SPSS

SPSS Advanced Statistics™ 6.1

Marija J. Norušis / SPSS Inc.

SPSS Inc.
444 N. Michigan Avenue
Chicago, Illinois 60611
Tel: (312) 329-2400
Fax: (312) 329-3668

SPSS Federal Systems (U.S.)
SPSS Asia Pacific Pte. Ltd.
SPSS Australasia Pty. Ltd.
SPSS Benelux BV
SPSS Central and Eastern Europe
SPSS France SARL
SPSS Germany
SPSS Hellas SA
SPSS Hispanoportuguesa S.L.
SPSS India Private Ltd.
SPSS Israel Ltd.
SPSS Italia srl
SPSS Japan Inc.
SPSS Latin America
SPSS Middle East and Africa
SPSS Scandinavia AB
SPSS UK Ltd.

For more information about SPSS® software products, please write or call

Marketing Department
SPSS Inc.
444 North Michigan Avenue
Chicago, IL 60611
Tel: (312) 329-2400
Fax: (312) 329-3668

4 5 6 7 8 9 0 96

ISBN 0-13-200065-2

Library of Congress Catalog Card Number: 94-068307

Preface

SPSS® 6.1 is a powerful software package for microcomputer data management and analysis. The Advanced Statistics option is an add-on enhancement that provides additional statistical analysis techniques. The procedures in Advanced Statistics must be used with the SPSS 6.1 Base system and are completely integrated into that system.

The Advanced Statistics option includes procedures for:

- Logistic regression
- Univariate and multivariate analysis of variance
- Model selection loglinear analysis
- General loglinear analysis
- Logit loglinear analysis
- Nonlinear regression
- Probit analysis
- Survival analysis, including life tables, Kaplan-Meier survival analysis, and Cox regression

The algorithms are identical to those used in SPSS software on mainframe computers, and the statistical results will be as precise as those computed on a mainframe.

What's New in SPSS 6.1. The information in this manual is essentially the same as the information in *SPSS for Windows Advanced Statistics Release 6.0* in conjunction with *SPSS 6.1 for Windows Update*.

Changes in procedures, incorporated in this manual and also described in *SPSS 6.1 for Windows Update*, include:

- **Kaplan-Meier survival analysis.** Censored cases are plotted on charts generated by this procedure.
- **General and logit loglinear analysis.** These new procedures, based on the Generalized Loglinear Model (GLM), replace the former General and Logit Loglinear Analysis procedures. The new procedures use a regression approach to the analysis and are available from the General Loglinear Analysis and Logit Loglinear Analysis dialog boxes. The former General Loglinear Analysis procedure, which uses a parameterized approach, is available only from syntax.
- **Model selection loglinear analysis.** This procedure was formerly called the Hierarchical Loglinear Analysis procedure. The Model Selection group for this procedure has been moved to the main dialog box.

New operational features of SPSS 6.1 are available for use with the Advanced Statistics option, including the following:

- Toolbar
- Exporting charts in various formats
- Variable information pop-up window
- Compatibility with Win32s

These features are described in the SPSS Base system documentation.

Installation

To install Advanced Statistics, follow the instructions for adding and removing features in the installation instructions supplied with the SPSS Base system. (To start, double-click on the SPSS Setup icon.)

Compatibility

The SPSS system is designed to operate on many computer systems. See the installation instructions that came with your system for specific information on minimum and recommended requirements.

Serial Numbers

Your serial number is your identification number with SPSS Inc. You will need this serial number when you call SPSS Inc. for information regarding support, payment, or an upgraded system. The serial number can be found on Disk 2, which came with your Base system. Before using the system, please copy this number to the registration card.

Registration Card

STOP! Before continuing on, *fill out and send us your registration card.* Until we receive your registration card, you have an unregistered system. Even if you have previously sent a card to us, please fill out and return the card enclosed in your Advanced Statistics package. Registering your system entitles you to:

- Technical support services
- Favored customer status
- New product announcements

Don't put it off—send your registration card now!

Customer Service

Contact Customer Service at 1-800-521-1337 if you have any questions concerning your shipment or account. Please have your serial number ready for identification when calling.

Training Seminars

SPSS Inc. provides both public and onsite training seminars for SPSS. All seminars feature hands-on workshops. SPSS seminars will be offered in major U.S. and European cities on a regular basis. For more information on these seminars, call the SPSS Inc. Training Department toll-free at 1-800-543-6607.

Technical Support

The services of SPSS Technical Support are available to registered customers. Customers may call Technical Support for assistance in using SPSS products or for installation help for one of the supported hardware environments.

To reach Technical Support, call 1-312-329-3410. Be prepared to identify yourself, your organization, and the serial number of your system.

If you are a Value Plus or Customer EXPress customer, use the priority 800 number you received with your materials. For information on subscribing to the Value Plus or Customer EXPress plan, call SPSS Software Sales at 1-800-543-2185.

Additional Publications

Additional copies of SPSS product manuals may be purchased from Prentice Hall, the exclusive distributor of SPSS publications. To order, fill out and mail the Publications order form included with your system or call toll-free. If you represent a bookstore or have an account with Prentice Hall, call 1-800-223-1360. If you are not an account customer, call 1-800-374-1200. In Canada, call 1-800-567-3800. Outside of North America, contact your local Prentice Hall office.

Lend Us Your Thoughts

Your comments are important. So send us a letter and let us know about your experiences with SPSS products. We especially like to hear about new and interesting applications using the SPSS system. Write to SPSS Inc. Marketing Department, Attn: Micro Software Products Manager, 444 N. Michigan Avenue, Chicago, IL 60611.

About This Manual

This manual is divided into two sections. The first section provides a guide to the various statistical techniques available with the Advanced Statistics option and how to obtain the appropriate statistical analyses with the dialog box interface. Illustrations of dialog boxes are taken from SPSS for Windows. Dialog boxes in other operating systems are similar. The second part of the manual is a Syntax Reference section that provides complete command syntax for all the commands included in the Advanced Statistics option. Most features of the system can be accessed through the dialog box interface, but some functionality can be accessed only through command syntax.

This manual contains two indexes: a subject index and a syntax index. The subject index covers both sections of the manual. The syntax index applies only to the Syntax Reference section.

Contacting SPSS Inc.

If you would like to be on our mailing list, contact one of our offices below. We will send you a copy of our newsletter and let you know about SPSS Inc. activities in your area.

SPSS Inc.
Chicago, Illinois, U.S.A.
Tel: 1.312.329.2400
Fax: 1.312.329.3668
Customer Service:
1.800.521.1337
Sales:
1.800.543.2185
sales@spss.com
Training:
1-800-543-6607
Technical Support:
1.312.329.3410
support@spss.com

SPSS Federal Systems
Arlington, Virginia, U.S.A.
Tel: 1.703.527.6777
Fax: 1.703.527.6866

SPSS Asia Pacific Pte. Ltd.
Singapore, Singapore
Tel: +65.221.2577
Fax: +65.221.9920

SPSS Australasia Pty. Ltd.
Sydney, Australia
Tel: +61.2.954.5660
Fax: +61.2.954.5616

SPSS Benelux BV
Gorinchem, The Netherlands
Tel: +31.0183.636711
Fax: +31.0183.635839

SPSS Central and Eastern Europe
Chertsey, Surrey, U.K.
Tel: +44.1932.566262
Fax: +44.1932.567020

SPSS France SARL
Boulogne, France
Tel: +33.1.4699.9670
Fax: +33.1.4684.0180

SPSS Germany
Munich, Germany
Tel: +49.89.4890740
Fax: +49.89.4483115

SPSS Hellas SA
Athens, Greece
Tel: +30.1.7251925
Fax: +30.1.7249124

SPSS Hispanoportuguesa S.L.
Madrid, Spain
Tel: +34.1.547.3703
Fax: +34.1.548.1346

SPSS India Private Ltd.
New Delhi, India
Tel: +91.11.600121 x1029
Fax: +91.11.6888851

SPSS Israel Ltd.
Herzlia, Israel
Tel: +972.9.598900
Fax: +972.9.598903

SPSS Italia srl
Bologna, Italy
Tel: +39.51.252573
Fax: +39.51.253285

SPSS Japan Inc.
Tokyo, Japan
Tel: +81.3.5474.0341
Fax: +81.3.5474.2678

SPSS Latin America
Chicago, Illinois, U.S.A.
Tel: 1.312.494.3226
Fax: 1.312. 494.3227

SPSS Middle East and Africa
Chertsey, Surrey, U.K.
Tel: +44.1753.622677
Fax: +44.1753.622644

SPSS Scandinavia AB
Stockholm, Sweden
Tel: +46.8.102610
Fax: +46.8.102550

SPSS UK Ltd.
Chertsey, Surrey, U.K.
Tel: +44.1932.566262
Fax: +44.1932.567020

Contents

1 Logistic Regression Analysis

Predicting whether an event will or will not occur, as well as identifying the variables useful in making the prediction, is important in most academic disciplines and in the "real" world. Why do some citizens vote and others do not? Why do some people develop coronary heart disease and others do not? Why do some businesses succeed and others fail?

A variety of multivariate statistical techniques can be used to predict a binary dependent variable from a set of independent variables. Multiple regression analysis and discriminant analysis are two related techniques that quickly come to mind. However, these techniques pose difficulties when the dependent variable can have only two values—an event occurring or not occurring.

When the dependent variable can have only two values, the assumptions necessary for hypothesis testing in regression analysis are necessarily violated. For example, it is unreasonable to assume that the distribution of errors is normal. Another difficulty with multiple regression analysis is that predicted values cannot be interpreted as probabilities. They are not constrained to fall in the interval between 0 and 1.

Linear discriminant analysis does allow direct prediction of group membership, but the assumption of multivariate normality of the independent variables, as well as equal variance-covariance matrices in the two groups, is required for the prediction rule to be optimal.

In this chapter, we will consider another multivariate technique for estimating the probability that an event occurs: the **logistic regression model**. This model requires far fewer assumptions than discriminant analysis; and even when the assumptions required for discriminant analysis are satisfied, logistic regression still performs well. (See Hosmer and Lemeshow, 1989, for an introduction to logistic regression.)

The Logistic Regression Model

In logistic regression, you directly estimate the probability of an event occurring. For the case of a single independent variable, the logistic regression model can be written as

$$\text{Prob (event)} = \frac{e^{B_0 + B_1 X}}{1 + e^{B_0 + B_1 X}} \qquad \textbf{Equation 1.1}$$

or equivalently

$$\text{Prob (event)} = \frac{1}{1 + e^{-(B_0 + B_1 X)}} \qquad \textbf{Equation 1.2}$$

where B_0 and B_1 are coefficients estimated from the data, X is the independent variable, and e is the base of the natural logarithms, approximately 2.718 .

For more than one independent variable, the model can be written as

$$\text{Prob (event)} = \frac{e^{Z}}{1 + e^{Z}} \qquad \textbf{Equation 1.3}$$

or equivalently

$$\text{Prob (event)} = \frac{1}{1 + e^{-Z}} \qquad \textbf{Equation 1.4}$$

where Z is the linear combination

$$Z = B_0 + B_1 X_1 + B_2 X_2 + \ldots + B_p X_p \qquad \textbf{Equation 1.5}$$

The probability of the event not occurring is estimated as

$$\text{Prob (no event)} = 1 - \text{Prob (event)} \qquad \textbf{Equation 1.6}$$

Figure 1.1 is a plot of a logistic regression curve when the values of Z are between -5 and $+5$. As you can see, the curve is S-shaped. It closely resembles the curve obtained when the cumulative probability of the normal distribution is plotted. The relationship between the independent variable and the probability is nonlinear. The probability estimates will always be between 0 and 1, regardless of the value of Z.

Figure 1.1 Plot of logistic regression curve

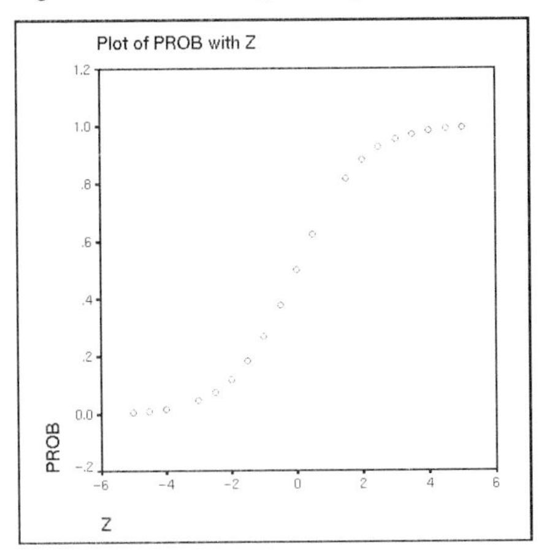

In linear regression, we estimate the parameters of the model using the **least-squares method**. That is, we select regression coefficients that result in the smallest sums of squared distances between the observed and the predicted values of the dependent variable.

In logistic regression, the parameters of the model are estimated using the **maximum-likelihood method**. That is, the coefficients that make our observed results most "likely" are selected. Since the logistic regression model is nonlinear, an iterative algorithm is necessary for parameter estimation.

An Example

The treatment and prognosis of cancer depends on how much the disease has spread. One of the regions to which a cancer may spread is the lymph nodes. If the lymph nodes are involved, the prognosis is generally poorer than if they are not. That's why it's desirable to establish as early as possible whether the lymph nodes are cancerous. For certain cancers, exploratory surgery is done just to determine whether the nodes are cancerous, since this will determine what treatment is needed. If we could predict whether the nodes are affected or not on the basis of data that can be obtained without performing surgery, considerable discomfort and expense could be avoided.

For this chapter, we will use data presented by Brown (1980) for 53 men with prostate cancer. For each patient, he reports the age, serum acid phosphatase (a laboratory value that is elevated if the tumor has spread to certain areas), the stage of the disease (an indication of how advanced the disease is), the grade of the tumor (an indication of aggressiveness), and x-ray results, as well as whether the cancer had spread to the regional lymph nodes at the time of surgery. The problem is to predict whether the nodes are positive for cancer based on the values of the variables that can be measured without surgery.

Coefficients for the Logistic Model

Figure 1.2 contains the estimated coefficients (under column heading *B*) and related statistics from the logistic regression model that predicts nodal involvement from a constant and the variables *age*, *acid*, *xray*, *stage*, and *grade*. The last three of these variables (*xray*, *stage*, and *grade*) are **indicator variables**, coded 0 or 1. The value of 1 for *xray* indicates positive x-ray findings, the value of 1 for *stage* indicates advanced stage, and the value of 1 for *grade* indicates a more aggressively spreading malignant tumor.

Figure 1.2 Parameter estimates for the logistic regression model

---------------------- Variables in the Equation ----------------------

Variable	B	S.E.	Wald	df	Sig	R	Exp(B)
AGE	-.0693	.0579	1.4320	1	.2314	.0000	.9331
ACID	.0243	.0132	3.4229	1	.0643	.1423	1.0246
XRAY	2.0453	.8072	6.4207	1	.0113	.2509	7.7317
GRADE	.7614	.7708	.9758	1	.3232	.0000	2.1413
STAGE	1.5641	.7740	4.0835	1	.0433	.1722	4.7783
Constant	.0618	3.4599	.0003	1	.9857		

Given these coefficients, the logistic regression equation for the probability of nodal involvement can be written as

$$\text{Prob (nodal involvement)} = \frac{1}{1 + e^{-Z}}$$ **Equation 1.7**

where

$$Z = 0.0618 - 0.0693\,(\text{age}) + 0.0243\,(\text{acid}) + 2.0453\,(\text{xray}) + 0.7614\,(\text{grade}) + 1.5641\,(\text{stage})$$ **Equation 1.8**

Applying this to a man who is 66 years old, with a serum acid phosphatase level of 48 and values of 0 for the remaining independent variables, we find

$$Z = 0.0618 - 0.0693\,(66) + 0.0243\,(48) = -3.346$$ **Equation 1.9**

The probability of nodal involvement is then estimated to be

$$\text{Prob (nodal involvement)} = \frac{1}{1 + e^{-(-3.346)}} = 0.0340$$ **Equation 1.10**

Based on this estimate, we would predict that the nodes are unlikely to be malignant. In general, if the estimated probability of the event is less than 0.5, we predict that the event will not occur. If the probability is greater than 0.5, we predict that the event will occur. (In the unlikely event that the probability is exactly 0.5, we can flip a coin for our prediction.)

Testing Hypotheses about the Coefficients

For large sample sizes, the test that a coefficient is 0 can be based on the **Wald statistic**, which has a chi-square distribution. When a variable has a single degree of freedom, the Wald statistic is just the square of the ratio of the coefficient to its standard error. For categorical variables, the Wald statistic has degrees of freedom equal to one less than the number of categories.

For example, the coefficient for age is -0.0693, and its standard error is 0.0579. (The standard errors for the logistic regression coefficients are shown in the column labeled *S.E.* in Figure 1.2.) The Wald statistic is $(-0.0693/0.0579)^2$, or about 1.432. The significance level for the Wald statistic is shown in the column labeled *Sig.* In this example, only the coefficients for *xray* and *stage* appear to be significantly different from 0, using a significance level of 0.05.

Unfortunately, the Wald statistic has a very undesirable property. When the absolute value of the regression coefficient becomes large, the estimated standard error is too large. This produces a Wald statistic that is too small, leading you to fail to reject the null hypothesis that the coefficient is 0, when in fact you should. Therefore, whenever you have a large coefficient, you should not rely on the Wald statistic for hypothesis testing. Instead, you should build a model with and without that variable and base your hypothesis test on the change in the log likelihood (Hauck & Donner, 1977).

Partial Correlation

As is the case with multiple regression, the contribution of individual variables in logistic regression is difficult to determine. The contribution of each variable depends on the other variables in the model. This is a problem, particularly when independent variables are highly correlated.

A statistic that is used to look at the partial correlation between the dependent variable and each of the independent variables is the R statistic, shown in Figure 1.2. R can range in value from -1 to $+1$. A positive value indicates that as the variable increases in value, so does the likelihood of the event occurring. If R is negative, the opposite is true. Small values for R indicate that the variable has a small partial contribution to the model.

The equation for the R statistic is

$$R = \pm\sqrt{\left(\frac{\text{Wald statistic} - 2K}{-2LL_{(0)}}\right)}$$

Equation 1.11

where K is the degrees of freedom for the variable (Atkinson, 1980). The denominator is -2 times the log likelihood of a base model that contains only the intercept, or a model with no variables if there is no intercept. (If you enter several blocks of variables, the base model for each block is the result of previous entry steps.) The sign of the corresponding coefficient is attached to R. The value of $2K$ in Equation 1.11 is an adjustment for the number of parameters estimated. If the Wald statistic is less than $2K$, R is set to 0.

Interpreting the Regression Coefficients

In multiple linear regression, the interpretation of the regression coefficient is straightforward. It tells you the amount of change in the dependent variable for a one-unit change in the independent variable.

To understand the interpretation of the logistic coefficients, consider a rearrangement of the equation for the logistic model. The logistic model can be rewritten in terms of the odds of an event occurring. (The **odds** of an event occurring are defined as the ratio of the probability that it will occur to the probability that it will not. For example, the odds of getting a head on a single flip of a coin are $0.5/0.5 = 1$. Similarly, the odds of getting a diamond on a single draw from a card deck are $0.25/0.75 = 1/3$. Don't confuse this technical meaning of odds with its informal usage to mean simply the probability.)

First let's write the logistic model in terms of the log of the odds, which is called a **logit**:

$$\log\left(\frac{\text{Prob (event)}}{\text{Prob (no event)}}\right) = B_0 + B_1X_1 + \ldots + B_pX_p$$

Equation 1.12

From Equation 1.12, you see that the logistic coefficient can be interpreted as the change in the log odds associated with a one-unit change in the independent variable. For example, from Figure 1.2, you see that the coefficient for *grade* is 0.76. This tells you that when the grade changes from 0 to 1 and the values of the other independent variables remain the same, the log odds of the nodes being malignant increase by 0.76.

Since it's easier to think of odds rather than log odds, the logistic equation can be written in terms of odds as

$$\frac{\text{Prob (event)}}{\text{Prob (no event)}} = e^{B_0 + B_1X_1 + \ldots + B_pX_p} = e^{B_0}e^{B_1X_1}\ldots e^{B_pX_p}$$

Equation 1.13

Then e raised to the power B_i is the factor by which the odds change when the ith independent variable increases by one unit. If B_i is positive, this factor will be greater than 1, which means that the odds are increased; if B_i is negative, the factor will be less than 1, which means that the odds are decreased. When B_i is 0, the factor equals 1, which leaves the odds unchanged. For example, when *grade* changes from 0 to 1, the odds are increased by a factor of 2.14, as is shown in the *Exp(B)* column in Figure 1.2.

As a further example, let's calculate the odds of having malignant nodes for a 60-year-old man with a serum acid phosphatase level of 62, a value of 1 for x-ray results, and values of 0 for stage and grade of tumor. First, calculate the probability that the nodes are malignant:

$$\text{Estimated prob (malignant nodes)} = \frac{1}{1 + e^{-Z}} \qquad \textbf{Equation 1.14}$$

where

$$Z = 0.0618 - 0.0693\,(60) + 0.0243\,(62) + 2.0453\,(1) + 0.7614\,(0) + (1.5641)\,(0) = -0.54 \qquad \textbf{Equation 1.15}$$

The estimated probability of malignant nodes is therefore 0.37. The probability of not having malignant nodes is 0.63 (that is, $1 - 0.37$). The *odds* of having a malignant node are then estimated as

$$\text{Odds} = \frac{\text{Prob (event)}}{\text{Prob (no event)}} = \frac{0.37}{1 - 0.37} = 0.59 \qquad \textbf{Equation 1.16}$$

and the log odds are -0.53 .

What would be the probability of malignant nodes if, instead of 0, the case had a value of 1 for *grade*? Following the same procedure as before, but using a value of 1 for *grade*, the estimated probability of malignant nodes is 0.554. Similarly, the estimated odds are 1.24, and the log odds are 0.22.

By increasing the value of *grade* by one unit, we have increased the log odds by about 0.75, the value of the coefficient for *grade*. (Since we didn't use many digits in our hand calculations, our value of 0.75 isn't exactly equal to the 0.76 value for *grade* shown in Figure 1.2. If we carried the computations out with enough precision, we would arrive at exactly the value of the coefficient.)

By increasing the value of *grade* from 0 to 1, the odds changed from 0.59 to 1.24. That is, they increased by a factor of about 2.1. This is the value of *Exp(B)* for *grade* in Figure 1.2.

Assessing the Goodness of Fit of the Model

There are various ways to assess whether or not the model fits the data. The sections "The Classification Table," below, through "Goodness of Fit with All Variables" on p. 10 discuss the goodness of fit of the model.

The Classification Table

One way to assess how well our model fits is to compare our predictions to the observed outcomes. Figure 1.3 is the classification table for this example.

Figure 1.3 Classification table

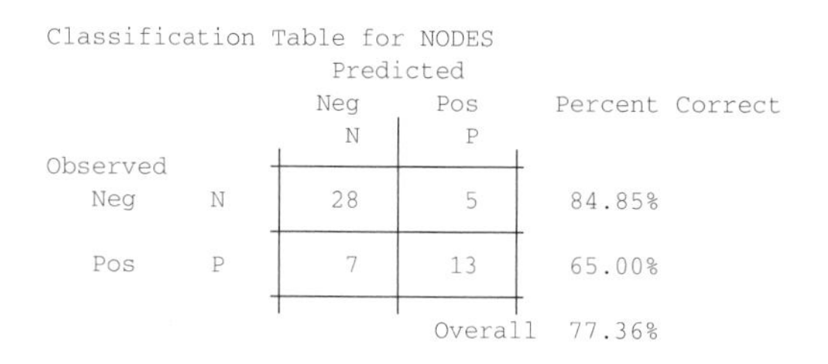

Classification Table for NODES

Observed		Predicted Neg N	Predicted Pos P	Percent Correct
Neg	N	28	5	84.85%
Pos	P	7	13	65.00%
			Overall	77.36%

From the table, you see that 28 patients without malignant nodes were correctly predicted by the model not to have malignant nodes. Similarly, 13 men with positive nodes were correctly predicted to have positive nodes. The off-diagonal entries of the table tell you how many men were incorrectly classified. A total of 12 men were misclassified in this example—5 men with negative nodes and 7 men with positive nodes. Of the men without diseased nodes, 84.85% were correctly classified. Of the men with diseased nodes, 65% were correctly classified. Overall, 77.36% of the 53 men were correctly classified.

The classification table doesn't reveal the distribution of estimated probabilities for men in the two groups. For each predicted group, the table shows only whether the estimated probability is greater or less than one-half. For example, you cannot tell from the table whether the seven patients who had false negative results had predicted probabilities near 50%, or low predicted probabilities. Ideally, you would like the two groups to have very different estimated probabilities. That is, you would like to see small estimated probabilities of positive nodes for all men without malignant nodes and large estimated probabilities for all men with malignant nodes.

Histogram of Estimated Probabilities

Figure 1.4 is a histogram of the estimated probabilities of cancerous nodes. The symbol used for each case designates the group to which the case actually belongs. If you have a model that successfully distinguishes the two groups, the cases for which the event has occurred should be to the right of 0.5, while the cases for which the event has not occurred should be to the left of 0.5. The more the two groups cluster at their respective ends of the plot, the better.

Figure 1.4 Histogram of estimated probabilities

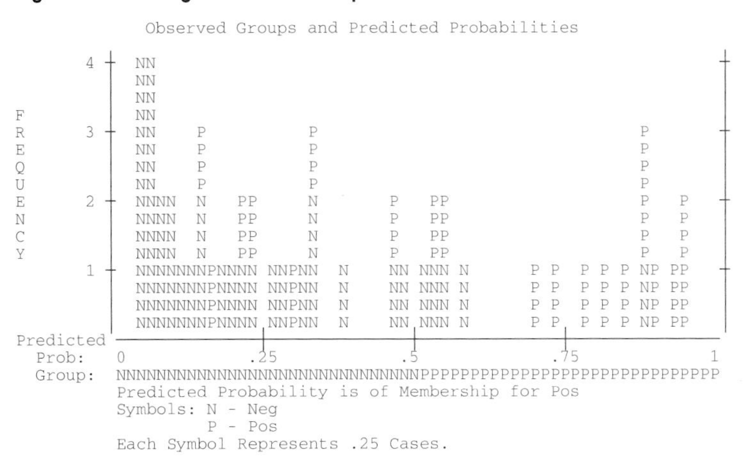

From Figure 1.4, you see that there is only one noncancerous case with a high estimated probability of having positive nodes (the case identified with the letter *N* at a probability value of about 0.88). However, there are four diseased cases with estimated probabilities less than 0.25.

By looking at this histogram of predicted probabilities, you can see whether a different rule for assigning cases to groups might be useful. For example, if most of the misclassifications occur in the region around 0.5, you might decide to withhold judgment for cases with values in this region. In this example, this means that you would predict nodal involvement only for cases where you were reasonably sure that the logistic prediction would be correct. You might decide to operate on all questionable cases.

If the consequences of misclassification are not the same in both directions (for example, calling nodes negative when they are really positive is worse than calling nodes positive when they are really negative), the classification rule can be altered to decrease the possibility of making the more severe error. For example, you might decide to call cases "negative" only if their estimated probability is less than 0.3. By looking at the histogram of the estimated probabilities, you can get some idea of how different classification rules might perform. (Of course, when you apply the model to new cases, you can't expect the classification rule to behave exactly the same.)

Goodness of Fit of the Model

Seeing how well the model classifies the observed data is one way of determining how well the logistic model performs. Another way of assessing the goodness of fit of the model is to examine how "likely" the sample results actually are, given the parameter estimates. (Recall that we chose parameter estimates that would make our observed results as likely as possible.)

The probability of the observed results, given the parameter estimates, is known as the **likelihood**. Since the likelihood is a small number less than 1, it is customary to use -2 times the log of the likelihood ($-2LL$) as a measure of how well the estimated model fits the data. A good model is one that results in a high likelihood of the observed results. This translates to a small value for $-2LL$. (If a model fits perfectly, the likelihood is 1, and -2 times the log likelihood is 0.)

For the logistic regression model that contains only the constant, $-2LL$ is 70.25, as shown in Figure 1.5.

Figure 1.5 –2LL for model containing only the constant

```
Dependent Variable..   NODES

Beginning Block Number  0.  Initial Log Likelihood Function

-2 Log Likelihood   70.252153

* Constant is included in the model.

No terms in the model.
```

Another measure of how well the model fits is the **goodness-of-fit statistic**, which compares the observed probabilities to those predicted by the model. The goodness-of-fit statistic is defined as

$$Z^2 = \sum \frac{\text{Residual}_i^2}{P_i(1-P_i)} \qquad \textbf{Equation 1.17}$$

where the residual is the difference between the observed value, Y_i, and the predicted value, P_i.

Goodness of Fit with All Variables

Figure 1.6 shows the goodness-of-fit statistics for the model with all of the independent variables. For the current model, the value of $-2LL$ is 48.126, which is smaller than the $-2LL$ for the model containing only a constant. The goodness-of-fit statistic is displayed in the second row of the table.

Figure 1.6 Statistics for model containing the independent variables

```
-2 Log Likelihood          48.126
Goodness of Fit            46.790

                       Chi-Square     df Significance

Model Chi-Square           22.126      5        .0005
Improvement                22.126      5        .0005
```

There are two additional entries in Figure 1.6. They are labeled *Model Chi-Square* and *Improvement*. In this example, the model chi-square is the difference between $-2LL$ for the model with only a constant and $-2LL$ for the current model. (If a constant is not included in the model, the likelihood for the model without any variables is used for comparison. If variables are already in the equation when variable selection begins, the model with these variables is used as the base model.) Thus, the model chi-square tests the null hypothesis that the coefficients for all of the terms in the current model, except the constant, are 0. This is comparable to the overall F test for regression.

In this example, $-2LL$ for the model containing only the constant is 70.25 (from Figure 1.5), while for the complete model, it is 48.126. The model chi-square, 22.126, is the difference between these two values. The degrees of freedom for the model chi-square are the difference between the number of parameters in the two models.

The entry labeled *Improvement* is the change in $-2LL$ between successive steps of building a model. It tests the null hypothesis that the coefficients for the variables added at the last step are 0. In this example, we considered only two models: the constant-only model and the model with a constant and five independent variables. Thus, the model chi-square and the improvement chi-square values are the same. If you sequentially consider more than just these two models, using either forward or backward variable selection, the model chi-square and improvement chi-square will differ. The improvement chi-square test is comparable to the F-change test in multiple regression.

Categorical Variables

In logistic regression, just as in linear regression, the codes for the independent variables must be meaningful. You cannot take a nominal variable like religion, assign arbitrary codes from 1 to 35, and then use the resulting variable in the model. In this situation, you must recode the values of the independent variable by creating a new set of variables that correspond in some way to the original categories.

If you have a two-category variable such as sex, you can code each case as 0 or 1 to indicate either female or not female. Or you could code it as being male or not male. This is called **dummy-variable** or **indicator-variable coding**. *Grade*, *stage*, and *xray* are all examples of two-category variables that have been coded as 0 and 1. The code of 1 indicates that the poorer outcome is present. The interpretation of the resulting coefficients for *grade*, *stage*, and *xray* is straightforward. It tells you the difference between the log odds when a case is a member of the "poor" category and when it is not.

When you have a variable with more than two categories, you must create new variables to represent the categories. The number of new variables required to represent a categorical variable is one less than the number of categories. For example, if instead of the actual values for serum acid phosphatase, you had values of 1, 2, or 3, depending on whether the value was low, medium, or high, you would have to create two new variables to represent the serum phosphatase effect. Two alternative coding schemes are described in "Indicator-Variable Coding Scheme," below, and "Another Coding Scheme" on p. 13.

Indicator-Variable Coding Scheme

One of the ways you can create two new variables for serum acid phosphatase is to use indicator variables to represent the categories. With this method, one variable would represent the low value, coded 1 if the value is low and 0 otherwise. The second variable would represent the medium value, coded 1 if the value is average and 0 otherwise. The value "high" would be represented by codes of 0 for both of these variables. The choice of the category to be coded as 0 for both variables is arbitrary.

With categorical variables, the only statement you can make about the effect of a particular category is in comparison to some other category. For example, if you have a variable that represents type of cancer, you can only make statements such as "lung cancer compared to bladder cancer decreases your chance of survival." Or you might say that "lung cancer compared to all the cancer types in the study decreases your chance of survival." You can't make a statement about lung cancer without relating it to the other types of cancer.

If you use indicator variables for coding, the coefficients for the new variables represent the effect of each category compared to a reference category. The coefficient for the reference category is 0. As an example, consider Figure 1.7. The variable *catacid1* is the indicator variable for low serum acid phosphatase, coded 1 for low levels and 0 otherwise. Similarly, the variable *catacid2* is the indicator variable for medium serum acid phosphatase. The reference category is high levels.

Figure 1.7 Indicator variables

```
---------------------- Variables in the Equation -----------------------

Variable           B       S.E.      Wald     df       Sig        R    Exp(B)

AGE            -.0522     .0630     .6862      1     .4075    .0000     .9492
CATACID1      -2.0079    1.0520    3.6427      1     .0563   -.1529     .1343
CATACID2      -1.0923     .9264    1.3903      1     .2384    .0000     .3355
XRAY           2.0348     .8375    5.9033      1     .0151    .2357    7.6503
GRADE           .8076     .8233     .9623      1     .3266    .0000    2.2426
STAGE          1.4571     .7683    3.5968      1     .0579    .1508    4.2934
Constant       1.7698    3.8088     .2159      1     .6422
```

The coefficient for *catacid1* is the change in log odds when you have a low value compared to a high value. Similarly, *catacid2* is the change in log odds when you have a medium value compared to a high value. The coefficient for the high value is necessarily 0, since it does not differ from itself. In Figure 1.7, you see that the coefficients for both of the indicator variables are negative. This means that compared to high values for serum acid phosphatase, low and medium values are associated with decreased log odds of malignant nodes. The low category decreases the log odds more than the medium category.

The SPSS Logistic Regression procedure will automatically create new variables for variables declared as categorical (see "Defining Categorical Variables" on p. 26). You can choose the coding scheme you want to use for the new variables.

Figure 1.8 shows the table that is displayed for each categorical variable. The rows of the table correspond to the categories of the variable. The actual value is given in the column labeled *Value*. The number of cases with each value is displayed in the column labeled *Freq*. Subsequent columns correspond to new variables created by the program. The number in parentheses indicates the suffix used to identify the variable in the output. The codes that represent each original category using the new variables are listed under the corresponding new-variable column.

Figure 1.8 Indicator-variable coding scheme

```
                      Parameter
         Value   Freq  Coding
                         (1)     (2)
CATACID
          1.00     15  1.000    .000
          2.00     20   .000   1.000
          3.00     18   .000    .000
```

From Figure 1.8, you see that there are 20 cases with a value of 2 for *catacid*. Each of these cases will be assigned a code of 0 for the new variable *catacid(1)* and a code of 1 for the new variable *catacid(2)*. Similarly, cases with a value of 3 for *catacid* will be given the code of 0 for both *catacid(1)* and *catacid(2)*.

Another Coding Scheme

The statement you can make based on the logistic regression coefficients depends on how you have created the new variables used to represent the categorical variable. As shown in the previous section, when you use indicator variables for coding, the coefficients for the new variables represent the effect of each category compared to a reference category. If, on the other hand, you wanted to compare the effect of each category to the average effect of all of the categories, you could have selected the default deviation coding scheme shown in Figure 1.9. This differs from indicator-variable coding only in that the last category is coded as -1 for each of the new variables.

With this coding scheme, the logistic regression coefficients tell you how much better or worse each category is compared to the average effect of all categories, as shown in Figure 1.10. For each new variable, the coefficients now represent the difference from the average effect over all categories. The value of the coefficient for the last category is not displayed, but it is no longer 0. Instead, it is the negative of the sum of the displayed coefficients. From Figure 1.10, the coefficient for "high" level is calculated as $-(-0.9745 - 0.0589) = 1.0334$.

Figure 1.9 Another coding scheme

```
                      Parameter
         Value   Freq  Coding
                         (1)     (2)
CATACID
          1.00     15  1.000    .000
          2.00     20   .000   1.000
          3.00     18 -1.000 -1.000
```

Figure 1.10 New coefficients

---------------------- Variables in the Equation -----------------------

Variable	B	S.E.	Wald	df	Sig	R	Exp(B)
AGE	-.0522	.0630	.6862	1	.4075	.0000	.9492
CATACID			3.8361	2	.1469	.0000	
CATACID(1)	-.9745	.6410	2.3116	1	.1284	-.0666	.3774
CATACID(2)	-.0589	.5727	.0106	1	.9181	.0000	.9428
XRAY	2.0348	.8375	5.9033	1	.0151	.2357	7.6503
GRADE	.8076	.8233	.9623	1	.3266	.0000	2.2426
STAGE	1.4571	.7683	3.5968	1	.0579	.1508	4.2934
Constant	.7364	3.7352	.0389	1	.8437		

Different coding schemes result in different logistic regression coefficients, but not in different conclusions. That is, even though the actual values of the coefficients differ between Figure 1.7 and Figure 1.10, they tell you the same thing. Figure 1.7 tells you the effect of category 1 compared to category 3, while Figure 1.10 tells you the effect of category 1 compared to the average effect of all of the categories. You can select the coding scheme to match the type of comparisons you want to make.

Interaction Terms

Just as in linear regression, you can include terms in the model that are products of single terms. For example, if it made sense, you could include a term for the *acid* by *age* interaction in your model.

Interaction terms for categorical variables can also be computed. They are created as products of the values of the new variables. For categorical variables, make sure that the interaction terms created are those of interest. If you are using categorical variables with indicator coding, the interaction terms generated as the product of the variables are generally not those that you are interested in. Consider, for example, the interaction term between two indicator variables. If you just multiply the variables together, you will obtain a value of 1 only if both of the variables are coded "present." What you would probably like is a code of 1 if both of the variables are present or both are absent. You will obtain this if you use the default coding scheme for category variables instead of specifying the scheme as indicator variables.

Selecting Predictor Variables

In logistic regression, as in other multivariate statistical techniques, you may want to identify subsets of independent variables that are good predictors of the dependent variable. All of the problems associated with variable selection algorithms in regression and discriminant analysis are found in logistic regression as well. None of the algorithms result in a "best" model in any statistical sense. Different algorithms for variable selection may result in different models. It is a good idea to examine several possible models and choose from among them on the basis of interpretability, parsimony, and ease of variable acquisition.

As always, the model is selected to fit a particular sample well, so there is no assurance that the same model will be selected if another sample from the same population is taken. The model will always fit the sample better than the population from which it is selected.

The SPSS Logistic Regression procedure has several methods available for model selection. You can enter variables into the model at will. You can also use forward stepwise selection and backward stepwise elimination for automated model building. The score statistic is always used for entering variables into a model. The Wald statistic, the change in likelihood, or the conditional statistic can be used for removing variables from a model (Lawless & Singhal, 1978). All variables that are used to represent the same categorical variable are entered or removed from the model together.

Forward Stepwise Selection

Forward stepwise variable selection in logistic regression proceeds the same way as in multiple linear regression. You start out with a model that contains only the constant unless the option to omit the constant term from the model is selected. At each step, the variable with the smallest significance level for the score statistic, provided it is less than the chosen cutoff value (by default 0.05), is entered into the model. All variables in the forward stepwise block that have been entered are then examined to see if they meet removal criteria. If the Wald statistic is used for deleting variables, the Wald statistics for all variables in the model are examined and the variable with the largest significance level for the Wald statistic, provided it exceeds the chosen cutoff value (by default 0.1), is removed from the model. If no variables meet removal criteria, the next eligible variable is entered into the model.

If a variable is selected for removal and it results in a model that has already been considered, variable selection stops. Otherwise, the model is estimated without the deleted variable and the variables are again examined for removal. This continues until no more variables are eligible for removal. Then variables are again examined for entry into the model. The process continues until either a previously considered model is encountered (which means the algorithm is cycling) or no variables meet entry or removal criteria.

The Likelihood-Ratio Test

A better criterion than the Wald statistic for determining variables to be removed from the model is the **likelihood-ratio (LR) test**. This involves estimating the model with each variable eliminated in turn and looking at the change in the log likelihood when each variable is deleted. The likelihood-ratio test for the null hypothesis that the coefficients of the terms removed are 0 is obtained by dividing the likelihood for the reduced model by the likelihood for the full model.

If the null hypothesis is true and the sample size is sufficiently large, the quantity −2 times the log of the likelihood-ratio statistic has a chi-square distribution with *r* degrees of freedom, where *r* is the difference between the number of terms in the full model and the reduced model. (The model chi-square and the improvement chi-square are both likelihood-ratio tests.)

When the likelihood-ratio test is used for removing terms from a model, its significance level is compared to the cutoff value. The algorithm proceeds as previously described but with the likelihood-ratio statistic, instead of the Wald statistic, being evaluated for removing variables.

You can also use the **conditional statistic** to test for removal. Like the likelihood-ratio test, the conditional statistic is based on the difference in the likelihood for the reduced and full models. However, the conditional statistic is computationally much less intensive since it does not require that the model be reestimated without each of the variables.

An Example of Forward Selection

To see what the output looks like for forward selection, consider Figure 1.11, which contains part of the summary statistics for the model when the constant is the only term included. First you see the previously described statistics for the constant. Then you see statistics for variables not in the equation. (The *R* for variables not in the equation is calculated using the score statistic instead of the Wald statistic.)

Figure 1.11 Variables not in the equation

---------------------- Variables in the Equation -----------------------

Variable	B	S.E.	Wald	df	Sig	R	Exp(B)
Constant	-.5008	.2834	3.1227	1	.0772		

-------------- Variables not in the Equation -----------------

Residual Chi Square 19.451 with 5 df Sig = .0016

Variable	Score	df	Sig	R
AGE	1.0945	1	.2955	.0000
ACID	3.1168	1	.0775	.1261
XRAY	11.2829	1	.0008	.3635
STAGE	7.4381	1	.0064	.2782
GRADE	4.0745	1	.0435	.1718

The residual chi-square statistic tests the null hypothesis that the coefficients for all variables not in the model are 0. (The residual chi-square statistic is calculated from the score statistics, so it is not exactly the same value as the improvement chi-square value that you see in Figure 1.6. In general, however, the two statistics should be similar in value.) If the observed significance level for the residual chi-square statistic is small (that is, if you have reason to reject the hypothesis that all of the coefficients are 0), it is sensible to proceed with variable selection. If you can't reject the hypothesis that the coefficients are 0, you should consider terminating variable selection. If you continue to build a model, there is a reasonable chance that your resulting model will not be useful for other samples from the same population.

In this example, the significance level for the residual chi-square is small, so we can proceed with variable selection. For each variable not in the model, the score statistic and its significance level, if the variable were entered next into the model, is shown. The score statistic is an efficient alternative to the Wald statistic for testing the hypothesis that a coefficient is 0. Unlike the Wald statistic, it does not require the explicit computation of parameter estimates, so it is useful in situations where recalculating parameter estimates for many different models would be computationally prohibitive. The likelihood-ratio statistic, the Wald statistic, and Rao's efficient score statistic are all equivalent in large samples, when the null hypothesis is true (Rao, 1973).

From Figure 1.11, you see that *xray* has the smallest observed significance level less than 0.05, the default value for entry, so it is entered into the model. Statistics for variables not in the model at this step are shown in Figure 1.12. You see that the *stage* variable has the smallest observed significance level and meets entry criteria, so it is entered next. Figure 1.13 contains logistic coefficients when *stage* is included in the model. Since the observed significance levels of the coefficients for both variables in the model are less than 0.1, the default criterion for removal, neither variable is removed from the model.

Figure 1.12 Variables not in the equation

```
--------------- Variables not in the Equation -----------------
Residual Chi Square       10.360 with       4 df       Sig =   .0348

Variable              Score     df        Sig         R

AGE                  1.3524      1      .2449     .0000
ACID                 2.0732      1      .1499     .0323
STAGE                5.6393      1      .0176     .2276
GRADE                2.3710      1      .1236     .0727
```

Figure 1.13 Logistic coefficients with variables stage and xray

```
---------------------- Variables in the Equation -----------------------

Variable           B        S.E.      Wald     df       Sig        R     Exp(B)

XRAY          2.1194       .7468    8.0537      1     .0045    .2935    8.3265
STAGE         1.5883       .7000    5.1479      1     .0233    .2117    4.8953
Constant     -2.0446       .6100   11.2360      1     .0008
```

The goodness-of-fit statistics for the model with *xray* and *stage* are shown in Figure 1.14. The model chi-square is the difference between $-2LL$ when only the constant is in the model and $-2LL$ when the constant, *xray*, and *stage* are in the model ($70.25 - 53.35 = 16.90$). The small observed significance level for the model chi-square indicates that you can reject the null hypothesis that the coefficients for *xray* and *stage* are zero. The improvement chi-square is the change in $-2LL$ when *stage* is added to a model containing *xray* and the constant. The small observed significance level indicates that the coefficient for *stage* is not zero. ($-2LL$ for the model with only the constant and *xray* is 59.001, so the improvement chi-square is $59.00 - 53.35 = 5.65$.)

Figure 1.14 Goodness-of-fit statistics with variables stage and xray

```
-2 Log Likelihood          53.353
Goodness of Fit            54.018

                        Chi-Square     df Significance

Model Chi-Square           16.899      2         .0002
Improvement                 5.647      1         .0175
```

The statistics for variables not in the model after *stage* is entered are shown in Figure 1.15. All three of the observed significance levels are greater than 0.05, so no additional variables are included in the model.

Figure 1.15 Variables not in the model after variable stage

```
-------------- Variables not in the Equation ----------------
Residual Chi Square        5.422 with       3 df      Sig =  .1434

Variable              Score     df        Sig        R

AGE                  1.2678      1      .2602    .0000
ACID                 3.0917      1      .0787    .1247
GRADE                 .5839      1      .4448    .0000
```

Forward Selection with the Likelihood-Ratio Criterion

If you select the likelihood-ratio statistic for deleting variables, the output will look slightly different from that previously described. For variables in the equation at a particular step, output similar to that shown in Figure 1.16 is produced in addition to the usual coefficients and Wald statistics.

Figure 1.16 Removal statistics

```
---------------- Model if Term Removed -----------------

Term         Log                                  Significance
Removed      Likelihood     -2 Log LR     df      of Log LR

XRAY            -31.276         9.199      1             .0024
STAGE           -29.500         5.647      1             .0175
```

For each variable in the model, Figure 1.16 contains the log likelihood for the model if the variable is removed from the model; -2 log LR, which tests the null hypothesis that the coefficient of the term is 0; and the observed significance level. If the observed significance level is greater than the cutoff value for remaining in the model, the term is removed from the model and the model statistics are recalculated to see if any other variables are eligible for removal.

Backward Elimination

Forward selection starts without any variables in the model. Backward elimination starts with all of the variables in the model. Then, at each step, variables are evaluated for entry and removal. The score statistic is always used for determining whether variables should be added to the model. Just as in forward selection, the Wald statistic, the likelihood-ratio statistic, or the conditional statistic can be used to select variables for removal.

Diagnostic Methods

Whenever you build a statistical model, it is important to examine the adequacy of the resulting model. In linear regression, we look at a variety of residuals, measures of influence, and indicators of collinearity. These are valuable tools for identifying points for which the model does not fit well, points that exert a strong influence on the coefficient estimates, and variables that are highly related to each other.

In logistic regression, there are comparable diagnostics that should be used to look for problems. The SPSS Logistic Regression procedure provides a variety of such statistics.

The **residual** is the difference between the observed probability of the event and the predicted probability of the event based on the model. For example, if we predict the probability of malignant nodes to be 0.80 for a man who has malignant nodes, the residual is $1 - 0.80 = 0.20$.

The **standardized residual** is the residual divided by an estimate of its standard deviation. In this case, it is

$$Z_i = \frac{\text{Residual}_i}{\sqrt{P_i(1 - P_i)}}$$ **Equation 1.18**

For each case, the standardized residual can also be considered a component of the chi-square goodness-of-fit statistic. If the sample size is large, the standardized residuals should be approximately normally distributed, with a mean of 0 and a standard deviation of 1.

For each case, the **deviance** is computed as

$$-2 \times \log(\text{predicted probability for the observed group})$$ **Equation 1.19**

The deviance is calculated by taking the square root of the above statistic and attaching a negative sign if the event did not occur for that case. For example, the deviance for a man without malignant nodes and a predicted probability of 0.8 for nonmalignant nodes is

$$\text{Deviance} = -\sqrt{-2\log(0.8)} = -0.668$$ **Equation 1.20**

Large values for deviance indicate that the model does not fit the case well. For large sample sizes, the deviance is approximately normally distributed.

The **Studentized residual** for a case is the change in the model deviance if the case is excluded. Discrepancies between the deviance and the Studentized residual may identify unusual cases. Normal probability plots of the Studentized residuals may be useful.

The **logit residual** is the residual for the model if it is predicted in the logit scale. That is,

$$\text{Logit residual}_i = \frac{\text{residual}_i}{P_i(1-P_i)} \qquad \textbf{Equation 1.21}$$

The **leverage** in logistic regression is in many respects analogous to the leverage in least-squares regression. Leverage values are often used for detecting observations that have a large impact on the predicted values. Unlike linear regression, the leverage values in logistic regression depend on both the dependent variable scores and the design matrix. Leverage values are bounded by 0 and 1. Their average value is p/n, where p is the number of estimated parameters in the model, including the constant, and n is the sample size.

Cook's distance is a measure of the influence of a case. It tells you how much deleting a case affects not only the residual for that case, but also the residuals of the remaining cases. Cook's distance (D) depends on the standardized residual for a case, as well as its leverage. It is defined as

$$D_i = \frac{Z_i^2 \times h_i}{(1-h_i)^2} \qquad \textbf{Equation 1.22}$$

where Z_i is the standardized residual and h_i is the leverage.

Another useful diagnostic measure is the change in the logistic coefficients when a case is deleted from the model, or **DfBeta**. You can compute this change for each coefficient, including the constant. For example, the change in the first coefficient when case i is deleted is

$$\text{DfBeta}(B_1^{(i)}) = B_1 - B_1^{(i)} \qquad \textbf{Equation 1.23}$$

where B_1 is the value of the coefficient when all cases are included and $B_1^{(i)}$ is the value of the coefficient when the ith case is excluded. Large values for change identify observations that should be examined.

Plotting Diagnostics

All of the diagnostic statistics described in this chapter can be saved for further analysis. If you save the values for the diagnostics, you can, when appropriate, obtain normal probability plots using the Examine procedure and plot the diagnostics using the Graph procedure (see the SPSS Base system documentation for more information on these procedures).

Figure 1.17 shows a normal probability plot and a detrended normal probability plot of the deviances. As you can see, the deviances do not appear to be normally distributed. That's because there are cases for which the model just doesn't fit well. In Figure 1.4, you see cases that have high probabilities for being in the incorrect group.

Figure 1.17 Normal probability of the deviances

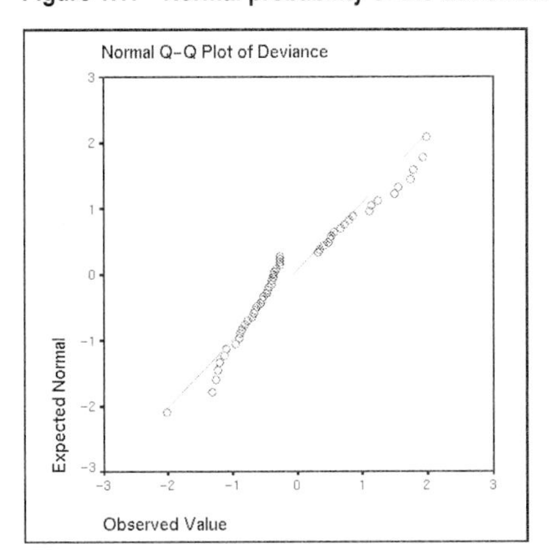

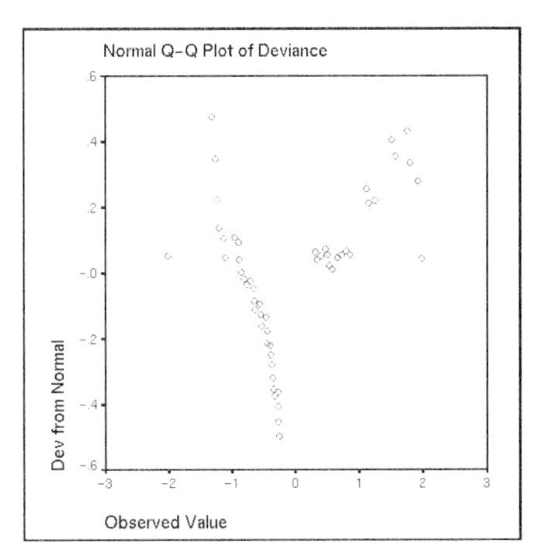

A plot of the standardized residuals against the case sequence numbers is shown in Figure 1.18. Again, you see cases with large values for the standardized residuals. Figure 1.19 shows that there is one case with a leverage value that is much larger than the rest. Similarly, Figure 1.20 shows that there is a case that has substantial impact on the estimation of the coefficient for *acid* (case 24). Examination of the data reveals that this case has the largest value for serum acid phosphatase and yet does not have malignant nodes. Since serum acid phosphatase was positively related to malignant nodes, as shown in Figure 1.2, this case is quite unusual. If we remove case 24 from the analysis, the coefficient for serum acid phosphatase changes from 0.0243 to 0.0490. A variable that was, at best, a very marginal predictor becomes much more important.

Figure 1.18 Plot of standardized residual with case ID

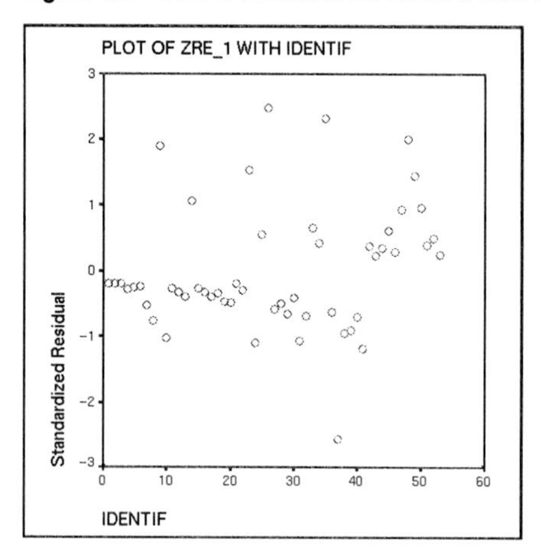

Figure 1.19 Plot of leverage with case ID

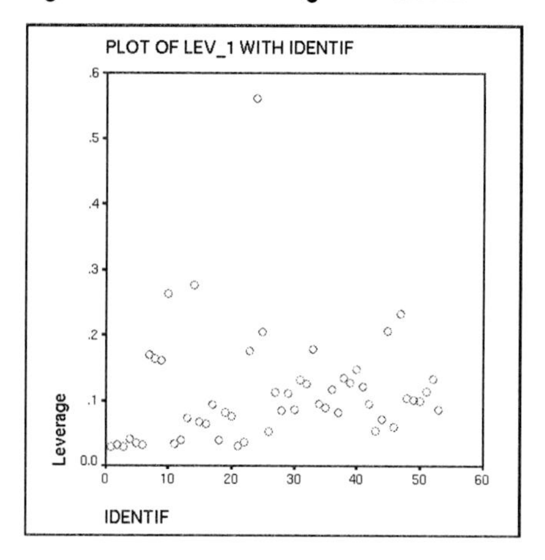

Figure 1.20 Plot of change in acid coefficient with case ID

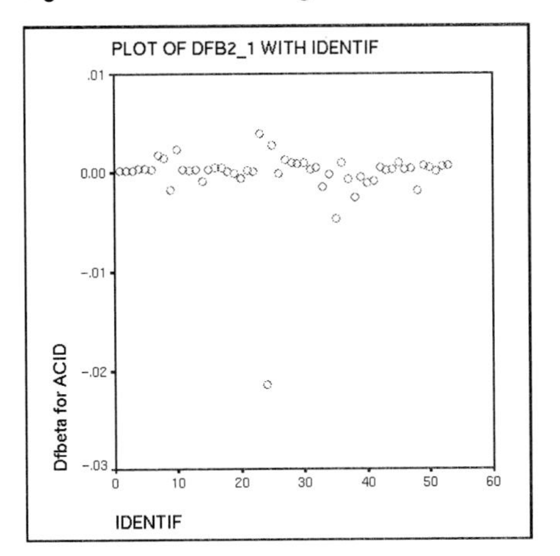

How to Obtain a Logistic Regression Analysis

The Logistic Regression procedure estimates the probability that an event will occur for a dichotomous dependent variable. You can force variables into the logistic regression model or use stepwise variable selection. Residual diagnostics, which help detect influential cases, outliers, and violations of model assumptions, are also available.

The minimum specifications are:

- A dichotomous dependent variable
- One or more predictor variables, or covariates

To obtain a logistic regression analysis, from the menus choose:

Statistics
  Regression ▸
    Logistic...

This opens the Logistic Regression dialog box, as shown in Figure 1.21.

Figure 1.21 Logistic Regression dialog box

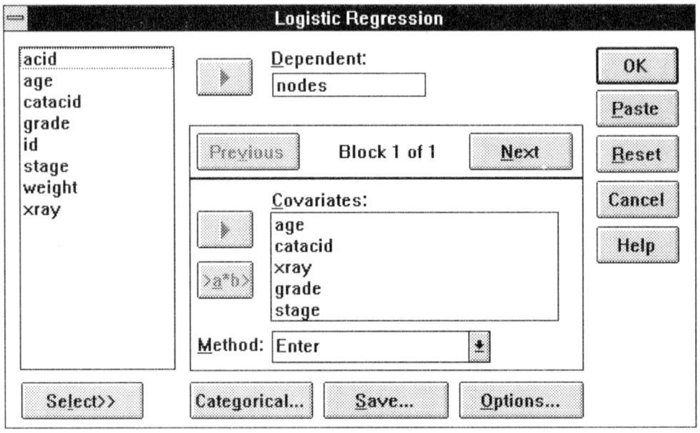

The numeric and short string variables in your working data file appear on the source list.

Dependent. Select a numeric or short string dependent variable that has only two values. If your dependent variable does not have exactly two nonmissing values, no analysis is produced.

Covariates. Select a block of one or more predictor variables. To create interaction terms, highlight two or more variables on the source list and click on a*b.

Each term in the block must be unique; however, a term can contain components of another term. For example, while *age*grade* can appear only once in a block, the same block can contain *age*, *grade*, and *age*grade*stage*.

You must also indicate which, if any, numeric covariates you want to treat as categorical (see "Defining Categorical Variables" on p. 26).

Macintosh: Use ⌘-click to select multiple factors or covariates, or a combination.

🡇 **Method**. Method selection controls the entry of the block of covariates into the model. For stepwise methods, the score statistic is used for entering variables; removal testing is based on the likelihood-ratio statistic or the Wald statistic. You can choose one of the following alternatives:

Enter. Forced entry. All variables in the block are entered in a single step. This is the default.

Forward: Conditional. Forward stepwise selection. Removal testing is based on the probability of the likelihood-ratio statistic based on conditional parameter estimates.

Forward: LR. Forward stepwise selection. Removal testing is based on the probability of the likelihood-ratio statistic based on the maximum-likelihood estimates.

Forward: Wald. Forward stepwise selection. Removal testing is based on the probability of the Wald statistic.

Backward: Conditional. Backward stepwise selection. Removal testing is based on the probability of the likelihood-ratio statistic based on conditional parameter estimates.

Backward: LR. Backward stepwise selection. Removal testing is based on the probability of the likelihood-ratio statistic based on the maximum-likelihood estimates.

Backward: Wald. Backward stepwise selection. Removal testing is based on the probability of the Wald statistic.

Covariates are entered into, or removed from, a single model. However, you can use different entry methods for different blocks of covariates. For example, you can enter one block using forced entry and another block using forward selection. After entering the first block of covariates into the model, click on Next to add a second block. If you do not want the default (forced entry), select an alternate entry method. To move back and forth between blocks of covariates, use Previous and Next.

Selecting a Subset of Cases for Analysis

By default, all cases in the working data file are used in the analysis. Optionally, you can estimate the logistic regression model for a subset of cases. Click on Select>> in the Logistic Regression dialog box. This expands the Logistic Regression dialog box, as shown in Figure 1.22.

Figure 1.22 Expanded Logistic Regression dialog box

Logistic Regression
acid
age
catacid
grade
id
stage
xray
Dependent:
nodes
Previous
Block 1 of 1
Next
Covariates:
age
catacid
xray
grade
stage
>a*b>
Method: Enter
OK
Paste
Reset
Cancel
Help
Select>>
Categorical...
Save...
Options...
Selection Variable:
weight
Set Value...

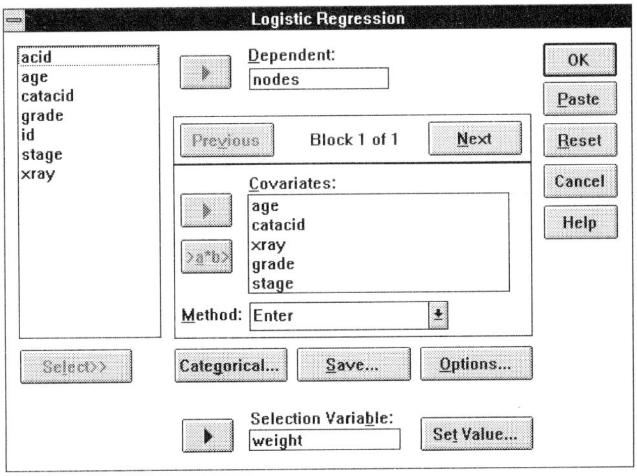

Case selection is based on values of a variable. Choose a numeric variable and then click on Set Value... to open the Logistic Regression Set Value dialog box, as shown in Figure 1.23.

Figure 1.23 Logistic Regression Set Value dialog box

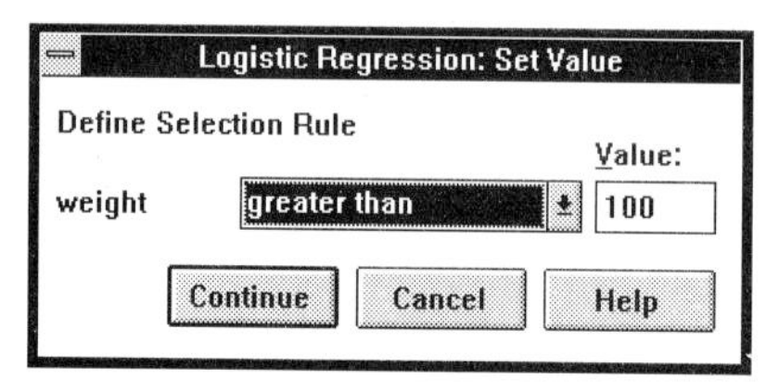

Select a relation from the drop-down list and enter a value for the variable you have specified. For example, if your selection variable is *weight* (measured in pounds), you can select individuals weighing more than 100 pounds by choosing greater than and entering 100. Only cases that satisfy the selection criterion—the selected cases—are used in model estimation. SPSS classifies selected and unselected cases separately, so you can use the case selection feature to build a model for one set of cases and apply the model to another set of cases.

Defining Categorical Variables

By default, SPSS treats string covariates as categorical and numeric variables as continuous. To treat one or more numeric covariates as categorical, click on Categorical... in the Logistic Regression dialog box. This opens the Logistic Regression Define Categorical Variables dialog box, as shown in Figure 1.24.

Figure 1.24 Logistic Regression Define Categorical Variables dialog box

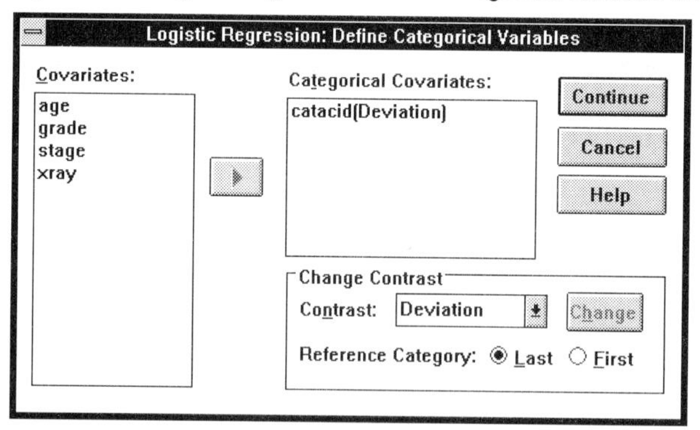

Categorical Covariates. Select numeric variables you want to treat as categorical from the list of numeric covariates. SPSS transforms categorical covariates and interaction terms containing one or more categorical covariates into contrast variables, which are entered or removed from the model as a set. String covariates are always treated as categorical (you can remove string variables from the list of categorical covariates only by removing them from the covariates list in the Logistic Regression dialog box).

Change Contrast. By default, each categorical covariate is transformed into a set of deviation contrasts. To obtain a different type of contrast, highlight one or more covariates, select a contrast type, and click on Change. Optionally, you can change the default reference category.

- **Contrast**. You can choose one of the following types of contrasts:

  Deviation. Each category of the predictor variable except the reference category is compared to the overall effect. This is the default.

  Simple. Each category of the predictor variable (except the reference category) is compared to the reference category.

  Difference. Also known as reverse Helmert contrasts. Each category of the predictor variable except the first category is compared to the average effect of previous categories.

  Helmert. Each category of the predictor variable except the last category is compared to the average effect of subsequent categories.

  Repeated. Each category of the predictor variable except the first category is compared to the category that precedes it.

  Polynomial. Orthogonal polynomial contrasts. Categories are assumed to be equally spaced. Polynomial contrasts are available for numeric variables only.

  Indicator. Contrasts indicate the presence or absence of category membership. The reference category is represented in the contrast matrix as a row of zeros.

  Reference Category. For deviation, simple, and indicator contrasts, you can override the default reference category. Choose one of the following alternatives:

  - **Last**. Uses the last category of the variable as the reference category. This is the default.
  - **First**. Uses the first category as the reference category.

Saving New Variables

To save predicted values, measures of influence, or residuals, click on Save... in the Logistic Regression dialog box. This opens the Logistic Regression Save New Variables dialog box, as shown in Figure 1.25.

Figure 1.25 Logistic Regression Save New Variables dialog box

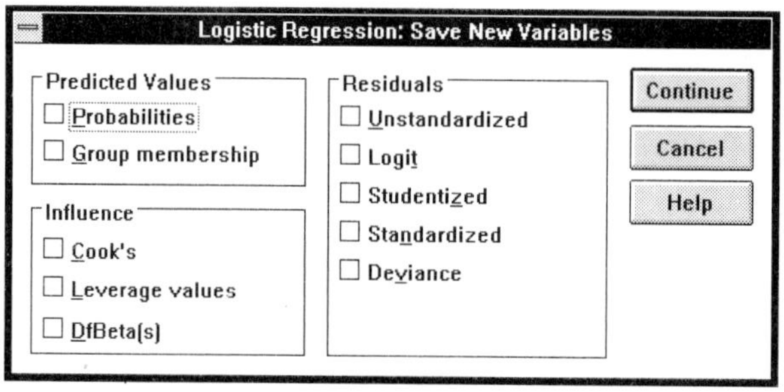

Logistic Regression: Save New Variables

Predicted Values
- ☐ Probabilities
- ☐ Group membership

Influence
- ☐ Cook's
- ☐ Leverage values
- ☐ DfBeta(s)

Residuals
- ☐ Unstandardized
- ☐ Logit
- ☐ Studentized
- ☐ Standardized
- ☐ Deviance

Continue | Cancel | Help

A table in the output displays names and contents of any new variables.

Predicted Values. You can choose one or both of the following predicted values:

- ❑ **Probabilities**. For each case, the predicted probability of occurrence of the event.
- ❑ **Group membership**. The group to which a case is assigned based on its predicted probability.

Influence. You can choose one or more of the following measures of influence:

- ❑ **Cook's**. The logistic regression analog of Cook's influence statistic.
- ❑ **Leverage values**. The relative influence of each observation on the model's fit.
- ❑ **DfBeta(s)**. The change in a regression coefficient when a case is omitted. One variable is saved for each term in the model.

Residuals. You can choose one or more of the following residuals:

- ❑ **Unstandardized**. The difference between observed and predicted values.
- ❑ **Logit**. Residuals in the logit scale.
- ❑ **Studentized**. The change in the model deviance if a case is excluded.
- ❑ **Standardized**. Residuals divided by their standard deviation.
- ❑ **Deviance**. Residuals based on the model deviance.

Options

To obtain optional statistics or plots, or to change model-building criteria, click on Options... in the Logistic Regression dialog box. This opens the Logistic Regression Options dialog box, as shown in Figure 1.26.

Figure 1.26 Logistic Regression Options dialog box

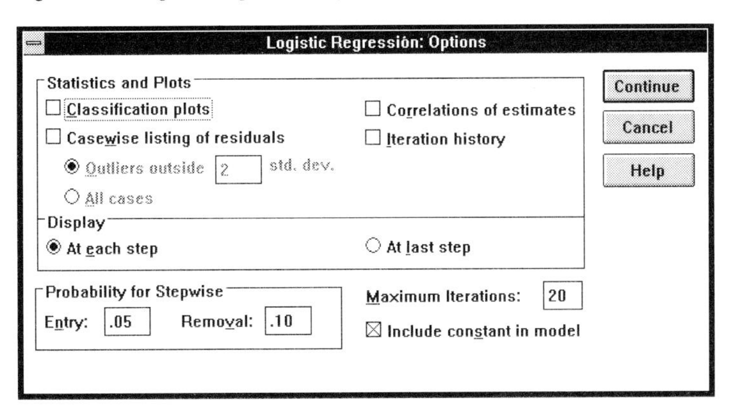

Statistics and Plots. The default output includes goodness-of-fit tests for the model, a classification table, and parameter estimates for terms in the equation. You can also choose one or more of the following plots and statistics:

- **Classification plots**. Histogram of the actual and predicted values of the dependent variable.
- **Casewise listing of residuals**. Displays unstandardized residuals, predicted probability, and observed and predicted group membership. You can choose one of the following alternatives for the casewise listing:
  - **Outliers outside n std. dev**. For the casewise plot, displays only cases with absolute Studentized residuals greater than a cutoff value. This is the default. To override the default cutoff value (2), enter a non-negative value.
  - **All cases**. Displays all cases in the casewise plot.
- **Correlations of estimates**. Displays a correlation matrix of parameter estimates for terms in the model.
- **Iteration history**. Displays coefficients and log likelihood at each iteration of the parameter estimation process.

Display. You can choose one of the following alternatives:

- **At each step**. Prints plots, tables, and statistics at each step. This is the default.
- **At last step**. Prints plots, tables, and statistics for the final model for a block. Summarizes intermediate steps.

Probability for Stepwise. Entry and removal criteria for stepwise variable selection. By default, a variable is entered into the model if the probability of its score statistic is less than 0.05. A variable is removed if its removal statistic (conditional, Wald, or likelihood ratio) has a probability greater than or equal to 0.10. To override the default stepwise criteria, enter entry and removal probabilities. The stepwise probabilities do not affect the direct-entry method.

Maximum Iterations. Maximum-likelihood coefficients are estimated in an iterative process. Convergence is attained if, from one iteration to the next, the change in parameter estimates is less than 0.001 or the log likelihood decreases less than 0.01%. Iteration terminates before convergence if the default maximum number of iterations (20) is reached. To specify a different maximum number of iterations, enter an integer value.

The following option is also available:

❑ **Include constant in model.** By default, the model includes a constant term (intercept). To suppress the constant term and obtain regression through the origin, deselect this item.

Additional Features Available with Command Syntax

You can customize your logistic regression analysis if you paste your selections into a syntax window and edit the resulting LOGISTIC REGRESSION command syntax. (For information on syntax windows, see the SPSS Base system documentation.) Additional features include:

- Additional contrast options, such as user-defined contrasts and alternate reference categories (using the CONTRAST subcommand).
- The ability to label casewise output with values of a variable (using the ID subcommand).
- Parameter estimates for every *n*th iteration (using the PRINT subcommand).
- Additional diagnostic measures for the casewise listing (using the CASEWISE subcommand).
- Additional criteria for estimation and model building (using the CRITERIA subcommand).
- User-specified names for new variables (using the SAVE subcommand).

See the Syntax Reference section of this manual for command syntax rules and for complete LOGISTIC REGRESSION command syntax.

2 General Factorial Analysis of Variance

In the SPSS Base system documentation, you saw analysis-of-variance techniques applied to some simple problems. In this and subsequent chapters, you'll see how analysis of variance can be used to analyze simple problems in more detail, as well as more complicated problems.

A Simple Factorial Design

When you want to examine the effect of several independent variables (factors)—for example, type of gasoline and car model—on a dependent variable, such as gas mileage or acceleration, you can use a factorial experimental design. For example, if you want to study four types of gasoline and three models of cars, you would take cars of each of the different model types and randomly assign them to the different gasolines. Each type of car is called a **level** of the car factor and, similarly, each brand of gasoline is a level of the gasoline variable. Schematically, this design is shown in Table 2.1.

Table 2.1 Two-factor model

		Cars		
		Model A	Model B	Model C
Gasoline	Brand 1			
	Brand 2			
	Brand 3			
	Brand 4			

Table 2.1 contains 12 cells, one for each combination of the two factor variables: car model and gasoline brand. If you have 80 cars of each model available for your experiment, you would assign 20 of them to each brand of gasoline. You would have 20 different cars in each cell. If the same car is tested using all four brands of gasoline, you have a special type of experiment called a **repeated measures design**. Special statistical techniques are needed for analysis of such data. (See Chapter 4 for a discussion of experiments where different levels of the same factor are applied to the same experi-

mental unit.) Why bother with this type of design? Why not do two separate experiments, one to assess the effect of different brands of gasoline on mileage and one to assess the effect of different models of cars on mileage? One of the advantages of a factorial design is that it can answer several important questions at once using the same cars. Not only can you draw conclusions about the effects of the different models and the different brands of gas, but you can also test whether there is an interaction of the two factors. That is, do particular brands of gasoline perform better in certain models of cars?

Factorial designs are by no means limited to two factors; you can have many more. For example, if you were also interested in the effect of five different climates on mileage, you would have a three-factor design with 60 cells. There would be a cell for each possible combination of auto model, gas brand, and climate. You could then test not only for the effects of auto, gas, and climate, but also for interactions between all pairs of factors (two-way interactions) and the interaction of all three variables (the three-way interaction).

Where Are We?

To see how you would go about analyzing a factorial design in SPSS, let's consider a simple example from Winer et al. (1991). You are interested in determining the effectiveness of three instructors who teach map reading using two different methods. You have two factors of interest: the method of teaching (variable *method*), which has two levels, and the instructor (variable *inst*), which has three levels. Five subjects are randomly assigned to each of the six cells in the design. For each, you record a map-reading achievement test score before and after the training. First, we'll analyze just the difference between the post-test and the pretest scores (variable *dif*). In the section "An Analysis-of-Covariance Model" on p. 44, we'll use analysis of covariance to analyze the same type of data.

In all of the examples in this chapter, we will consider what are known as **fixed-effects designs**. These are designs in which the factors and levels under study are the only ones that you are interested in drawing conclusions about. This means that the two methods for teaching are not considered to be a sample from many possible methods for teaching map reading, and that the three instructors are not considered to be a sample from a possible pool of instructors. Models in which the factor levels under study are considered a sample of possible levels are called **random-effects designs** (also known as **variance-component designs**). See Winer et al. (1991) for a discussion of such models.

Describing the Data

Before proceeding to a formal analysis of data, it's always a good idea to calculate descriptive statistics and displays. These will help you plan your analyses and interpret your results. Clustered boxplots showing each combination of method and instructor are shown in Figure 2.1 (Clustered boxplots are available in the Boxplot procedure. For more information, see the Base system documentation.) From the boxplots, you see that the median change scores for method 1 are a little lower than for method 2 and that instructor 3 has the smallest change score for both methods.

Figure 2.1 Boxplots for cells

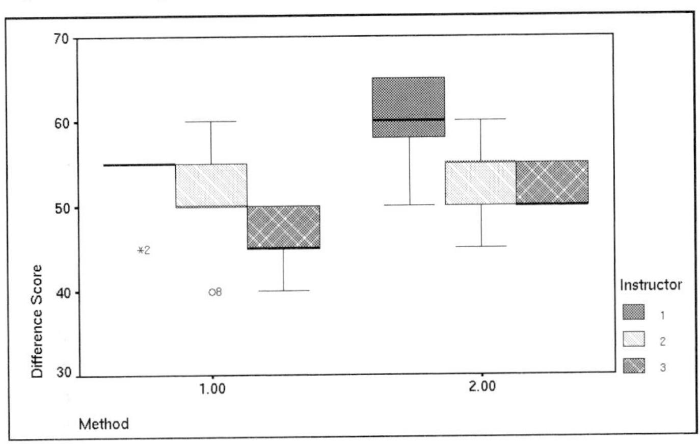

Means and standard deviations for each of the cells are shown in Figure 2.2. You see that all of the instructors have larger average change scores for method 2 than for method 1. You also see that the first instructor has the largest average scores for both methods, while the third instructor has the smallest average scores for both methods. The distribution of means for the six cells is shown in Figure 2.3. You see that three of the cell means are very close to each other and that two are somewhat distant from the rest.

Figure 2.2 Cell statistics from SPSS Means procedure

```
- - Description of Subpopulations - -

Summaries of      DIF
By levels of      METHOD
                  INST

Variable         Value  Label                          Mean     Std Dev    Cases

For Entire Population                               52.4333      6.3283       30

METHOD            1.00                              50.0000      5.9761       15
  INST            1.00                              53.0000      4.4721        5
  INST            2.00                              51.0000      7.4162        5
  INST            3.00                              46.0000      4.1833        5

METHOD            2.00                              54.8667      5.8781       15
  INST            1.00                              59.6000      6.1887        5
  INST            2.00                              53.0000      5.7009        5
  INST            3.00                              52.0000      2.7386        5

  Total Cases = 30
```

Figure 2.3 Histogram of cell means

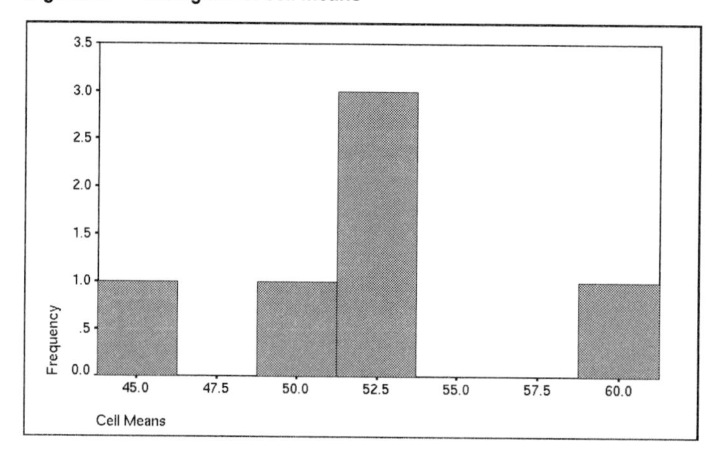

Equality of Variances

The assumptions needed for analysis of variance are that for each cell, the data are a random sample for a normal population and that in the population, all of the cell variances are the same. There are several ways to check the equality-of-variance assumption. You can compute the homogeneity-of-variance tests shown in Figure 2.4. For this example, neither test leads you to reject the null hypothesis that all population cell variances are equal. Both of the homogeneity-of-variance tests are sensitive to departures from normality, so you should keep this in mind when interpreting the results.

Figure 2.4 Homogeneity-of-variance tests

```
Univariate Homogeneity of Variance Tests

 Variable .. DIF

       Cochrans C(4,6) =                    .32201, P =  .520 (approx.)
       Bartlett-Box F(5,741) =              .79055, P =  .557
```

One of the common ways in which the equality-of-variance assumption is violated is when the variances or standard deviations are proportional to the cell means. That is, larger variances are associated with larger values of the dependent variable. To look for a possible relationship between the variability and the cell means, you can plot the variances and standard deviations against the cell means, as shown in Figure 2.5 and Figure 2.6. In this example, there doesn't appear to be a relationship between the measures of variability and the cell means.

Figure 2.5 Plot of cell variances against cell means

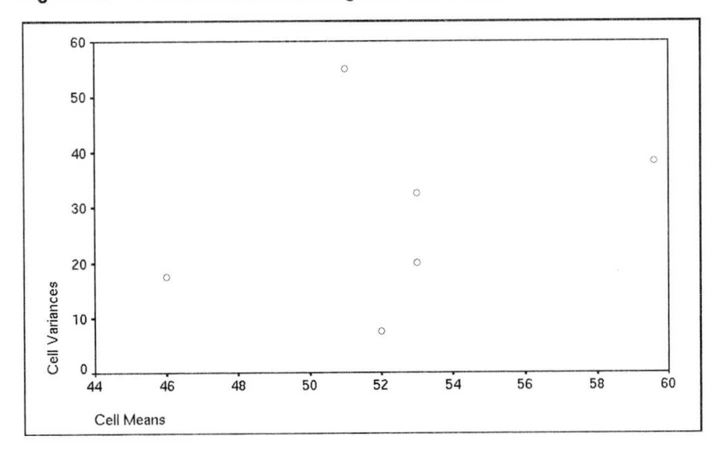

Figure 2.6 Plot of cell standard deviations against cell means

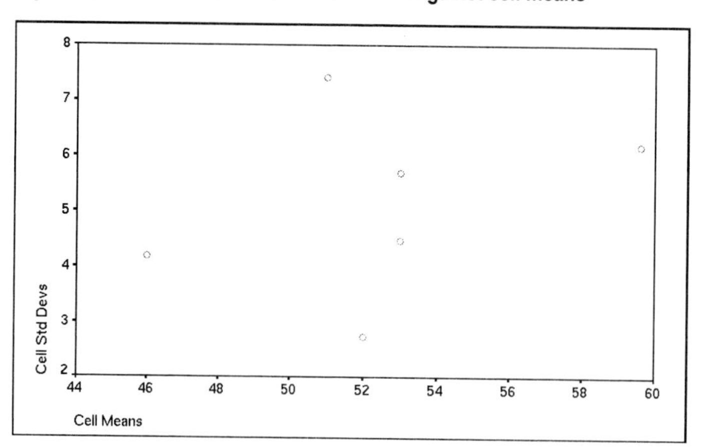

The Analysis-of-Variance Table

When you have a factorial design, you can test hypotheses about the **main effects** (the factor variables individually) and the **interactions** (various combinations of factor variables). Consider Figure 2.7, which is the analysis-of-variance table for the map-reading experiment. (For general discussion of the analysis-of-variance table, see the Base system documentation.)

Figure 2.7 ANOVA table for full factorial model

Tests of Significance for DIF using UNIQUE sums of squares

Source of Variation	SS	DF	MS	F	Sig of F
WITHIN CELLS	683.20	24	28.47		
METHOD	177.63	1	177.63	6.24	.020
INST	269.27	2	134.63	4.73	.019
METHOD BY INST	31.27	2	15.63	.55	.585

The first row of the table is labeled *WITHIN CELLS*. The mean square for this term is the estimate of variability derived from the individual cells. The within-cells means square will be the denominator for the *F* tests in the ANOVA table.

The last row of the table, *METHOD BY INST*, provides a test of the method-by-instructor interaction. This tests the hypothesis that the effect of the methods is the same across the instructors. (Interaction terms are discussed in more detail in the Base system documentation.) This means that there are not particular combinations of methods and instructors that perform better or worse than we would expect based on considering the two effects, method and instructor, individually. In this example, the observed significance level for the interaction term is large ($p = 0.585$), so you cannot reject the null hypothesis that there is no interaction between the two variables. When you find a significant interaction term, you should not perform tests of the main effects, since it makes no sense to say that, overall, the methods or instructors differ or do not. In this situation, effectiveness depends on the combinations of instructors and methods, and you must consider both of the variables together when describing results.

The row labeled *METHOD* tests whether the two methods are equally effective for teaching map reading. Since the observed significance level is small ($p = 0.020$), you can reject the null hypothesis that the two methods are equally effective. The next row, labeled *INST*, is a test of the instructor effect. Again, since the observed significance level is small ($p = 0.019$), you can reject the null hypothesis that the three instructors are equally effective.

Full Factorial Model

In the example we've considered, there were two main-effect terms (method and instructor) and one interaction term (method by instructor). Since these three terms represent all of the possible main effects and the interaction, the model is said to be a **full factorial model**. All possible main-effect and interaction terms are included. When analyzing a factorial design, you need not include all possible terms in a model. Although you will usually include all main-effect terms, when the number of factors is large, you may want to suppress some of the higher-order interactions.

For example, if you have reason to believe that there is no method-by-instructor interaction, you can fit a model that includes only main effects. Figure 2.8 shows the analysis-of-variance table when the method-by-instructor interaction is not tested. Instead, it is included in the within+residual error term, which is now used for the denominator of the F tests.

Figure 2.8 ANOVA table for main-effects model

```
Tests of Significance for DIF using UNIQUE sums of squares
Source of Variation          SS      DF        MS         F  Sig of F

WITHIN+RESIDUAL          714.47      26     27.48
METHOD                   177.63       1    177.63      6.46      .017
INST                     269.27       2    134.63      4.90      .016
```

Examining Contrasts

An analysis-of-variance table provides summary results. It tells you whether there is a factor or interaction effect, but it doesn't pinpoint which factor levels are different. One way to examine factor level differences is to construct linear contrasts of their means. For example, you can compare the average of each factor level to the grand mean (the average overall factor levels). These are known as **deviation contrasts**.

To see how this is done, consider Figure 2.9, which shows the overall mean (from the SPSS Descriptives procedure), and Figure 2.10, which shows the observed means for each of the method and instructor factor levels. (You'll note that two entries are printed for each level, one labeled *WGT.* for weighted mean and the other *UNWGT.* for unweighted mean. In this example, both entries are the same, since the same number of subjects is observed in each cell of the design. If there are different numbers of cases in the cells, unweighted means are obtained simply by averaging the means of the individual cells that are part of that effect. All cells contribute equally, regardless of how many cases are in a cell. Weighted means take into account the number of cases in a cell and average the cell means by weighting them by the number of cases in a cell.)

Figure 2.9 Descriptive statistics for change score

```
                                                                  Valid
Variable          Mean      Std Dev    Minimum    Maximum          N   Label

DIF               52.43        6.33      40.00      65.00         30
```

Figure 2.10 Observed means by method and instructor

```
Combined Observed Means for METHOD
 Variable .. DIF
        METHOD
             1          WGT.      50.00000
                      UNWGT.      50.00000
             2          WGT.      54.86667
                      UNWGT.      54.86667

- - - - - - - - - - - - - - - - - - - - - - - - - - - - - - - - - - - - - - -
 Combined Observed Means for INST
 Variable .. DIF
          INST
             1          WGT.      56.30000
                      UNWGT.      56.30000
             2          WGT.      52.00000
                      UNWGT.      52.00000
             3          WGT.      49.00000
                      UNWGT.      49.00000
```

From Figure 2.9 and Figure 2.10, you can calculate the deviation contrasts, which are the differences between the mean for each level of a factor and the overall mean. For example, for method 1, the value of the deviation contrast is 50 (its value) minus 52.43 (the overall mean), or –2.43. For method 2, the deviation contrast is 54.86 minus 52.43, or 2.43. Similarly, for instructor 1, the deviation contrast is 56.30 minus 52.43, or 3.87.

The values for the deviation contrasts (labeled *Estimates for DIF*) are shown in Figure 2.11. The number of parameter estimates printed is one less than the number of levels of the factor. That's because you can always calculate what the values for the omitted factor are, based on the printed values. You can test the null hypothesis that the value of a parameter is 0, using the *t* value and its associated significance level. You can also compute confidence intervals for each of the parameters.

Figure 2.11 Deviation contrasts

```
Effect Size Measures and Observed Power at the .0500 Level
                              Partial Noncen-
 Source of Variation          ETA Sqd trality       Power

 METHOD                          .206    6.240       .667
 INST                            .283    9.459       .736
 METHOD BY INST                  .044    1.098       .132

 Estimates for DIF
 --- Individual univariate .9500 confidence intervals
 --- two-tailed observed power taken at .0500 level
 METHOD

  Parameter      Coeff.  Std. Err.     t-Value    Sig. t Lower -95%  CL- Upper   Noncent.   Power

        2   -2.4333333      .97411    -2.49801    .01974   -4.44380    -.42287    6.24005    .667
 INST

  Parameter      Coeff.  Std. Err.     t-Value    Sig. t Lower -95%  CL- Upper   Noncent.   Power

        3   3.86666667     1.37760     2.80682    .00977    1.02344    6.70989    7.87822    .766
        4   -.43333333     1.37760     -.31456    .75582   -3.27656    2.40989     .09895    .053
```

From the parameter estimates in Figure 2.11, you see that method 1 (labeled *Parameter 2*) has a change score significantly worse than average. Similarly, instructor 1 (*Parameter 3*) has an average change score significantly better than the average for the sample. Instructor 2, however, has a change score that does not differ significantly from the average of the change scores. The parameter estimate for instructor 3 is –3.433, since the coefficients for deviation parameter estimates must sum to 0 over all levels of a factor.

Simple Contrasts

The parameter estimates in Figure 2.11 are computed by comparing each level of the factor to the overall mean. This is but one of many types of contrasts that can be computed. (See "Specifying Contrasts" on p. 50 for other types of contrasts available in SPSS.) Another type of contrast that is often used is called the **simple contrast**. With simple contrasts, you compare each level of a factor not to the average of all levels, but to a reference level. For example, if you choose the last category as the reference category (the default), you will obtain the parameter estimates shown in Figure 2.12. The values for the coefficients are the differences between each factor level and the last factor level. The coefficients for the last factor level, which are not printed, are all 0, since it is the reference to which the levels are compared.

Figure 2.12 Simple contrasts

```
Estimates for DIF
 --- Joint univariate .9500 BONFERRONI confidence intervals
 --- two-tailed observed power taken at .0500 level
 METHOD

  Parameter      Coeff.  Std. Err.    t-Value   Sig. t Lower -95%  CL- Upper   Noncent.     Power

          2  -4.8666667    1.94822   -2.49801   .01974   -8.88759    -.84574    6.24005      .667
 INST

  Parameter      Coeff.  Std. Err.    t-Value   Sig. t Lower -95%  CL- Upper   Noncent.     Power

          3  7.30000000    2.38607    3.05942   .00539    1.59503   13.00497    9.36007      .834
          4  3.00000000    2.38607    1.25730   .22074   -2.70497    8.70497    1.58080      .225
```

The Problem of Multiple Comparisons

In the Base system documentation, we discussed the problems associated with making many nonindependent comparisons. As the number of comparisons increases, so does the probability that you will call a difference "statistically significant" when, in fact, it is not. The same problem occurs when you examine a set of contrasts. As the number of contrasts you examine increases, so does the likelihood that you will call differences "significant" when they are not. Multiple comparison procedures similar to those used for comparing all possible pairs of means can be used for contrasts as well. The goal is the same—to provide control over the Type 1 error rate. (Remember, a Type 1 error occurs when you reject the null hypothesis when, in fact, it is true.)

The SPSS General Factorial ANOVA procedure can be used to calculate confidence intervals for parameters using a Bonferroni or Scheffé correction. These intervals are known as **simultaneous confidence intervals** (also called **joint confidence intervals**), since they are constructed for all of the parameters together. Both of these methods result in wider confidence intervals than when no protection for multiple comparisons is made. The Bonferroni method is based on the number of comparisons actually made, while the Scheffé method is based on all possible contrasts. For a large number of contrasts, Bonferroni intervals will be wider than the Scheffé intervals. Both the Scheffé and Bonferroni intervals are computed separately for each term in the design. See Timm (1975) for a discussion of the comparative merits of the intervals.

Once again, consider Figure 2.12. The confidence intervals in the figure are simultaneous Bonferroni confidence intervals. To compute Bonferroni protected tests for each of the parameters, do not use the *Sig. t* entry, but instead determine whether the 95% confidence interval for a parameter includes 0. If it does, you cannot reject the null hypothesis that in the population the parameter value is 0, using a 5% significance level.

Measuring Effect Size

In a one-way analysis-of-variance design, the **eta-squared statistic** is sometimes used to describe the proportion of the total variability "explained" by the grouping or factor variable. A value close to 1 indicates that all of the total variability is attributable to differences between the groups, while a value close to 0 indicates that the grouping variable explains little of the total variability. Eta squared is sometimes called an **effect-size measure**.

For a factorial design, you can compute several effect-size measures based on the eta-squared statistic. The **partial eta-squared statistic** has the same numerator as the eta-squared statistic: the sum of squares for the effect of interest. But instead of the total sum of squares in the denominator, it has the sum of the sum of squares for error and the sum of squares for the effect of interest. From Figure 2.11, you see that the partial eta squared for the method effect is 0.21 $(177.63/(177.63 + 683.20))$.

Power Computations

Whenever you perform an experiment or conduct a study, you should be concerned with **power** (your ability to reject the null hypothesis when it is false). In general, power depends on the magnitude of the true differences and the sample sizes. If the true differences are very large, even small sample sizes should detect them. If, on the other hand, the true differences are small, you will need large sample sizes to detect them. Your decision on how many experimental units to include in your study should be based on power considerations. That is, you want your sample to be large enough so that you stand a good chance of detecting differences you consider important. See Cohen (1977) for a discussion of power analysis.

Sometimes it is also useful to evaluate power after an experiment has been completed. You calculate the probability that you would call your observed difference statistically significant at a chosen alpha level based on the sample size used. For example, consider the simple situation of two independent groups with 10 cases in each. If you found that the difference in average change scores was 7 and the pooled estimate of the standard deviation within a group was 10, you could calculate, based on tables available in Cohen (1977), that your power of detecting this difference, using a 5% significance level for the *t* test, would be about 0.31. This means that more than half of the time you conducted such an experiment, you would fail to reject the null hypothesis when, in fact, it was false. Thus, even though on the basis of your sample data you failed to reject the null hypothesis that the true difference is 0 (your *t* value is only 1.57), you know that your experiment had limited power to detect a true difference of 7 points or less.

For analysis-of-variance models, the computation of power is considerably more complicated. However, the basic idea is the same as for the *t* test. You assume that the observed effect size is a true difference, and, based on the sample size you have used and the alpha level you have selected, you calculate the probability that you would have rejected the null hypothesis. If the power is small, you're not very confident about a decision not to reject the null hypothesis. Figure 2.11 shows the power values for the terms in the map-reading example.

Examining Residuals

Once you have estimated an analysis-of-variance model, you can examine the residuals to see how well it fits the data. For each case, you have an observed value and can calculate a predicted value based on the model. (To calculate a predicted value, you estimate parameters for each of the terms in your model.) Figure 2.13 is a listing of the residuals. For each of the cases, you see the observed or actual change score, the score predicted from the model, the raw residual, and the standardized residual. The raw residual is the difference between the observed and predicted values. The standardized residual is the raw residual divided by an estimate of its standard deviation. In this example, the standard deviation is the square root of the within-cells mean square.

Figure 2.13 Residuals listing

Observed and Predicted Values for Each Case
Dependent Variable.. DIF

Case No.	Observed	Predicted	Raw Resid.	Std Resid.
1	55.000	53.000	2.000	.375
2	45.000	53.000	-8.000	-1.499
3	55.000	53.000	2.000	.375
4	55.000	53.000	2.000	.375
5	55.000	53.000	2.000	.375
6	55.000	51.000	4.000	.750
7	60.000	51.000	9.000	1.687
8	40.000	51.000	-11.000	-2.062
9	50.000	51.000	-1.000	-.187
10	50.000	51.000	-1.000	-.187
11	40.000	46.000	-6.000	-1.125
12	45.000	46.000	-1.000	-.187
13	50.000	46.000	4.000	.750
14	50.000	46.000	4.000	.750
15	45.000	46.000	-1.000	-.187
16	50.000	59.600	-9.600	-1.799
17	65.000	59.600	5.400	1.012
18	60.000	59.600	.400	.075
19	65.000	59.600	5.400	1.012
20	58.000	59.600	-1.600	-.300
21	50.000	53.000	-3.000	-.562
22	60.000	53.000	7.000	1.312
23	55.000	53.000	2.000	.375
24	45.000	53.000	-8.000	-1.499
25	55.000	53.000	2.000	.375
26	50.000	52.000	-2.000	-.375
27	55.000	52.000	3.000	.562
28	50.000	52.000	-2.000	-.375
29	55.000	52.000	3.000	.562
30	50.000	52.000	-2.000	-.375

You can also look at various plots of the residuals to see if the analysis-of-variance assumptions have been violated. Figure 2.14 contains a scatterplot matrix of the observed and predicted values and the standardized residuals. You should not see a pattern in the plot of residuals against predicted values (the second plot in the last row of the matrix). If the spread of the residuals increases with the magnitude of the predicted values, you have reason to suspect that the variance may not be constant in all cells. If the model fits and the assumption of normality is not violated, the distribution of residuals should be approximately normal. Figure 2.15 is a plot of observed residuals against those expected from a normal distribution. If the normality assumption is not violated, the points should fall more or less on the normal line. Figure 2.16 is a detrended normal plot that shows the distances between the observed points and the expected line. You should not see any pattern in a detrended plot if the assumption of normality is met.

Figure 2.14 Scatterplot matrix of observed and predicted values and residuals

Observed

Predicted

Residuals

Figure 2.15 Normal probability plot of standardized residuals

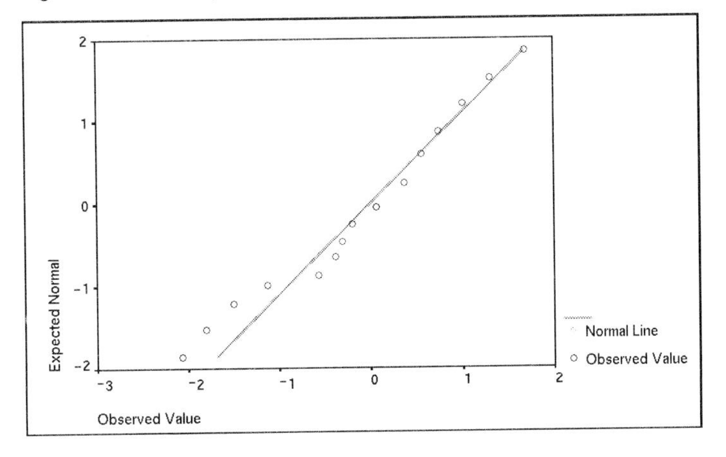

Figure 2.16 Detrended normal plot of standardized residuals

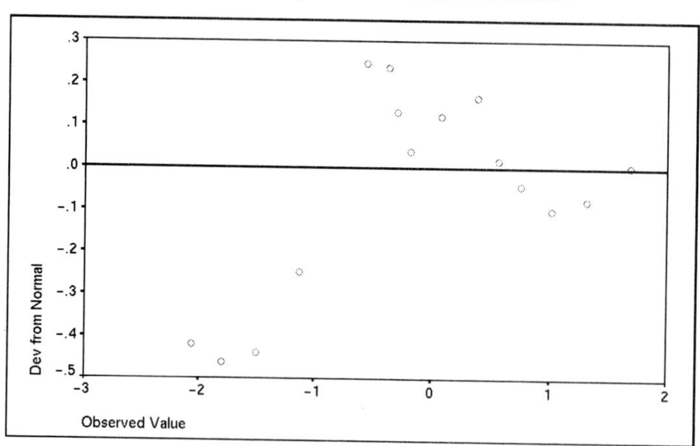

An Analysis-of-Covariance Model

In the map-reading experiment, we had two variables for each subject—the score before the training program and the score after the training program. We had two variables because we knew that differences in final scores are attributable not only to differences between the methods and the instructors but also to differences among the subjects. Everyone did not read maps equally well before the training started. By considering the differences between the two scores, we controlled for some of the initial differences among subjects. In this experiment, the two scores were obtained from similar tests, so computing differences between the two measurements was reasonable.

If we didn't use a pretest but instead recorded a variable, such as IQ or years of education, thought to be related to map-reading aptitude, we couldn't analyze the data by simply computing differences. Instead, we'd need a statistical technique that would allow us to incorporate initial differences among subjects in our analysis.

Analysis of covariance is a statistical technique that can be thought of as a combination of linear regression analysis and analysis of variance. The analysis-of-variance results are adjusted for the linear relationships between the dependent variable and the covariates.

To see how analysis of covariance is used, consider the following example from Winer et al. (1991). Seven subjects are randomly assigned to each of three methods for teaching a course. The dependent variable is a measure of achievement after completing the course. The covariate is an aptitude score obtained prior to the course. Figure 2.17 is a plot of the achievement and aptitude scores for subjects in the three groups. You see that there is a fairly strong linear relationship between the two variables. This is important, since if there were no linear relationship between the two variables, it would make little sense to try to adjust achievement scores based on aptitude scores using a linear regression model. From the plot, it also appears that the relationship between the two variables is similar in the three teaching groups. That is, the three groups have almost parallel regression lines.

Figure 2.17 Plot of achievement against aptitude score

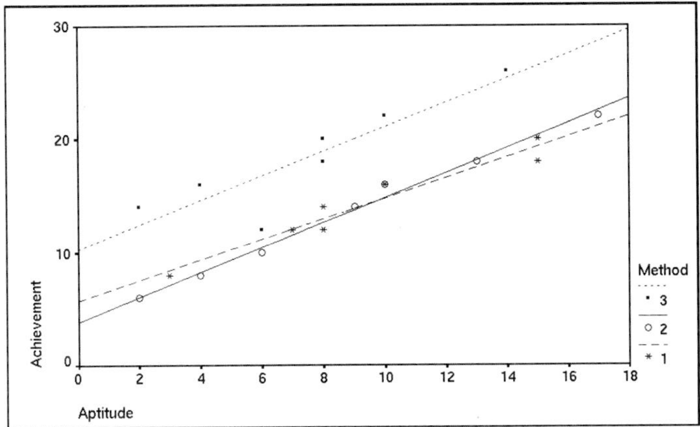

As a first step, let's see what the analysis-of-variance table for achievement looks like when we ignore the aptitude scores (see Figure 2.18). You see that, based on the observed significance level for the method effect, there is not sufficient evidence to reject the null hypothesis that all three teaching methods are equally effective.

Figure 2.18 Analysis-of-variance table for achievement

Tests of Significance for ACHIEVE using UNIQUE sums of squares

Source of Variation	SS	DF	MS	F	Sig of F
WITHIN CELLS	436.57	18	24.25		
METHOD	94.10	2	47.05	1.94	.173

Now let's see what happens when we adjust for initial differences between the aptitudes of the subjects in the three groups. Our goal will be to decrease the error sums of squares by removing the variability due to differences in aptitudes.

Testing for Equal Slopes

Before we perform the analysis of covariance, we have to determine whether it is reasonable to assume that the slopes of the lines relating achievement and aptitude are the same in the three groups. From visual inspection of Figure 2.17, that seemed plausible, but we need a formal test. Figure 2.19 contains the analysis-of-variance table that can be used to test whether the slopes of the regression line are the same in all three groups. The term of interest is the method-by-aptitude interaction. If this interaction is significant, you reject the hypothesis that the slope is the same for all three methods. In this example, the interaction is not significant, so we will fit an analysis-of-covariance model with a common slope. (If the assumption of a common slope is rejected, you can fit separate slopes for each of the cells).

Figure 2.19 Test of homogeneity of slopes

Tests of Significance for ACHIEVE using UNIQUE sums of squares

Source of Variation	SS	DF	MS	F	Sig of F
WITHIN+RESIDUAL	38.11	15	2.54		
APTITUDE	370.97	1	370.97	146.00	.000
METHOD	31.24	2	15.62	6.15	.011
METHOD BY APTITUDE	2.62	2	1.31	.52	.607

The Analysis-of-Covariance Table

The analysis-of-covariance table for this example is shown in Figure 2.20. You see that there is a row of the table that is labeled *REGRESSION*. This is a test of the hypothesis that the common slope is 0. Since the observed significance level is very small, you can reject the null hypothesis that the slope is 0.

Figure 2.20 Analysis-of-covariance table for achievement

Tests of Significance for ACHIEVE using UNIQUE sums of squares

Source of Variation	SS	DF	MS	F	Sig of F
WITHIN CELLS	40.74	17	2.40		
REGRESSION	395.84	1	395.84	165.19	.000
METHOD	169.52	2	84.76	35.37	.000

Consider again the row labeled *METHOD*. Looking at the observed significance level, you see that you can now reject the null hypothesis that all three teaching methods are equally effective. Why has the conclusion changed from that obtained in Figure 2.18? The reason is that we have dramatically decreased the within-cells variability, the denominator for the *F* test, by eliminating the variability due to differences in aptitude. Our estimate of variability in each of the cells is now "purer."

From Figure 2.21, you see that the estimate of the common slope is 1.033. This estimate can be used to compute what are called **adjusted means** for each of the cells of the design. The basic idea is to estimate what the mean achievement score for each group would be if the aptitude score were the same for all groups. To compute adjusted means, we must know what the average value of the covariate is in each of the teaching methods. These values, from the SPSS Means procedure, are shown in Figure 2.22. For method 1, the average aptitude score was 9.43; for method 2, 8.71; and for method 3, 7.43. The overall average aptitude score was 8.52. So you see that for method 1, the average aptitude score is quite a bit higher than average, while for method 3, it is somewhat less than average. The observed mean achievement scores for each of the groups is shown in Figure 2.23 in the column labeled *Obs. Mean*. To calculate what the mean for each group would be if the average aptitude score was 8.52 (the adjusted mean), you must calculate how much above or below average each group is compared to the overall average and then adjust the observed mean accordingly. The formula is

$$\text{adjusted mean}_j = \text{observed mean}_j - \text{slope}\,(\text{mean}_j - \text{unweighted grand mean})$$

Equation 2.1

Using Equation 2.1, the adjusted mean for method 1 is

$$\text{adjusted mean}_{\text{method 1}} = 14.29 - 1.03\,(9.43 - 8.52) = 13.35$$

Equation 2.2

The adjusted mean is less than the observed mean, since the average value of the aptitude score was higher for method 1 than for the other methods. The adjusted means for the three methods are shown in Figure 2.23 in the column labeled *Adj. Mean*. Comparisons between groups should be based on these adjusted means, not on the original means.

Figure 2.21 Regression statistics for aptitude score

```
Regression analysis for WITHIN CELLS error term
--- Individual Univariate .9500 confidence intervals
Dependent variable .. ACHIEVE

COVARIATE          B       Beta    Std. Err.    t-Value    Sig. of t   Lower -95%   CL- Upper

APTITUDE     1.03313    .95220          .080     12.853         .000         .864       1.203
```

Figure 2.22 Means for aptitude score

```
                 - - Description of Subpopulations - -

Summaries of     APTITUDE
By levels of     METHOD

Variable        Value  Label                          Mean     Std Dev     Cases

For Entire Population                               8.5238      4.3888        21

METHOD           1.00                               9.4286      4.3534         7
METHOD           2.00                               8.7143      5.2190         7
METHOD           3.00                               7.4286      3.9521         7

  Total Cases = 21
```

Figure 2.23 Observed and adjusted means

```
Adjusted and Estimated Means
Variable .. ACHIEVE
 CELL           Obs. Mean    Adj. Mean    Est. Mean   Raw Resid. Std. Resid.

    1             14.286       13.351       14.286         .000         .000
    2             13.429       13.232       13.429         .000         .000
    3             18.286       19.417       18.286         .000         .000
```

More Than One Factor

Although analysis of covariance was illustrated using only one covariate and one grouping factor, its use is not restricted to these situations. You can include any number of covariates that are linearly related to the dependent variable. Similarly, you can use analysis of variance for any type of factorial design. When you have several factors, however, testing for a common slope is more complicated. For further discussion, see MANOVA: Univariate in the Syntax Reference section of this manual.

Types of Sums of Squares

The examples in this chapter had the same number of cases in each of the cells of the design. This greatly simplifies the interpretation of the ANOVA hypotheses. When the number of cases in all of the cells is not equal—that is, when the design is **unbalanced**—several different types of sums of squares can be computed. Different sums of squares correspond to tests of different hypotheses. Two frequently used methods for calculating sums of squares are the **regression method** and the **sequential method**. In the regression method, all effects are adjusted for all other effects in the model. In the sequential method, an effect is adjusted only for effects that precede it in the model.

For any connected design, the hypotheses associated with sequential sums of squares are weighted functions of the population cell means, with weights depending on cell frequencies (Searle, 1971). For designs in which every cell contains observations, the hypotheses corresponding to the regression sums of squares are hypotheses about unweighted cell means. With empty cells, the hypotheses depend on the pattern of missing cells.

When your design contains empty cells, the analysis is greatly complicated. Hypotheses that involve parameters corresponding to empty cells usually cannot be tested. The output from an analysis involving empty cells should be treated with caution (Milliken & Johnson, 1984).

How to Obtain a General Factorial Analysis of Variance

The General Factorial ANOVA procedure is a generalized analysis-of-variance procedure for balanced and unbalanced univariate models. While the One-Way ANOVA and Simple Factorial ANOVA procedures in the SPSS Base system also perform univariate analysis of variance, the General Factorial ANOVA procedure can analyze a wider variety of models than the other procedures, including models containing covariate interactions. The General Factorial ANOVA procedure also provides additional statistical displays, including adjusted means, collinearity diagnostics, effect sizes, and approximate power values.

To analyze models containing several dependent variables, use the Multivariate ANOVA procedure (see Chapter 3). For repeated measures designs, see Chapter 4.

The minimum specification is one dependent variable. This produces a test of whether the mean of the dependent variable is equal to 0 (when the dependent variable is a difference variable, this is equivalent to a paired *t* test). Other models require additional specifications. For example:

- For independent-samples designs, specify at least one factor variable.
- For one-sample models involving continuous predictor variables (such as linear regression models), specify one or more covariates.
- For analysis of covariance (ANCOVA), specify a factor variable and at least one covariate; to test the parallelism assumption, also specify a factor-by-covariate interaction.

To obtain a general factorial analysis of variance, from the menus choose:

Statistics
 ANOVA Models ▸
  General Factorial...

This opens the General Factorial ANOVA dialog box, as shown in Figure 2.24.

Figure 2.24 General Factorial ANOVA dialog box

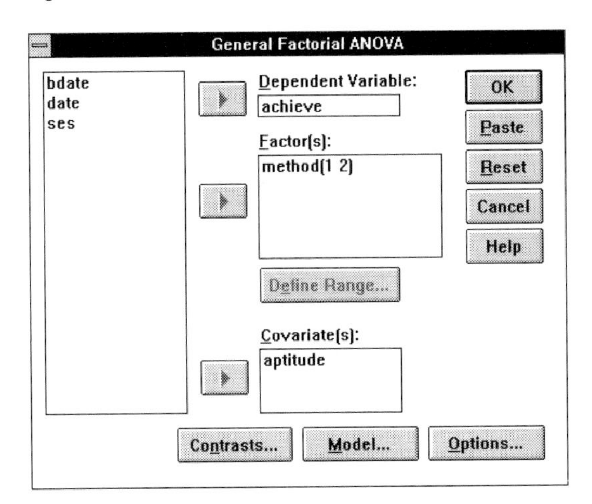

The numeric variables in your working data file appear on the source list.

Dependent Variable. Select a continuous dependent variable.

Factor(s). Optionally, you can select one or more categorical predictor variables. You must indicate which factor levels you want to use in the analysis (see "Defining Levels of Factor Variables," below).

Covariate(s). Optionally, you can select one or more continuous predictor variables.

Defining Levels of Factor Variables

For each factor variable, you must indicate which factor levels you want to use in the analysis. Highlight one or more factors and click on Define Range... in the General Factorial ANOVA dialog box. This opens the General Factorial ANOVA Define Range dialog box, as shown in Figure 2.25.

Figure 2.25 General Factorial ANOVA Define Range dialog box

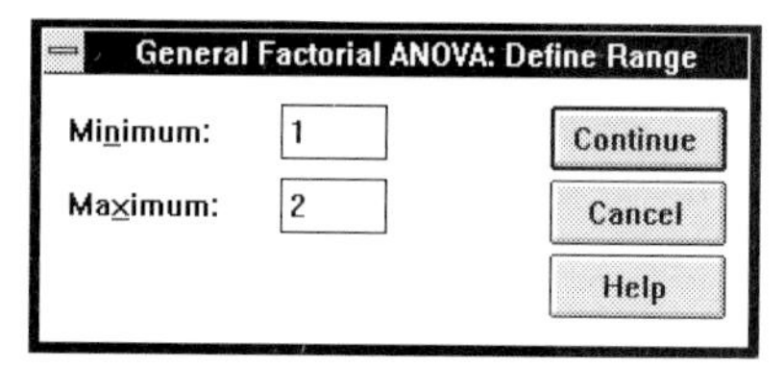

Enter integer values corresponding to the lowest and highest factor levels you want to use. Each integer category in the inclusive range is used as a factor level in the analysis (non-integer codes are truncated for the analysis, so cases with the same integer portion are combined into the same factor level). Cases with values outside the range are excluded from the analysis. To avoid empty cells, factor levels within the range must be consecutive. If a variable has empty categories, use the Automatic Recode facility on the Transform menu to create consecutive categories. (For more information on recoding variables, see the Base system documentation.)

Specifying Contrasts

Factor variables are transformed into contrasts for the analysis. To display parameter estimates and confidence intervals for contrasts, or to override the default contrast type (deviation), click on Contrasts... in the General Factorial ANOVA dialog box. This opens the General Factorial ANOVA Contrasts dialog box, as shown in Figure 2.26.

Figure 2.26 General Factorial ANOVA Contrasts dialog box

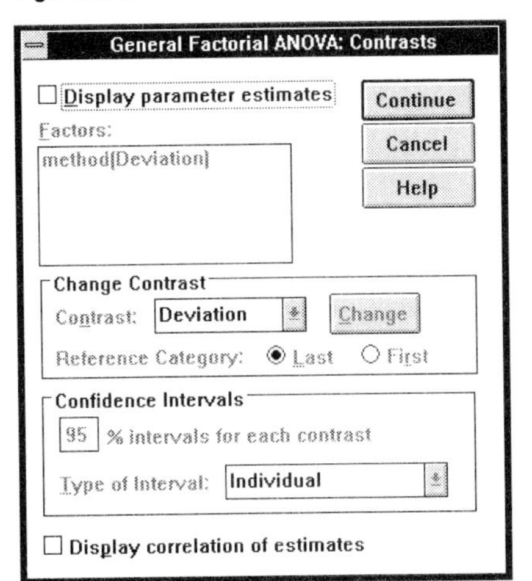

❑ **Display parameter estimates**. Displays parameter estimates, standard errors, t statistics, and confidence intervals. If you select this option, you can also change the contrast type and confidence intervals.

Change Contrast. By default, each factor is transformed into deviation contrasts. To obtain a different type of contrast, highlight one or more factor(s), select a contrast type, and click on Change. Optionally, you can change the default reference category.

- **Contrast**. You can choose one of the following contrast types:

  Deviation. Each category of the factor except the reference category is compared to the overall effect. This is the default.

  Simple. Each category of the factor (except the reference category) is compared to the reference category.

  Difference. Also known as reverse Helmert contrasts. Each category of the factor except the first category is compared to the average effect of previous categories.

  Helmert. Each category of the factor except the last category is compared to the mean effect of subsequent categories.

  Repeated. Each category of the factor except the first category is compared to the category that precedes it.

  Polynomial. The factor is partitioned into linear, quadratic, and cubic effects, and so on (depending on the number of categories). Categories are assumed to be equally spaced.

Reference Category. For deviation and simple contrasts, you can specify which category is to be omitted from the comparisons. Choose one of the following alternatives:

❍ **Last.** Uses the highest category of the factor as the reference category. This is the default.

❍ **First.** Uses the lowest category of the factor as the reference category.

Confidence Intervals. If you request parameter estimates for contrasts, SPSS displays confidence intervals for the estimates. To override the default confidence level (95%), enter a value for n% intervals for each contrast. The value must be greater than 0 and less than 100.

➧ **Type of Interval.** You can also choose between individual and joint confidence intervals. Choose one of the following alternatives:

Individual. Displays a separate interval for each parameter. These are the default confidence intervals.

Joint Bonferroni. Displays simultaneous Bonferroni intervals for parameters. These intervals are based on the number of contrasts actually made.

Joint Scheffé. Displays simultaneous Scheffé intervals for parameters. These intervals are based on all possible contrasts.

The following option is also available:

❑ **Display correlation of estimates.** Displays correlations and covariance factors between parameter estimates.

Specifying the MANOVA Model

By default, SPSS analyzes a full factorial model. To define a different model or control model options, click on Model... in the General Factorial ANOVA dialog box. This opens the General Factorial ANOVA Model dialog box, as shown in Figure 2.27.

Figure 2.27 General Factorial ANOVA Model dialog box

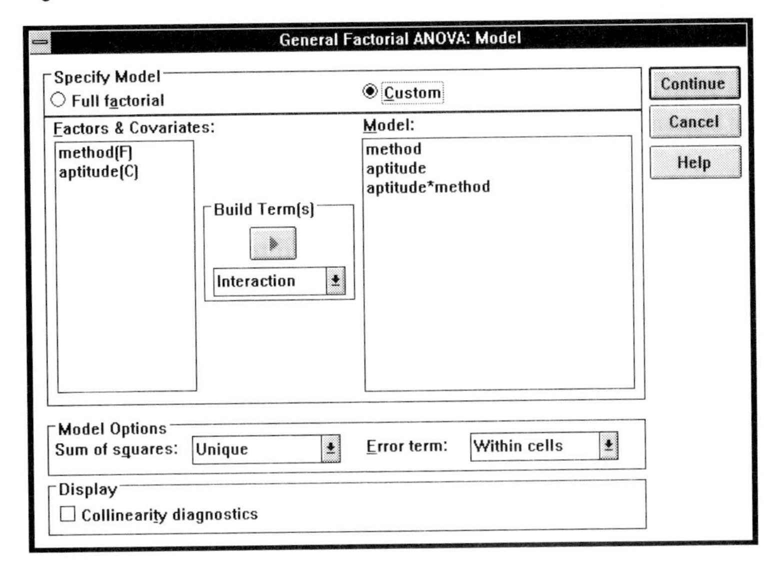

Specify Model. You can choose one of the following alternatives:

❍ **Full factorial**. Model contains all factor and covariate main effects and all factor-by-factor interactions. This is the default. Covariate interactions are not included in the model.

❍ **Custom**. Select this to define a model containing factor-by-covariate interactions or a model containing a subset of possible factor-by-factor interactions. You must indicate which terms you want to include in the model.

Model. To include a term in a custom model, select one or more factors or covariates, or a combination. If you don't want to create the highest-order interaction for the selected variables, select an alternative method for building the term. To add more terms to the model, repeat this process. Do not use the same term more than once in the model.

For example, to define a model containing main effects for method and aptitude and the aptitude-by-method interaction, first highlight method(F) and aptitude(C), select Main effects from the Build Term(s) list, and click on [▶]. Then highlight method(F) and aptitude(C), select Interaction (the default) from the Build Term(s) list, and click on [▶].

Macintosh: Use ⌘-click to select multiple factors or covariates, or a combination.

➧ **Build Term(s)**. You can build main effects or interactions for the selected variables. If you request an interaction of a higher order than the number of variables, SPSS creates a term for the highest-order interaction for the variables. If only one variable is selected, a main-effect term is added to the model, regardless of your Build Term(s) selection. Choose one of the following alternatives:

Interaction. Creates the highest-level interaction term for the variables. This is the default.

Main effects. Creates a main-effect term for each variable.

All 2-way. Creates all possible two-way interactions for the variables.

All 3-way. Creates all possible three-way interactions for the variables.

All 4-way. Creates all possible four-way interactions for the variables.

All 5-way. Creates all possible five-way interactions for the variables.

Model Options. SPSS includes a constant term in all models. You can control the computation of sums of squares and the error term used for the model.

- **Sum of squares**. Select one of the following methods for decomposing sums of squares:

  Unique. Uses the regression method to partition sums of squares. This is the default. Each effect is adjusted for all other effects in the model.

  Sequential. Decomposes sums of squares hierarchically. For default (full factorial) models, SPSS first adjusts covariates for all factors and interactions. Then each factor is adjusted for all covariates and for factors that precede it in the model specification. Each interaction is adjusted for all covariates, factors, and preceding interactions.

  For custom models, SPSS first adjusts covariates that appear only as main effects for all other terms in the model. Next, all other terms in the model are adjusted for covariates that appear only as main effects. Then SPSS evaluates all remaining terms (factor main effects, factor interactions, covariate interactions, and main effects for covariates that also appear in interactions) in the order in which they appear in the model specification.

- **Error term**. You can control the error term used in tests of significance. Select one of the following alternatives:

  Within cells. Tests model terms against the within-cells sum of squares. This is the default.

  Residual. Tests model terms against the residual sum of squares.

  Within+Residual. Tests model terms against the pooled within-cells and residual sum of squares.

Display. You can also choose the following statistical display:

- ❑ **Collinearity diagnostics**. Displays collinearity diagnostics for the design matrix, including singular values of the normalized decomposition (which are the same as the normalized design), condition indexes, and the proportion of variance of parameters accounted for by each principal component.

Options

To obtain optional statistics, click on **Options...** in the General Factorial ANOVA dialog box. This opens the General Factorial ANOVA Options dialog box, as shown in Figure 2.28.

Figure 2.28 General Factorial ANOVA Options dialog box

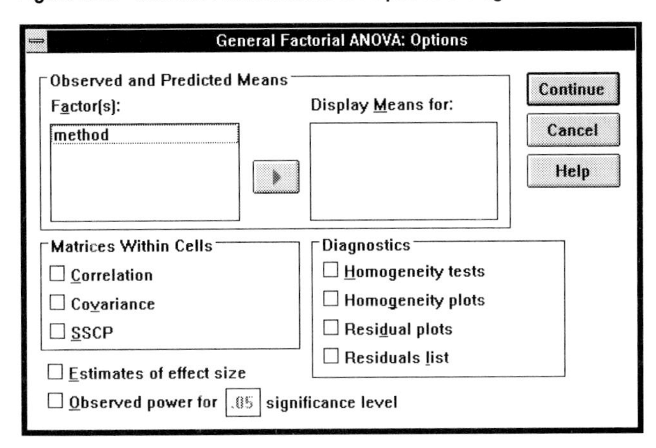

Observed and Predicted Means. You can obtain observed means (of dependent variables and covariates) and predicted means (of dependent variables) for subgroups defined by factors.

Factor(s). The list contains the variables you have selected as factors.

Display Means for. Select the factors you want to use to define subgroups. This displays means for each factor level. For observed means, SPSS shows weighted means (which give equal weight to all cases) and unweighted means (which weight cells equally regardless of the number of cases they contain). For predicted means, SPSS displays cell means that are adjusted (adjusted means) and unadjusted (estimated means) for all covariates that appear only as main effects. Raw and standardized residuals are also shown.

Matrices Within Cells. Matrices within cells are computed for the dependent variable and covariates. The matrices are produced only if you have at least one factor and covariate. You can choose one or more of the following:

- **Correlation.** Displays a correlation matrix for each cell. Standard deviations are shown on the diagonal.
- **Covariance.** Displays a variance-covariance matrix for each cell. The pooled within-cells variance-covariance matrix is also shown.
- **SSCP.** Displays a matrix of sums-of-squares and cross-products for each cell.

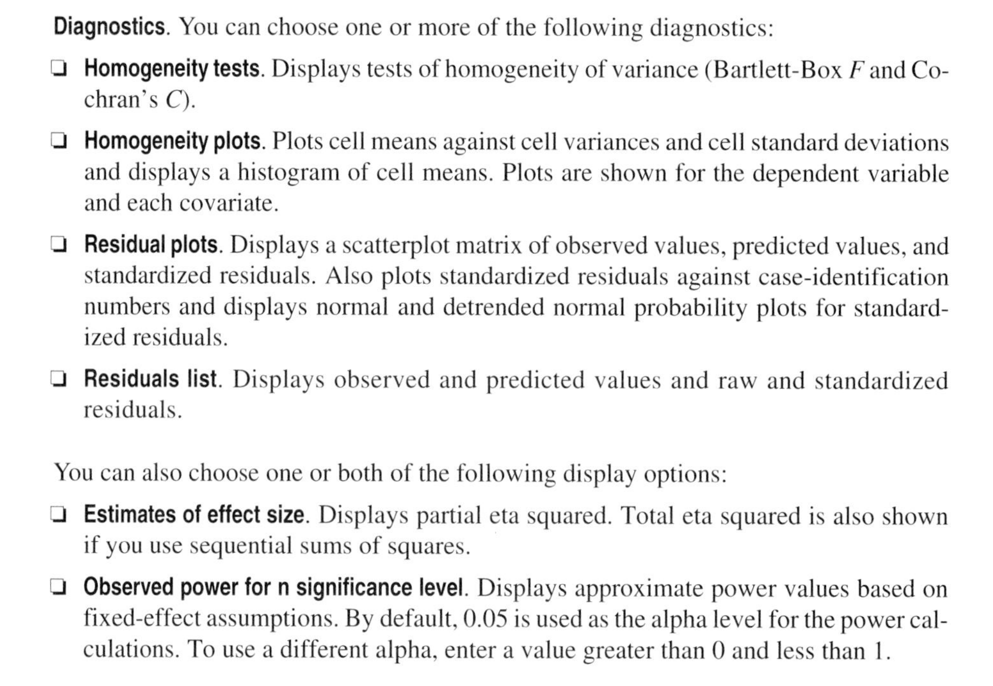

Diagnostics. You can choose one or more of the following diagnostics:

❑ **Homogeneity tests.** Displays tests of homogeneity of variance (Bartlett-Box *F* and Cochran's *C*).

❑ **Homogeneity plots.** Plots cell means against cell variances and cell standard deviations and displays a histogram of cell means. Plots are shown for the dependent variable and each covariate.

❑ **Residual plots.** Displays a scatterplot matrix of observed values, predicted values, and standardized residuals. Also plots standardized residuals against case-identification numbers and displays normal and detrended normal probability plots for standardized residuals.

❑ **Residuals list.** Displays observed and predicted values and raw and standardized residuals.

You can also choose one or both of the following display options:

❑ **Estimates of effect size.** Displays partial eta squared. Total eta squared is also shown if you use sequential sums of squares.

❑ **Observed power for n significance level.** Displays approximate power values based on fixed-effect assumptions. By default, 0.05 is used as the alpha level for the power calculations. To use a different alpha, enter a value greater than 0 and less than 1.

Additional Features Available with Command Syntax

You can customize your general factorial analysis of variance if you paste your selections into a syntax window and edit the resulting MANOVA command syntax. (For information on syntax windows, see the SPSS Base system documentation.) Additional features include:

- Additional contrast options, such as user-defined contrasts (using the CONTRAST subcommand).
- Additional design information, including basis matrices (using the PRINT subcommand).
- Linear transformations of covariates and dependent variables (using the TRANSFORM subcommand).
- The ability to read matrix data and to write matrix files that can be used in subsequent analyses (using the MATRIX subcommand).
- The ability to partition degrees of freedom associated with a factor (using the PARTITION subcommand).
- The ability to analyze nested models and to specify separate slopes for each cell in a model (using WITHIN on the DESIGN subcommand).
- Predicted means for combinations of factor levels (using the PMEANS subcommand).

See the Syntax Reference section of this manual for command syntax rules and for complete MANOVA command syntax.

3 Multivariate Analysis of Variance

Tilted houses featured in amusement parks capitalize on the challenge of navigating one's way in the presence of misleading visual cues. It's difficult to maintain balance when walls are no longer parallel and rooms assume strange shapes. We are all dependent on visual information to guide movement, but the extent of this dependence has been found to vary considerably.

Based on a series of experiments, Witkin et al. (1954) classified individuals into two categories: those who can ignore misleading visual cues, termed field-independent, and those who cannot, termed field-dependent. Field dependence has been linked to a variety of psychological characteristics such as self-image and intelligence. Psychologists theorize that it derives from childhood socialization patterns—field-dependent children learn to depend on highly structured environments while field-independent children learn to cope with ambiguous situations.

In this chapter, the relationship between field dependence, sex, and various motor abilities is examined using data reported by Barnard (1973). Students (63 female and 71 male) from the College of Southern Idaho were administered a test of field independence: the rod and frame test. On the basis of this test, subjects were classified as field-dependent, field-independent, or intermediate. Four tests of motor ability were also conducted: two tests of balance, a test of gross motor skills, and a test of fine motor skills.

For the balance test, subjects were required to maintain balance while standing on one foot on a rail. Two trials for each of two conditions, eyes open and eyes closed, were administered, and the average number of seconds a subject maintained balance under each condition was recorded. Gross motor coordination was assessed with the sidestepping test. For this test, three parallel lines are drawn four feet apart on the floor and a subject stands on the middle line. At the start signal, the subject must sidestep to the left until the left foot crosses the left line. He or she then sidesteps to the right until the right line is crossed. A subject's score is the average number of lines crossed in three ten-second trials. The Purdue Pegboard Test was used to quantify fine motor skills. Subjects are required to place small pegs into holes using only the left hand, only the right hand, and then both hands simultaneously. The number of pegs placed in two 30-second trials for each condition was recorded and the average over six trials calculated.

The experiment described above is fairly typical of many investigations. There are several classification, or independent, variables—sex and field independence in this case—and a dependent variable. The goal of the experiment is to examine the relationship between the classification variables and the dependent variable. For example, is motor ability related to field dependence? Does the relationship differ for men and women? Analysis-of-variance techniques are usually used to answer these questions, as discussed in Chapter 2.

In this experiment, however, the dependent variable is not a single measure but four different scores obtained for each student. Although ANOVA tests can be computed separately for each of the dependent variables, this approach ignores the interrelation among the dependent variables. Substantial information may be lost when correlations between variables are ignored. For example, several bivariate regression analyses cannot substitute for a multiple regression model, which considers the independent variables jointly. Only when the independent variables are uncorrelated with each other are the bivariate and multivariate regression results equivalent. Similarly, analyzing multiple two-dimensional tables cannot substitute for an analysis that considers the variables simultaneously.

Multivariate Analysis of Variance

The extension of univariate analysis of variance to the case of multiple dependent variables is termed **multivariate analysis of variance**, abbreviated as MANOVA. Univariate analysis of variance is just a special case of MANOVA, the case with a single dependent variable. The hypotheses tested with MANOVA are similar to those tested with ANOVA. The difference is that sets of means (sometimes called a **vector**) replace the individual means specified in ANOVA. In a one-way design, for example, the hypothesis tested is that the populations from which the groups are selected have the same means for all dependent variables. Thus, the hypothesis might be that the population means for the four motor-ability variables are the same for the three field-dependence categories.

Assumptions

For the case of a single dependent variable, the following assumptions are necessary for the proper application of the ANOVA test: the groups must be random samples from normal populations with the same variance. Similar assumptions are necessary for MANOVA. Since we are dealing with several dependent variables, however, we must make assumptions about their joint distribution—that is, the distribution of the variables considered together. The extension of the ANOVA assumptions to MANOVA requires that the dependent variables have a multivariate normal distribution with the same variance-covariance matrix in each group. A **variance-covariance matrix**, as its name indicates, is a square arrangement of elements with the variances of the variables on the diagonal, and the covariances of pairs of variables off the diagonal. A variance-covariance matrix can be transformed into a correlation matrix by dividing each covariance by the standard deviations of the two variables. Later in this chapter, we present tests for these assumptions.

One-Sample Hotelling's T^2

Before considering more complex generalizations of ANOVA techniques, let's consider the simple one-sample t test and its extension to the case of multiple dependent variables. As you will recall, the one-sample t test is used to test the hypothesis that the sample originates from a population with a known mean. For example, you might want to test the hypothesis that schizophrenics do not differ in mean IQ from the general population, which is assumed to have a mean IQ of 100. If additional variables such as reading comprehension, mathematical aptitude, and motor dexterity are also to be considered, a test that allows comparison of several observed means to a set of constants is required.

A test developed by Hotelling, called Hotelling's T^2, is often used for this purpose. It is the simplest example of MANOVA. To illustrate this test and introduce some of the SPSS Multivariate ANOVA procedure output, we will use the field-dependence and motor-ability data to test the hypothesis that the observed sample comes from a population with specified values for the means of the four tests. That is, we will assume that normative data are available for the four tests, and we will test the hypothesis that our sample is from a population having the normative means.

For illustrative purposes, the standard values are taken to be 13 seconds for balancing with eyes open, 3 seconds for balancing with eyes closed, 18 lines for the sidestepping test, and 10 pegs for the pegboard test. Since the Multivariate ANOVA procedure automatically tests the hypothesis that a set of means is equal to 0, we subtract the normative values from the observed scores prior to the MANOVA analysis to test the hypothesis that the differences are 0.

Figure 3.1 contains the message displayed by SPSS when the cases are processed. It indicates the number of cases to be used in the analysis as well as the number of cases to be excluded. In this example, 134 cases will be included in the analysis. SPSS also indicates whether any cases contain missing values for the variables being analyzed or have independent-variable (factor) values outside the designated range. Such cases are excluded from the analysis.

Since the hypothesis being tested involves only a test of a single sample, all observations are members of one **cell**, in ANOVA terminology. If two independent samples, for example, males and females, were compared, two cells would exist. The last line of Figure 3.1 indicates how many different MANOVA models have been specified.

Figure 3.1 Case information

```
134 cases accepted.
 0 cases rejected because of out-of-range factor values.
 0 cases rejected because of missing data.
 1 non-empty cell.

 1 design will be processed.
```

Descriptive Statistics

One of the first steps in any statistical analysis, regardless of how simple or complex it may be, is examination of the individual variables. This preliminary screening provides information about a variable's distribution and permits identification of unusual or outlying values.

Of course, when multivariate analyses are undertaken, it is not sufficient just to look at the characteristics of the variables individually. Information about their joint distribution must also be obtained. Similarly, identification of outliers must be based on the joint distribution of variables. For example, a height of six feet is not very unusual, and neither is a weight of 100 pounds, nor being a man. A six-foot-tall male who weighs 100 pounds, however, is fairly atypical and needs to be identified to ascertain that the values have been correctly recorded, and if so, to gauge the effect of such a lean physique on subsequent analyses. (See the discussion on Mahalanobis distance and Cook's distance and deleted residuals in the SPSS Base system documentation.)

Figure 3.2 contains means and confidence intervals for each of the four motor-ability variables after the normative values have been subtracted (means are shown in the column labeled *Coeff.*, and confidence intervals appear in the last two columns). The sample exceeds the norm for balancing with eyes closed (*balcmean*) and peg insertion (*pp*) and is poorer than the norm for balancing with eyes open (*balomean*) and sidestepping (*sstmean*). The only confidence interval that includes 0 is for the balancing-with-eyes-closed variable.

Figure 3.2 Cell means and confidence intervals

```
Estimates for BALOMEAN
--- Individual univariate .9500 confidence intervals
CONSTANT

 Parameter      Coeff.  Std. Err.      t-Value      Sig. t Lower -95%  CL- Upper

         1  -1.5399254      .50623    -3.04193      .00283   -2.54123    -.53862

- - - - - - - - - - - - - - - - - - - - - - - - - - - - - - - - - - - - - - - - -
Estimates for BALCMEAN
--- Individual univariate .9500 confidence intervals
CONSTANT

 Parameter      Coeff.  Std. Err.      t-Value      Sig. t Lower -95%  CL- Upper

         1  .142910448      .12136     1.17754      .24108    -.09714     .38296

- - - - - - - - - - - - - - - - - - - - - - - - - - - - - - - - - - - - - - - - -
Estimates for SSTMEAN
--- Individual univariate .9500 confidence intervals
CONSTANT

 Parameter      Coeff.  Std. Err.      t-Value      Sig. t Lower -95%  CL- Upper

         1  -2.5970149      .23164   -11.21165      .00000   -3.05518   -2.13885

- - - - - - - - - - - - - - - - - - - - - - - - - - - - - - - - - - - - - - - - -
Estimates for PP
--- Individual univariate .9500 confidence intervals
CONSTANT

 Parameter      Coeff.  Std. Err.      t-Value      Sig. t Lower -95%  CL- Upper

         1  4.97263682      .12912    38.51214      .00000    4.71725    5.22803
```

The 95% confidence intervals that are displayed are individual confidence intervals. This means that no adjustment has been made for the fact that the confidence intervals for several variables have been computed. We have 95% confidence that each of the individual intervals contains the unknown parameter value. We do not have 95% confidence that *all* intervals considered jointly contain the unknown parameters. The distinction here is closely related to the problem of multiple comparisons in ANOVA.

When many tests are done, the chance that some observed differences appear to be statistically significant when there are no true differences in the populations increases with the number of comparisons made. To protect against calling too many differences "real" when in fact they are not, the criterion for how large a difference must be before it is considered significant is made more stringent. That is, larger differences are required, depending on the number of comparisons made. The larger the number of comparisons, the greater the observed difference must be. Similarly, if a confidence region that simultaneously contains values of observed population parameters with a specified overall confidence level is to be constructed, the confidence interval for each variable must be wider than that needed if only one variable is considered.

Further Displays for Checking Assumptions

Although the summary statistics presented in Figure 3.2 provide some information about the distributions of the dependent variables, more detailed information is often desirable. Figure 3.3 is a stem-and-leaf plot of the pegboard variable from the SPSS Explore procedure after the normative values have been subtracted. Stem-and-leaf plots provide a convenient way to examine the distribution of a variable.

In Figure 3.3, the numbers to the left of the dotted line are called the **stem**, while those to the right are the **leaves**. Each case is represented by a leaf. For example, the first line of the plot is for a case with a value of 1.1 and a case with a value of 1.3 (the actual values for these cases are 1.17 and 1.33, but the plot uses only the first decimal place). The stem (1) is the same for both cases, while the values of the leaves (1 and 3) differ. When there are several cases with the same values, the leaf value is repeated. For example, there are two cases with a value of 2.6 and four cases with a value of 2.8. In this example, each stem value occurs twice—once for cases with leaves 0 through 4 and once for cases with leaves 5 through 9. This is not always the case, since the stem values depend on the actual data. In this example, the decimal point for each case occurs between the value of the stem and the leaf. This also is not always the case. The SPSS Explore procedure scales the variables so that the stem-and-leaf plot is based on the number of significant digits. Since the purpose of the plot is to display the distribution of the variable, the actual scale is not important. The final row of the stem-and-leaf plot, labeled *Extremes*, shows cases with values far removed from the rest. In the frequency column, we see that there is one extreme case that has a pegboard test value of 9.3.

Figure 3.3 Stem-and-leaf plot for pegboard test

```
Frequency     Stem &  Leaf

     2.00        1 *  13
     1.00        1 .  8
     3.00        2 *  033
     6.00        2 .  668888
     5.00        3 *  01333
    18.00        3 .  555555566666668888
    12.00        4 *  001111111113
    19.00        4 .  5555555556666668888
    14.00        5 *  00000111111333
    17.00        5 .  55555555666688888
    13.00        6 *  0000111113333
    10.00        6 .  5566666668
     8.00        7 *  00000113
     4.00        7 .  5668
     1.00        8 *  1
     1.00 Extremes    (9.3)

Stem width:       1.00
Each leaf:        1 case(s)
```

One assumption needed for hypothesis testing in MANOVA is the assumption that the dependent variables have a multivariate normal distribution. If variables have a multivariate normal distribution, each one taken individually must be normally distributed. (However, variables that are normally distributed individually will not necessarily have a multivariate normal distribution when considered together.) The stem-and-leaf plots for each variable allow us to assess the reasonableness of the normality assumption, since if any distribution appears to be markedly non-normal, the assumption of multivariate normality is likely to be violated.

Normal Plots

Although the stem-and-leaf plot gives a rough idea of the normality of the distribution of a variable, other plots that are especially designed for assessing normality can also be obtained. For example, we can assess normality using a normal probability plot, which is obtained by ranking the observed values of a variable from the smallest to the largest and then pairing each value with an expected normal value for a sample of that size from a standard normal distribution.

Figure 3.4 is a normal probability plot of the pegboard variable. If the observed scores are from a normal distribution, points in the plot should be approximately in a straight line. Since the distribution of the pegboard scores appeared fairly normal in the stem-and-leaf plot (Figure 3.3), the normal probability plot should be fairly linear, and it is. (Additional tests for normality are available in the Explore procedure; see the SPSS Base system documentation.)

Figure 3.4 Normal probability plot for pegboard test

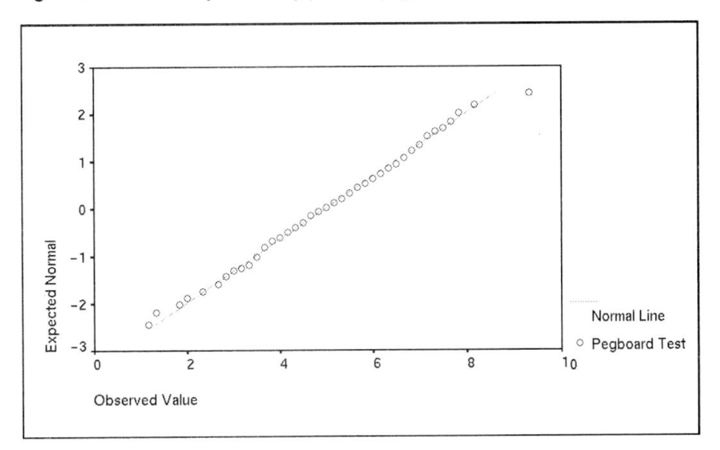

To further assess the linearity of the normal probability plot, we can calculate the difference between the observed point and the expected point under the assumption of normality and plot this difference for each case. If the observed sample is from a normal distribution, these differences should be fairly close to 0 and be randomly distributed.

Figure 3.5 is a plot of the differences for the normal probability plot shown in Figure 3.4. This plot is called a **detrended normal plot**, since the trend (or straight line) in Figure 3.4 has been removed. Note that the values fall roughly in a horizontal band around 0, though there appears to be some pattern. Also notice the two outliers, one in the lower left corner and the other in the upper right corner. These correspond to the smallest and largest observations in the sample and indicate that the observed distribution doesn't have quite as much spread in the tails as expected. For the smallest value, the deviation from the expected line is negative, indicating that the smallest value is not quite as large as would be expected. For the largest value, the value is not as small as would be expected. Since most of the points cluster nicely around 0, this small deviation is probably not of too much concern. Nonetheless, it is usually a good idea to check outlying points to make sure that they have been correctly recorded and entered. If the distribution of a variable appears markedly non-normal, transformation of the data should be considered.

Figure 3.5 Detrended normal plot for pegboard test

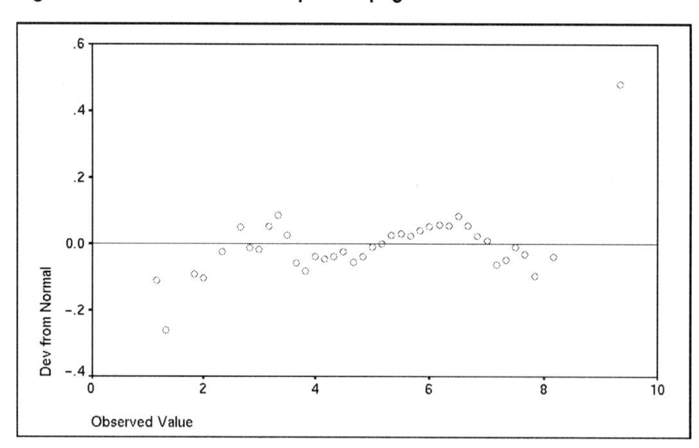

Another Plot

To get a little more practice in interpreting the previously described plots, consider Figure 3.6, which is the plot for the balancing-with-eyes-closed variable. From this stem-and-leaf plot (from the Explore procedure), you can see that the distribution of the data is skewed to the right. That is, there are several large values that are quite removed from the rest.

Figure 3.6 Stem-and-leaf plot for balance variable

```
Frequency    Stem &  Leaf

    2.00       -2 *  00
    6.00       -1 .  555668
   19.00       -1 *  0000112222222334444
   22.00       -0 .  5555556666667777788889
   19.00       -0 *  0011111222223334444
   21.00        0 *  000000111222233444444
   17.00        0 .  55555556778899999
   14.00        1 *  00000112222344
    6.00        1 .  555889
    3.00        2 *  244
     .00        2 .
    1.00        3 *  2
    4.00 Extremes    (4.8), (6.2), (6.5)

Stem width:      1.00
Each leaf:       1 case(s)
```

Figure 3.7 is the corresponding normal probability plot. Note that the plot is no longer linear but is curved, especially for larger values of the variable. This downward curve indicates that the observed values are larger than predicted by the corresponding expected normal values. This is also seen in the stem-and-leaf plot. The detrended normal plot for this variable is shown in Figure 3.8, which shows that there is a definite pattern to the deviations. The values no longer cluster in a horizontal band around 0. A transformation might be considered.

Figure 3.7 Normal probability plot for balance variable

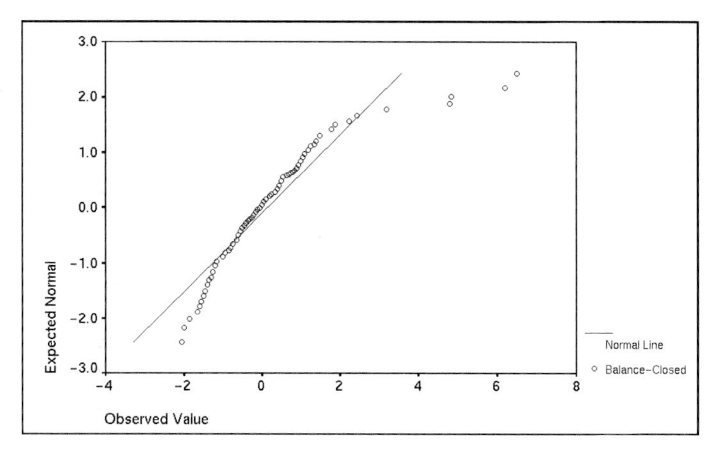

Figure 3.8 Detrended normal plot for balance variable

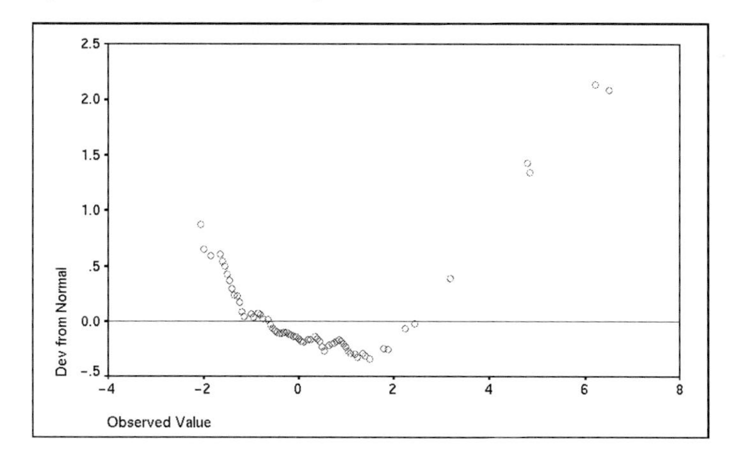

Bartlett's Test of Sphericity

Since there is no reason to use the multivariate analysis-of-variance procedure if the dependent variables are not correlated, it is useful to examine the correlation matrix of the dependent variables. If the variables are independent, the observed correlation matrix is expected to have small off-diagonal elements. Bartlett's test of sphericity can be used to test the hypothesis that the population correlation matrix is an **identity matrix**—that is, all diagonal terms are 1 and all off-diagonal terms are 0.

The test is based on the determinant of the error correlation matrix. A determinant that is close in value to 0 indicates that one or more of the variables can almost be expressed as a linear function of the other dependent variables. Thus, the hypothesis that the variables are independent is rejected if the determinant is small. Figure 3.9 contains output from Bartlett's test of sphericity. The log of the determinant is displayed first, followed by a transformation of the determinant (which has a chi-square distribution). Since the observed significance level is small (less than 0.0005), the hypothesis that the population correlation matrix is an identity matrix is rejected.

Figure 3.9 Bartlett's test of sphericity

```
Statistics for WITHIN CELLS correlations

Log(Determinant) =                  -.18703
Bartlett test of sphericity =     24.46993 with 6 D. F.
Significance =                        .000
```

A graphical test of the hypothesis that the correlation matrix is an identity matrix is described by Everitt (1978). The observed correlation coefficients are transformed using Fisher's Z-transform, and then a half-normal plot of the transformed coefficients is obtained. A half-normal plot is very similar to the normal plot described above. The only difference is that both positive and negative values are treated identically. If the population correlation matrix is an identity matrix, the plot should be fairly linear and the line should pass through the origin.

Figure 3.10 is the plot of the transformed correlation coefficients for the four motor-ability variables. The plot shows deviations from linearity, suggesting that the dependent variables are not independent, a result also indicated by Bartlett's test.

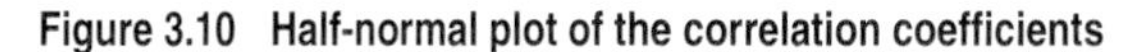

Figure 3.10 Half-normal plot of the correlation coefficients

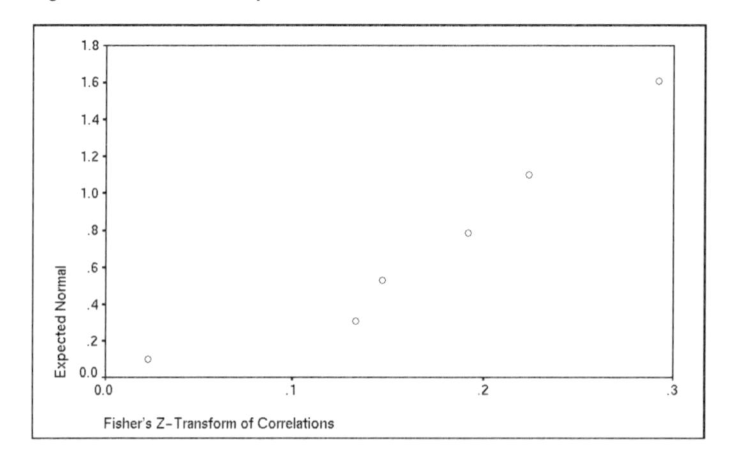

Testing Hypotheses

Once the distributions of the variables have been examined, we are ready to test the hypothesis that there is no difference between the population means and the hypothesized values. To test the various hypotheses of interest, the Multivariate ANOVA procedure computes a design matrix whose columns correspond to the various effects in the model (see "Parameter Estimates" on p. 77).

To understand how the statistic for testing the hypothesis that all differences are 0 is constructed, recall the one-sample *t* test. The statistic for testing the hypothesis that the population mean is some known constant, which we will call μ_0, is

$$t = \frac{\bar{X} - \mu_0}{S / \sqrt{N}}$$ **Equation 3.1**

where $\bar{X}$ is the sample mean, *S* is the sample standard deviation, and *N* is the number of cases. The numerator of the *t* value is a measure of how much the sample mean differs from the hypothesized value, while the denominator is a measure of the variability of the sample mean.

When you simultaneously test the hypothesis that several population means do not differ from a specified set of constants, a statistic that considers all variables together is required. Hotelling's T^2 statistic is usually used for this purpose. It is computed as

$$T^2 = N\left(\bar{X} - \mu_0\right)' S^{-1}\left(\bar{X} - \mu_0\right)$$ **Equation 3.2**

where S^{-1} is the inverse of the variance-covariance matrix of the dependent variables and $\bar{X} - \mu_0$ is the vector of differences between the sample means and the hypothesized constants. The two matrices upon which Hotelling's T^2 is based can be displayed by the Multivariate ANOVA procedure. The matrix that contains the differences between the observed sample means and the hypothesized values is shown in Figure 3.11.

Figure 3.11 Hypothesis sum-of-squares and cross-products matrix

```
EFFECT .. CONSTANT
Adjusted Hypothesis Sum-of-Squares and Cross-Products

                  BALOMEAN         BALCMEAN          SSTMEAN               PP

BALOMEAN         317.76360
BALCMEAN         -29.48957          2.73674
SSTMEAN          535.89403        -49.73284        903.76119
PP             -1026.10361         95.22600      -1730.47761       3313.43367
```

The diagonal elements are just the squares of the sample means minus the hypothesized values, multiplied by the sample size. For example, from Figure 3.2, the difference between the eyes-open balance score and the hypothesized value is -1.54. Squaring this difference and multiplying it by the sample size of 134, we get 317.76, the entry for *balomean* in Figure 3.11. The off-diagonal elements are the product of the differences for the two variables, multiplied by the sample size. For example, the cross-products entry for the *balomean* and *balcmean* is, from Figure 3.2, $-1.54 \times 0.14 \times 134 = -29.49$. For a fixed sample size, as the magnitude of the differences between the sample means and hypothesized values increases, so do the entries of this sum-of-squares and cross-products matrix.

The matrix whose entries indicate how much variability there is in the dependent variables is called the **within-cells sums-of-squares and cross-products matrix**. It is designated as S in the previous formula and is shown in Figure 3.12. The diagonal entries are $(N-1)$ times the variance of the dependent variables. For example, the entry for the pegboard variable is, from Figure 3.2, $1.49^2 \times 133 = 297.12$. The off-diagonal entries are the sums of the cross-products for the two variables. For example, for variables X and Y, the cross-product is

$$CPSS_{xy} = \sum_{i=1}^{N} \left(X_i - \bar{X}\right)\left(Y_i - \bar{Y}\right)$$ **Equation 3.3**

To compute Hotelling's T^2, the inverse of this within-cells sums-of-squares matrix is required, as well as the hypothesis sums-of-squares matrix in Figure 3.11. The significance level associated with T^2 can be obtained from the F distribution. Figure 3.13 contains the value of Hotelling's T^2 divided by $(N-1)$, its transformation to a variable that has an F distribution, and the degrees of freedom associated with the F statistic. Since there are four dependent variables in this example, the hypothesis degrees of freedom are 4, while the remaining degrees of freedom, 130, are associated with the error sums of squares. Since the observed significance level is small (less than 0.0005), the null hypothesis that the population means do not differ from the hypothesized constants is rejected.

Figure 3.12 Within-cells sums-of-squares and cross-products matrix

WITHIN CELLS Sum-of-Squares and Cross-Products

	BALOMEAN	BALCMEAN	SSTMEAN	PP
BALOMEAN	4567.27390			
BALCMEAN	311.07957	262.50076		
SSTMEAN	459.65597	94.83284	956.23881	
PP	26.79527	40.78234	70.42206	297.12189

Figure 3.13 Hotelling's statistic

Multivariate Tests of Significance (S = 1, M = 1 , N = 64)

Test Name	Value	Exact F	Hypoth. DF	Error DF	Sig. of F
Hotellings	13.20588	429.19095	4.00	130.00	.000

Univariate Tests

When the hypothesis of no difference is rejected, it is often informative to examine the univariate test results to get some idea of where the differences may be. Figure 3.14 contains the univariate results for the four dependent variables. The hypothesis and error sums of squares are the diagonal terms in Figure 3.11 and Figure 3.12. The mean squares are obtained by dividing the sums of squares by their degrees of freedom, 1 for the hypothesis sums of squares and 133 for the error sums of squares. The ratio of the two mean squares is displayed in the column labeled *F*. These *F* values are nothing more than the squares of one-sample *t* values. Thus, from Figure 3.2, the mean difference for the *balomean* variable is −1.54 and the standard deviation of the difference is 5.86. The corresponding *t* value is

$$t = \frac{-1.54}{5.86/\sqrt{134}} = -3.04$$ **Equation 3.4**

Squaring this produces the value 9.25, which is the entry in Figure 3.14. From Figure 3.2, we can see that the balancing-with-eyes-closed variable is the only one for which the *t* value is not significant. This is to be expected, since it has a 95% confidence interval that includes 0. The significance levels for the univariate statistics are not adjusted for the fact that several comparisons are being made and thus should be used with a certain amount of caution. For a discussion of the problem of multiple comparisons, see Miller (1981) or Burns (1984).

Figure 3.14 Univariate F tests

Univariate F-tests with (1,133) D. F.

Variable	Hypoth. SS	Error SS	Hypoth. MS	Error MS	F	Sig. of F
BALOMEAN	317.76360	4567.27390	317.76360	34.34041	9.25334	.003
BALCMEAN	2.73674	262.50076	2.73674	1.97369	1.38661	.241
SSTMEAN	903.76119	956.23881	903.76119	7.18977	125.70107	.000
PP	3313.43367	297.12189	3313.43367	2.23400	1483.18482	.000

The Two-Sample Multivariate T Test

In the previous sections, we were concerned with testing the hypothesis that the sample was drawn from a population with a particular set of means. There was only one sample involved, though there were several dependent variables. In this section, we will consider the multivariate generalization of the two-sample *t* test. The hypothesis that men and women do not differ on the four motor-ability variables will be tested.

Figure 3.15 contains descriptive statistics from the Means procedure for each variable according to sex. The female subjects are coded as 1's, and the males are coded as 2's. Males appear to maintain balance with eyes open longer than females and cross more lines in the stepping test.

Figure 3.15 Cell means and standard deviations

```
Summaries of     BALOMEAN   Balance Test - Eyes Open
By levels of     SEX        Sex of Subject

Variable       Value  Label                       Mean     Std Dev     Cases

For Entire Population                          11.4601      5.8601       134

SEX                 1  Female                   9.7484      5.7072        63
SEX                 2  Male                    12.9789      5.6054        71


Summaries of     BALCMEAN   Balance Test - Eyes Closed
By levels of     SEX        Sex of Subject

Variable       Value  Label                       Mean     Std Dev     Cases

For Entire Population                           3.1429      1.4049       134

SEX                 1  Female                   3.1913      1.5179        63
SEX                 2  Male                     3.1000      1.3059        71


Summaries of     SSTMEAN    Sidestepping Test
By levels of     SEX        Sex of Subject

Variable       Value  Label                       Mean     Std Dev     Cases

For Entire Population                          15.4030      2.6814       134

SEX                 1  Female                  14.0952      2.2548        63
SEX                 2  Male                    16.5634      2.5005        71


Summaries of     PP         Purdue Pegboard Test
By levels of     SEX        Sex of Subject

Variable       Value  Label                       Mean     Std Dev     Cases

For Entire Population                          14.9726      1.4947       134

SEX                 1  Female                  15.4656      1.4527        63
SEX                 2  Male                    14.5352      1.4009        71
```

Another way to visualize the distribution of scores in each of the groups is with box-and-whiskers plots from the Explore procedure, as shown in Figure 3.16 for *balomean* and Figure 3.17 for *balcmean*. The upper and lower boundaries of the boxes are the upper and lower quartiles. The box length is the interquartile distance, and the box contains the middle 50% of values in a group. The horizontal line inside the box identifies the group median. The larger the box, the greater the spread of the observations. Any points between 1.5 and 3 interquartile ranges from the end of the box (outliers) are marked with circles. The lines emanating from each box (the whiskers) extend to the smallest and largest observations in a group that are not outliers. Points more than 3 interquartile distances away from the box (extreme values) are marked with asterisks.

Figure 3.16 Box-and-whiskers plots for balomean

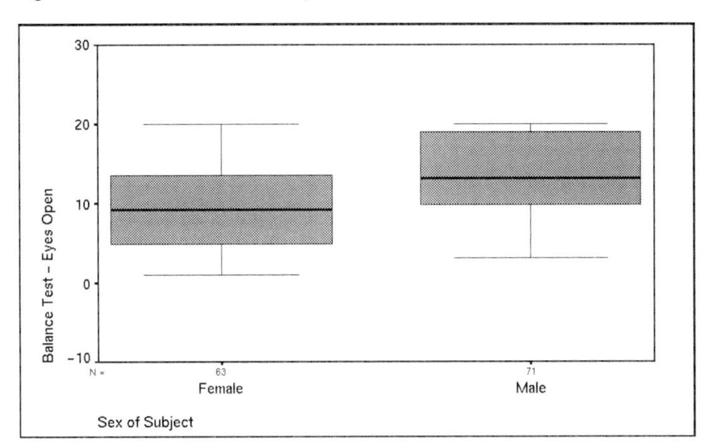

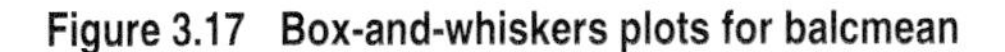

Figure 3.17 Box-and-whiskers plots for balcmean

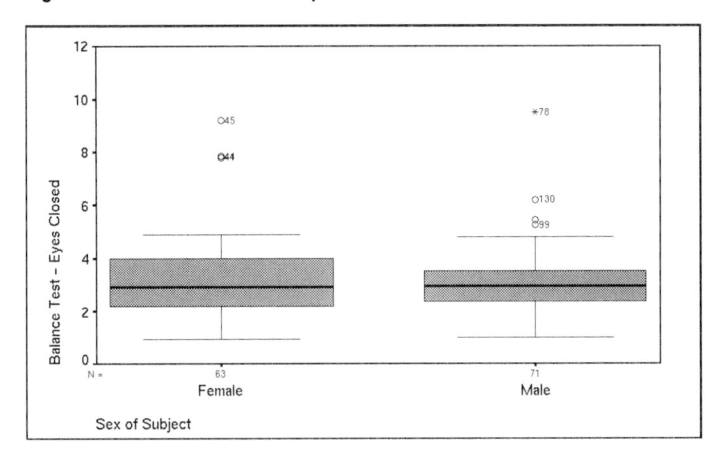

Tests of Homogeneity of Variance

In the one-sample Hotelling's T^2 test, there was no need to worry about the homogeneity of the variance-covariance matrices, since there was only one matrix. In the two-sample test, there are two matrices (one for each group), and tests for their equality are necessary.

The variance-covariance matrices are shown in Figure 3.18. They are computed for each group using the means of the variables within that group. Thus, each matrix indicates how much variability there is in a group. Combining these individual matrices into a common variance-covariance matrix results in the pooled matrix displayed in Figure 3.19.

Figure 3.18 Group variance-covariance matrices for females and males

Cell Number .. 1

Variance-Covariance matrix

	BALOMEAN	BALCMEAN	SSTMEAN	PP
BALOMEAN	32.57185			
BALCMEAN	3.39103	2.30400		
SSTMEAN	.42704	.55622	5.08397	
PP	1.43677	.36703	1.30888	2.11036

Cell Number .. 2

Variance-Covariance matrix

	BALOMEAN	BALCMEAN	SSTMEAN	PP
BALOMEAN	31.42090			
BALCMEAN	1.58111	1.70536		
SSTMEAN	2.38612	.96952	6.25267	
PP	.54349	.21702	.94178	1.96263

Figure 3.19 Pooled variance-covariance matrix

Pooled within-cells Variance-Covariance matrix

	BALOMEAN	BALCMEAN	SSTMEAN	PP
BALOMEAN	31.96150			
BALCMEAN	2.43122	1.98654		
SSTMEAN	1.46594	.77540	5.70374	
PP	.96306	.28748	1.11421	2.03202

Figure 3.20 contains two homogeneity-of-variance tests (Cochran's C and the Bartlett-Box F) for each variable individually. The significance levels indicate that there is no reason to reject the hypotheses that the variances in the two groups are equal. Although these univariate tests are a convenient starting point for examining the equality of the covariance matrices, they are not sufficient. A test that simultaneously considers both the variances and covariances is required.

Box's M test, which is based on the determinants of the variance-covariance matrices in each cell as well as of the pooled variance-covariance matrix, provides a multivariate test for the homogeneity of the matrices. However, Box's M test is very sensitive to departures from normality. The significance level can be based on either an F or a chi-square statistic, and both approximations are given in the output, as shown in Figure 3.21. Given the result of Box's M test, there appears to be no reason to suspect the homogeneity-of-dispersion-matrices assumption.

Figure 3.20 Univariate homogeneity-of-variance tests

```
Univariate Homogeneity of Variance Tests

Variable .. BALOMEAN            Balance Test - Eyes Open

     Cochrans C(66,2) =                          .50899, P =  .884 (approx.)
     Bartlett-Box F(1,51762) =                   .02113, P =  .884

Variable .. BALCMEAN            Balance Test - Eyes Closed

     Cochrans C(66,2) =                          .57466, P =  .224 (approx.)
     Bartlett-Box F(1,51762) =                  1.48033, P =  .224

Variable .. SSTMEAN             Sidestepping Test

     Cochrans C(66,2) =                          .55155, P =  .403 (approx.)
     Bartlett-Box F(1,51762) =                   .69438, P =  .405

Variable .. PP                  Purdue Pegboard Test

     Cochrans C(66,2) =                          .51813, P =  .769 (approx.)
     Bartlett-Box F(1,51762) =                   .08603, P =  .769
```

Figure 3.21 Homogeneity-of-dispersion matrices

```
Cell Number .. 1
Variance-Covariance matrix

Determinant of Covariance matrix of dependent variables =        538.89812
LOG(Determinant) =                                                 6.28953

- - - - - - - - - -

Cell Number .. 2

Determinant of Covariance matrix of dependent variables =        522.34851
LOG(Determinant) =                                                 6.25834

- - - - - - - - - -

Determinant of pooled Covariance matrix of dependent vars. =       557.40735
LOG(Determinant) =                                                   6.32330

 - - - - - - - - - - - - - - - - - - - - - - - - - - - - - - - - - - - - -

Multivariate test for Homogeneity of Dispersion matrices

Boxs M =                        6.64102
F WITH (10,80608) DF =           .64228, P =   .779 (Approx.)
Chi-Square with 10 DF =         6.42362, P =   .779 (Approx.)
```

Hotelling's T^2 for Two Independent Samples

The actual statistic for testing the equality of several means with two independent samples is also based on Hotelling's T^2. The formula is

$$T^2 = \frac{N_1 N_2}{N_1 + N_2} (\bar{X}_1 - \bar{X}_2)' S^{-1} (\bar{X}_1 - \bar{X}_2) \qquad \textbf{Equation 3.5}$$

where $\bar{X}_1$ is the vector of means for the first group (females), $\bar{X}_2$ is the vector for the second group (males), and S^{-1} is the inverse of the pooled within-groups covariance matrix. The statistic is somewhat similar to the one described for the one-sample test in "Testing Hypotheses" on p. 67.

Again, two matrices are used in the computation of the statistic. The pooled within-groups covariance matrix has been described previously and is displayed in Figure 3.19. The second matrix is the adjusted hypothesis sums-of-squares and cross-products matrix. It is displayed in Figure 3.22. The entries in this matrix are the weighted squared differences of the group means from the combined mean. For example, the entry for the balancing-with-eyes-open variable is

$$SS = 63\,(9.75 - 11.46)^2 + 71\,(12.98 - 11.46)^2 = 348.36 \quad \textbf{Equation 3.6}$$

where, from Figure 3.15, 9.75 is the mean value for the 63 females, 12.98 is the value for the 71 males, and 11.46 is the mean for the entire sample. The first off-diagonal term for balancing with eyes open (*balomean*) and eyes closed (*balcmean*) is similarly

$$\begin{aligned} SS = {} & (9.75 - 11.46)\,(3.19 - 3.14)\,63 \\ & + (12.98 - 11.46)\,(3.10 - 3.14)\,71 = -9.84 \end{aligned} \quad \textbf{Equation 3.7}$$

The diagonal terms should be recognizable to anyone familiar with analysis-of-variance methodology. They are the sums of squares due to groups for each variable.

Figure 3.22 Adjusted hypothesis sums-of-squares and cross-products matrix

```
EFFECT .. SEX
Adjusted Hypothesis Sum-of-Squares and Cross-Products

                  BALOMEAN        BALCMEAN          SSTMEAN            PP

BALOMEAN         348.35575
BALCMEAN          -9.84206          .27807
SSTMEAN          266.15138        -7.51955        203.34545
PP              -100.32910         2.83459        -76.65362        28.89554
```

Figure 3.23 contains the value of Hotelling's T^2 statistic divided by $N - 2$ for the test of the hypothesis that men and women do not differ on the motor-ability test scores. The significance level is based on the F distribution, with 4 and 129 degrees of freedom. The observed significance level is small (less than 0.0005), so the null hypothesis that men and women perform equally well on the motor-ability tests is rejected.

Figure 3.23 Adjusted Hotelling's statistic

```
Multivariate Tests of Significance (S = 1, M = 1 , N = 63 1/2)

Test Name          Value   Approx. F  Hypoth. DF    Error DF   Sig. of F

Hotellings        .66953    21.59226        4.00      129.00       .000
```

Univariate Tests

To get some idea of where the differences between men's and women's scores occur, the univariate tests for the individual variables may be examined. These are the same as the *F* values from one-way analyses of variance. In the case of two groups, the *F* values are just the squares of the two-sample *t* values.

Figure 3.24 Univariate tests

EFFECT .. SEX (Cont.)
Univariate F-tests with (1,132) D. F.

Variable	Hypoth. SS	Error SS	Hypoth. MS	Error MS	F	Sig. of F
BALOMEAN	348.35575	4218.91815	348.35575	31.96150	10.89923	.001
BALCMEAN	.27807	262.22270	.27807	1.98654	.13998	.709
SSTMEAN	203.34545	752.89336	203.34545	5.70374	35.65126	.000
PP	28.89554	268.22635	28.89554	2.03202	14.22012	.000

From Figure 3.24, we can see that there are significant univariate tests for all variables except balancing with eyes closed. Again, the significance levels are not adjusted for the fact that four tests, rather than one, are being performed.

Discriminant Analysis

In *SPSS Professional Statistics*, we considered the problem of finding the best linear combination of variables for distinguishing among several groups. Coefficients for the variables are chosen so that the ratio of between-groups sums of squares to total sums of squares is as large as possible. Although the equality-of-means hypotheses tested in MANOVA may initially appear quite unrelated to the discriminant problem, the two procedures are closely related. In fact, MANOVA can be viewed as a problem of first finding linear combinations of the dependent variables that best separate the groups and then testing whether these new variables are significantly different for the groups. For this reason, the usual discriminant analysis statistics can be obtained as part of the SPSS Multivariate ANOVA procedure output.

Figure 3.25 contains the eigenvalues and canonical correlation for the canonical discriminant function that separates males and females. Remember that the **eigenvalue** is the ratio of the between-groups sum of squares to the within-groups sum of squares, while the **square of the canonical correlation** is the ratio of the between-groups sums of squares to the total sum of squares. Thus, about 40% of the variability in the discriminant scores is attributable to between-group differences ($0.633^2 = 0.401$).

Figure 3.25 Eigenvalue and canonical correlation

Eigenvalues and Canonical Correlations

Root No.	Eigenvalue	Pct.	Cum. Pct.	Canon Cor.
1	.66953	100.00000	100.00000	.63327

When there are two groups, Wilks' lambda can be interpreted as a measure of the proportion of total variability not explained by group differences. As shown in Figure 3.26, almost 60% of the observed variability is not explained by the group differences. The hypothesis that in the population there are no differences between the group means can be tested using Wilks' lambda. Lambda is transformed to a variable that has an *F* distribution. For the two-group situation, the *F* value for Wilks' lambda is identical to that given for Hotelling's T^2 (Figure 3.23). Both raw and standardized discriminant function coefficients can be displayed by the SPSS Multivariate ANOVA procedure. The **raw coefficients** are the multipliers of the dependent variables in their original units, while the **standardized coefficients** are the multipliers of the dependent variables when the latter have been standardized to a mean of 0 and a standard deviation of 1. Both sets of coefficients are displayed in Figure 3.27.

Figure 3.26 Wilks' lambda

```
Multivariate Tests of Significance (S = 1, M = 1 , N = 63 1/2)

Test Name              Value        Exact F      Hypoth. DF      Error DF       Sig. of F

Wilks                 .59897       21.59226           4.00        129.00            .000
```

Figure 3.27 Raw and standardized discriminant function coefficients

```
EFFECT .. SEX (Cont.)
Raw discriminant function coefficients
         Function No.

Variable              1

BALOMEAN         -.07465
BALCMEAN          .19245
SSTMEAN          -.36901
PP                .49189


Standardized discriminant function coefficients
         Function No.

Variable              1

BALOMEAN         -.42205
BALCMEAN          .27125
SSTMEAN          -.88128
PP                .70118
```

The sidestepping and pegboard scores have the largest standardized coefficients and, as you will recall from Figure 3.24, they also have the largest univariate *F*'s, suggesting that they are important for separating the two groups. Of course, when variables are correlated, the discriminant function coefficients must be interpreted with care, since highly correlated variables "share" the discriminant weights.

The correlation coefficients for the discriminant scores and each dependent variable, sometimes called **structure coefficients**, are displayed in Figure 3.28. Once again, the sidestepping test and the pegboard test are most highly correlated with the discriminant function, while the correlation coefficient between balancing with eyes closed and the discriminant function is near 0. This is not surprising, since balancing with eyes closed had a nonsignificant univariate F.

Figure 3.28 Structure coefficients

```
Correlations between DEPENDENT and canonical variables
          Canonical Variable

Variable                  1

BALOMEAN              -.35118
BALCMEAN               .03980
SSTMEAN               -.63514
PP                     .40113
```

An additional statistic displayed as part of the discriminant output in the SPSS Multivariate ANOVA procedure is an estimate of the effect for the canonical variable. Consider Figure 3.29 (from the Discriminant Analysis procedure described in *SPSS Professional Statistics*), which gives the average canonical function scores for the two groups. We can estimate the effect of a canonical variable by measuring its average distance from 0 across all groups. In this example, the average distance is $(0.862 + 0.765) / 2 = 0.814$, as shown in Figure 3.30. Canonical variables with small effects do not contribute much to separation between groups.

Figure 3.29 Average canonical function scores

```
Canonical discriminant functions evaluated at group means (group centroids)

   Group       Func  1

     1        -.86214
     2         .76500
```

Figure 3.30 Estimate of effect for canonical variable

```
Estimates of effects for canonical variables
          Canonical Variable

 Parameter                1

       2              .81357
```

Parameter Estimates

As for most other statistical techniques, there is for MANOVA a mathematical model that expresses the relationship between the dependent variable and the independent variables. Recall, for example, that in a univariate analysis-of-variance model with four groups, the mean for each group can be expressed as

$$\begin{aligned} y_1 &= \mu + 1\alpha_1 + 0\alpha_2 + 0\alpha_3 + 0\alpha_4 + e_1 \\ y_2 &= \mu + 0\alpha_1 + 1\alpha_2 + 0\alpha_3 + 0\alpha_4 + e_2 \\ y_3 &= \mu + 0\alpha_1 + 0\alpha_2 + 1\alpha_3 + 0\alpha_4 + e_3 \\ y_4 &= \mu + 0\alpha_1 + 0\alpha_2 + 0\alpha_3 + 1\alpha_4 + e_4 \end{aligned}$$

Equation 3.8

where y_j is the mean for group j. In matrix form, this can be written as

$$\begin{bmatrix} y_1 \\ y_2 \\ y_3 \\ y_4 \end{bmatrix} = \begin{bmatrix} 1 & 1 & 0 & 0 & 0 \\ 1 & 0 & 1 & 0 & 0 \\ 1 & 0 & 0 & 1 & 0 \\ 1 & 0 & 0 & 0 & 1 \end{bmatrix} \begin{bmatrix} \mu \\ \alpha_1 \\ \alpha_2 \\ \alpha_3 \\ \alpha_4 \end{bmatrix} + \begin{bmatrix} e_1 \\ e_2 \\ e_3 \\ e_4 \end{bmatrix}$$

Equation 3.9

or

$$y = A\theta^* + e$$

Equation 3.10

Since the θ matrix has more columns than rows, it does not have a unique inverse. We are unable to estimate five parameters (μ and α_1 to α_4) on the basis of four sample means. Instead, we can estimate four linear combinations, termed **contrasts**, of the parameters (see Finn, 1974).

Several types of contrasts are available in the SPSS Multivariate ANOVA procedure, resulting in different types of parameter estimates. **Deviation contrasts**, the default, estimate each parameter as its difference from the overall average. This results in parameter estimates of the form $\mu_j - \mu$. Deviation contrasts do not require any particular ordering of the factor levels.

Simple contrasts are useful when one of the factor levels is a comparison or control group. All parameter estimates are then expressed as a deviation from the value of the control group. When factor levels have an underlying metric, **orthogonal polynomial contrasts** may be used to determine whether group means are related to the values of the factor level. For example, if three doses of an agent are administered (10 units, 20 units, and 30 units), you can test whether response is related to dose in a linear or quadratic fashion.

Figure 3.31 contains parameter estimates corresponding to the default deviation contrasts.There are two estimates for the sex parameter, one for females and one for males.

The output includes only values for the first parameter (females), since the value for the second parameter is just the negative of the value for the first. The sex effect for females is estimated as the difference between the mean score of females and the overall unweighted mean. For balancing with eyes open, it is $9.75 - 11.36 = -1.61$, the value displayed in Figure 3.31. Confidence intervals and t tests for the null hypothesis that a parameter value is 0 can also be calculated. SPSS can calculate either individual or joint confidence intervals for all dependent variables. Note that the t values displayed for each parameter are equal to the square root of the F values (ignoring sign) displayed for the univariate F tests in Figure 3.24.

Figure 3.31 Parameter estimates

```
Estimates for BALOMEAN
--- Individual univariate .9500 confidence intervals
SEX

 Parameter       Coeff.  Std. Err.     t-Value     Sig. t Lower -95%  CL- Upper

        2   -1.6152303      .48926    -3.30140     .00124   -2.58303    -.64743

- - - - - - - - - - - - - - - - - - - - - - - - - - - - - - - - - - - - - - -
Estimates for BALCMEAN
--- Individual univariate .9500 confidence intervals
SEX

 Parameter       Coeff.  Std. Err.     t-Value     Sig. t Lower -95%  CL- Upper

        2    .045634921      .12198      .37413     .70891    -.19564     .28691

- - - - - - - - - - - - - - - - - - - - - - - - - - - - - - - - - - - - - - -
Estimates for SSTMEAN
--- Individual univariate .9500 confidence intervals
SEX

 Parameter       Coeff.  Std. Err.     t-Value     Sig. t Lower -95%  CL- Upper

        2   -1.2340711      .20668    -5.97087     .00000   -1.64291    -.82523

- - - - - - - - - - - - - - - - - - - - - - - - - - - - - - - - - - - - - - -
Estimates for PP
--- Individual univariate .9500 confidence intervals
SEX

 Parameter       Coeff.  Std. Err.     t-Value     Sig. t Lower -95%  CL- Upper

        2    .465198599      .12336     3.77096     .00024     .22117     .70922
```

A Multivariate Factorial Design

So far we have considered two very simple multivariate designs, generalizations of the one- and two-sample t tests to the case of multiple dependent variables. We are now ready to examine a more complex design. Recall that in the experiment conducted by Barnard, the hypothesis of interest concerned the relationship among field dependence, sex, and motor ability. Since subjects are classified into one of three field-dependence categories—low (1), intermediate (2), and high (3)—we have a two-way multivariate factorial design with three levels of field dependence and two categories of sex. The four motor-ability variables are the dependent variables.

Some additional plots for cells in the model may be useful. Plotting the mean of a variable for all the cells in the design, as shown in Figure 3.32, gives an idea of the spread of the means. You can see from the plot that there are two cells with means close to 14 seconds, three cells with means less than 11 seconds, and one cell with a mean in between.

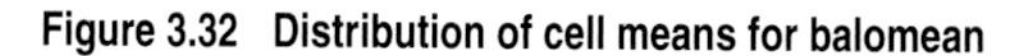

Figure 3.32 Distribution of cell means for balomean

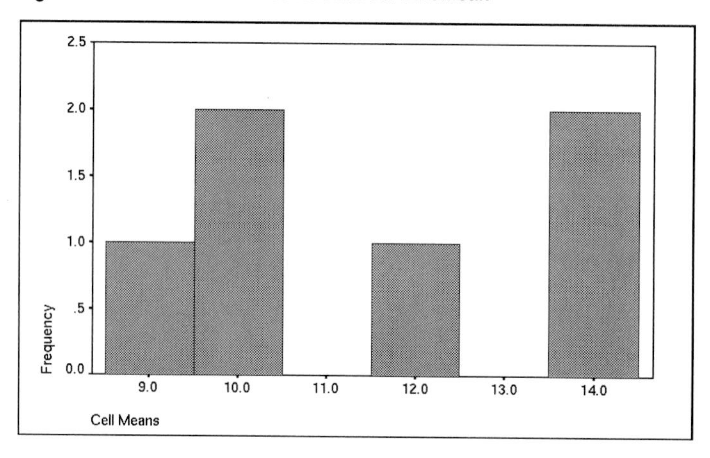

Although both univariate and multivariate analyses of variance require equal variances in the cells, there are many situations in which cell means and standard deviations, or cell means and variances, are proportional. Figure 3.33 shows plots of cell means versus cell variances and cell standard deviations. There appears to be no relationship between the means and the measures of variability. If patterns were evident, transformations of the dependent variables might be used to stabilize the variances.

Figure 3.33 Plots of cell means, cell variances, and cell standard deviations

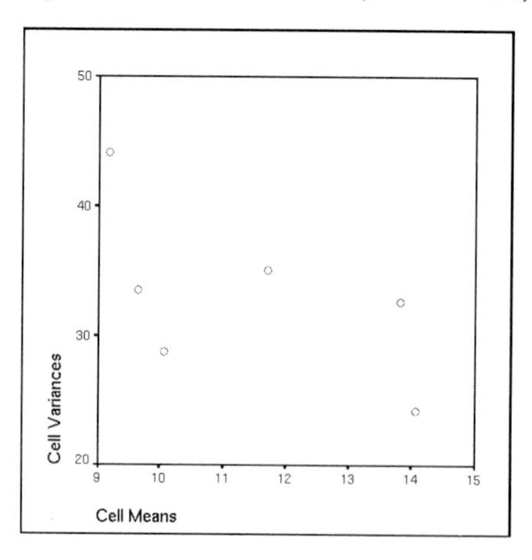

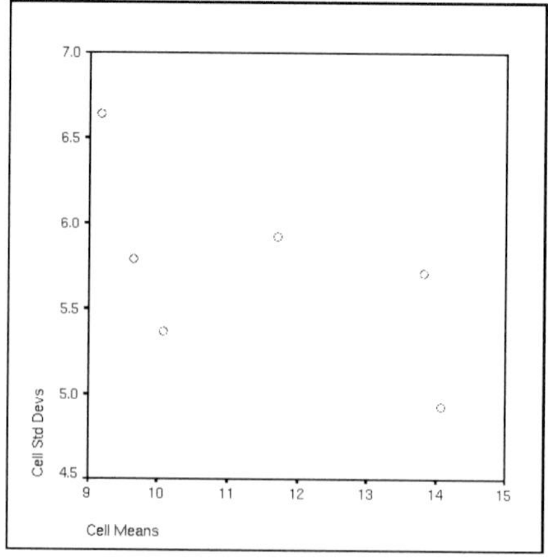

Principal Components Analysis

Bartlett's test of sphericity provides information about the correlations among the dependent variables by testing the hypothesis that the correlation matrix is an identity matrix. Another way to examine dependencies among the dependent variables is to perform a principal components analysis of their within-cells correlation matrix (see *SPSS Professional Statistics* for a discussion of principal components analysis). If principal components analysis reveals that one of the variables can be expressed as a linear combination of the others, the error sums-of-squares and cross-products matrix will be singular and a unique inverse cannot be obtained.

Figure 3.34 contains the within-cells correlation matrix for the two-way factorial design. The eigenvalues and percentage of variance explained by each are shown in Figure 3.35. The first two principal components account for about two-thirds of the total variance, and the remaining two components account for the rest. None of the eigenvalues is close enough to 0 to cause concern about the error matrix being singular.

Figure 3.34 Within-cells correlation matrix

WITHIN CELLS Correlations with Std. Devs. on Diagonal

	BALOMEAN	BALCMEAN	SSTMEAN	PP
BALOMEAN	5.67488			
BALCMEAN	.30019	1.41485		
SSTMEAN	.11506	.24341	2.39327	
PP	.11100	.14334	.34698	1.43842

Figure 3.35 Eigenvalues and percentage of variance

Eigenvalues of WITHIN CELLS correlation matrix

	Eigenvalue	Pct of Var	Cum Pct
1	1.63610	40.90257	40.90257
2	1.02780	25.69499	66.59757
3	.72258	18.06448	84.66205
4	.61352	15.33795	100.00000

Figure 3.36 contains the loadings, which in this case are equivalent to correlations, between the principal components and dependent variables. The components can be rotated, as shown in Figure 3.37, to increase interpretability. Each variable loads highly on only one of the components, suggesting that none is redundant or highly correlated with the others. When there are many dependent variables, a principal components analysis may indicate how the variables are related to each other. This information is useful for establishing the number of unique dimensions being measured by the dependent variables.

Figure 3.36 Correlations between principal components and dependent variables

Normalized principal components
Components

Variables	1	2	3	4
BALOMEAN	-.55200	-.63056	-.48683	.24635
BALCMEAN	-.66950	-.41517	.49417	-.36770
SSTMEAN	-.69880	.41598	.27636	.51212
PP	-.62837	.53366	-.40619	-.39416

Figure 3.37 Rotated correlations between components and dependent variables

VARIMAX rotated correlations between components and DEPENDENT variable
Can. Var.

DEP. VAR.	1	2	3	4
BALOMEAN	.98696	.04774	.14699	.04503
BALCMEAN	.15005	.05958	.98016	.11501
SSTMEAN	.04631	.17390	.11606	.97680
PP	.04835	.98230	.05905	.17104

Tests of Multivariate Differences

Once the preliminary steps of examining the distribution of the variables for outliers, non-normality, and inequality of variances have been taken and no significant violations have been found, hypothesis testing can begin. In the one- and two-sample tests, Hotelling's T^2, a multivariate generalization of the univariate t value, is used. For more complicated designs, an extension of the familiar analysis-of-variance F test to the multivariate case is needed.

The univariate F tests in ANOVA are the ratios of the hypothesis mean squares to the error mean squares. When there is more than one dependent variable, there is no longer a single number that represents the hypothesis and error sums of squares. Instead, as shown in "Testing Hypotheses" on p. 67, there are matrices for the hypothesis and error sums of squares and cross-products. These matrices must be combined into some type of test statistic.

Most multivariate test statistics are based on the determinant of HE^{-1}, where H is the hypothesis sums-of-squares and cross-products matrix and E^{-1} is the inverse of the error sums-of-squares and cross-products matrix. The determinant is a measure of the generalized variance, or dispersion, of a matrix. This determinant can be calculated as the product of the eigenvalues of a matrix, since each eigenvalue represents a portion of the generalized variance. In fact, the process of extracting eigenvalues can be viewed as a principal components analysis on the HE^{-1} matrix.

There are a variety of test statistics for evaluating multivariate differences based on the eigenvalues of HE^{-1}. Four of the most commonly used tests are displayed by the SPSS Multivariate ANOVA procedure:

Pillai's trace:

$$V = \sum_{i=1}^{s} \frac{1}{1+\lambda_i}$$ **Equation 3.11**

Wilks' lambda:

$$W = \prod_{i=1}^{s} \frac{1}{1+\lambda_i}$$ **Equation 3.12**

Hotelling's trace:

$$T = \Sigma\lambda_i$$ **Equation 3.13**

Roy's largest root:

$$R = \frac{\lambda_{MAX}}{1+\lambda_{MAX}}$$ **Equation 3.14**

where λ_{MAX} is the largest eigenvalue, λ_i is the *i*th eigenvalue, and *s* is the number of non-zero eigenvalues of HE^{-1}.

Although the exact distributions of the four criteria differ, they can be transformed into statistics that have approximately an *F* distribution. Tables of the exact distributions of the statistics are also available.

When there is a single dependent variable, all four criteria are equivalent to the ordinary ANOVA *F* statistic. When there is a single sample or two independent samples with multiple dependent variables, they are all equivalent to Hotelling's T^2. In both situations, the transformed statistics are distributed exactly as *F*'s.

Two concerns dictate the choice of the multivariate criterion—**power** and **robustness**. That is, the test statistic should detect differences when they exist and not be much affected by departures from the assumptions. For most practical situations, when differences among groups are spread along several dimensions, the ordering of the test criteria in terms of decreasing power is Pillai's, Wilks', Hotelling's, and Roy's. Pillai's trace is also the most robust criterion. That is, the significance level based on it is reasonably correct even when the assumptions are violated. This is important, since a test that results in distorted significance levels in the presence of mild violations of homogeneity of covariance matrices or multivariate normality is of limited use (Olsen, 1976).

Testing the Effects

Since our design is a two-by-three factorial (two sexes and three categories of field dependence), there are three effects to be tested: the sex and field-dependence main effects and the sex-by-field-dependence interaction. As in univariate analysis of variance, the terms are tested in reverse order. That is, higher-order effects are tested before lower-order ones, since it is difficult to interpret lower-order effects in the presence of higher-order interactions. For example, if there is a sex-by-field-dependence interaction, testing for sex and field-dependence main effects is not particularly useful and can be misleading.

The SPSS Multivariate ANOVA procedure displays separate output for each effect. Figure 3.38 shows the label displayed on each page to identify the effect being tested. The hypothesis sums-of-squares and cross-products matrix can be displayed for each effect. The same error matrix (the pooled within-cells sums-of-squares and cross-products matrix) is used to test all effects and is displayed only once before the effect-by-effect output. Figure 3.39 contains the error matrix for the factorial design. Figure 3.40 is the hypothesis sums-of-squares and cross-product matrix. These two matrices are the ones involved in the computation of the test statistics displayed in Figure 3.41.

Figure 3.38 Label for effect being tested

```
EFFECT .. SEX BY FIELD
```

Figure 3.39 Error sums-of-squares and cross-products matrix

WITHIN CELLS Sum-of-Squares and Cross-Products

	BALOMEAN	BALCMEAN	SSTMEAN	PP
BALOMEAN	4122.15200			
BALCMEAN	308.51395	256.22940		
SSTMEAN	200.02104	105.49895	733.14908	
PP	115.98133	37.33883	152.89508	264.83786

Figure 3.40 Hypothesis sums-of-squares and cross-products matrix

Adjusted Hypothesis Sum-of-Squares and Cross-Products

	BALOMEAN	BALCMEAN	SSTMEAN	PP
BALOMEAN	18.10215			
BALCMEAN	8.23621	3.77316		
SSTMEAN	2.49483	.55197	13.52286	
PP	-.00368	.21323	-4.85735	1.78988

Figure 3.41 Multivariate tests of significance

Multivariate Tests of Significance (S = 2, M = 1/2, N = 61 1/2)

Test Name	Value	Approx. F	Hypoth. DF	Error DF	Sig. of F
Pillais	.05245	.84827	8.00	252.00	.561
Hotellings	.05410	.83848	8.00	248.00	.570
Wilks	.94813	.84339	8.00	250.00	.565
Roys	.03668				

Note.. F statistic for WILKS' Lambda is exact.

The first line of Figure 3.41 contains the values of the parameters (*S*, *M*, *N*) used to find significance levels in tables of the exact distributions of the statistics. For the first three tests, the value of the test statistic is given, followed by its transformation to a statistic that has approximately an *F* distribution. The next two columns contain the numerator (hypothesis) and denominator (error) degrees of freedom for the *F* statistic. The observed significance level (the probability of observing a difference at least as large as the one found in the sample when there is no difference in the populations) is given in the last column. All of the observed significance levels are large, causing us not to reject the hypothesis that there is no sex-by-field-dependence interaction. There is no straightforward transformation for Roy's largest root criterion to a statistic with a known distribution, so only the value of the largest root is displayed.

Since the multivariate results are not statistically significant, there is no reason to examine the univariate results shown in Figure 3.42. When the multivariate results are significant, however, the univariate statistics may help determine which variables contribute to the overall differences. The univariate *F* tests for the sex-by-field interaction are the same as the *F*'s for sex by field in a two-way ANOVA. Figure 3.43 contains a two-way analysis of variance for the balancing-with-eyes-open (*balomean*) variable from the General Factorial ANOVA procedure. The within-cells sum of squares is identical to the diagonal entry for *balomean* in Figure 3.39. Similarly, the sex-by-field sum of squares is identical to the diagonal entry for *balomean* in Figure 3.40. The *F* value for the interaction term, 0.28, is the same as in the first line of Figure 3.42.

Figure 3.42 Univariate F tests

```
EFFECT .. SEX BY FIELD (CONT.)
Univariate F-tests with (2,128) D. F.

Variable     Hypoth. SS     Error SS  Hypoth. MS     Error MS           F   Sig. of F

BALOMEAN       18.10215   4122.15200     9.05107     32.20431      .28105        .755
BALCMEAN        3.77316    256.22940     1.88658      2.00179      .94244        .392
SSTMEAN        13.52286    733.14908     6.76143      5.72773     1.18047        .310
PP              1.78988    264.83786      .89494      2.06905      .43254        .650
```

Figure 3.43 Two-way analysis of variance

```
Tests of Significance for BALOMEAN using UNIQUE sums of squares
Source of Variation          SS      DF        MS          F  Sig of F

WITHIN CELLS            4122.15     128     32.20
SEX                      389.33       1    389.33      12.09      .001
FIELD                     55.19       2     27.59        .86      .427
SEX BY FIELD              18.10       2      9.05        .28      .755
```

Discriminant Analysis for the Interaction Effect

As discussed in "Discriminant Analysis" on p. 75, for the two-group situation, the MANOVA problem can also be viewed as one of finding the linear combinations of the dependent variables that best separate the categories of the independent variables. For main effects, the analogy with discriminant analysis is clear: what combinations of the variables distinguish men from women and what combinations distinguish the three categories of field dependence? When interaction terms are considered, we must distinguish among the six (two sex and three field-dependence) categories jointly. This is done by finding the linear combination of variables that maximizes the ratio of the hypothesis to error sums of squares. Since the interaction effect is not significant, there is no particular reason to examine the discriminant analysis results. However, we will consider them for illustrative purposes.

Figure 3.44 contains the standardized discriminant function coefficients for the interaction term. The number of functions that can be derived is equal to the degrees of freedom for that term if it is less than the number of dependent variables. The two variables that have the largest standardized coefficients for function 1 are the sidestepping test and the Purdue Pegboard Test. All warnings concerning the interpretation of coefficients when variables are correlated apply in this situation as well. However, the magnitude of the coefficients may give us some idea of the variables contributing most to group differences.

Figure 3.44 Standardized discriminant function coefficients

```
EFFECT .. SEX BY FIELD
Standardized discriminant function coefficients
          Function No.

Variable               1            2

BALOMEAN          .02621       -.27281
BALCMEAN         -.19519       -.90129
SSTMEAN           .98773        .02420
PP               -.73895        .15295
```

A measure of the strength of the association between the discriminant functions and the grouping variables is the **canonical correlation coefficient**. Its square is the proportion of variability in the discriminant function scores explained by the independent variables. All multivariate significance tests, which were expressed as functions of the eigenvalues in "Tests of Multivariate Differences" on p. 82, can also be expressed as functions of the canonical correlations.

Figure 3.45 contains several sets of statistics for the discriminant functions. The entry under *Eigenvalue* is the dispersion associated with each function. The next column is the percentage of the total dispersion associated with each function. (It is obtained by dividing each eigenvalue by the sum of all eigenvalues and multiplying by 100.) The last column contains the canonical correlation coefficients. The results in Figure 3.45 are consistent with the results of the multivariate significance tests. Both the eigenvalues and canonical correlation coefficients are small, indicating that there is no interaction effect.

Figure 3.45 Eigenvalues and canonical correlations

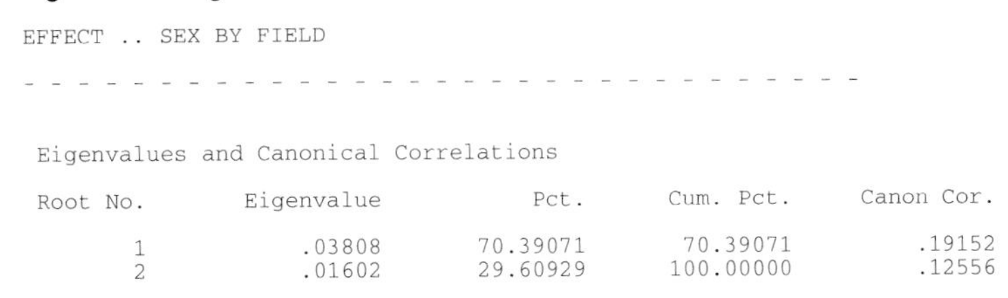

EFFECT .. SEX BY FIELD

Eigenvalues and Canonical Correlations

Root No.	Eigenvalue	Pct.	Cum. Pct.	Canon Cor.
1	.03808	70.39071	70.39071	.19152
2	.01602	29.60929	100.00000	.12556

When more than one discriminant function can be derived, you should examine how many functions contribute to group differences. The same tests for successive eigenvalues available in the Discriminant Analysis procedure can be obtained from the SPSS Multivariate ANOVA procedure. The first line in Figure 3.46 is a test of the hypothesis that all eigenvalues are equal to 0. The value for Wilks' lambda in the first line of Figure 3.46 is equal to the test of multivariate differences in Figure 3.41. Successive lines in Figure 3.46 correspond to tests of the hypothesis that all remaining functions are equal in the groups. These tests allow you to assess the number of dimensions on which the groups differ.

Figure 3.46 Dimension reduction analysis

EFFECT .. SEX BY FIELD

Dimension Reduction Analysis

Roots	Wilks L.	F	Hypoth. DF	Error DF	Sig. of F
1 TO 2	.94813	.84339	8.00	250.00	.565
2 TO 2	.98424	.67273	3.00	126.00	.570

Testing for Differences among Field Dependence and Sex Categories

Since the field-dependence-by-sex interaction term is not significant, the main effects can be tested. Figure 3.47 contains the multivariate tests of significance for the field-dependence variable. All four criteria indicate that there is not sufficient evidence to reject the null hypothesis that the means of the motor-ability variables do not differ for the three categories of field dependence.

Figure 3.47 Multivariate tests of significance for field

```
Multivariate Tests of Significance (S = 2, M = 1/2, N = 61 1/2)

Test Name         Value       Approx. F    Hypoth. DF      Error DF      Sig. of F

Pillais          .04152        .66780          8.00         252.00          .720
Hotellings       .04244        .65789          8.00         248.00          .728
Wilks            .95889        .66285          8.00         250.00          .724
Roys             .02533
Note.. F statistic for WILKS' Lambda is exact.
```

The multivariate tests of significance for the sex variable are shown in Figure 3.48. All four criteria indicate that there are significant differences between men and women on the motor-ability variables. Compare Figure 3.48 with Figure 3.23, which contains Hotelling's statistic for the two-group situation. Note that although the two statistics are close in value—0.669 when *sex* is considered alone and 0.609 when *sex* is included in a model containing field dependence and the sex-by-field-dependence interaction—they are not identical. The reason is that in the second model the sex effect is adjusted for the other effects in the model (see "Different Types of Sums of Squares" on p. 89).

Figure 3.48 Multivariate tests of significance for sex

```
EFFECT .. SEX

Multivariate Tests of Significance (S = 1, M = 1 , N = 61 1/2)

Test Name         Value       Exact F    Hypoth. DF      Error DF      Sig. of F

Pillais          .37871      19.04872        4.00         125.00          .000
Hotellings       .60956      19.04872        4.00         125.00          .000
Wilks            .62129      19.04872        4.00         125.00          .000
Roys             .37871
Note.. F statistics are exact.
```

Stepdown F Tests

If the dependent variables are ordered in some fashion, it is possible to test for group differences of variables adjusting for effects of other variables. This is termed a **stepdown procedure**. Consider Figure 3.49, which contains the stepdown tests for the motor-ability variables. The first line is just a univariate F test for the balancing-with-eyes-open variable. The value is the same as that displayed in Figure 3.50, which contains the univariate F tests. The next line in Figure 3.49 is the univariate F test for balancing with eyes closed when balancing with eyes open is taken to be a covariate. That is, differences in balancing with eyes open are eliminated from the comparison of balancing with eyes closed. Figure 3.51 is the ANOVA table from the General Factorial ANOVA procedure for the *balcmean* variable when *balomean* is treated as the covariate. Note that the F value of 1.77 for *sex* in Figure 3.51 is identical to that for *balcmean* in Figure 3.49.

Figure 3.49 Stepdown tests

EFFECT .. SEX (Cont.)
Roy-Bargman Stepdown F - tests

Variable	Hypoth. MS	Error MS	StepDown F	Hypoth. DF	Error DF	Sig. of F
BALOMEAN	389.32561	32.20431	12.08924	1	128	.001
BALCMEAN	3.24111	1.83574	1.76556	1	127	.186
SSTMEAN	149.80346	5.46262	27.42335	1	126	.000
PP	44.19767	1.84900	23.90361	1	125	.000

Figure 3.50 Univariate F tests

EFFECT .. SEX (Cont.)
Univariate F-tests with (1,128) D. F.

Variable	Hypoth. SS	Error SS	Hypoth. MS	Error MS	F	Sig. of F
BALOMEAN	389.32561	4122.15200	389.32561	32.20431	12.08924	.001
BALCMEAN	.16537	256.22940	.16537	2.00179	.08261	.774
SSTMEAN	172.11258	733.14908	172.11258	5.72773	30.04902	.000
PP	23.56120	264.83786	23.56120	2.06905	11.38747	.001

Figure 3.51 ANOVA table for balcmean with balomean as the covariate

Tests of Significance for BALCMEAN using UNIQUE sums of squares

Source of Variation	SS	DF	MS	F	Sig of F
WITHIN CELLS	233.14	127	1.84		
REGRESSION	23.09	1	23.09	12.58	.001
SEX	3.24	1	3.24	1.77	.186
FIELD	2.40	2	1.20	.65	.522
SEX BY FIELD	2.63	2	1.32	.72	.490

The third line of Figure 3.49, the test for the sidestepping variable, is adjusted for both the balancing-with-eyes-open variable and the balancing-with-eyes-closed variable. Similarly, the last line, the Purdue Pegboard Test score, has balancing with eyes open, balancing with eyes closed, and the sidestepping test as covariates. Thus, each variable in Figure 3.49 is adjusted for variables that precede it in the table.

The order in which variables are displayed on the stepdown tests in the SPSS Multivariate ANOVA procedure output depends only on the order in which the variables are specified for the procedure. If this order is not meaningful, the stepdown tests that result will not be readily interpretable or meaningful.

Different Types of Sums of Squares

When there is more than one factor and unequal numbers of cases in each cell in univariate analysis of variance, the total sums of squares cannot be partitioned into additive components for each effect. That is, the sums of squares for all effects do not add up to the total sums of squares. Differences between factor means are "contaminated" by the effects of the other factors.

There are many different algorithms for calculating the sums of squares for unbalanced data. Different types of sums of squares correspond to tests of different hypotheses. Two frequently used methods for calculating the sums of squares are the **unique method**, also known as the **regression method**, in which an effect is adjusted for all other effects in the model, and the **sequential method**, in which an effect is adjusted only for effects that precede it in the model.

In multivariate analysis of variance, unequal sample sizes in the cells lead to similar problems. Again, different procedures for calculating the requisite statistics are available. The SPSS Multivariate ANOVA procedure offers two options: the unique (regression) solution and the sequential solution. All output displayed in this chapter is obtained from the regression solution, the default (see Milliken & Johnson, 1984).

For any connected design, the hypotheses associated with the sequential sums of squares are weighted functions of the population cell means, with weights depending on the cell frequencies (see Searle, 1971). For designs in which every cell is filled, it can be shown that the hypotheses corresponding to the regression model sums of squares are the hypotheses about the unweighted cell means. With empty cells, the hypotheses will depend on the pattern of empty cells.

Problems with Empty Cells

When there are no observations in one or more cells in a design, the analysis is greatly complicated. This is true for both univariate and multivariate designs. Empty cells result in the inability to estimate uniquely all of the necessary parameters. Hypotheses that involve parameters corresponding to the empty cells usually cannot be tested. Thus, the output from an analysis involving empty cells should be treated with caution (see Freund, 1980; Milliken & Johnson, 1984).

Examining Residuals

Residuals—the differences between observed values and those predicted from a model—provide information about the adequacy of fit of the model and the assumptions. Residual analysis for regression models is discussed in the SPSS Base system documentation. The same techniques are appropriate for analysis-of-variance models as well.

In the two-factor model with interactions, the equation is

$$\hat{y}_{ij} = \hat{\mu} + \hat{\alpha}_i + \hat{\beta}_j + \hat{\delta}_{ij}$$ **Equation 3.15**

where $\hat{y}_{ij}$ is the predicted value for the cases in the ith category of the first variable and the jth category of the second. As before, $\hat{\mu}$ is the grand mean, $\hat{\alpha}_i$ is the effect of the ith category of the first variable, $\hat{\beta}_j$ is the effect of the jth category of the second variable, and $\hat{\delta}_{ij}$ is their interaction.

Consider Figure 3.52, which contains deviation parameter estimates for the balancing-with-eyes-open variable. From this table and the grand mean (11.42), the predicted value for females in the low field-dependence category is

$$\hat{y}_{11} = 11.42 - 1.78 - 0.97 + 0.51 = 9.18 \qquad \textbf{Equation 3.16}$$

Similarly, the predicted value for males in the high field-dependence category is

$$\hat{y}_{23} = 11.42 + 1.78 + 0.53 + 0.09 = 13.82 \qquad \textbf{Equation 3.17}$$

Note that only independent parameter estimates are displayed and that the other estimates must be derived from these. For example, the parameter estimate displayed for the sex variable is for the first category, females. The estimate for males is the negative of the estimate for females, since the two values must sum to 0 for the default deviation contrasts. The value for the third category of field dependence is 0.53, the negative of the sum of the values for the first two categories. The two parameter estimates displayed for the interaction effects are for females with low field dependence and females with medium field dependence. The remaining estimates can be easily calculated. For example, the value for males in the low field-dependence category is –0.51 , the negative of that for females. Similarly, the value for females in the high field-dependence category is –0.087 , the negative of the sum of the values for females in low and medium field-dependence categories.

Figure 3.52 Parameter estimates for deviation contrasts

```
Estimates for BALOMEAN
--- Individual univariate .9500 confidence intervals
SEX

 Parameter        Coeff.  Std. Err.      t-Value      Sig. t Lower -95%  CL- Upper

         2   -1.7790199      .51166     -3.47696      .00069    -2.79143     -.76661
FIELD

 Parameter        Coeff.  Std. Err.      t-Value      Sig. t Lower -95%  CL- Upper

         3   -.97446425      .74485     -1.30827      .19313    -2.44828      .49935
         4   .445727448      .70743       .63006      .52978     -.95405     1.84551
SEX BY FIELD

 Parameter        Coeff.  Std. Err.      t-Value      Sig. t Lower -95%  CL- Upper

         5   .514565812      .74485       .69083      .49092     -.95925     1.98838
         6   -.42730420      .70743      -.60402      .54690    -1.82709      .97248
```

Figure 3.53 contains an excerpt of some cases in the study and their observed and predicted values for the balancing-with-eyes-open variable. The fourth column is the residual: the difference between the observed and predicted values. Standardized residuals shown in the fifth column are obtained by dividing each raw residual by the error standard deviation.

Figure 3.53 Observed and predicted values

Observed and Predicted Values for Each Case

Dependent Variable.. BALOMEAN   Balance Test - Eyes Open

Case No.	Observed	Predicted	Raw Resid.	Std Resid.
1	4.000	10.082	-6.082	-1.072
2	8.850	9.181	-.331	-.058
3	10.000	9.659	.341	.060
4	3.550	10.082	-6.532	-1.151
5	2.800	9.659	-6.859	-1.209
6	14.500	9.659	4.841	.853
.	.	.	.	.
.	.	.	.	.
.	.	.	.	.
127	5.000	13.815	-8.815	-1.553
128	20.000	11.710	8.290	1.461
129	20.000	13.815	6.185	1.090
130	19.000	11.710	7.290	1.285
131	16.150	14.072	2.078	.366
132	20.000	14.072	5.928	1.045
.	.	.	.	.
.	.	.	.	.
.	.	.	.	.

As in regression analysis, a variety of plots is useful for checking the assumptions. Figure 3.54 is a scatterplot of the observed, predicted, and standardized residual values for the *balomean* variable. For observed versus predicted, the cases fall into six rows for the predicted values, since all cases in the same cell have the same predicted value. Residuals are plotted against the case numbers in Figure 3.55. The plot against the case numbers is useful if the data are gathered and entered into the file sequentially. Any patterns in this plot lead to suspicions that the data are not independent of each other. Of course, if the data have been sorted before being entered, a pattern is to be expected.

Figure 3.54 Scatterplot matrix of observed, predicted, and residual values for balance test

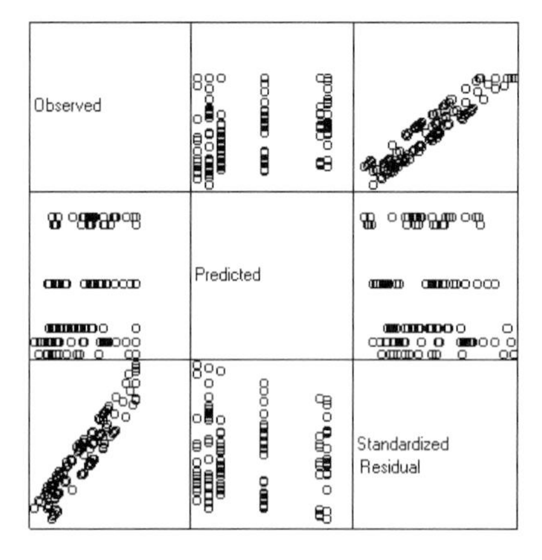

Figure 3.55 Plot of case numbers and residuals for balance test (balomean)

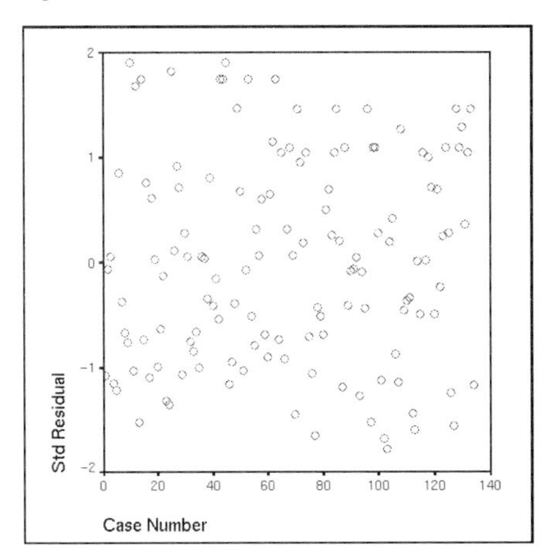

If the assumption of multivariate normality is met, the distribution of the residuals for each variable should be approximately normal. Figure 3.56 is a normal plot of the residuals for the balancing-with-eyes-open variable, while Figure 3.57 is the detrended normal plot of the same variable. Both of these plots suggest that there may be reason to suspect that the distribution of the residuals is not normal.

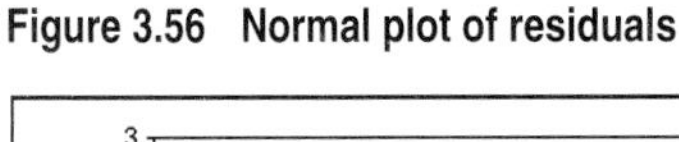

Figure 3.56 Normal plot of residuals

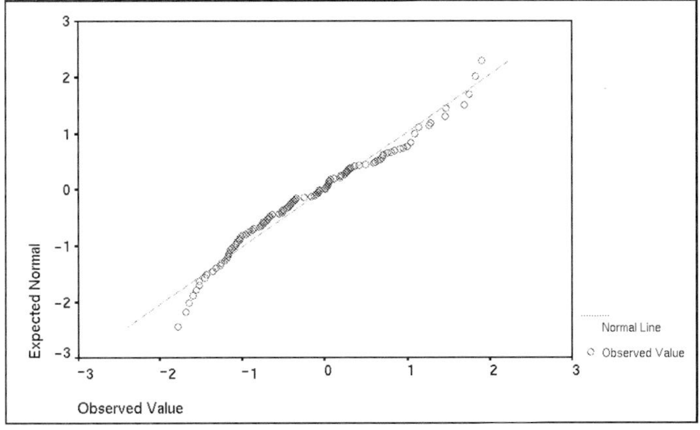

Figure 3.57 Detrended normal plot

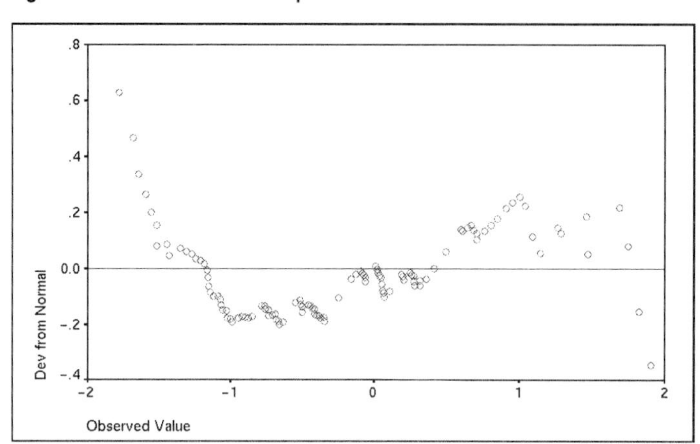

Predicted Means

Figure 3.58 shows a table that contains observed and predicted means as well as residuals for all cells in the design for the balancing-with-eyes-open variable (predicted means for combinations of factor levels are available using MANOVA command syntax. See MANOVA: Univariate in the Syntax Reference section of this manual for more information). The first column of the table, labeled *Obs. Mean,* is the observed mean for that cell. The next value, labeled *Adj. Mean*, is the mean predicted from the model adjusted for the covariates. The column labeled *Est. Mean* contains the predicted means without correcting for covariates. When covariates are present, the differences between the adjusted and estimated means provide an indication of the effectiveness of the covariate adjustment (see Finn, 1974). For a complete factorial design without covariates, the observed, adjusted, and estimated means will always be equal. The difference between them, the residual, will also be equal to 0.

Figure 3.58 Table of predicted means

```
Adjusted and Estimated Means
Variable .. BALOMEAN              Balance Test - Eyes Open
      Factor          Code          Obs. Mean      Adj. Mean      Est. Mean    Raw Resid.    Std. Resid.


 SEX              Female
  FIELD                  1          9.18077        9.18077        9.18077        .00000         .00000
  FIELD                  2          9.65909        9.65909        9.65909        .00000         .00000
  FIELD                  3         10.08214       10.08214       10.08214        .00000         .00000

 SEX              Male
  FIELD                  1         11.70968       11.70968       11.70968        .00000         .00000
  FIELD                  2         14.07174       14.07174       14.07174        .00000         .00000
  FIELD                  3         13.81471       13.81471       13.81471        .00000         .00000
```

If the design is not a full-factorial model, the observed means and those predicted using the parameter estimates will differ. Consider Figure 3.59, which contains a table of means for the main-effects-only model. The observed cell means are no longer equal to those predicted by the model. The difference between the observed and estimated mean is shown in the column labeled *Raw Resid.* The residual divided by the error standard deviation is shown in the column labeled *Std. Resid.* From this table it is possible to identify cells for which the model does not fit well.

It is also possible to obtain various combinations of the adjusted means. For example, Figure 3.59 contains the combined adjusted means for the *sex* and *field* variables for the main-effects design. The means are labeled as unweighted, since the sample sizes in the cells are not used when means are combined over the categories of a variable.

Figure 3.59 Table of predicted means

```
Adjusted and Estimated Means
Variable .. BALOMEAN            Balance Test - Eyes Open
      Factor            Code              Obs. Mean      Adj. Mean       Est. Mean     Raw Resid.    Std. Resid.


 SEX              Female
  FIELD                    1              9.18077        8.40954        8.40954         .77123          .13590
  FIELD                    2              9.65909       10.06241       10.06241       -.40331         -.07107
  FIELD                    3             10.08214       10.12332       10.12332       -.04118         -.00726

 SEX              Male
  FIELD                    1             11.70968       12.03310       12.03310       -.32342         -.05699
  FIELD                    2             14.07174       13.68596       13.68596        .38578          .06798
  FIELD                    3             13.81471       13.74688       13.74688        .06783          .01195

- - - - - - - - - - - - - - - - - - - - - - - - - - - - - - - - - - - - - - - - - - - - - - - - - - - - -
Combined Adjusted Means for SEX
Variable .. BALOMEAN
          SEX
       Female      UNWGT.      9.53176
         Male      UNWGT.     13.15531

- - - - - - - - - - - - - - - - - - - - - - - - - - - - - - - - - - - - - - - - - - - - - - - - - - - - -
Combined Adjusted Means for FIELD
Variable .. BALOMEAN
        FIELD
            1      UNWGT.     10.22132
            2      UNWGT.     11.87418
            3      UNWGT.     11.93510
```

Computation of Power and Effect Size

Whenever you evaluate data, it is important to consider both the magnitude of the observed effect and the power of detecting an effect of the observed magnitude. (Recall that the power of an experiment is the probability of rejecting the null hypothesis when it is false.) The SPSS Multivariate ANOVA procedure will compute power estimates for both univariate and multivariate tests.

Some Final Comments

In this chapter, only the most basic aspects of multivariate analysis of variance have been covered. The SPSS Multivariate ANOVA procedure is capable of testing more elaborate models of several types. Chapter 4 considers a special class of designs called **repeated measures designs**. In such designs, the same variable or variables are measured on several occasions. For further discussion of multivariate analysis of variance, consult Morrison (1976) and Tatsuoka (1971).

How to Obtain a Multivariate Analysis of Variance

SPSS Multivariate ANOVA is a procedure for analysis of variance and covariance for models containing two or more dependent variables. The procedure also produces principal components, discriminant function coefficients, canonical correlations, and other statistics for balanced and unbalanced designs. To analyze models containing a single dependent variable, use the General Factorial ANOVA procedure (see Chapter 2). For repeated measures designs, see Chapter 4.

The minimum specification is two or more dependent variables (this produces one-sample Hotelling's T^2). Most other models require additional specifications. For example:

- For independent-samples designs, specify at least one factor variable.
- For one-sample models involving continuous predictor variables (such as multivariate regression analysis), specify one or more covariates.
- For multivariate analysis of covariance, specify a factor variable and at least one covariate.

To obtain a multivariate analysis of variance, from the menus choose:

Statistics
 ANOVA Models ▸
  Multivariate...

This opens the Multivariate ANOVA dialog box, as shown in Figure 3.60.

Figure 3.60 Multivariate ANOVA dialog box

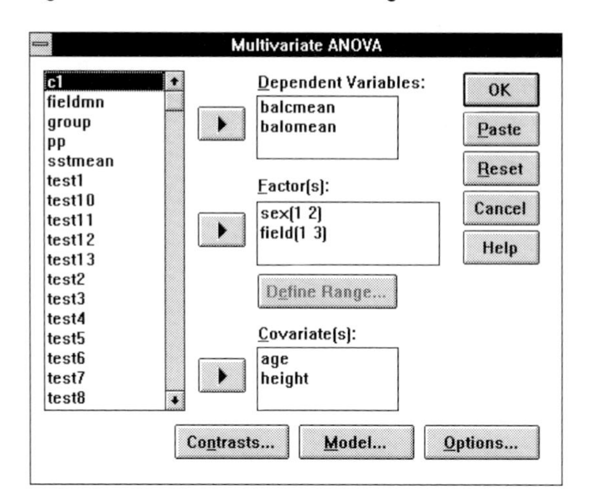

The numeric variables in your working data file appear on the source list.

Dependent Variables. Select two or more continuous dependent variables. SPSS treats the variables as jointly dependent and analyzes a multivariate design.

Factor(s). Optionally, you can select one or more categorical predictor variables. You must indicate which factor levels you want to use in the analysis (see "Defining Levels of Factor Variables," below).

Covariate(s). Optionally, you can select one or more continuous predictor variables.

Defining Levels of Factor Variables

For each factor variable, you must indicate which factor levels you want to use in the analysis. Highlight one or more factors and click on Define Range... in the Multivariate ANOVA dialog box. This opens the Multivariate ANOVA Define Range dialog box, as shown in Figure 3.61.

Figure 3.61 Multivariate ANOVA Define Range dialog box

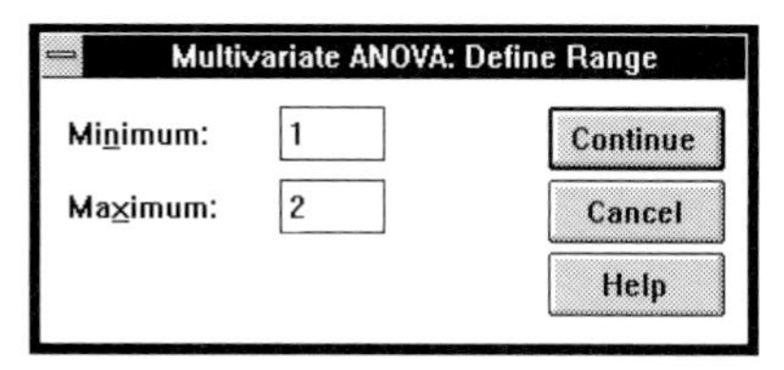

Enter integer values corresponding to the lowest and highest factor levels you want to use. Each integer category in the inclusive range is used as a factor level in the analysis (non-integer codes are truncated for the analysis, so cases with the same integer portion are combined into the same factor level). Cases with values outside the range are excluded from the analysis. To avoid empty cells, factor levels within the range must be consecutive. If a variable has empty categories, use the Automatic Recode facility on the Transform menu to create consecutive categories (for more information on recoding variables, see the SPSS Base system documentation).

Specifying Contrasts

Factor variables are transformed into contrasts for the analysis. To display parameter estimates or override the default contrast type (deviation), click on Contrasts... in the Multivariate ANOVA dialog box. This opens the Multivariate ANOVA Contrasts dialog box, as shown in Figure 3.62.

Figure 3.62 Multivariate ANOVA Contrasts dialog box

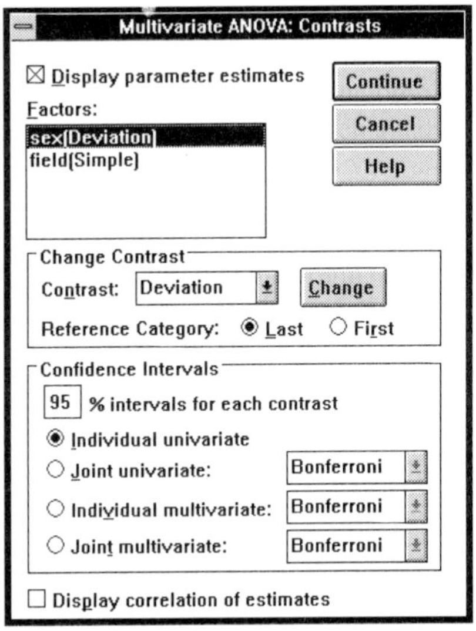

- **Display parameter estimates.** Displays parameter estimates for contrasts as well as their standard errors, *t* statistics, and confidence intervals. If you select this option, you can also change the contrast type and confidence intervals.

Change Contrast. By default, each factor is transformed into deviation contrasts. To obtain a different type of contrast, highlight one or more factors, select a contrast type, and click on Change. Optionally, you can change the default reference category.

- **Contrast**. You can choose one of the following contrast types:

  Deviation. Each category of the factor except the reference category is compared to the overall effect. This is the default.

  Simple. Each category of the factor (except the reference category) is compared to the reference category.

  Difference. Also known as reverse Helmert contrasts. Each category of the factor except the first category is compared to the average effect of previous categories.

  Helmert. Each category of the factor except the last category is compared to the mean effect of subsequent categories.

  Repeated. Each category of the factor except the first category is compared to the category that precedes it.

  Polynomial. The factor is partitioned into linear, quadratic, and cubic effects, and so forth (depending on the number of categories). Categories are assumed to be equally spaced.

  Reference Category. For deviation and simple contrasts, you can specify which category is to be omitted from the comparisons. Choose one of the following alternatives:

  - **Last**. Uses the highest category of the factor as the reference category. This is the default.
  - **First**. Uses the lowest category of the factor as the reference category.

Confidence Intervals. If you request parameter estimates for contrasts, SPSS displays confidence intervals for the estimates. To override the default confidence level (95%), enter a value for n% intervals for each contrast. The value must be greater than 0 and less than 100.

You can also choose between individual and joint confidence intervals and between univariate and multivariate intervals. Choose one of the following alternatives:

- **Individual univariate**. Displays a separate interval for each parameter within each dependent variable. These are the default confidence intervals.
- **Joint univariate**. Displays simultaneous intervals for parameters within each dependent variable. By default, SPSS displays Bonferroni intervals, which are based on the number of contrasts actually made. To obtain intervals based on all possible contrasts, choose Scheffé from the drop-down list.
- **Individual multivariate**. Displays separate intervals for each parameter across all dependent variables. You can choose one of the following methods for computing the multivariate intervals:

  Bonferroni. Displays confidence intervals based on Student's *t* distribution.

  Roy. Displays confidence intervals based on Roy's largest root.

Pillai. Displays confidence intervals based on Pillai's trace.

Hotelling. Displays confidence intervals based on Hotelling's trace.

Wilks. Displays confidence intervals based on Wilks' lambda.

❍ **Joint multivariate**. Displays simultaneous intervals for parameters across all dependent variables. You can choose one of the following methods for computing the multivariate intervals:

Bonferroni. Displays confidence intervals based on Student's t distribution.

Roy. Displays confidence intervals based on Roy's largest root.

Pillai. Displays confidence intervals based on Pillai's trace.

Hotelling. Displays confidence intervals based on Hotelling's trace.

Wilks. Displays confidence intervals based on Wilks' lambda.

The following option is also available:

❑ **Display correlation of estimates**. Displays correlations and covariances between parameter estimates.

Specifying the MANOVA Model

By default, SPSS analyzes a full factorial model. To define a different model or control model options, click on Model... in the Multivariate ANOVA dialog box. This opens the Multivariate ANOVA Model dialog box, as shown in Figure 3.63.

Figure 3.63 Multivariate ANOVA Model dialog box

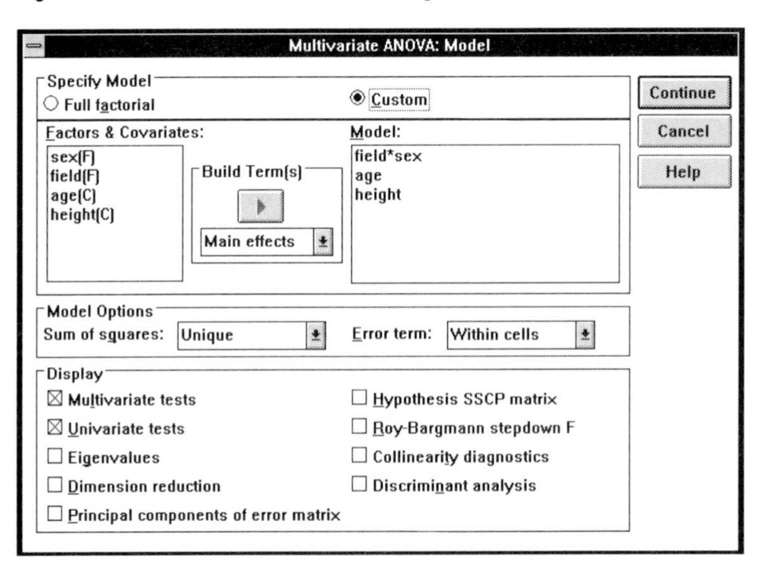

Specify Model. You can choose one of the following alternatives:

❍ **Full factorial.** Model contains all factor and covariate main effects and all factor-by-factor interactions. This is the default. Covariate interactions are not included in the model.

❍ **Custom.** Select this item to define a model containing factor-by-covariate interactions, or a model containing a subset of possible factor-by-factor interactions. You must indicate which terms you want to include in the model.

Model. To include a term in a custom model, select one or more factors or covariates, or a combination. If you don't want to create the highest-order interaction for the selected variables, select an alternate method for building the term. To add more terms to the model, repeat this process. Do not use the same term more than once in the model.

For example, to define a model containing the factor-by-factor interaction *field*sex* and covariates *age* and *height*, highlight field(F) and sex(F), select Interaction (the default) from the Build Term(s) drop-down list, and click on ▸. Then highlight age(C) and height(C), select Main effects from the Build Term(s) drop-down list, and click on ▸.

If you remove a variable from the Factor(s) or Covariate(s) list in the main dialog box, any terms that include that variable are removed from the model.

Macintosh: Use ⌘-click to select multiple factors or covariates, or a combination.

⬇ **Build Term(s).** You can build main effects or interactions for the selected variables. If you request an interaction of a higher order than the number of variables, SPSS creates a term for the highest-order interaction for the variables. If only one variable is selected, a main-effect term is added to the model, regardless of your Build Term(s) selection. Choose one of the following alternatives:

Interaction. Creates the highest-level interaction term for the variables. This is the default.

Main effects. Creates a main-effect term for each variable.

All 2-way. Creates all possible two-way interactions for the variables.

All 3-way. Creates all possible three-way interactions for the variables.

All 4-way. Creates all possible four-way interactions for the variables.

All 5-way. Creates all possible five-way interactions for the variables.

Model Options. SPSS includes a constant term in all models. You can control the computation of sums of squares and the error term used for the model.

⬇ **Sum of squares.** Select one of the following methods for decomposing sums of squares:

Unique. Uses the regression method to partition sums of squares. This is the default. Each effect is adjusted for all other effects in the model.

Sequential. Decomposes sums of squares hierarchically. For default (full factorial) models, SPSS first adjusts covariates for all factors and interactions. Then each factor is adjusted for all covariates and for factors that precede it in the model specification. Each interaction is adjusted for all covariates, factors, and preceding interactions.

For custom models, SPSS first adjusts covariates that appear only as main effects for all other terms in the model. Next, all other terms in the model are adjusted for covariates that appear only as main effects. Then SPSS evaluates all remaining terms (factor main effects, factor interactions, covariate interactions, and main effects for covariates that also appear in interactions) in the order in which they appear in the model specification.

- **Error term**. You can control the error term used in tests of significance. Select one of the following alternatives:

  Within cells. Tests model terms against the within-cells sum of squares. This is the default.

  Residual. Tests model terms against the residual sum of squares.

  Within+Residual. Tests model terms against the pooled within-cells and residual sum of squares.

Display. At least one of the following statistical displays must be selected:

- **Multivariate tests**. Displays multivariate test statistics and their F values, degrees of freedom, and probabilities. Shown by default.
- **Univariate tests**. Displays univariate F tests and their probabilities. Also shows hypothesis and error sums of squares and mean squares. Shown by default.
- **Eigenvalues**. Displays eigenvalues of HE^{-1} and canonical correlations.
- **Dimension reduction**. Displays a table showing Wilks' lambda and its transformed F value, degrees of freedom, and probability for each discriminant function and for functions that remain when successive functions are removed.
- **Principal components of error matrix**. Displays principal components analysis of within-cells correlation matrix. These principal components are corrected for the effects of factors and covariates in the analysis. Also shows principal component loadings after orthogonal (varimax) rotation. All principal components are extracted.
- **Hypothesis SSCP matrix**. Displays a matrix of hypothesis sums of squares and cross-products.
- **Roy-Bargmann stepdown F**. Displays analyses of covariance that test for a dependent variable's contribution to group differences while controlling for other dependent variables. SPSS controls for dependent variables in the order in which they are specified in the Multivariate ANOVA dialog box.
- **Collinearity diagnostics**. Displays collinearity diagnostics for the design matrix, including singular values of the normalized decomposition (which are the same as the normalized design), their condition indices, and the proportion of variance of parameters accounted for by each principal component.

❑ **Discriminant analysis**. Produces a canonical discriminant or canonical correlation analysis for each effect in the model. Output includes raw and standardized discriminant (or canonical) coefficients, effect estimates for canonical variables, and structure coefficients. All possible canonical functions are extracted. If two or more functions are extracted, SPSS rotates structure coefficients using the varimax method and displays the transformation matrix.

Options

To obtain optional statistics, click on Options... in the Multivariate ANOVA dialog box. This opens the Multivariate ANOVA Options dialog box, as shown in Figure 3.64.

Figure 3.64 Multivariate ANOVA Options dialog box

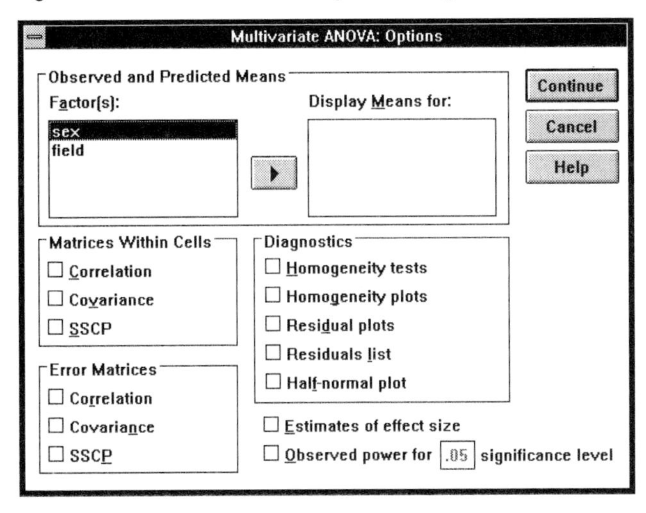

Observed and Predicted Means. You can obtain observed means (of dependent variables and covariates) and predicted means (of dependent variables) for subgroups defined by factors.

Factor(s). The list contains the variables you have selected as factors.

Display Means for. Select the factors you want to use to define subgroups. This displays means for each factor level. For observed means, SPSS shows weighted means (which give equal weight to all cases) and unweighted means (which weight cells equally regardless of the number of cases they contain). For predicted means, SPSS displays cell means that are adjusted (adjusted means) and unadjusted (estimated means) for all covariates that appear only as main effects. Raw and standardized residuals are also shown.

Matrices Within Cells. Matrices within cells are produced only if you have at least one factor variable. You can choose one or more of the following:

- ❑ **Correlation.** Displays a correlation matrix for each cell. Standard deviations are shown on the diagonal.
- ❑ **Covariance.** Displays a variance-covariance matrix for each cell. The pooled within-cells variance-covariance matrix is also shown.
- ❑ **SSCP.** Displays a matrix of sums-of-squares and cross-products for each cell.

Error Matrices. You can choose one or more of the following error matrices:

- ❑ **Correlation.** Displays an error correlation matrix. Also shows the log of the determinant of the matrix and Bartlett's test of sphericity, which tests the hypothesis that the population correlation matrix is an identity matrix.
- ❑ **Covariance.** Displays an error variance-covariance matrix.
- ❑ **SSCP.** Displays a matrix of error sums-of-squares and cross-products.

Diagnostics. You can choose one or more of the following diagnostics:

- ❑ **Homogeneity tests.** Displays homogeneity tests for univariate (Bartlett-Box F and Cochran's C) and multivariate (Box's M) models.
- ❑ **Homogeneity plots.** Plots cell means against cell variances and cell standard deviations and displays a histogram of cell means. Plots are shown for each dependent variable and covariate.
- ❑ **Residual plots.** Displays a scatterplot matrix of observed values, predicted values, and standardized residuals. Also plots standardized residuals against case-identification numbers and displays normal and detrended normal probability plots for standardized residuals. Plots are shown for each dependent variable.
- ❑ **Residuals list.** Displays observed and predicted values and raw and standardized residuals. A separate listing is shown for each dependent variable.
- ❑ **Half-normal plot.** Plots the within-cell correlations between the dependent variables against their expected values. Correlations shown in the plot are transformed using Fisher's Z transformation.

You can also choose one or both of the following display options:

- ❑ **Estimates of effect size.** Displays effect-size estimates for multivariate tests. For univariate models, η^2 (eta squared) is shown.
- ❑ **Observed power for n significance level.** Displays approximate power values for univariate and multivariate F and t tests based on fixed-effect assumptions. By default, 0.05 is used as the alpha level for the power calculations. To use a different alpha, enter a value greater than 0 and less than 1.

Additional Features Available with Command Syntax

You can customize your multivariate analysis of variance if you paste your selections into a syntax window and edit the resulting MANOVA command syntax. (For information on syntax windows, see the SPSS Base system documentation.) Additional features include:

- Additional contrast options, such as user-defined contrasts (using the CONTRAST subcommand).
- Additional design information, including basis matrices (using the PRINT subcommand).
- Linear transformations of covariates and dependent variables (using the TRANSFORM subcommand).
- Extraction of a subset of discriminant functions (using the DISCRIM subcommand).
- Extraction and rotation options for principal components analysis (using the PCOMPS subcommand).
- The ability to read matrix data and to write matrix files that can be used in subsequent analyses (using the MATRIX subcommand).
- The ability to partition degrees of freedom associated with a factor (using the PARTITION subcommand).
- The ability to analyze nested models and to specify separate slopes for each cell in a model (using WITHIN on the DESIGN subcommand).
- Predicted means for combinations of factor levels (using the PMEANS subcommand).

See the Syntax Reference section of this manual for command syntax rules and for complete MANOVA command syntax.

4 Repeated Measures Analysis of Variance

Anyone who experiences difficulties sorting through piles of output, stacks of bills, or assorted journals must be awed by the brain's ability to organize, update, and maintain the memories of a lifetime. It seems incredible that someone can instantly recall the name of his or her first-grade teacher. On the other hand, that same person might spend an entire morning searching for a misplaced shoe.

Memory has two components: storage and retrieval. Many different theories explaining its magical operation have been proposed (see, for example, Eysenck, 1977). In this chapter, an experiment concerned with latency—the length of time required to search a list of items in memory—is examined.

Bacon (1980) conducted an experiment in which subjects were instructed to memorize a number. They were then given a "probe" digit and told to indicate whether it was included in the memorized number. One hypothesis of interest is the relationship between the number of digits in the memorized number and the latency. Is more time required to search through a longer number than a shorter one? Twenty-four subjects were tested on 60 memorized numbers, 20 each of two, three, and four digits in random order. The average latencies, in milliseconds, were calculated for each subject on the two-, three-, and four-digit numbers. Thus, three scores are recorded for each subject, one for each of the number lengths. The probe digit was included in the memorized number in a random position.

Repeated Measures

When the same variable is measured on several occasions for each subject, it is a **repeated measures design**. The simplest repeated measures design is one in which two measurements are obtained for each subject—such as pretest and post-test scores. These types of data are usually analyzed with a paired *t* test.

The advantages of repeated measurements are obvious. Besides requiring fewer experimental units (in this study, human subjects), they provide a control on the differences among units. That is, variability due to differences between subjects can be eliminated from the experimental error. Less attention has been focused on some of the difficulties that may be encountered with repeated measurements. Broadly, these problems can be classified as the carry-over effect, the latent effect, and the order or learning effect.

The carry-over effect occurs when a new treatment is administered before the effect of a previous treatment has worn off. For example, Drug B is given while Drug A still has an effect. The carry-over effect can usually be controlled by increasing the time between treatments. In addition, special designs that allow you to assess directly the carry-over effects are available (Cochran & Cox, 1957).

The latent effect—in which one treatment may activate the dormant effect of the previous treatment or interact with the previous treatment—is not so easily countered. This effect is especially problematic in drug trials and should be considered before using a repeated measures experimental design. Usually, if a latency effect is suspected, a repeated measures design should not be used.

The learning effect occurs when the response may improve merely by repetition of a task, independent of any treatment. For example, subjects' scores on a test may improve each time they take the test. Thus, treatments that are administered later may appear to improve performance, even though they have no effect. In such situations, it is important to pay particular attention to the sequencing of treatments. Learning effects can be assessed by including a control group that performs the same tasks repeatedly without receiving any treatment.

Describing the Data

In a repeated measures experiment, as well as any other, the first step is to obtain descriptive statistics. These provide some idea of the distributions of the variables, as well as their average values and dispersions. Figure 4.1 contains means and standard deviations for the latency times (obtained with the Descriptives procedure). The shortest average latency time (520 milliseconds) was observed for the two-digit numbers (*p2digit*). The longest (581 milliseconds) was observed for the four-digit numbers. A plot of the mean latency times against the number of digits is shown in Figure 4.2. Note that there appears to be a linear relationship between the latency time and the number of digits in the memorized number.

Figure 4.1 Means and standard deviations

Variable	Mean	Std Dev	N	Label
P2DIGIT	520.58	131.37	24	
P3DIGIT	560.00	118.78	24	
P4DIGIT	581.25	117.32	24	

Figure 4.2 Plot of means from Graph procedure

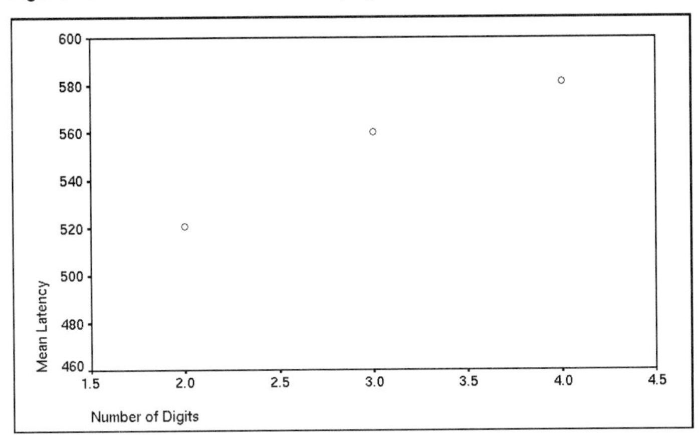

The stem-and-leaf plot of latencies for the *p4digit* is shown in Figure 4.3 (stem-and-leaf plots are available from the Explore procedure). The corresponding normal probability plot is shown in Figure 4.4. From this, it appears that the values are somewhat more "bunched" than you would expect if they were normally distributed. The bunching of the data might occur because of limitations in the accuracy of the measurements. Similar plots can be obtained for the other two variables.

Figure 4.3 Stem-and-leaf plot for p4digit

```
Frequency    Stem &  Leaf

     1.00       3 .  9
     3.00       4 .  247
    11.00       5 .  00011114558
     4.00       6 .  1888
     4.00       7 .  0015
     1.00       8 .  5

Stem width:    100.00
Each leaf:       1 case(s)
```

Figure 4.4 Normal probability plot for p4digit

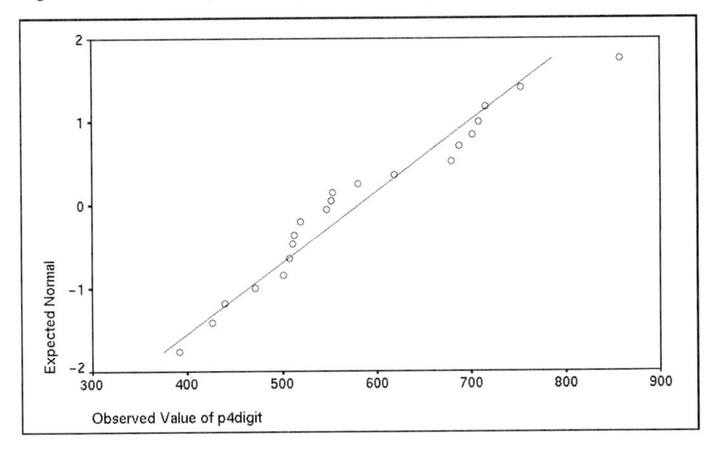

Analyzing Differences

Since multiple observations are made on the same experimental unit in a repeated measures design, special procedures that incorporate dependencies within an experimental unit must be used. For example, in the paired *t* test design, instead of analyzing each score separately, we analyze the difference between the two scores. When there are more than two scores for a subject, the analysis becomes somewhat more complicated. For example, if each subject receives three treatments, there are three pairwise differences: the difference between the first two treatments, the difference between the second two treatments, and the difference between the first and third treatments.

Performing three paired *t* tests of the differences may seem to be the simplest analysis, but it is not, for several reasons, the best strategy. First, the three *t* tests are not statistically independent, since they involve the same means in overlapping combinations. Some overall protection against calling too many differences significant is needed, especially as the number of treatments increases. Second, since there are three separate *t* tests, a single test of the hypothesis that there is no difference between the treatments is not available.

There are several approaches for circumventing such problems in the analysis of data from repeated measures experiments. These approaches are described in the remainder of this chapter.

Transforming the Variables

To test the null hypothesis that the mean latencies are the same for the three digit lengths, the original three variables must be transformed. That is, instead of analyzing the original three variables, we analyze linear combinations of their differences. (In the paired *t* test, the transformation is the difference between the values for each subject or pair.) For certain methods of analysis, these linear combinations, sometimes called **contrasts**, must be chosen so that they are statistically independent (orthogonal) and so that the sum of the squared coefficients is 1 (normalized). Such contrasts are termed **orthonormalized**. The number of statistically independent contrasts that can be formed for a factor is one less than the number of levels of the factor. In addition, a contrast corresponding to the overall mean (the constant term in the model) is always formed. For this example, the contrast for the overall mean is

$$\text{Contrast 1} = \text{p2digit} + \text{p3digit} + \text{p4digit} \qquad \textbf{Equation 4.1}$$

There are many types of transformations that can be used to form contrasts among the means. One is the **difference contrast**, which compares each level of a factor to the average of the levels that precede it. (Another transformation, orthogonal polynomials, is discussed in "Selecting Other Contrasts" on p. 117.)

The first difference contrast for the digit factor (the second contrast for the experiment) is

$$\text{Contrast 2} = \text{p3digit} - \text{p2digit}$$ **Equation 4.2**

The second difference contrast is

$$\text{Contrast 3} = (2 \times \text{p4digit}) - \text{p3digit} - \text{p2digit}$$ **Equation 4.3**

The contrasts can be normalized by dividing each contrast by the square root of the sum of the squared coefficients. The contrast for the overall mean is divided by

$$\sqrt{1^2 + 1^2 + 1^2} = \sqrt{3}$$ **Equation 4.4**

and the second difference contrast is divided by

$$\sqrt{2^2 + 1^2 + 1^2} = \sqrt{6}$$ **Equation 4.5**

Figure 4.5 shows the orthonormalized transformation matrix from MANOVA for creating new variables from *p2digit*, *p3digit*, and *p4digit* with the difference transformation. Each column contains the coefficients for a particular contrast. Each row corresponds to one of the original variables. Thus, the first linear combination is

$$\text{Constant} = 0.577\,(\text{p2digit} + \text{p3digit} + \text{p4digit})$$ **Equation 4.6**

This new variable is, for each case, the sum of the values of the three original variables, multiplied by 0.577. Again, the coefficients are chosen so that the sum of their squared values is 1. The next two variables are the normalized difference contrasts for the digit effect.

Figure 4.5 Orthonormalized transformation matrix for difference contrasts

```
Orthonormalized Transformation Matrix (Transposed)

                    T1          T2          T3

P2DIGIT            .577       -.707       -.408
P3DIGIT            .577        .707       -.408
P4DIGIT            .577        .000        .816
```

What do these new variables represent? The first variable, the sum of the original variables, is the average response over all treatments. It measures latency times over all digit lengths. The hypothesis that the average response, the constant, is equal to 0 is based on this variable. The second two contrasts together represent the treatment (digit) effect. These two contrasts are used to test hypotheses about differences in latencies for the three digit lengths.

Testing for Differences

In Chapter 3, analysis of variance models with more than one dependent variable are described. The same techniques can be used for repeated measures data. For example, we can use the single-sample tests to determine whether particular sets of the transformed variables in the Bacon experiment have means of 0. If the number of digits does not affect latency times, the mean values of the two contrasts for the digit effect are expected to be 0. Different hypotheses are tested using different transformed variables.

Testing the Constant Effect

In Repeated Measures ANOVA, several hypotheses are automatically tested. The first hypothesis is that the overall mean latency time is 0. It is based on the first transformed variable, which corresponds to the constant effect.

To help identify results, Repeated Measures ANOVA displays the names of the transformed variables used to test a hypothesis. The names for the transformed variables begin with the letter *t* and are followed by a sequential number. For example, in Figure 4.5, the transformed variables are *t1*, *t2*, and *t3*.

Figure 4.6 shows the explanation displayed when the constant effect is tested. The column labeled *Variates* indicates which transformed variables are involved in the test of a particular effect. Since the test for the constant is based only on variable *t1*, its name appears in that column. When there are no covariates in the analysis, the column labeled *Covariates* is empty, as shown. The note below the table is a reminder that analyses are based on the transformed variables and that the particular analysis is for the constant effect.

Figure 4.6 Renamed variable used in the analysis

```
Order of Variables for Analysis

  Variates     Covariates

  T1

   1 Dependent Variable
   0 Covariates

- - - - - - - - - - - - - - - - - - - - - - - - - - - - - - - - - - -

     Note..  TRANSFORMED variables are in the variates column.
             These TRANSFORMED variables correspond to the
             Between-subject effects.
```

The Analysis-of-Variance Table

Since the test for the constant is based on a single variable, the results are displayed in the usual univariate analysis-of-variance table (Figure 4.7). The large F value and the small observed significance level indicate that the hypothesis that the constant is 0 is rejected. This finding is of limited importance here, since we do not expect the time required to search a number to be 0. Tests about the constant might be of interest if the original variables are difference scores—change from baseline, for example—since the test of the constant would then correspond to the test that there has been no overall change from baseline.

Figure 4.7 Analysis-of-variance table

```
Tests of Significance for T1 using UNIQUE sums of squares
Source of Variation              SS        DF         MS            F  Sig of F

WITHIN CELLS                 995411.11     23   43278.74
CONSTANT                   22093520.22      1   22093520     510.49       .000
```

After the analysis-of-variance table, MANOVA displays parameter estimates and tests of the hypotheses that the individual transformed variables have means of 0. These are shown in Figure 4.8. For the constant effect, the parameter estimate is nothing more than

$$0.57735 \times (520.58 + 560.00 + 581.25) = 959.46$$ **Equation 4.7**

where the numbers within the parentheses are just the means of the original three variables shown in Figure 4.1. The 0.57735 is the value used to normalize the contrast (the rounded value, 0.577, is shown in Figure 4.5). The test of the hypothesis that the true value of the first parameter is 0 is equivalent to the test of the hypothesis that the constant is 0. Therefore, the t value displayed for the test is the square root of the F value from the analysis-of-variance table (the square root of 510.49 is 22.59). (When the t statistic is squared, it is equal to an F statistic with one degree of freedom for the numerator and the same degrees of freedom for the denominator as the t statistic.)

Figure 4.8 Parameter estimates

```
Estimates for T1
--- Individual univariate .9500 confidence intervals
CONSTANT

 Parameter      Coeff.  Std. Err.     t-Value     Sig. t Lower -95%   CL- Upper

         1  959.459922   42.46506    22.59410     .00000  871.61426 1047.30558
```

Testing the Digit Effect

The hypothesis of interest in this study is whether latency time depends on the number of digits in the memorized number. As shown in Figure 4.9, this test is based on the two transformed variables, labeled *t2* and *t3*. Figure 4.10 contains the multivariate tests of the hypothesis that the means of these two variables are 0. In this situation, all multivariate criteria are equivalent and lead to rejection of the hypothesis that the number of digits does not affect latency time.

Figure 4.9 Transformed variables used

```
Order of Variables for Analysis

  Variates      Covariates

  T2
  T3

   2 Dependent Variables
   0 Covariates

- - - - - - - - - - - - - - - - - - - - - - - - - - - - - -

     Note..  TRANSFORMED variables are in the variates column.
             These TRANSFORMED variables correspond to the
             'DIGIT' WITHIN-SUBJECT effect.
```

Figure 4.10 Multivariate hypothesis tests

```
EFFECT .. DIGIT
 Multivariate Tests of Significance (S = 1, M = 0, N = 10 )

 Test Name          Value     Exact F Hypoth. DF   Error DF  Sig. of F

 Pillais           .60452    16.81439       2.00      22.00       .000
 Hotellings       1.52858    16.81439       2.00      22.00       .000
 Wilks             .39548    16.81439       2.00      22.00       .000
 Roys              .60452
 Note.. F statistics are exact.
```

Univariate *F* tests for the individual transformed variables are shown in Figure 4.11. The first row corresponds to a test of the hypothesis that there is no difference in average latency times for numbers consisting of two digits and those consisting of three. (This is equivalent to a one-sample *t* test that the mean of the second transformed variable *t2* is 0.) The second row of Figure 4.11 provides a test of the hypothesis that there is no difference between the average response to numbers with two and three digits and numbers with four digits. The average value of this contrast is also significantly different from 0, since the observed significance level is less than 0.0005. (See "Selecting Other Contrasts" on p. 117 for an example of a different contrast type.)

Figure 4.11 Univariate hypothesis tests

```
Univariate F-tests with (1,23) D. F.

 Variable   Hypoth. SS    Error SS Hypoth. MS    Error MS            F  Sig. of F

 T2         18644.0833 19800.9167 18644.0833  860.90942   21.65627       .000
 T3         26841.3611 22774.3056 26841.3611  990.18720   27.10736       .000
```

To estimate the magnitudes of differences among the digit lengths, we can examine the values of each contrast. These are shown in Figure 4.12. The column labeled *Coeff.* is the average value for the normalized contrast. As shown in Figure 4.5, the second contrast is 0.707 times the difference between three- and two-digit numbers, which is $0.707 \times (560 - 520.58) = 27.87$, the value shown in Figure 4.12. To obtain an estimate of the absolute difference between mean response to two and three digits, the parameter estimate must be divided by 0.707. This value is 39.42. Again, the t value for the hypothesis that the contrast is 0 is equivalent to the square root of the F value from the univariate analysis-of-variance table.

Figure 4.12 Estimates for contrasts

```
Estimates for T2
 --- Individual univariate .9500 confidence intervals

 DIGIT

  Parameter       Coeff.  Std. Err.      t-Value      Sig. t Lower -95%  CL- Upper

         1   27.8717923     5.98926     4.65363      .00011    15.48207    40.26152

 - - - - - - - - - - - - - - - - - - - - - - - - - - - - - - - - - - - - - - - -
 Estimates for T3
 --- Individual univariate .9500 confidence intervals

 DIGIT

  Parameter       Coeff.  Std. Err.      t-Value      Sig. t Lower -95%  CL- Upper

         1   33.4423391     6.42322     5.20647      .00003    20.15489    46.72979
```

Similarly, the parameter estimate for the third contrast, the normalized difference between the average of two and three digits and four digits, is 33.44. The actual value of the difference is 40.96 (33.44 divided by 0.816). The t value, the ratio of the parameter estimate to its standard error, is 5.206, which, when squared, equals the F value in Figure 4.11. The t values are the same for normalized and non-normalized contrasts.

The parameter estimates and univariate F tests help to identify which individual contrasts contribute to overall differences. However, the observed significance levels for the individual parameters are not adjusted for the fact that several comparisons are being made (Miller, 1981; Burns, 1984). Thus, the significance levels should serve only as guides for identifying potentially important differences.

Averaged Univariate Results

The individual univariate tests, since they are orthogonal, can be pooled to obtain the averaged F test shown in Figure 4.13. The entries in Figure 4.13 are obtained from Figure 4.11 by summing the hypothesis and error sums of squares and the associated degrees of freedom. In this example, the averaged F test also leads us to reject the hypothesis that average latency times do not differ for numbers of different lengths. This is the same F statistic as that obtained by specifying a repeated measures design as a mixed-model univariate analysis of variance (Winer et al., 1991).

Figure 4.13 Averaged univariate hypothesis test

```
AVERAGED Tests of Significance for MEAS.1 using UNIQUE sums of squares
Source of Variation              SS       DF          MS          F  Sig of F

WITHIN CELLS               42575.22       46      925.55
DIGIT                      45485.44        2    22742.72      24.57      .000
```

Choosing Multivariate or Univariate Results

In the previous section, we saw that hypothesis tests for the digit effect could be based on multivariate criteria such as Wilks' lambda, or on the averaged univariate *F* tests. When both approaches lead to similar results, choosing between them is not of much importance. However, there are situations in which the multivariate and univariate approaches lead to different results, and the question of which is appropriate arises.

The **multivariate approach** considers the measurements on a subject to be a sample from a multivariate normal distribution and makes no assumption about the characteristics of the variance-covariance matrix. The **univariate approach** (sometimes called the mixed-model approach) requires certain assumptions about the variance-covariance matrix. If these conditions are met, especially for small sample sizes, the univariate approach is more powerful than the multivariate approach. That is, it is more likely to detect differences when they exist.

Modifications of the univariate results when the assumptions are violated have also been proposed. These corrected results are approximate and are based on the adjustment of the degrees of freedom of the *F* ratio (Greenhouse & Geisser, 1959; Huynh & Feldt, 1976). The significance levels for the corrected tests will always be larger than for the uncorrected. Thus, if the uncorrected test is not significant, there is no need to calculate corrected values.

Assumptions Needed for the Univariate Approach

Since subjects in the current example are not subdivided by any grouping characteristics, the only assumption required for using the univariate results is that the variances of all the transformed variables for an effect be equal and that their covariances be 0. (Assumptions required for more complicated designs are described in "Additional Univariate Assumptions" on p. 124.)

Mauchly's test of sphericity is available in the Repeated Measures ANOVA procedure for testing the hypothesis that the covariance matrix of the transformed variables has a constant variance on the diagonal and zeros off the diagonal (Morrison, 1976). For small sample sizes, this test is not very powerful. For large sample sizes, the test may be significant even when the impact of the departure on the analysis-of-variance results may be small.

Figure 4.14 contains the correlation matrix with standard deviations on the diagonal for the two transformed variables corresponding to the digit effect, Mauchly's test of sphericity, and the observed significance level based on a chi-square approximation. The observed significance level is 0.150, so the hypothesis of sphericity is not rejected. If the observed significance level is small and the sphericity assumption appears to be violated, an adjustment to the numerator and denominator degrees of freedom can be made. Two estimates of this adjustment, called **epsilon**, are available in the Repeated Measures ANOVA procedure. These are also shown in Figure 4.14. Both the numerator and denominator degrees of freedom must be multiplied by epsilon, and the significance of the F ratio must be evaluated with the new degrees of freedom. The **Huynh-Feldt epsilon** is an attempt to correct the **Greenhouse-Geisser epsilon**, which tends to be overly conservative, especially for small sample sizes. The lowest value possible for epsilon is also displayed. The Huynh-Feldt epsilon sometimes exceeds the value of 1. When this occurs, Repeated Measures ANOVA displays a value of 1.

Figure 4.14 Mauchly's test of sphericity

```
WITHIN CELLS Correlations with Std. Devs. on Diagonal

                    T2          T3

 T2              29.341
 T3                .393      31.467

 - - - - - - - - - - - - - - - - - - - - - - - - - - - - - - - - - - - -

Tests involving 'DIGIT' Within-Subject Effect.

 Mauchly sphericity test, W =       .84172
 Chi-square approx. =              3.79088 with 2 D. F.
 Significance =                       .150

 Greenhouse-Geisser Epsilon =       .86335
 Huynh-Feldt Epsilon =              .92700
 Lower-bound Epsilon =              .50000
```

Selecting Other Contrasts

Based on the orthonormalized transformation matrix shown in Figure 4.5 and the corresponding parameter estimates in Figure 4.12, hypotheses about particular combinations of the means were tested. Remember that the second contrast compared differences between the two- and three-digit numbers, while the third contrast compared the average of the two- and three-digit number to the four-digit number. A variety of other hypotheses can be tested by selecting different orthogonal contrasts.

For example, to test the hypothesis that latency time increases linearly with the number of digits in the memorized number, orthogonal polynomial contrasts can be used. (In fact, polynomial contrasts should have been the first choice for data of this type. Difference contrasts were used for illustrative purposes.) When polynomial contrasts are used, the first contrast for the digit effect represents the linear component, and the second contrast represents the quadratic component. Figure 4.15 contains parameter estimates corresponding to the polynomial contrasts. You can see that there is a significant linear trend, but the quadratic trend is not significant. This is also shown in the plot in Figure 4.2, since the means fall more or less on a straight line, which does not appear to curve upward or downward.

Figure 4.15 Parameter estimates for polynomial contrasts

```
Estimates for T2
--- Individual univariate .9500 confidence intervals

DIGIT

 Parameter       Coeff.   Std. Err.     t-Value      Sig. t Lower -95%  CL- Upper

         1   42.8978114     7.27957     5.89290      .00001   27.83887   57.95675

- - - - - - - - - - - - - - - - - - - - - - - - - - - - - - - - - - - - - - - -
Estimates for T3
--- Individual univariate .9500 confidence intervals

DIGIT

 Parameter       Coeff.   Std. Err.     t-Value      Sig. t Lower -95%  CL- Upper

         1   -7.4165106     4.91293    -1.50959      .14476  -17.57968    2.74666
```

The requirement that contrasts be orthonormalized is necessary for the averaged *F* tests. It is not required for the multivariate approach. Repeated Measures ANOVA, however, requires all contrasts for the within-subjects factors to be orthonormal. If nonorthogonal contrasts, such as simple or deviation, are requested, they are orthonormalized prior to the actual analysis. The transformation matrix should always be displayed so that the parameter estimates and univariate *F* ratios can be properly interpreted.

Adding Another Factor

The experimental design discussed so far is a very simple one. Responses to all levels of one factor (number of digits) were measured for all subjects. However, repeated measures designs can be considerably more complicated. Any factorial design can be applied to a single subject. For example, we can administer several different types of medication at varying dosages and times of day. Such a design has three factors (medication, dosage, and time) applied to each subject.

All of the usual analysis-of-variance hypotheses for factorial designs can be tested when each subject is treated as a complete replicate of the design. However, since observations from the same subject are not independent, the usual analysis-of-variance method is inappropriate. Instead, we need to extend the previously described approach to analyzing repeated measures experiments.

To illustrate the analysis of a two-factor repeated measures design, consider the Bacon data again. The experiment was actually more involved than first described. The single "probe" digit was not always present in the memorized number. Instead, each subject was tested under two conditions—probe digit present and probe digit absent. Presence and absence of the probe digit were randomized. The two conditions were included to test the hypothesis that it takes longer to search through numbers when the probe digit is not present than when it is present. If memory searches are performed sequentially, you would expect that when the probe digit is encountered, searching stops. When the probe digit is not present in a number, searching must continue through all digits of the memorized number.

Thus, each subject has in fact six observations: latency times for the three number lengths when the probe is present, and times for the numbers when the probe is absent. This design has two factors: number of digits, and the probe presence or absence condition. The digit factor has three levels (two, three, and four digits) and the condition factor has two levels (probe present and probe absent).

Testing a Two-Factor Model

The analysis of this modified experiment proceeds similarly to the analysis described for the single-factor design. However, instead of testing only the digit effect, tests for the digit effect, the condition effect, and the digit-by-condition interaction are required. Figure 4.16 is the orthonormalized transformation matrix for the two-factor design.

Figure 4.16 Orthonormalized transformation matrix for the two-factor design

Orthonormalized Transformation Matrix (Transposed)

	T1	T2	T3	T4	T5	T6
P2DIGIT	.408	-.408	-.500	-.289	.500	.289
P3DIGIT	.408	-.408	.500	-.289	-.500	.289
P4DIGIT	.408	-.408	.000	.577	.000	-.577
NP2DIGIT	.408	.408	-.500	-.289	-.500	-.289
NP3DIGIT	.408	.408	.500	-.289	.500	-.289
NP4DIGIT	.408	.408	.000	.577	.000	.577

The Transformed Variables

The coefficients of the transformation matrix indicate that the first contrast is an average of all six variables. The second contrast is the average response under the absent condition compared to the average response under the present condition. The third contrast is the difference between two and three digits averaged over the two conditions. The fourth contrast is the average of two and three digits compared to four digits, averaged over both conditions. As before, these two contrasts jointly provide a test of the digit effect. The last two contrasts are used for the test of interaction. If there is no interaction effect, the difference between the two- and three-digit numbers should be the same for the two probe conditions. Contrast *t5* is the difference between two and three digits for condition 2, minus two and three digits for condition 1. Similarly, if there is no interaction between probe presence and the number of digits, the average of two and three digits compared to four should not differ for the two probe conditions. Contrast *t6* is used to test this hypothesis.

Testing Hypotheses

Hypothesis testing for this design proceeds similarly to the single factor design. Each effect is tested individually. Both multivariate and univariate results can be obtained for tests of each effect. (When a factor has only two levels, there is one contrast for the effect, and the multivariate and univariate results are identical.) Since the test of the constant is not of interest, we will proceed to the test of the condition effect.

The Condition Effect

The table in Figure 4.17 explains that variable *t2* is being used in the analysis of the condition effect. The analysis-of-variance table in Figure 4.18 indicates that the condition effect is significant. The F value of 52 has an observed significance level of less than 0.0005. The parameter estimate for the difference between the two conditions is -75 , as shown in Figure 4.19. The estimate of the actual difference between mean response under the two conditions is obtained by dividing -75 by 0.408, since the contrast is actually

$$\text{Contrast} = 0.408 \times (\text{mean present} - \text{mean absent})$$ **Equation 4.8**

Since the contrast value is negative, the latency times for the absent condition are larger than the latency times for the present condition. This supports the notion that memory searching may be sequential, terminating when an item is found rather than continuing until all items are examined.

Figure 4.17 Test of the condition effect

```
Order of Variables for Analysis

   Variates      Covariates

    T2

    1 Dependent Variable
    0 Covariates

 - - - - - - - - - - - - - - - - - - - - - - - - - - - - - - - - - - - - - - -
 Note..  TRANSFORMED variables are in the variates column.
         These TRANSFORMED variables correspond to the
         'COND' WITHIN-SUBJECT effect.
```

Figure 4.18 Analysis-of-variance table for the condition effect

```
Tests involving 'COND' Within-Subject Effect.

 Tests of Significance for T2 using UNIQUE sums of squares
 Source of Variation          SS      DF        MS          F  Sig of F

 WITHIN CELLS             58810.08      23   2556.96
 COND                    134322.25       1 134322.25     52.53      .000
```

Figure 4.19 Parameter estimates for the condition effect

```
Estimates for T2
 --- Individual univariate .9500 confidence intervals

 COND

  Parameter       Coeff.  Std. Err.     t-Value      Sig. t Lower -95%  CL- Upper

         1   74.8114992   10.32182     7.24790      .00000    53.45918   96.16381
```

The Number of Digits

The next effect to be tested is the number of digits in the memorized number. As shown in Figure 4.20, the test is based on the two contrasts labeled *t3* and *t4*. To use the univariate approach, the assumption of sphericity is necessary.

Figure 4.20 Test of the digit effect

```
Order of Variables for Analysis

   Variates      Covariates

    T3
    T4

    2 Dependent Variables
    0 Covariates

 - - - - - - - - - - - - - - - - - - - - - - - - - - - - - - - - - - - - - - -
 Note..  TRANSFORMED variables are in the variates column.
         These TRANSFORMED variables correspond to the
         'DIGIT' WITHIN-SUBJECT effect.
```

Based on the multivariate criteria in Figure 4.21, the hypothesis that there is no digit effect should be rejected. From Figure 4.22, we see that both of the contrasts are also individually different from 0. This can also be seen from Figure 4.23, which contains the parameter estimates for the contrasts. Note that the tests that the parameter values are 0 are identical to the corresponding univariate F tests. The averaged tests of significance for the digit effect, as shown in Figure 4.24, lead to the same conclusion as the multivariate results in Figure 4.21.

Figure 4.21 Multivariate tests of significance

```
EFFECT .. DIGIT

Multivariate Tests of Significance (S = 1, M = 0, N = 10 )

Test Name           Value       Approx. F      Hypoth. DF        Error DF       Sig. of F

Pillais            .64216        19.73989            2.00           22.00            .000
Hotellings        1.79454        19.73989            2.00           22.00            .000
Wilks              .35784        19.73989            2.00           22.00            .000
Roys               .64216
```

Figure 4.22 Univariate tests of significance

```
Univariate F-tests with (1,23) D. F.

Variable   Hypoth. SS   Error SS Hypoth. MS   Error MS          F  Sig. of F

T3         41002.6667 34710.8333 41002.6667 1509.16667   27.16908       .000
T4         53682.7222 30821.4444 53682.7222 1340.06280   40.05986       .000
```

Figure 4.23 Parameter estimates for the digit contrasts

```
Estimates for T3
 --- Individual univariate .9500 confidence intervals

 DIGIT

  Parameter      Coeff.  Std. Err.     t-Value     Sig. t Lower -95%  CL- Upper

        1   41.3333333     7.92981     5.21240     .00003   24.92926   57.73740

 - - - - - - - - - - - - - - - - - - - - - - - - - - - - - - - - - - - -
 Estimates for T4
 --- Individual univariate .9500 confidence intervals

 DIGIT

  Parameter      Coeff.  Std. Err.     t-Value     Sig. t Lower -95%  CL- Upper

        1   47.2946096     7.47235     6.32929     .00000   31.83688   62.75233
```

Figure 4.24 Averaged tests of significance

```
AVERAGED Tests of Significance for MEAS.1 using UNIQUE sums of squares
Source of Variation          SS      DF        MS         F  Sig of F

WITHIN CELLS             65532.28    46   1424.61
DIGIT                    94685.39     2  47342.69     33.23      .000
```

The Interaction

The interaction between the number of digits in the memorized number and the presence or absence of the probe digit is based on the last two contrasts, labeled *tint1* and *tint2*, as indicated in Figure 4.25. Based on the multivariate criteria shown in Figure 4.26 and the averaged univariate results in Figure 4.27, the hypothesis that there is no interaction is not rejected.

Figure 4.25 Test of the interaction effect

```
Order of Variables for Analysis

  Variates     Covariates

  T5
  T6

   2 Dependent Variables
   0 Covariates
- - - - - - - - - -

Note..  TRANSFORMED variables are in the variates column.
        These TRANSFORMED variables correspond to the
        'COND BY DIGIT' WITHIN-SUBJECT effect.
```

Figure 4.26 Multivariate tests of significance

```
EFFECT .. COND BY DIGIT

Multivariate Tests of Significance (S = 1, M = 0, N = 10 )

Test Name          Value      Exact F   Hypoth. DF    Error DF    Sig. of F

Pillais           .00890      .09873        2.00        22.00        .906
Hotellings        .00898      .09873        2.00        22.00        .906
Wilks             .99110      .09873        2.00        22.00        .906
Roys              .00890
```

Figure 4.27 Averaged tests of significance

```
AVERAGED Tests of Significance for MEAS.1 using UNIQUE sums of squares
Source of Variation          SS      DF        MS         F  Sig of F

WITHIN CELLS           20751.50      46    451.12
COND BY DIGIT             88.17       2     44.08       .10      .907
```

Putting It Together

Consider Figure 4.28, which contains a plot of the average latencies for each of the three number lengths and conditions. Means and standard deviations are shown in Figure 4.29. Note that the average latency times for the absent condition are always higher than those for the present condition. The test for the statistical significance of this observation is based on the condition factor. Figure 4.28 shows that as the number of digits in a number increases, so does the average latency time. The test of the hypothesis that latency time is the same regardless of the number of digits is based on the digit factor. Again, the hypothesis that there is no digit effect is rejected. The relationship between latency time and the number of digits appears to be fairly similar for the two probe conditions. This is tested by the condition-by-digit interaction, which was not found to be statistically significant. If there is a significant interaction, the relationship between latency time and the number of digits would differ for the two probe conditions.

Figure 4.28 Plot of average latencies from Graph procedure

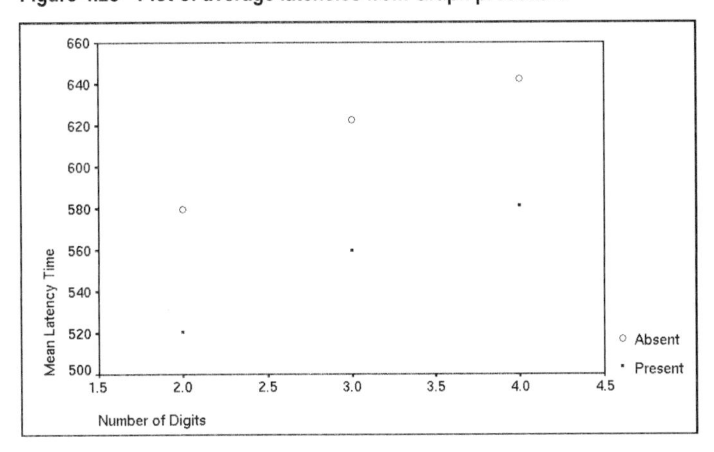

Figure 4.29 Means and standard deviations from SPSS Tables option

	2 DIGIT		3 DIGIT		4 DIGIT	
	Mean	Standard Deviation	Mean	Standard Deviation	Mean	Standard Deviation
PRESENT	520.58	131.37	560.00	118.78	581.25	117.32

	2 DIGIT		3 DIGIT		4 DIGIT	
	Mean	Standard Deviation	Mean	Standard Deviation	Mean	Standard Deviation
ABSENT	579.75	132.17	623.00	135.14	642.33	144.81

Within-Subjects and Between-Subjects Factors

Both number of digits and probe status are called **within-subjects factors**, since all combinations occur within each of the subjects. It is also possible to have between-subjects factors in repeated measures designs. **Between-subjects factors** subdivide the sample into discrete subgroups. Each subject has only one value for a between-subjects factor. For example, if cases in the previously described study are subdivided into males and females, sex is a between-subjects factor. Similarly, if cases are classified as those who received "memory enhancers" and those who did not, the memory enhancement factor is a between-subjects factor. If the same subject is tested with and without memory enhancers, memory enhancement would be a within-subjects factor. Thus, the same factor can be either a within-subjects or between-subjects factor, depending on the experimental design. Some factors, such as sex and race, can be only between-subjects factors, since the same subject can be of only one sex or race.

Additional Univariate Assumptions

In a within-subjects design, a sufficient condition for the univariate model approach to be valid is that, for each effect, the variance-covariance matrix of the transformed variables used to test the effect has covariances of 0 and equal variances. Including between-subjects factors in a design necessitates an additional assumption. The variance-covariance matrices for the transformed variables for a particular effect must be equal for all levels of the between-subjects factors. These two assumptions are often called the **symmetry conditions**. If they are not tenable, the *F* ratios from the averaged univariate results may not be correct.

Back to Memory

In addition to the two within-subjects factors, Bacon's experiment also included a between-subjects factor—the hand subjects used to press the instrument that signaled the presence or absence of the probe digit. All subjects were right-handed, but half were required to signal with the right hand and half with the left. (If a subject had been tested under both conditions, right hand and left hand, hand would be a within-subjects factor.) The hypothesis of interest was whether latency would increase for subjects using the left hand.

The hypothesis that the variance-covariance matrices are equal across all levels of the between-subjects factor can be examined using the multivariate generalization of Box's *M* test. It is based on the determinants of the variance-covariance matrices for all between-subjects cells in the design. Figure 4.30 contains the multivariate test for equality of the variance-covariance matrices for the two levels of the hand factor (which hand the subject used). Note that in Repeated Measures ANOVA, this test is based on all the original variables for the within-subjects effects.

Figure 4.30 Box's M

```
Multivariate test for Homogeneity of Dispersion matrices

Boxs M =                       25.92434
F with (21,1780) DF =            .86321, P =    .641 (Approx.)
Chi-Square with 21 DF =        18.43322, P =    .621 (Approx.)
```

Adding a between-subjects factor to the experiment introduces more terms into the analysis-of-variance model. Besides the digit, condition, and condition-by-digit effects, the model includes the main effect, hand, and the interaction terms hand by digit, hand by condition, and hand by digit by condition.

The analysis proceeds as before. Variables corresponding to the within-subjects factors are again transformed using the transformation matrix shown in Figure 4.31. The between-subjects factors are not transformed.

Figure 4.31 Transformation matrix

Orthonormalized Transformation Matrix (Transposed)

	T1	T2	T3	T4	T5	T6
P2DIGIT	.408	-.408	-.500	-.289	.500	.289
P3DIGIT	.408	-.408	.500	-.289	-.500	.289
P4DIGIT	.408	-.408	.000	.577	.000	-.577
NP2DIGIT	.408	.408	-.500	-.289	-.500	-.289
NP3DIGIT	.408	.408	.500	-.289	.500	-.289
NP4DIGIT	.408	.408	.000	.577	.000	.577

The within-subjects factors and their interactions are tested as when there were no between-subject factors in the design. Tests of the between-subjects factors and the interactions of the between- and within-subjects factors treat the transformed within-subjects variables as dependent variables. For example, the test of the hand effect is identical to a two-sample *t* test, with the transformed variable corresponding to the constant as the dependent variable. The test of the condition-by-hand interaction treats the transformed variable corresponding to the condition effect as the dependent variable. Similarly, the digit-by-hand interaction considers the two transformed variables for the digit effect as dependent variables. The test of the three-way interaction hand by digit by condition treats the two variables corresponding to the interaction of digit by condition as the dependent variables. *Hand* is always the grouping variable.

The SPSS output for a design with both within- and between-subjects factors looks much like before. The variables used for each analysis are first identified, as shown for the constant effect in Figure 4.32.

Figure 4.32 Test of the constant effect

```
Order of Variables for Analysis

  Variates     Covariates

   T1

   1 Dependent Variable
   0 Covariates

Note..  TRANSFORMED variables are in the variates column.
        These TRANSFORMED variables correspond to the
        Between-subject effects.
```

All tests based on the same transformed variables are presented together. Since the test of the hand effect is based on the same variable as the test for the constant effect, the tests are displayed together, as shown in Figure 4.33. This is the usual analysis-of-variance table, since there is only one dependent variable in the analyses. From the analysis-of-variance table, it appears that the constant is significantly different from 0, an uninteresting finding that we have made before. The hand effect, however, is not statistically significant, since its observed significance level is very close to 1. This means that there is insufficient evidence to reject the null hypothesis that there is no difference in average latency times for subjects who used their right hand and subjects who used their left.

Figure 4.33 Tests of significance using unique sums of squares

```
Tests of Between-Subjects Effects.

 Tests of Significance for T1 using UNIQUE sums of squares
 Source of Variation          SS       DF         MS          F  Sig of F

 WITHIN CELLS         2196663.86       22   99848.36
 CONSTANT            49193858.03        1   49193858     492.69      .000
 HAND                     300.44        1     300.44        .00      .957
```

For the transformed variable used to test the hand effect, the overall mean is 1432. The mean is 1435 for the right-hand subjects and 1428 for the left-hand subjects. The parameter estimate in Figure 4.34 is based on these means. The parameter estimate for hand is the deviation of the right-hand group from the overall mean. (For between-subjects variables such as *hand*, deviation parameter estimates are the default.)

Figure 4.34 Parameter estimates for the hand contrasts

```
Estimates for T1
 --- Individual univariate .9500 confidence intervals

 HAND

  Parameter       Coeff.  Std. Err.      t-Value      Sig. t Lower -95%  CL- Upper

          2   3.53815185   64.50076       .05485      .95675 -130.22824  137.30454
```

Figure 4.35 is the analysis-of-variance table based on the transformed variable for the condition effect. Again, there is a significant effect for condition, but the hand-by-condition effect is not significant.

Figure 4.35 Analysis of variance for the interaction

```
Tests of Significance for T2 using UNIQUE sums of squares
Source of Variation          SS       DF        MS           F  Sig of F

WITHIN CELLS             57889.97     22    2631.36
COND                    134322.25      1 134322.25      51.05      .000
HAND BY COND               920.11      1    920.11        .35      .560
```

Since the digit effect has two degrees of freedom, its test is based on two transformed variables, and both univariate and multivariate results are displayed for hypotheses involving digit. The multivariate results for the hand-by-digit interaction are shown in Figure 4.36. It appears that there is no interaction between number of digits and the hand used to signal the response. The multivariate results for the digit effect are shown in Figure 4.37. Again, they are highly significant. Figure 4.38 shows the averaged univariate results for terms involving the digit effect.

Figure 4.36 Multivariate tests of significance for the hand-by-digit interaction

```
EFFECT .. HAND BY DIGIT
Multivariate Tests of Significance (S = 1, M = 0, N = 9 1/2)

Test Name          Value    Exact F Hypoth. DF   Error DF  Sig. of F

Pillais           .08477     .97258       2.00      21.00       .394
Hotellings        .09263     .97258       2.00      21.00       .394
Wilks             .91523     .97258       2.00      21.00       .394
Roys              .08477
```

Figure 4.37 Multivariate tests of significance for digit effect

```
EFFECT .. DIGIT
Multivariate Tests of Significance (S = 1, M = 0, N = 9 1/2)

Test Name         Value    Exact F Hypoth. DF   Error DF  Sig. of F

Pillais          .64219   18.84488       2.00      21.00       .000
Hotellings      1.79475   18.84488       2.00      21.00       .000
Wilks            .35781   18.84488       2.00      21.00       .000
Roys             .64219
```

Figure 4.38 Averaged tests of significance

```
AVERAGED Tests of Significance for MEAS.1 using UNIQUE sums of squares
Source of Variation          SS      DF        MS         F  Sig of F

WITHIN CELLS           64376.89      44   1463.11
DIGIT                  94685.39       2  47342.69     32.36      .000
HAND BY DIGIT           1155.39       2    577.69       .39      .676
```

Figure 4.39 shows the multivariate results for the hand-by-condition-by-digit effect. Again, these are not significant. The univariate tests, shown in Figure 4.40, agree with the multivariate results.

Figure 4.39 Multivariate tests of significance for the three-way interaction

```
EFFECT .. HAND BY COND BY DIGIT
Multivariate Tests of Significance (S = 1, M = 0, N = 9 1/2)

Test Name         Value    Exact F Hypoth. DF   Error DF  Sig. of F

Pillais          .07632     .86760       2.00      21.00       .434
Hotellings       .08263     .86760       2.00      21.00       .434
Wilks            .92368     .86760       2.00      21.00       .434
Roys             .07632
```

Figure 4.40 Averaged tests of significance for the three-way interaction

```
AVERAGED Tests of Significance for MEAS.1 using UNIQUE sums of squares
Source of Variation          SS      DF        MS         F  Sig of F

WITHIN CELLS           20138.11      44    457.68
COND BY DIGIT             88.17       2     44.08       .10      .908
HAND BY COND BY DIGIT    613.39       2    306.69       .67      .517
```

Summarizing the Results

The hand used by the subject to signal the response does not seem to affect overall latency times. It also does not appear to interact with the number of digits in the memorized number or with the presence or absence of the probe digit. This is an interesting finding, since it was conceivable that using a nonpreferred hand would increase the time required to signal. Or subjects using their right hands might have had shorter latency times because of the way different activities are governed by the hemispheres of the brain. Memory is thought to be a function of the left hemisphere, which also governs the activity of the right hand. Thus, using the right hand would not necessitate a "switch" of hemispheres and might result in shorter latency times.

Analysis of Covariance with a Constant Covariate

The speed with which a subject signals that the probe digit is or is not present in the memorized number may depend on various characteristics of the subject. For example, some subjects may generally respond more quickly to stimuli than others. The reaction time or "speed" of a subject may influence performance in the memory experiments. To control for differences in responsiveness, we might administer a reaction-time test to each subject. This time can then be used as a covariate in the analysis. If subjects responding with the right hand are generally slower than subjects responding with the left, we may be able to account for the fact that no differences between the two groups were found.

To adjust for differences in covariates, the regression between the dependent variable and the covariate is calculated. For each subject, the response that would have been obtained if they had the same average speed is then calculated. Further analyses are based on these corrected values.

In a repeated measures design, the general idea is the same. The between-subjects effects are adjusted for the covariates. There is no need to adjust the within-subjects effects for covariates whose values do not change during the course of an experiment, since within-subjects factor differences are obtained from the same subject. That is, a subject's quickness or slowness is the same for all within-subjects factors.

To see how this is done, consider a hypothetical extension of Bacon's experiment. Let's assume that prior to the memory tests, each subject was given a series of trials in which he or she pressed a bar as soon as a light appeared. The interval between the time the light appeared and the time the subject pressed the bar will be termed the reaction time. For each subject, the average reaction time over a series of trials is calculated. This reaction time (variable *react*) will be considered a covariate in the analysis.

As in the previous examples, all analyses are based on the transformed within-subjects variables. Figure 4.41 shows the orthonormalized transformation matrix. It is similar to the one used before, except that there are now six additional rows and columns that are used to transform the covariates. Although there is only one covariate in this example, it is repeated six times in the matrix, once for each within-subjects variable. The transformation applied to the covariates is the same as the transformation for the within-subjects factors.

Figure 4.41 Transformation matrix with covariates

```
Orthonormalized Transformation Matrix (Transposed)

                  T1          T2          T3          T4          T5          T6

P2DIGIT          .408       -.408       -.500       -.289        .500        .289
P3DIGIT          .408       -.408        .500       -.289       -.500        .289
P4DIGIT          .408       -.408        .000        .577        .000       -.577
NP2DIGIT         .408        .408       -.500       -.289       -.500       -.289
NP3DIGIT         .408        .408        .500       -.289        .500       -.289
NP4DIGIT         .408        .408        .000        .577        .000        .577
REACT            .000        .000        .000        .000        .000        .000
REACT            .000        .000        .000        .000        .000        .000
REACT            .000        .000        .000        .000        .000        .000
REACT            .000        .000        .000        .000        .000        .000
REACT            .000        .000        .000        .000        .000        .000
REACT            .000        .000        .000        .000        .000        .000

                  T7          T8          T9         T10         T11         T12

P2DIGIT          .000        .000        .000        .000        .000        .000
P3DIGIT          .000        .000        .000        .000        .000        .000
P4DIGIT          .000        .000        .000        .000        .000        .000
NP2DIGIT         .000        .000        .000        .000        .000        .000
NP3DIGIT         .000        .000        .000        .000        .000        .000
NP4DIGIT         .000        .000        .000        .000        .000        .000
REACT            .408       -.408       -.500       -.289        .500        .289
REACT            .408       -.408        .500       -.289       -.500        .289
REACT            .408       -.408        .000        .577        .000       -.577
REACT            .408        .408       -.500       -.289       -.500       -.289
REACT            .408        .408        .500       -.289        .500       -.289
REACT            .408        .408        .000        .577        .000        .577
```

Figure 4.42 is a description of the variables used in the analysis of the constant and the between-subjects variable. The dependent variable is the constant effect, the covariate *t7*. All of the other variables, including the five transformed replicates of the same covariate labeled *t8* to *t12*, are not used for this analysis.

Figure 4.42 Test of the constant effect

```
Order of Variables for Analysis

   Variates     Covariates

    T1             T7

    1 Dependent Variable
    1 Covariate

 - - - - - - - - - - - - - - - - - - - - - - - - - - - - - - - - - - - -
 Note..  TRANSFORMED variables are in the variates column.
         These TRANSFORMED variables correspond to the
         Between-subject effects.
```

The analysis-of-variance table for the hand and constant effects is shown in Figure 4.43. Notice how the table has changed from Figure 4.33, the corresponding table without the reaction-time covariate. In Figure 4.43, the within-cells error term is subdivided into two components—error sums of squares and sums of squares due to the regression. In an analysis-of-covariance model, we are able to explain some of the variability within a cell of the design by the fact that cases have different values for the covariates. The regression sum of squares is the variability attributable to the covariate. If you were to calculate a regression between the transformed constant term and the transformed covariate, the regression sums of squares would be identical to those in Figure 4.43. The test for the hand effect is thus no longer based on the constant but on the constant adjusted for the covariate. Thus, differences between the groups in overall reaction time are eliminated. The hand effect is still not significant, however. The sums of squares attributable to hand have also changed. This is because the hypothesis tested is no longer that the overall mean is 0 but that the intercept in the regression equation for the constant and the reaction time is 0.

Figure 4.43 Analysis of covariance

```
Tests of Significance for T1 using UNIQUE sums of squares
 Source of Variation          SS      DF        MS         F  Sig of F

 WITHIN CELLS            656243.80     21  31249.70
 REGRESSION             1540420.06      1 1540420.1    49.29      .000
 HAND                      3745.89      1   3745.89      .12      .733
```

Figure 4.44 shows that the regression coefficient for the transformed reaction time is 97.5746. The test of the hypothesis that the coefficient is 0 is identical to the test of the regression effect in the analysis-of-variance table.

Figure 4.44 Regression coefficient for reaction time

```
Regression analysis for WITHIN CELLS error term
 --- Individual Univariate .9500 confidence intervals
 Dependent variable .. T1

 COVARIATE            B        Beta   Std. Err.     t-Value   Sig. of t  Lower -95%  CL- Upper

 T7            97.57466      .83788      13.898       7.021        .000      68.673     126.476
```

When constant covariates are used, messages such as that shown in Figure 4.45 are displayed. All the message says is that since all the covariates are identical, they are linearly dependent. You can simply ignore this message.

Figure 4.45 Linear dependency warning message

```
* * * * * * * * * * * * * * * * * * * * * * * * * * * * * * * * * * *
*                *                                                   *
*   W A R N I N G  * For WITHIN CELLS error matrix, these covariates *
*                * appear LINEARLY DEPENDENT on preceding            *
*                * variables ...                                     *
*                *   REACT                                           *
*                * 1 D.F. will be returned to this error term.       *
*                *                                                   *
* * * * * * * * * * * * * * * * * * * * * * * * * * * * * * * * * * *
```

Doubly Multivariate Repeated Measures Designs

Only one dependent variable in Bacon's experiment, latency time, was measured for all subjects. In some situations, more than one dependent variable may be measured for each factor combination. For example, to test the effect of different medications on blood pressure, both systolic and diastolic blood pressures may be recorded for each treatment. This is sometimes called a doubly multivariate repeated measures design, since each subject has multiple variables measured at multiple times.

Analysis of doubly multivariate repeated measures data is similar to the analyses above. Each dependent variable is transformed using the same orthonormalized transformation. All subsequent analyses are based on these transformed variables. The tests for each effect are based on the appropriate variables for all dependent variables. For example, in the memory experiment, if two dependent variables had been measured for the three number lengths, the test of the digit effect would be based on four transformed variables, two for the first dependent variable and two for the second. Similarly, the test for constant would have been based on two variables corresponding to averages for each dependent variable.

How to Obtain a Repeated Measures ANOVA

With the Repeated Measures ANOVA procedure, you can analyze a wide variety of repeated measures designs using the univariate (mixed-model) or multivariate approach. You can also specify transformations of within- and between-subjects factors and obtain tests of symmetry conditions.

To obtain Repeated Measures ANOVA, from the menus choose:

Statistics
 ANOVA Models ▸
  Repeated Measures...

This opens the Repeated Measures Define Factor(s) dialog box, as shown in Figure 4.46.

Figure 4.46 Repeated Measures Define Factor(s) dialog box

Repeated Measures Define Factor(s)
Within-Subject Factor Name:
Number of Levels:
Add
Change
Remove
cond(2)
digit(3)
Define
Reset
Cancel
Help
Measure>>

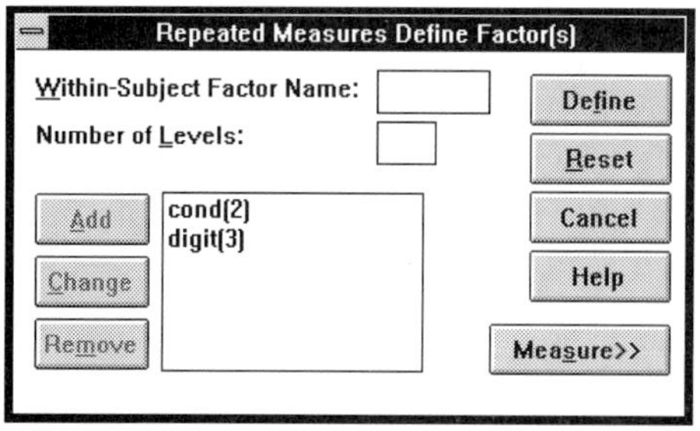

You must specify at least one within-subjects factor to distinguish measurements made on the same subject or case. You can specify multiple factors. Enter a unique name for each factor, enter the number of factor levels, and click on Add.

For example, in Figure 4.46 there are two within-subjects factors: the length of the numbers memorized by each subject (*digit*) and the presence or absence of a test digit in the number (*cond*). The total number of observations (dependent variables) for each subject is the product of the number of levels for each factor. For example, with two levels for *cond* and three levels for *digit*, there should be six observations for each subject.

Within-Subject Factor Name. Enter a unique name for each factor. The names cannot exceed eight characters and should not duplicate any variable names in the working data file. Each additional factor indicates factor levels within levels of the previous factor. For example, in Figure 4.46, there are three levels of *digit* within each level of *cond*.

Number of Levels. Enter the number of levels for each factor. Each level represents a different observation for the same subject.

Doubly Multivariate Repeated Measures

To specify a doubly multivariate repeated measures design (in which more than one variable is measured at each combination of the factor levels), click on Measure>>. This expands the Repeated Measures Define Factor(s) dialog box, as shown in Figure 4.47.

Figure 4.47 Expanded Repeated Measures Define Factor(s) dialog box

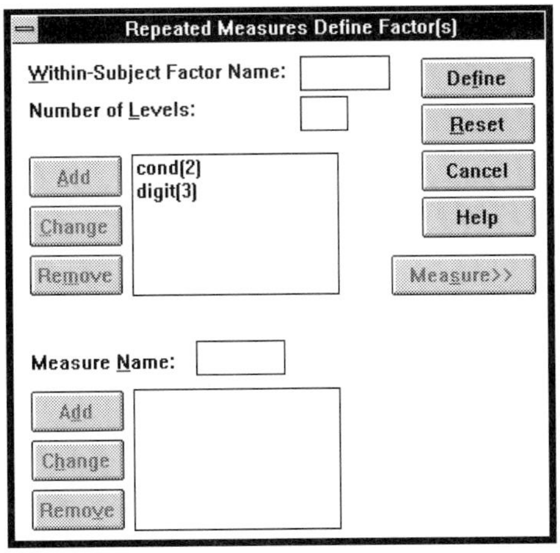

Measure Name. Enter a unique name for each measure. The names cannot exceed eight characters and should not duplicate any variable names in the working data file. The number of names entered indicates the number of variables measured at each combination of factor levels.

The total number of observations for each subject is determined by the number of measures multiplied by the number of levels for each factor. For example, if you enter two measure names, one factor with three levels, and one factor with two levels, there should be 12 observations for each subject.

Within-Subject Variables and Between-Subjects Factors

After specifying within-subject factors and optional measure names, click on Define in the Repeated Measures Define Factor(s) dialog box to open the Repeated Measures ANOVA dialog box, as shown in Figure 4.48.

Figure 4.48 Repeated Measures ANOVA dialog box

Repeated Measures ANOVA
c1
c2
c3
c4
c5
c6
hand
nhand
react
sr1
Within-Subject Variables: [cond,digit]
p2digit[1,1]
p3digit[1,2]
p4digit[1,3]
np2digit[2,1]
np3digit[2,2]
np4digit[2,3]
OK
Paste
Reset
Cancel
Help
Between-Subject Factor(s):
Define Range...
Model contains 0 covariates
Covariates...
Contrasts...
Model...
Options...

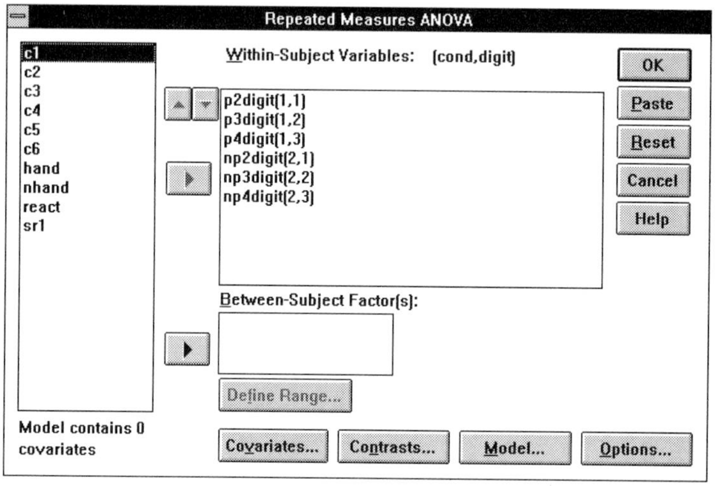

The numeric variables in the working data file are displayed on the source variable list.

Within-Subject Variables. Select a dependent variable that corresponds to each combination of within-subject factors (and optionally, measure) on the list. The numbers displayed in parentheses indicate the factor level for each within-subject factor. For example, in Figure 4.48, p3digit (1,2) indicates that the variable *p3digit* represents the observation for level 1 of *cond* and level 2 of *digit*. For doubly multivariate designs, the measure name is displayed after the factor levels in parentheses.

You can change the position (and consequently the corresponding factor levels) of a variable on the list with the ▲ and ▼ pushbuttons. If you want to define different within-subjects factors or measures or change the order of within-subjects factors, you can re-open the Repeated Measures Define Factor(s) dialog box and make the changes without closing the Repeated Measures ANOVA dialog box.

Between-Subjects Factor(s). Optionally, you can select one or more variables that divide the sample into discrete subgroups. Each subject has only one value for a between-subjects factor. For example, a variable for gender could be used as a between-subjects factor to divide the sample into males and females.

Defining Levels for Between-Subjects Factor Variables

For each between-subjects factor variable, you must indicate which factor levels you want to use in the analysis. Highlight one or more factors and click on Define Range... in the Repeated Measures ANOVA dialog box. This opens the Repeated Measures ANOVA Define Range dialog box, as shown in Figure 4.49.

Figure 4.49 Repeated Measures ANOVA Define Range dialog box

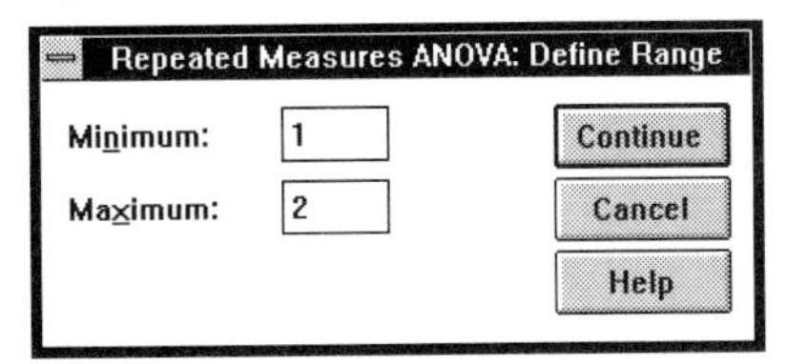

Enter integer values corresponding to the lowest and highest factor levels you want to use. Each integer category in the inclusive range is used as a factor level in the analysis (non-integer codes are truncated for the analysis, so cases with the same integer portion are combined into the same factor level). Cases with values outside the range are excluded from the analysis. To avoid empty cells, factor levels within the range must be consecutive. If a variable has empty categories, use the Automatic Recode facility on the Transform menu to create consecutive categories (for more information on recoding variables, see the SPSS Base system documentation).

Covariates

To specify covariates for the analysis, click on Covariates... in the Repeated Measures ANOVA dialog box. This opens the Repeated Measures ANOVA Covariates dialog box, as shown in Figure 4.50.

Figure 4.50 Repeated Measures ANOVA Covariates dialog box

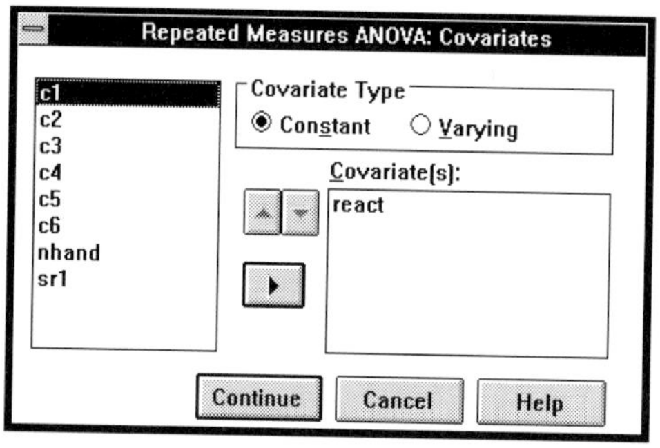

Covariate Type. You can select one of the following alternatives:

- **Constant.** The value of each covariate is constant for all within-subjects factor levels. For example, in Figure 4.50, variable *react* is a measure of each subject's average reaction time at the beginning of the study.
- **Varying.** The value of each covariate can vary for different within-subjects factor levels. You must specify a set of variables that represent the value of each covariate for each combination of within-subjects factor levels. (See Figure 4.51.)

Covariate(s). You can select multiple variables as covariates. If you want to specify more than one varying covariate, select the set of variables for the first covariate, and then select the set of variables for the next covariate.

Figure 4.51 Selecting varying covariates

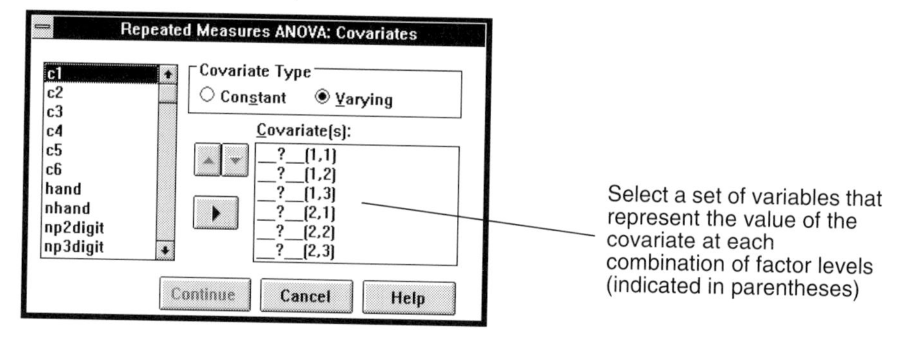

Contrasts

Within-subjects factors and between-subjects factors are transformed into contrasts for the analysis. To display parameter estimates and confidence intervals or to override the default contrast type (polynomial for within-subjects and deviation for between-subjects), click on Contrasts... in the Repeated Measures ANOVA dialog box. This opens the Repeated Measures ANOVA Contrasts dialog box, as shown in Figure 4.52.

Figure 4.52 Repeated Measures ANOVA Contrasts dialog box

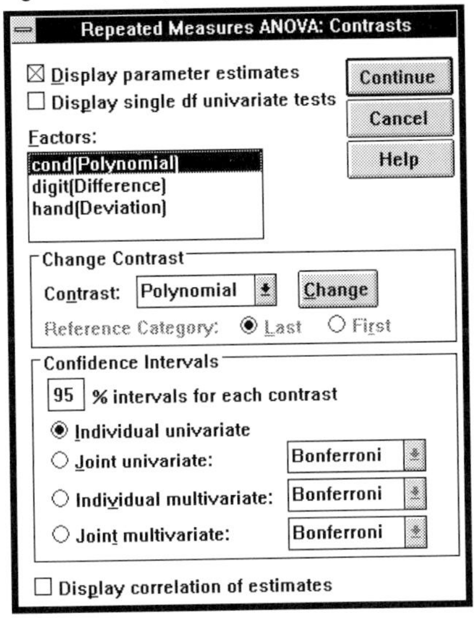

- **Display parameter estimates**. Displays parameter estimates for contrasts as well as their standard errors, *t* statistics, and confidence intervals. If you select this option, you can also change the contrast type and confidence intervals.
- **Display single df univariate tests.** Displays univariate *F* tests and their probabilities. Also shows hypothesis sums of squares and mean squares. If you select this option, you can also change the contrast type and confidence intervals.

Change Contrast. To obtain a different type of contrast, highlight one or more factors, select a contrast type, and click on Change. Optionally, you can change the default reference category.

- **Contrast.** You can choose one of the following contrast types:

  Deviation. Each category of the factor except the reference category is compared to the overall effect. This is the default for between-subjects factors. Available only for between-subjects factors.

  Simple. Each category of the factor (except the reference category) is compared to the reference category. Available only for between-subjects factors.

  Difference. Also known as reverse Helmert contrasts. Each category of the factor except the first category is compared to the average effect of previous categories.

  Helmert. Each category of the factor except the last category is compared to the mean effect of subsequent categories.

  Repeated. Each category of the factor except the first category is compared to the category that precedes it. Available only for between-subjects factors.

  Polynomial. The factor is partitioned into linear, quadratic, and cubic effects, and so on (depending on the number of categories). Categories are assumed to be equally spaced. This is the default for within-subjects factors.

  Reference Category. For deviation and simple contrasts, you can specify which category is to be omitted from the comparisons. Choose one of the following alternatives:

  - **Last.** Uses the highest category of the factor as the reference category. This is the default.
  - **First.** Uses the lowest category of the factor as the reference category.

Confidence Intervals. If you request parameter estimates for contrasts, SPSS displays confidence intervals for the estimates. To override the default confidence level (95%), enter a value for n% intervals for each contrast. The value must be greater than 0 and less than 100.

You can also choose between individual and joint confidence intervals and between univariate and multivariate intervals. Choose one of the following alternatives:

- **Individual univariate.** Displays a separate interval for each parameter within each dependent variable. These are the default confidence intervals.
- **Joint univariate.** Displays simultaneous intervals for parameters within each dependent variable. By default, SPSS displays Bonferroni intervals, which are based on the number of contrasts actually made. To obtain intervals based on all possible contrasts, choose Scheffé from the drop-down list.
- **Individual multivariate.** Displays separate intervals for each parameter across all dependent variables. You can choose one of the following methods for computing the multivariate intervals:

  Bonferroni. Displays confidence intervals based on Student's t distribution.

Roy. Displays confidence intervals based on Roy's largest root.

Pillai. Displays confidence intervals based on Pillai's trace.

Hotelling. Displays confidence intervals based on Hotelling's trace.

Wilks. Displays confidence intervals based on Wilks' lambda.

❍ **Joint multivariate**. Displays simultaneous intervals for parameters across all dependent variables. You can choose one of the following methods for computing the multivariate intervals:

Bonferroni. Displays confidence intervals based on Student's *t* distribution.

Roy. Displays confidence intervals based on Roy's largest root.

Pillai. Displays confidence intervals based on Pillai's trace.

Hotelling. Displays confidence intervals based on Hotelling's trace.

Wilks. Displays confidence intervals based on Wilks' lambda.

The following option is also available:

❑ **Display correlation of estimates**. Displays correlations and covariances between parameter estimates.

Specifying the Model

By default, SPSS analyzes a full factorial model. To define a different model or control model options, click on Model... in the Repeated Measures ANOVA dialog box. This opens the Repeated Measures ANOVA Model dialog box, as shown in Figure 4.53.

Figure 4.53 Repeated Measures ANOVA Model dialog box

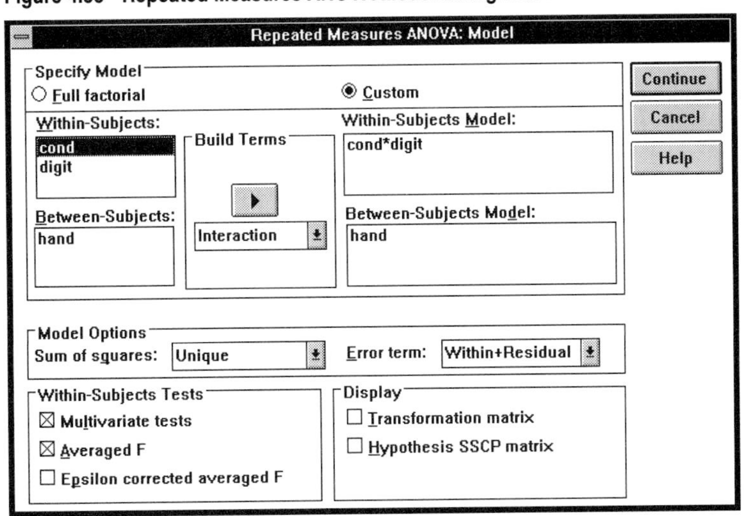

Specify Model. You can choose one of the following alternatives:

❍ **Full Factorial.** Model contains all within-subjects and between-subjects factor main effects and all factor-by-factor interactions. This is the default.

❍ **Custom.** Select this item to define a model containing subsets of possible factor-by-factor interactions. You must indicate which terms you want to include in the model.

Within-Subjects Model/Between-Subjects Model. To include a term in a custom model, select one or more factors. If you don't want to create the highest-order interaction for the selected factors, select an alternate method for building the term. To add more terms to the model, repeat this process. Do not use the same term more than once in the model.

Macintosh: Use ⌘-click to select multiple factors or covariates, or a combination.

🠻 **Build Terms.** You can build main effects or interactions for the selected factors. If you request an interaction of a higher order than the number of factors, SPSS creates a term for the highest-order interaction for the factors. If only one factor is selected, a main-effect term is added to the model, regardless of your Build Terms selection. Choose one of the following alternatives:

  Interaction. Creates the highest-level interaction term for the variables. This is the default.

  Main effects. Creates a main-effect term for each variable.

  All 2-way. Creates all possible two-way interactions for the variables.

  All 3-way. Creates all possible three-way interactions for the variables.

  All 4-way. Creates all possible four-way interactions for the variables.

  All 5-way. Creates all possible five-way interactions for the variables.

  Simple within. Creates the mean at all levels of the selected variable (or all levels of the interaction of selected variables) to be used for simple effects analysis. If the term is a within-subjects term, all between-subjects terms will be tested within each of the levels. If the term is a between-subjects term, all within-subjects terms will be tested within each of the levels. A separate term is displayed on the list for each factor level or combination of levels. You can selectively remove factor levels from the model.

Model Options. SPSS includes a constant term in all models. You can control the computation of sums of squares and the error term used for the model.

🠻 **Sum of squares.** Select one of the following methods for decomposing sums of squares:

  Unique. Uses the regression method to partition sums of squares. This is the default. Each effect is adjusted for all other effects in the model.

Sequential. Decomposes sums of squares hierarchically. For default (full factorial) models, SPSS first adjusts covariates for all factors and interactions. Then each factor is adjusted for all covariates and for factors that precede it in the model specification. Each interaction is adjusted for all covariates, factors, and preceding interactions.

For custom models, SPSS first adjusts covariates for all terms in the model. Next, all terms in the model are adjusted for covariates. Then SPSS evaluates all remaining terms (factor main effects and factor interactions) in the order in which they appear in the model specification.

- **Error term.** You can control the error term used in tests of significance. Select one of the following alternatives:

  Within cells. Tests model terms against the within-cells sum of squares.

  Residual. Tests model terms against the residual sum of squares.

  Within+Residual. Tests model terms against the pooled within-cells and residual sum of squares. This is the default.

Within-Subjects Tests. At least one of the following tests or one of the Display options must be selected:

- ❑ **Multivariate tests.** Displays multivariate test statistics and their *F* values, degrees of freedom, and probabilities. Shown by default.
- ❑ **Averaged F.** Displays averaged *F* tests for repeated measures. Shown by default.
- ❑ **Epsilon corrected averaged F.** Displays the Huynh-Feldt, Greenhouse-Geisser, and lowerbound corrected significance values for averaged univariate *F* tests.

Display. At least one of the following display options or one Within-Subjects Tests must be selected:

- ❑ **Transformation matrix.** Displays the orthonormalized transformation matrix of the transformed within-subject variables.
- ❑ **Hypothesis SSCP matrix.** Displays a matrix of hypothesis sums of squares and cross-products.

Options

To obtain optional statistics, click on Options... in the Repeated Measures ANOVA dialog box. This opens the Repeated Measures ANOVA Options dialog box, as shown in Figure 4.54.

Figure 4.54 Repeated Measures ANOVA Options dialog box

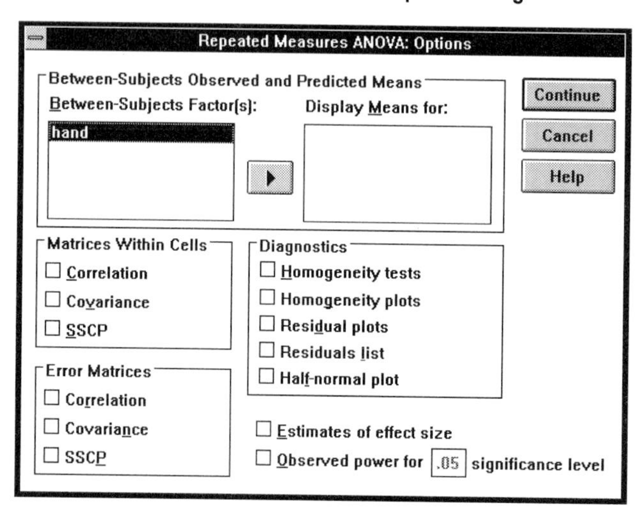

Between-Subjects Observed and Predicted Means. You can obtain observed means (of dependent variables and covariates) and predicted means (of dependent variables) for subgroups defined by between-subjects factors.

Between-Subjects Factor(s). The list contains the variables you have selected as between-subjects factors.

Display Means for. Select the between-subjects factors you want to use to define subgroups. This displays means for each factor level. For observed means, SPSS shows weighted means (which give equal weight to all cases) and unweighted means (which weight cells equally regardless of the number of cases they contain). For predicted means, SPSS displays cell means that are adjusted (adjusted means) and unadjusted (estimated means) for all covariates that appear only as main effects. Raw and standardized residuals are also shown.

Matrices Within Cells. Matrices within cells are produced only if you have at least one between-subjects factor variable. You can choose one or more of the following:

- **Correlation.** Displays a correlation matrix for each cell. Standard deviations are shown on the diagonal.
- **Covariance.** Displays a variance-covariance matrix for each cell. The pooled within-cells variance-covariance matrix is also shown.
- **SSCP.** Displays a matrix of sums of squares and cross-products for each cell.

Error Matrices. You can choose one or more of the following error matrices:

- ❑ **Correlation.** Displays an error correlation matrix. Also shows the log of the determinant of the matrix and Bartlett's test of sphericity, which tests the hypothesis that the population correlation matrix is an identity matrix.
- ❑ **Covariance.** Displays an error variance-covariance matrix.
- ❑ **SSCP.** Displays a matrix of error sums of squares and cross-products.

Diagnostics. You can choose one or more of the following diagnostics:

- ❑ **Homogeneity tests.** Displays homogeneity tests for univariate (Bartlett-Box *F* and Cochran's *C*) and multivariate (Box's *M*) models.
- ❑ **Homogeneity plots.** Plots cell means against cell variances and cell standard deviations and displays a histogram of cell means. Plots are shown for each within-subject variable (dependent variable) and covariate.
- ❑ **Residual plots.** Displays a scatterplot matrix of observed values, predicted values, and standardized residuals. Also plots standardized residuals against case-identification numbers and displays normal and detrended normal probability plots for standardized residuals. Plots are shown for each within-subject variable (dependent variable).
- ❑ **Residuals list.** Displays observed and predicted values and raw and standardized residuals. A separate listing is shown for each within-subject variable (dependent variable).
- ❑ **Half-normal plot.** Plots the within-cell correlations between the dependent variables against their expected values. Correlations shown in the plot are transformed using Fisher's *Z* transformation.

You can also choose one or both of the following display options:

- ❑ **Estimates of effect size.** Displays effect-size estimates for multivariate tests. For univariate models, η^2 (eta squared) is shown. For sequential models, η^2 and Ω^2 (omega squared) are shown.
- ❑ **Observed power for n significance level.** Displays approximate power values for univariate and multivariate *F* and *t* tests based on fixed-effect assumptions. By default, 0.05 is used as the alpha level for the power calculations. To use a different alpha, enter a value greater than 0 and less than 1.

Additional Features Available with Command Syntax

You can customize your repeated measures analysis of variance if you paste your selections into a syntax window and edit the resulting MANOVA command syntax. (For information on syntax windows, see the SPSS Base system documentation.) Additional features include:

- Additional contrast options, such as user-defined contrasts (using the CONTRAST subcommand).
- Additional design information, including basis matrices (using the PRINT subcommand).
- The ability to read matrix data and to write matrix files that can be used in subsequent analyses (using the MATRIX subcommand).
- The ability to partition degrees of freedom associated with a factor (using the PARTITION subcommand).
- Predicted means for combinations of factor levels (using the PMEANS subcommand).

See the Syntax Reference section of this manual for command syntax rules and for complete MANOVA command syntax.

5 Model Selection Loglinear Analysis

There is only one way to achieve
happiness on this terrestrial ball,
And that is to have either a clear
conscience or, none at all.

Ogden Nash

Ignoring Nash's warning, many of us continue to search for happiness in terrestrial institutions and possessions. Marriage, wealth, and health are all hypothesized to contribute to happiness. But do they really? And how can you investigate possible associations?

Consider Figure 5.1, which is a two-way crosstabulation of marital status and score on a happiness scale. The data are from the 1982 General Social Survey conducted by the National Opinion Research Center. Of the 854 currently married respondents, 92% indicated that they were very happy or pretty happy, while only 79% of the 383 divorced, separated, or widowed people classified themselves as very happy or pretty happy.

Figure 5.1 Two-way crosstabulation

HAPPY by MARITAL

	Count / Col Pct	MARITAL: MARRIED 1.00	SINGLE 2.00	SPLIT 3.00	Row Total
HAPPY					Page 1 of 1
YES	1.00	787 92.2	221 82.5	301 78.6	1309 87.0
NO	2.00	67 7.8	47 17.5	82 21.4	196 13.0
	Column Total	854 56.7	268 17.8	383 25.4	1505 100.0

Although the results in Figure 5.1 are interesting, they suggest many new questions. What role does income play in happiness? Are poor married couples happier than affluent singles? What about health? Determining the relationship among such variables is potentially complicated. If family income is recorded in three categories and condition of health in two, a separate two-way table of happiness score and marital status is obtained for each of the six possible combinations of health and income. As additional variables are included in the cross-classification tables, the number of cells rapidly increases and it is difficult, if not impossible, to unravel the associations among the variables by examining only the cell entries.

The usual response of researchers faced with crosstabulated data is to compute a chi-square test of independence for each subtable. This strategy is fraught with problems and usually does not result in a systematic evaluation of the relationship among the variables. The classical chi-square approach also does not provide estimates of the effects of the variables on each other, and its application to tables with more than two variables is complicated.

Loglinear Models

The advantages of statistical models that summarize data and test hypotheses are well recognized. Regression analysis, for example, examines the relationship between a dependent variable and a set of independent variables. Analysis-of-variance techniques provide tests for the effects of various factors on a dependent variable. But neither technique is appropriate for categorical data, where the observations are not from populations that are normally distributed with constant variance.

A special class of statistical techniques, called **loglinear models**, has been formulated for the analysis of categorical data (Haberman, 1978; Bishop, Fienberg & Holland, 1975). These models are useful for uncovering the potentially complex relationships among the variables in a multiway crosstabulation. Loglinear models are similar to multiple regression models. In loglinear models, all variables that are used for classification are independent variables, and the dependent variable is the number of cases in a cell of the crosstabulation.

A Fully Saturated Model

Consider Figure 5.1 again. Using a loglinear model, the number of cases in each cell can be expressed as a function of marital status, degree of happiness, and the interaction between degree of happiness and marital status. To obtain a linear model, the natural logs of the cell frequencies, rather than the actual counts, are used. The natural logs of the cell frequencies in Figure 5.1 are shown in Table 5.1. (Recall that the natural log of a number is the power to which the number e (approximately 2.718) is raised to give that number. For example, the natural log of the first cell entry is 6.668, since $e^{6.668} = 787$.)

Table 5.1 Natural logs

Happy	Married	Single	Split	Average
Yes	6.668	5.398	5.707	5.924
No	4.205	3.850	4.407	4.154
Average	5.436	4.624	5.057	5.039

The loglinear model for the first cell in Table 5.1 is

$$\log (787) = \mu + \lambda_{\text{yes}}^{\text{happy}} + \lambda_{\text{married}}^{\text{marital}} + \lambda_{\text{yes}}^{\text{happy}}{}_{\text{married}}^{\text{marital}}$$ **Equation 5.1**

The term denoted as μ is comparable to the grand mean in the analysis of variance. It is the average of the logs of the frequencies in all table cells. The lambda (λ) parameters represent the increments or decrements from the base value (μ) for particular combinations of values of the row and column variables.

Each individual category of the row and column variables has an associated lambda. The term $\lambda_{\text{married}}^{\text{marital}}$ indicates the effect of being in the *married* category of the marital status variable, and similarly the effect of being in the *yes* category of the happy variable is represented by $\lambda_{\text{yes}}^{\text{happy}}$. The term $\lambda_{\text{yes}}^{\text{happy}}{}_{\text{married}}^{\text{marital}}$ represents the interaction of being happy and married. Thus, the number of cases in a cell is a function of the values of the row and column variables and their interactions.

In general, the model for the log of the observed frequency in the *i*th row and the *j*th column is given by

$$\ln (n_{ij}) = \mu + \lambda_i^{\text{H}} + \lambda_j^{\text{S}} + \lambda_{ij}^{\text{HS}}$$ **Equation 5.2**

where n_{ij} is the observed frequency in the cell, λ_i^{H} is the effect of the *i*th happiness category, λ_j^{S} is the effect of the *j*th marital status category, and λ_{ij}^{HS} is the interaction effect for the *i*th value of the happiness category and the *j*th value of the marital status variable.

The lambda parameters and μ are estimated from the data. The estimate for μ is simply the average of the logs of the frequencies in all table cells. From Table 5.1, the estimated value of μ is 5.039. Estimates for the lambda parameters are obtained in a manner similar to analysis of variance. For example, the effect of the *happy* category is estimated as

$$\lambda_{\text{yes}}^{\text{happy}} = 5.924 - 5.039 = 0.885$$ **Equation 5.3**

where 5.924 is the average of the logs of the observed counts in the *happy* cells. The lambda parameter is just the average log of the frequencies in a particular category minus the grand mean. In general, the effect of the *i*th category of a variable, called a **main effect**, is estimated as

$$\lambda_i = \mu_i - \mu$$ **Equation 5.4**

where μ_i is the mean of the logs in the *i*th category and μ is the grand mean. Positive values of lambda occur when the average number of cases in a row or a column is larger than the overall average. For example, since there are more married people in the sample than single or separated people, the lambda for *married* is positive. Similarly, since there are fewer unhappy people than happy people, the lambda for *not happy* is negative.

The interaction parameters indicate how much difference there is between the sums of the effects of the variables taken individually and collectively. They represent the "boost" or "interference" associated with particular combinations of the values. For example, if marriage does result in bliss, the number of cases in the *happy* and *married* cell would be larger than the number expected based only on the frequency of married people ($\lambda^{\text{marital}}_{\text{married}}$) and the frequency of happy people ($\lambda^{\text{happy}}_{\text{yes}}$). This excess would be represented by a positive value for $\lambda^{\text{happy}}_{\text{yes}}{}^{\text{marital}}_{\text{married}}$. If marriage decreases happiness, the value for the interaction parameter would be negative. If marriage neither increases nor decreases happiness, the interaction parameter would be 0.

The estimate for the interaction parameter is the difference between the log of the observed frequency in a particular cell and the log of the predicted frequency using only the lambda parameters for the row and column variables. For example,

$$\begin{aligned}\lambda^{\text{happy}}_{\text{yes}}{}^{\text{marital}}_{\text{married}} &= \ln(n_{11}) - (\mu + \lambda^{\text{happy}}_{\text{yes}} + \lambda^{\text{marital}}_{\text{married}}) \\ &= 6.668 - (5.039 + 0.885 + 0.397) = 0.347\end{aligned}$$

Equation 5.5

where n_{11} is the observed frequency in the *married* and *happy* cell. Table 5.2 contains the estimates of the lambda parameters for the main effects (marital status and happiness) and their interactions.

To uniquely estimate the lambda parameters, we need to impose certain constraints on them. The lambdas must sum to 0 across the categories of a variable. For example, the sum of the lambdas for marital status is $0.397 + (-0.415) + 0.018 = 0$. Similar constraints are imposed on the interaction terms. They must sum to 0 over all categories of a variable.

Table 5.2 Estimates of lambda parameters

$$\lambda_{\text{yes}}^{\text{happy}} = 5.924 - 5.039 = 0.885$$

$$\lambda_{\text{no}}^{\text{happy}} = 4.154 - 5.039 = -0.885$$

$$\lambda_{\text{married}}^{\text{marital}} = 5.436 - 5.039 = 0.397$$

$$\lambda_{\text{single}}^{\text{marital}} = 4.624 - 5.039 = -0.415$$

$$\lambda_{\text{split}}^{\text{marital}} = 5.057 - 5.039 = 0.018$$

$$\lambda_{\text{YM}}^{\text{HM}} = 6.668 - (5.039 + 0.885 + 0.397) = 0.347$$

$$\lambda_{\text{NM}}^{\text{HM}} = 4.205 - (5.039 - 0.885 + 0.397) = -0.347$$

$$\lambda_{\text{YSi}}^{\text{HM}} = 5.398 - (5.039 + 0.885 - 0.415) = -0.111$$

$$\lambda_{\text{NSi}}^{\text{HM}} = 3.850 - (5.039 - 0.885 - 0.415) = 0.111$$

$$\lambda_{\text{YSp}}^{\text{HM}} = 5.707 - (5.039 + 0.885 + 0.018) = -0.235$$

$$\lambda_{\text{NSp}}^{\text{HM}} = 4.407 - (5.039 - 0.885 + 0.018) = 0.235$$

Each of the observed cell frequencies is reproduced exactly by a model that contains all main-effect and interaction terms. This type of model is called a **saturated model**. For example, the observed log frequency in cell 1 of Table 5.1 is given by

$$\begin{aligned} \ln(n_{11}) &= \mu + \lambda_{\text{yes}}^{\text{happy}} + \lambda_{\text{married}}^{\text{marital}} + \lambda_{\text{yes}}^{\text{happy}}{}_{\text{married}}^{\text{marital}} \\ &= 5.039 + 0.885 + 0.397 + 0.346 = 6.67 \end{aligned}$$

Equation 5.6

All other observed cell frequencies can be similarly expressed as a function of the lambdas and the grand mean.

Output for the Cells

Consider Figure 5.2, which contains the observed and expected (predicted from the model) counts for the data shown in Figure 5.1. The first column gives the names of the variables used in the analysis, and the second contains the value labels for the cells in the table. The next column indicates the number of cases in each of the cells. For example, 787 people who classify themselves as *happy* are married, which is 52.29% of all respondents in the survey (787 out of 1505). This percentage is listed in the next column. (If the output width is not large enough, percentages are not displayed.) Since the number of cases in each cell is expressed as a percentage of the total cases, the sum of all the percentages is 100. The saturated model reproduces the observed cell frequencies exactly, so the expected and observed cell counts and percentages are equal. For the same reason, the next two columns, which compare the observed and expected counts, are all zeros. (Models that do not exactly reproduce the observed cell counts are examined later.)

Figure 5.2 Observed and expected frequencies for saturated model

Observed, Expected Frequencies and Residuals.

Factor	Code	OBS. count	& PCT.	EXP. count	& PCT.	Residual	Std. Resid.
HAPPY	YES						
MARITAL	MARRIED	787.00	(52.29)	787.00	(52.29)	.000	.000
MARITAL	SINGLE	221.00	(14.68)	221.00	(14.68)	.000	.000
MARITAL	SPLIT	301.00	(20.00)	301.00	(20.00)	.000	.000
HAPPY	NO						
MARITAL	MARRIED	67.00	(4.45)	67.00	(4.45)	.000	.000
MARITAL	SINGLE	47.00	(3.12)	47.00	(3.12)	.000	.000
MARITAL	SPLIT	82.00	(5.45)	82.00	(5.45)	.000	.000

Parameter Estimates

The estimates of the loglinear model parameters can also be displayed by the Model Selection Loglinear Analysis procedure. Figure 5.3 contains parameter estimates for the data in Figure 5.1. (Since there are no zero-frequency cells, delta is set to 0.) The parameter estimates are displayed in blocks for each effect.

Parameter estimates for the interaction effects are displayed first. Because estimates must sum to 0 across the categories of each variable, only two parameter estimates need to be displayed for the interaction effects: those for $\lambda_{yes\ \ married}^{happy\ marital}$ and for $\lambda_{yes\ \ single}^{happy\ marital}$. All other interaction parameter estimates can be derived from these.

After the interaction terms, parameter estimates for the main effects of the variables are displayed. The first estimate is for λ_{yes}^{happy} and is identical to the value given in Table 5.2 (0.885). Again, only one parameter estimate is displayed, since we can infer that the parameter estimate for λ_{no}^{happy} is –0.885 (the two estimates must sum to 0). Two parameter estimates and associated statistics are displayed for the marital status variable, since it has three categories. The estimate for the third parameter, $\lambda_{split}^{marital}$, is the negative of the sum of the estimates for $\lambda_{married}^{marital}$ and $\lambda_{single}^{marital}$. Thus, $\lambda_{split}^{marital}$ is estimated to be 0.0178, as shown in Table 5.2.

Figure 5.3 Estimates for parameters

Estimates for Parameters.

HAPPY*MARITAL

Parameter	Coeff.	Std. Err.	Z-Value	Lower 95 CI	Upper 95 CI
1	.3464441901	.05429	6.38147	.24004	.45285
2	-.1113160745	.06122	-1.81833	-.23131	.00867

HAPPY

Parameter	Coeff.	Std. Err.	Z-Value	Lower 95 CI	Upper 95 CI
1	.8853236244	.03997	22.14946	.80698	.96367

MARITAL

Parameter	Coeff.	Std. Err.	Z-Value	Lower 95 CI	Upper 95 CI
1	.3972836534	.05429	7.31793	.29088	.50369
2	-.4150216289	.06122	-6.77930	-.53501	-.29503

Since the individual parameter estimates in Figure 5.3 are not labeled, the following rules for identifying the categories to which they correspond may be helpful. For main effects, the parameter estimates correspond to the first $K-1$ categories of the variable, where K is the total number of categories. For interaction parameters, the number of estimates displayed is the product of the number of categories, minus 1, of each variable in the interaction, multiplied together. For example, marital status has three categories and happiness has two, so the number of estimates displayed is $(3-1) \times (2-1) = 2$.

To identify individual parameters, first look at the order in which the variable labels are listed in the heading. In this case, the heading is *HAPPY*MARITAL*. The first estimate corresponds to the first category of both variables, which is happy–yes and marital–married. The next estimate corresponds to the first category of the first variable (*happy*) and the second category of the second variable (*marital*). In general, the categories of the last variable rotate most quickly and those of the first variable most slowly. Terms involving the last categories of the variables are omitted.

Figure 5.3 also displays the standard error for each estimate. For the $\lambda_{\text{yes}}^{\text{happy}}$ parameter, the standard error is 0.03997. The ratio of the parameter estimate to its standard error is given in the column labeled *Z-Value*. For sufficiently large sample sizes, the test of the null hypothesis that lambda is 0 can be based on this Z value, since the standardized lambda is approximately normally distributed with a mean of 0 and a standard deviation of 1 if the model fits the data. Lambdas with Z values greater than 1.96 in absolute value can be considered significant at the 0.05 level. When tests for many lambdas are calculated, however, the usual problem of multiple comparisons arises. That is, when many comparisons are made, the probability that some are found to be significant when there is no effect increases rapidly. Special multiple comparison procedures for testing the lambdas are discussed in Goodman (1984).

Individual confidence intervals can be constructed for each lambda. The 95% confidence interval for $\lambda_{\text{yes}}^{\text{happy}}$ is $0.885 \pm (1.96 \times 0.03997)$, which results in a lower limit of 0.80698 and an upper limit of 0.96367. Since the confidence interval does not include 0, the hypothesis that the population value is 0 can be rejected. These values are displayed in the last two columns of Figure 5.3.

The Independence Model

Representing an observed-frequency table with a loglinear model that contains as many parameters as there are cells (a saturated model) does not result in a parsimonious description of the relationship between the variables. It may, however, serve as a good starting point for exploring other models that could be used to represent the data. Parameters that have small values can be excluded from subsequent models.

To illustrate the general procedure for fitting a model that does not contain all possible parameters (an unsaturated or custom model), consider the familiar independence hypothesis for a two-way table. If variables are independent, they can be represented by a loglinear model that does not have any interaction terms. For example, if happiness and marital status are independent,

$$\log (m_{ij}) = \mu + \lambda_i^{\text{happy}} + \lambda_j^{\text{marital}} \quad \textbf{Equation 5.7}$$

Note that m_{ij} is no longer the observed frequency in the (i,j)th cell, but it is now the expected frequency based on the model. The estimates for the expected frequencies must be obtained using an iterative algorithm. Each time an estimate is obtained, it is called an **iteration**, while the largest amount by which successive estimates differ is called the **convergence criterion**.

The message shown in Figure 5.4 gives the number of iterations required for convergence. For this example, two iterations were required for convergence. The observed and expected counts in each of the cells of the table are shown in Figure 5.5.

Figure 5.4 Message indicating number of iterations required for convergence

```
DESIGN 1 has generating class

    HAPPY
    MARITAL

The Iterative Proportional Fit algorithm converged at iteration 2.
The maximum difference between observed and fitted marginal totals is      .000
and the convergence criterion is      .787
```

Figure 5.5 Observed and expected frequencies for unsaturated model

```
Observed, Expected Frequencies and Residuals.

     Factor          Code       OBS. count  & PCT.    EXP. count  & PCT.      Residual   Std. Resid.

 HAPPY           YES
  MARITAL         MARRIED           787.00 (52.29)        742.78 (49.35)        44.219         1.622
  MARITAL         SINGLE            221.00 (14.68)        233.10 (15.49)       -12.098         -.792
  MARITAL         SPLIT             301.00 (20.00)        333.12 (22.13)       -32.121        -1.760

 HAPPY           NO
  MARITAL         MARRIED            67.00 ( 4.45)        111.22 ( 7.39)       -44.219        -4.193
  MARITAL         SINGLE             47.00 ( 3.12)         34.90 ( 2.32)        12.098         2.048
  MARITAL         SPLIT              82.00 ( 5.45)         49.88 ( 3.31)        32.121         4.548
```

The expected values are identical to those obtained from the usual formulas for expected values in a two-way crosstabulation, as shown in Figure 5.6.

Figure 5.6 Crosstabulation of happiness by marital status

```
HAPPY  by  MARITAL

                   MARITAL                   Page 1 of 1
            Count |
          Exp Val |MARRIED  SINGLE   SPLIT
                  |                               Row
                  |    1.00|    2.00|    3.00|   Total
HAPPY             +--------+--------+--------+
             1.00 |    787 |    221 |    301 |   1309
  YES             |  742.8 |  233.1 |  333.1 |   87.0%
                  +--------+--------+--------+
             2.00 |     67 |     47 |     82 |    196
  NO              |  111.2 |   34.9 |   49.9 |   13.0%
                  +--------+--------+--------+
           Column      854      268      383     1505
            Total    56.7%    17.8%    25.4%   100.0%

        Chi-Square                   Value            DF            Significance
--------------------          -----------          ----            ------------

Pearson                           48.81639            2                .00000
Likelihood Ratio                  48.01183            2                .00000
Mantel-Haenszel test for          47.18692            1                .00000
      linear association

Minimum Expected Frequency -    34.902
```

For example, from Figure 5.1, the estimated probability of an individual being happy is $1309/1505$, and the estimated probability of an individual being married is $854/1505$. If marital status and happiness are independent, the probability of being a happy, married person is estimated to be

$$\frac{1309}{1505} \times \frac{854}{1505} = 0.4935$$

Equation 5.8

The expected number of happy, married people in a sample of 1505 is then

$$0.4935 \times 1505 = 742.78 \qquad \textbf{Equation 5.9}$$

This is the value displayed in the expected-count column in Figure 5.5. Since the independence model is not saturated, the observed and expected counts are no longer equal, as was the case in Figure 5.2.

Chi-square Goodness-of-Fit Tests

The test of the hypothesis that a particular model fits the observed data can be based on the familiar **Pearson chi-square statistic**, which is calculated as

$$\chi^2 = \sum_i \frac{(O_i - E_i)^2}{E_i} \qquad \textbf{Equation 5.10}$$

where the subscripts i and j include all cells in the table. An alternative statistic is the **likelihood-ratio chi-square**, which is calculated as

$$G^2 = 2\sum_i O_i \ln\left(\frac{O_i}{E_i}\right) \qquad \textbf{Equation 5.11}$$

For large sample sizes, these statistics are equivalent. The advantage of the likelihood-ratio chi-square statistic is that it, like the total sums of squares in analysis of variance, can be subdivided into interpretable parts that add up to the total (see "Partitioning the Chi-square Statistic" on p. 157).

Figure 5.7 shows that the value of the Pearson chi-square statistic is 48.82 and the likelihood-ratio chi-square is 48.01. The degrees of freedom associated with a particular model equal the number of cells in the table minus the number of independent parameters in the model. In this example, there are six cells and four independent parameters to be estimated (the grand mean, $\lambda_{\text{yes}}^{\text{happy}}$, $\lambda_{\text{married}}^{\text{marital}}$, $\lambda_{\text{split}}^{\text{marital}}$), so there are two degrees of freedom. (There are only four independent parameters because of the constraint that parameter estimates must sum to 0 over the categories of a variable. Therefore, the value for one of the categories is determined by the values of the others and is not, in a statistical sense, independent.)

Because the observed significance level associated with both chi-square statistics is very small (less than 0.0005), the independence model is rejected. Note that both the Pearson and likelihood-ratio chi-square statistics displayed for the independence model are the same as the chi-square value displayed by the SPSS Crosstabs procedure, as shown in Figure 5.6.

Figure 5.7 Chi-square goodness-of-fit test

```
Goodness-of-fit test statistics

   Likelihood ratio chi square =     48.01186     DF = 2  P =  .000
            Pearson chi square =     48.81639     DF = 2  P =  .000
```

Residuals

Another way to assess how well a model fits the data is to examine the differences between the observed and expected cell counts based on the model. If the model fits the observed data well, these differences, called **residuals**, should be fairly small in value and should not have any discernible pattern. The column labeled *Residual* in Figure 5.5 shows the differences between the observed and expected counts in each cell. For example, 787 individuals were found to be very happy and married, while 742.78 are expected to fall into this category if the independence model is correct. The residual is $787 - 742.78 = 44.22$.

As in regression analysis, it is useful to standardize the residuals by dividing them by an estimate of the standard deviation of the observed count—in this case, the square root of the expected cell count. For example, the standardized residual for the first cell in Figure 5.5 is

$$\frac{44.2}{\sqrt{742.8}} = 1.6$$ **Equation 5.12**

This value is displayed in the column labeled *Std. Resid.* in Figure 5.5. If the model is adequate, the standardized residuals are approximately normally distributed with a mean of 0 and a standard deviation of less than 1. Standardized residuals greater than 1.96 or less than -1.96 suggest important discrepancies, since they are unlikely to occur if the model is adequate. Particular combinations of cells with large standardized residuals may suggest which other models might be more appropriate.

The same types of diagnostics for residuals used in regression analysis can be used in loglinear models. (For more information on regression, see the Base system documentation.) Figure 5.8 is a scatterplot matrix of the observed and expected counts and the standardized residuals. If the model is adequate, there should be no discernible pattern in the plot. Patterns suggest that the chosen loglinear model, or the loglinear representation in general, may not be appropriate for the data.

Figure 5.8 Scatterplot matrix of observed and expected counts and standardized residuals

Observed Counts

Expected Counts

Std. Residual

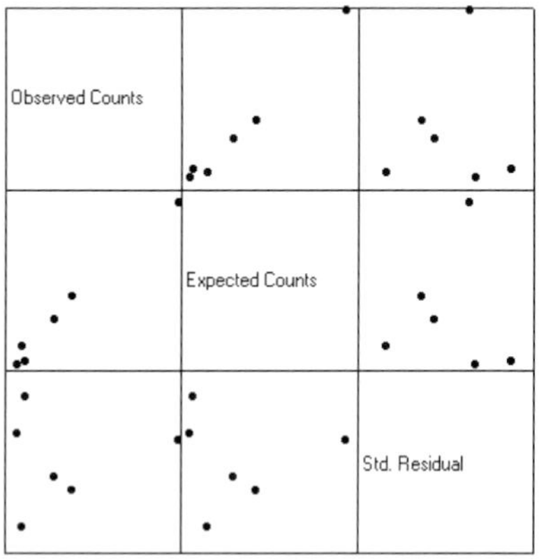

If the standardized residuals are normally distributed, the normal probability plot (Figure 5.9) should be approximately linear. In this plot, the standardized residuals are plotted against expected residuals from a normal distribution.

Figure 5.9 Plot of standardized residuals versus expected residuals

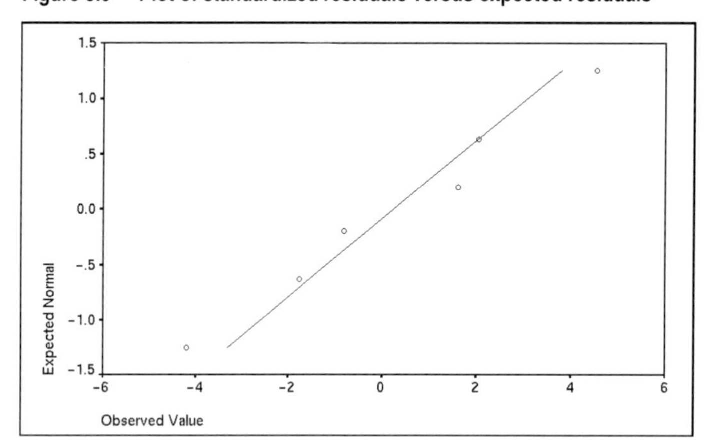

Hierarchical Models

A saturated loglinear model contains all possible effects. For example, a saturated model for a two-way table contains terms for the row main effects, the column main effects, and their interaction. Different models can be obtained by deleting terms from a saturated model. The independence model is derived by deleting the interaction effect. Although it is possible to delete any particular term from a model, in loglinear analysis attention is often focused on a special class of models called **hierarchical models**.

In a hierarchical model, if a term exists for the interaction of a set of variables, there must be lower-order terms for all possible combinations of these variables. For a two-variable model, this means that the interaction term can be included only if both main effects are present. For a three-variable model, if the term λ^{ABC} is included in a model, the terms λ^{A}, λ^{B}, λ^{C}, λ^{AB}, λ^{BC}, and λ^{AC} must also be included.

To describe a hierarchical model, it is sufficient to list the highest-order terms in which variables appear. This is called the **generating class** of a model. For example, the specification A*B*C indicates that a model contains the term λ^{ABC} and all its lower-order relations. (Terms are "relatives" if all variables that are included in one term are also included in the other. For example, the term λ^{ABCD} is a higher-order relative of the terms λ^{ABC}, λ^{BCD}, λ^{ACD}, λ^{ABD}, as well as all other lower-order terms involving variables *A*, *B*, *C*, or *D*. Similarly, λ^{AB} is a lower-order relative of both λ^{ABC} and λ^{ABD}.) The model

$$\ln(m_{ijk}) = \mu + \lambda_i^A + \lambda_j^B + \lambda_k^C + \lambda_{ij}^{AB}$$ **Equation 5.13**

can be represented by the generating class (A*B)(C), since *AB* is the highest-order term in which *A* and *B* occur, and *C* is included in the model only as a main effect.

Model Selection

Even if attention is restricted to hierarchical models, many different models are possible for a set of variables. How do you choose among them? The same guidelines discussed for model selection in regression analysis apply to loglinear models. (For more information on regression, see the Base system documentation.) A model should fit the data and be substantively interpretable and as simple (parsimonious) as possible. For example, if models with and without higher-order interaction terms fit the data well, the simpler models are usually preferable, since higher-order interaction terms are difficult to interpret.

A first step in determining a suitable model might be to fit a saturated model and examine the standardized values for the parameter estimates. Effects with small estimated values can usually be deleted from a model. Another strategy is to systematically test the contribution to a model made by terms of a particular order. For example, you might fit a model with interaction terms and then with main effects only. The change in the chi-square value between the two models is attributable to the interaction effects.

Partitioning the Chi-square Statistic

In regression analysis, the change in multiple R^2 when a variable is added to a model indicates the additional information conveyed by the variable. Similarly, in loglinear analysis, the decrease in the value of the likelihood-ratio chi-square statistic when terms are added to the model signals their contribution to the model. (Remember that R^2 increases when additional variables are added to a model, since large values of R^2 are associated with good models. Chi-square decreases when terms are added, since small values of chi-square are associated with good models.)

As an example, consider the happiness and marital status data when two additional variables, total income in 1982 and the condition of one's health, are included. Three designs generated by custom models will be considered. The statistics for a main-effects-only model are shown in Figure 5.10. The design in Figure 5.11 contains main effects and second-order interactions only. The design in Figure 5.12 contains all terms except the four-way interaction of *happy*, *marital*, *income82*, and *health*.

Figure 5.10 Goodness-of-fit statistics for main effects

```
DESIGN 1 has generating class

    HAPPY
    HEALTH
    INCOME82
    MARITAL

The Iterative Proportional Fit algorithm converged at iteration 2.
The maximum difference between observed and fitted marginal totals is     .000
and the convergence criterion is     .269
- - - - - - - - - - - - - - - - - - - - - - - - - - - - - - - - - - - - - - - - - -
Goodness-of-fit test statistics

    Likelihood ratio chi square =    404.07144    DF = 29  P =  .000
             Pearson chi square =    592.88559    DF = 29  P =  .000
```

The first design (Figure 5.10) has a large chi-square value and an observed significance level of less than 0.0005, so it definitely does not fit well. To judge the adequacy of the other two designs, consider the changes in the chi-square goodness-of-fit statistic as terms are removed from the model.

Figure 5.11 Goodness-of-fit statistics for second-order interactions

```
DESIGN 1 has generating class

    HAPPY*HEALTH
    HAPPY*INCOME82
    HAPPY*MARITAL
    HEALTH*INCOME82
    HEALTH*MARITAL
    INCOME82*MARITAL

The Iterative Proportional Fit algorithm converged at iteration 5.
The maximum difference between observed and fitted marginal totals is     .246
and the convergence criterion is     .269

- - - - - - - - - - - - - - - - - - - - - - - - - - - - - - - - - - - - - - - - -

 Goodness-of-fit test statistics

    Likelihood ratio chi square =     12.60526    DF = 16  P =  .701
             Pearson chi square =     12.32150    DF = 16  P =  .722
```

Figure 5.12 Goodness-of-fit statistics for third-order interactions

```
DESIGN 1 has generating class

    HAPPY*HEALTH*INCOME82
    HAPPY*HEALTH*MARITAL
    HAPPY*INCOME82*MARITAL
    HEALTH*INCOME82*MARITAL

The Iterative Proportional Fit algorithm converged at iteration 3.
The maximum difference between observed and fitted marginal totals is     .176
and the convergence criterion is     .269

- - - - - - - - - - - - - - - - - - - - - - - - - - - - - - - - - - - - - - - -

 Goodness-of-fit test statistics

    Likelihood ratio chi square =      3.99394    DF = 4  P =  .407
             Pearson chi square =      3.97293    DF = 4  P =  .410
```

For a saturated model, the value of the chi-square statistic is always 0. Eliminating the fourth-order interaction (Figure 5.12) results in a likelihood-ratio chi-square value of 3.99. The change in chi-square from 0 to 3.99 is attributable to the fourth-order interaction. The change in the degrees of freedom between the two models equals 4, since a saturated model has 0 degrees of freedom and the third-order interaction model has 4. The change in the chi-square value can be used to test the hypothesis that the fourth-order interaction term is 0. If the observed significance level for the change is small, the hypothesis that the fourth-order term is 0 is rejected, since this indicates that the model without the fourth-order term does not fit well. A chi-square value of 3.99 has an observed significance of 0.41, so the hypothesis that the fourth-order term is 0 is not rejected.

The likelihood-ratio chi-square value for the second-order interaction model (Figure 5.11) is 12.61. This value provides a test of the hypothesis that all third- and fourth-order interaction terms are 0. The difference between 12.61 and 3.99 (8.62) provides a test of the hypothesis that all third-order terms are 0. In general, the test of the hypothesis that the *k*th order terms are 0 is based on

$$\chi^2 = \chi^2_{k-1} - \chi^2_k \quad \textbf{Equation 5.14}$$

where χ^2_k is the value for the model that includes the *k*th-order effect or effects, and χ^2_{k-1} is the chi-square value for the model without the *k*th-order effects.

The SPSS Model Selection Loglinear Analysis procedure calculates tests of two types of hypotheses: the hypothesis that all *k*th- and higher-order effects are 0 and the hypothesis that the *k*th-order effects are 0. Figure 5.13 contains the test for the hypothesis that *k*th- and higher-order effects are 0.

Figure 5.13 Tests that k-way and higher-order effects are 0

Tests that K-way and higher order effects are zero.

K	DF	L.R. Chisq	Prob	Pearson Chisq	Prob	Iteration
4	4	3.994	.4069	3.973	.4097	3
3	16	12.605	.7014	12.317	.7219	5
2	29	404.071	.0000	592.885	.0000	2
1	35	2037.780	.0000	2938.739	.0000	0

The first line of Figure 5.13 is a test of the hypothesis that the fourth-order interaction is 0. Note that the likelihood-ratio chi-square value of 3.99 is the same as the value displayed for design 3 in Figure 5.10. This is the goodness-of-fit statistic for a model without the fourth-order interaction. Similarly, the entry for the k of 3 is the goodness-of-fit test for a model without third- and fourth-order effects, as shown in design 2 in Figure 5.10. The last line, $k = 1$, corresponds to a model that has no effects except the grand mean. That is, the expected value for all cells is the same—the average of the logs of the observed frequencies in all cells.

The column labeled *Prob* in Figure 5.13 gives the observed significance levels for the tests that kth- and higher-order effects are 0. Small observed significance levels indicate that the hypothesis that terms of particular orders are 0 should be rejected. Note in Figure 5.13 that the hypotheses that all effects are 0 and that second-order and higher effects are 0 should be rejected. Since the observed significance level for the test that third- and higher-order terms are 0 is large (0.70), the hypothesis that third- and fourth-order interactions are 0 should not be rejected. Thus, it appears that a model with first- and second-order effects is adequate to represent the data.

It is sometimes also of interest to test whether interaction terms of a particular order are 0. For example, rather than asking if all effects greater than two-way are 0, the question is whether two-way effects are 0. Figure 5.14 gives the tests for the hypothesis that k-way effects are 0.

Figure 5.14 Tests that k-way effects are 0

Tests that K-way effects are zero.

K	DF	L.R. Chisq	Prob	Pearson Chisq	Prob	Iteration
1	6	1633.708	.0000	2345.854	.0000	0
2	13	391.466	.0000	580.568	.0000	0
3	12	8.612	.7357	8.345	.7576	0
4	4	3.994	.4069	3.973	.4097	0

From Figure 5.13, the likelihood-ratio chi-square for a model with only the mean is 2037.78. The value for a model with first-order effects is 404.07. The difference between these two values, 1633.71, is displayed on the first line of Figure 5.14. The difference is an indication of how much the model improves when first-order effects are included. The observed significance level for a chi-square value of 1634 with six degrees of freedom ($35 - 29$) is small, less than 0.00005, so the hypothesis that first-order effects are 0 is rejected. The remaining entries in Figure 5.14 are obtained in a similar fashion. The test that third-order effects are 0 is the difference between a model without third-order terms (χ^2_{LR} = 12.61) and a model with third-order terms (χ^2_{LR} = 3.99). The resulting chi-square value of 8.61 with 12 degrees of freedom has a large observed significance level (0.76), so the hypothesis that third-order terms are 0 is not rejected.

Testing Individual Terms in the Model

The two tests described in the previous section provide an indication of the collective importance of effects of various orders. They do not, however, test the individual terms. That is, although the overall hypothesis that second-order terms are 0 may be rejected, that does not mean that every second-order effect is present.

One strategy for testing individual terms is to fit two models differing only in the presence of the effect to be tested. The difference between the two likelihood-ratio chi-square values, sometimes called the **partial chi-square**, also has a chi-square distribution and can be used to test the hypothesis that the effect is 0. For example, to test that the *happy*-by-*marital*-by-*income82* effect is 0, a model with all three-way interactions can be fit. From Figure 5.10, the likelihood-ratio chi-square value for this model is 3.99. When a model without the *happy*-by-*marital*-by-*income82* effect is fit (Figure 5.15), the likelihood ratio is 7.36. Thus, the partial chi-square value with four (8 – 4) degrees of freedom is 3.37 (7.36 – 3.99).

Figure 5.15 Model without happy by marital by income82

```
DESIGN 1 has generating class

    HAPPY*MARITAL*HEALTH
    MARITAL*INCOME82*HEALTH
    HAPPY*INCOME82*HEALTH


The Iterative Proportional Fit algorithm converged at iteration 4.
The maximum difference between observed and fitted marginal totals is     .050
and the convergence criterion is     .269

- - - - - - - - - - - - - - - - - - - - - - - - - - - - - - - - - - - - - - - -


 Goodness-of-fit test statistics

    Likelihood ratio chi square =      7.36366    DF = 8  P =  .498
             Pearson chi square =      7.50538    DF = 8  P =  .483
```

Figure 5.16 contains the partial chi-square values and their observed significance levels for all effects in the *happy*-by-*marital*-by-*health*-by-*income82* table. Note that the observed significance levels are large for all three-way effects, confirming that first- and second-order effects are sufficient to represent the data. The last column indicates the number of iterations required to achieve convergence.

Figure 5.16 Partial chi-squares for happy by marital by health by income82

Tests of PARTIAL associations.

Effect Name	DF	Partial Chisq	Prob	Iter
HAPPY*MARITAL*INCOME82	4	3.370	.4979	4
HAPPY*MARITAL*HEALTH	2	.458	.7955	4
HAPPY*INCOME82*HEALTH	2	.955	.6205	3
MARITAL*INCOME82*HEALTH	4	3.652	.4552	5
HAPPY*MARITAL	2	15.050	.0005	5
HAPPY*INCOME82	2	16.120	.0003	5
MARITAL*INCOME82	4	160.738	.0000	4
HAPPY*HEALTH	1	55.696	.0000	5
MARITAL*HEALTH	2	8.391	.0151	5
INCOME82*HEALTH	2	35.600	.0000	4
HAPPY	1	849.537	.0000	2
MARITAL	2	343.395	.0000	2
INCOME82	2	86.115	.0000	2
HEALTH	1	354.662	.0000	2

Model Selection Using Backward Elimination

As in regression analysis, another way to arrive at a "best" model is by using variable-selection algorithms. Forward selection adds effects to a model, while backward elimination starts with all effects in a model and then removes those that do not satisfy the criterion for remaining in the model. Since backward elimination appears to be the better procedure for model selection in hierarchical loglinear models (Benedetti & Brown, 1978), it is the only procedure described here.

The initial model for backward elimination need not be saturated but can be any hierarchical model. At the first step, the effect whose removal results in the least significant change in the likelihood-ratio chi-square is eligible for elimination, provided that the observed significance level is larger than the criterion for remaining in the model. To ensure a hierarchical model, only effects corresponding to the generating class are examined at each step. For example, if the generating class is *marital*happy*income*health*, the first step examines only the fourth-order interaction.

Figure 5.17 shows output at the first step. Elimination of the fourth-order interaction results in a chi-square change of 3.99, which has an associated significance level of 0.41. Since this significance level is not less than 0.05 (the default criterion for remaining in the model), the effect is removed. The new model has all three-way interactions as its generating class.

Figure 5.17 First step in backward elimination

```
Backward Elimination (p = .050) for DESIGN 1 with generating class

  HAPPY*MARITAL*INCOME82*HEALTH

 Likelihood ratio chi square =      .00000    DF = 0  P = 1.000

- - - - - - - - - - - - - - - - - - - - - - - - - - - - - - - - - - - - - - - - - - -

If Deleted Simple Effect is                     DF   L.R. Chisq Change    Prob  Iter

 HAPPY*MARITAL*INCOME82*HEALTH                   4               3.994   .4069     3

Step 1

  The best model has generating class

      HAPPY*MARITAL*INCOME82
      HAPPY*MARITAL*HEALTH
      HAPPY*INCOME82*HEALTH
      MARITAL*INCOME82*HEALTH

  Likelihood ratio chi square =     3.99367    DF = 4  P =  .407
```

Figure 5.18 contains the statistics for the second step. The effects eligible for removal are all three-way interactions. The *happy*marital*health* interaction has the largest observed significance level for the change in the chi-square if it is removed, so it is eliminated from the model. The likelihood-ratio chi-square for the resulting model is 4.45.

Figure 5.18 Statistics used to eliminate second effect

```
If Deleted Simple Effect is                     DF   L.R. Chisq Change    Prob  Iter

 HAPPY*MARITAL*INCOME82                          4               3.370   .4979     4
 HAPPY*MARITAL*HEALTH                            2                .458   .7955     4
 HAPPY*INCOME82*HEALTH                           2                .955   .6205     3
 MARITAL*INCOME82*HEALTH                         4               3.652   .4552     5

Step 2

  The best model has generating class

      HAPPY*MARITAL*INCOME82
      HAPPY*INCOME82*HEALTH
      MARITAL*INCOME82*HEALTH

  Likelihood ratio chi square =     4.45127    DF = 6  P =  .616
```

At the next three steps, the remaining third-order interactions are removed from the model. The sixth step begins with all second-order items in the model, as shown in Figure 5.19.

Figure 5.19 Sixth step in backward elimination

```
If Deleted Simple Effect is                  DF   L.R. Chisq Change    Prob  Iter

 HAPPY*HEALTH                                 1              55.697   .0000     5
 HAPPY*MARITAL                                2              15.051   .0005     5
 HAPPY*INCOME82                               2              16.120   .0003     5
 MARITAL*INCOME82                             4             160.738   .0000     5
 MARITAL*HEALTH                               2               8.392   .0151     5
 INCOME82*HEALTH                              2              35.601   .0000     4

Step 6

  The best model has generating class

      HAPPY*HEALTH
      HAPPY*MARITAL
      HAPPY*INCOME82
      MARITAL*INCOME82
      MARITAL*HEALTH
      INCOME82*HEALTH

  Likelihood ratio chi square =     12.60482    DF = 16   P =  .701

- - - - - - - - - - - - - - - - - - - - - - - - - - - - - - - - - - - - - - - -

The final model has generating class

    HAPPY*HEALTH
    HAPPY*MARITAL
    HAPPY*INCOME82
    MARITAL*INCOME82
    MARITAL*HEALTH
    INCOME82*HEALTH
```

Since the observed significance level for removal of any of the two-way interactions is smaller than 0.05, no more effects are removed from the model. The final model contains all second-order interactions and has a chi-square value of 12.60. This is the same model suggested by the partial-association table.

At this point, it is a good idea to examine the residuals to see if any anomalies are apparent. The largest standardized residual is 1.34, which suggests that there are no cells with bad fits. The residual plots also show nothing suspicious.

SPSS Loglinear Analysis Procedures

SPSS provides three procedures for loglinear analysis. For hierarchical models, the Model Selection Loglinear Analysis procedure, described in this chapter, is most efficient, but it produces parameter estimates for saturated models only. The General Loglinear Analysis procedure can estimate parameters for a variety of models, including nonhierarchical models. For logit models, use the Logit Loglinear Analysis procedure. The General and Logit Loglinear Analysis procedures are discussed in Chapter 6 and Chapter 7.

How to Obtain a Model Selection Loglinear Analysis

The Model Selection Loglinear Analysis procedure fits hierarchical loglinear models to multidimensional contingency tables using an iterative proportional-fitting algorithm. You can use forced entry or backward elimination to build models. Parameter estimates are available for saturated models.

The minimum specifications are:

- Two or more numeric factor variables.
- Factor levels you want to tabulate in the analysis.

To obtain a model selection loglinear analysis, from the menus choose:

Statistics
 Loglinear ▸
  Model Selection...

This opens the Model Selection Loglinear Analysis dialog box, as shown in Figure 5.20.

Figure 5.20 Model Selection Loglinear Analysis dialog box

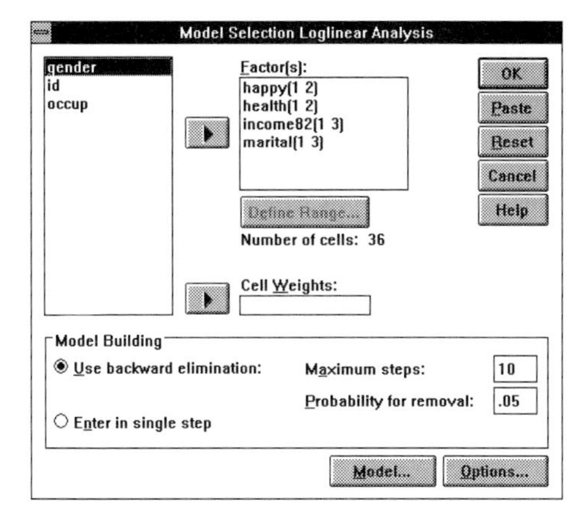

The numeric variables in your data file appear on the source list.

Factor(s). Select two or more categorical variables to use in the analysis. For each factor variable, you must also define the factor levels you want to tabulate (see "Defining Factor Ranges" on p. 166). You can select a maximum of 10 factor variables.

Cell Weights. By default, a weight of 1 is used for each cell. To provide your own cell weights for use in an unsaturated model, select a variable whose values are the cell weights. A cell is weighted by the mean weight for cases in the cell. You can use this feature to define logically empty cells (structural zeros). Do not use this facility to weight aggregate data; instead, use the Weight Cases facility on the Data menu (for more information on case weighting, see the Base system documentation).

Model Building. You can choose one of the following methods for building the model:

- **Use backward elimination.** Selects factors using backward elimination. All terms are entered into the model and then are tested for removal. This is the default. The following options are available for backward elimination:

  Maximum steps. By default, the maximum number of steps for backward elimination is 10. To override the default, enter a positive integer.

  Probability for removal. By default, a variable is removed from the model if the probability of the change in the likelihood ratio is greater than 0.05. To override the default probability for removal, enter a value greater than 0 and less than 1.

- **Enter in single step.** Enters all factors into the model simultaneously.

Defining Factor Ranges

To indicate which factor levels you want to tabulate, highlight one or more factor variables and click on Define Range... in the Model Selection Loglinear Analysis dialog box. This opens the Loglinear Analysis Define Range dialog box, as shown in Figure 5.21.

Figure 5.21 Loglinear Analysis Define Range dialog box

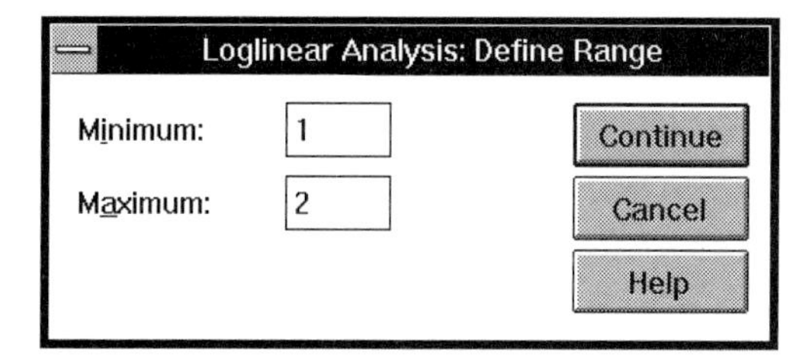

Enter integer values corresponding to the lowest and highest categories you want to use in the analysis. SPSS tabulates each integer category in the inclusive range (non-integer data values in the range are truncated before tabulation, so cases with the same integer portion are combined into the same category). Cases with values outside the range are excluded from analysis. To avoid empty cells, values of a factor variable should be consecutive. If a variable has empty categories, you can use the Automatic Recode facility on the Transform menu to create consecutive categories (for more information on recoding variables, see the Base system documentation).

Model Specification

By default, SPSS analyzes a saturated model. To specify an unsaturated model, click on Model... in the Model Selection Loglinear Analysis dialog box. This opens the Loglinear Analysis Model dialog box, as shown in Figure 5.22.

Figure 5.22 Loglinear Analysis Model dialog box

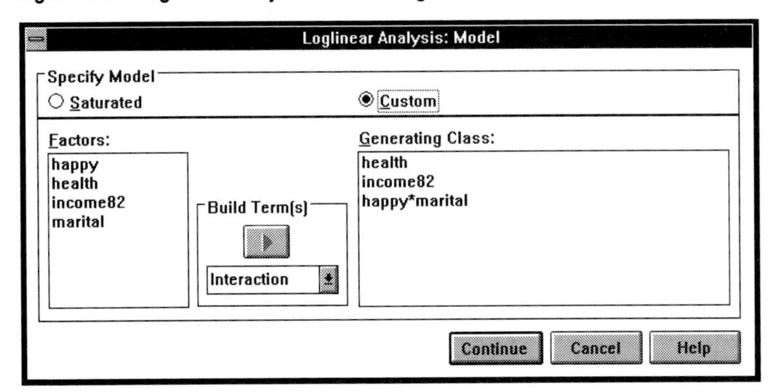

Specify Model. You can choose one of the following models:

❍ **Saturated**. The saturated model contains all main effects and interactions for factor variables. This is the default. For example, if you have factor variables *happy*, *health*, and *marital*, the model contains the third-order interaction term *happy*health*marital*, second-order effects *happy*health*, *health*marital*, and *happy*marital*, and main effects *happy*, *health*, and *marital*.

❍ **Custom**. Select this item to obtain an unsaturated model. If you request a custom model, you must build a generating class for the model.

For custom models, if you remove a factor in the main dialog box, any terms in the model that contain that factor will also be deleted.

Generating Class. A generating class is a list of the highest-order terms in which factors appear (see "Hierarchical Models" on p. 156). SPSS builds a hierarchical model containing the terms that define the generating class and all lower-order relatives.

To add an element to the generating class, select one or more factors. If you don't want to create the highest-order interaction for the factors, select an alternate method of treating the factors on the Build Term(s) drop-down list. To add more terms to the generating class, repeat this process. Each element of the generating class must be unique. Do not specify lower-order relatives in the generating class.

For example, to define the generating class *(happy*marital)*(*income82)(health)*, first highlight variables *happy* and *marital*, select Interaction (the default) from the Build Term(s) drop-down list, and click on [▶]. Next, highlight *income82* and *health* (together or separately), select Main effects from the Build Term(s) drop-down list, and click on [▶]. (In this example, SPSS analyzes a model containing *happy*marital*, *happy*, *marital*, *income82*, and *health*).

Macintosh: Use ⌘-click to select multiple factors.

- **Build Term(s)**. You can treat factors as main effects or interactions. If you request an interaction whose order is higher than the number of factors, SPSS creates a term for the highest-order interaction possible for the factors. If only one factor is selected, SPSS adds a main-effect term for that factor to the generating class. You can choose one of the following alternatives:

  Interaction. Creates the highest-level interaction. This is the default.

  Main effects. Creates a main-effect term for each variable.

  All 2-way. Creates all possible two-way interactions.

  All 3-way. Creates all possible three-way interactions.

  All 4-way. Creates all possible four-way interactions.

  All 5-way. Creates all possible five-way interactions.

Options

To obtain optional statistics or plots, or to control estimation criteria, click on Options... in the Model Selection Loglinear Analysis dialog box. This opens the Loglinear Analysis Options dialog box, as shown in Figure 5.23.

Figure 5.23 Loglinear Analysis Options dialog box

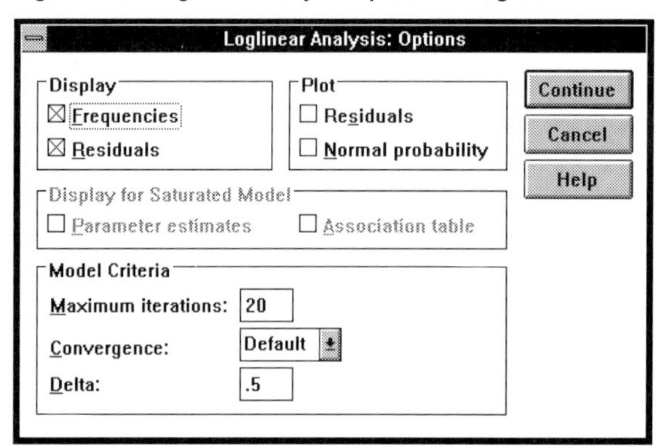

Display. You can choose one or both of the following displays:

- **Frequencies**. Observed and expected cell frequencies. Frequencies are displayed by default. To suppress frequencies, deselect this item.
- **Residuals**. Standardized and unstandardized residuals. Residuals are displayed by default. To suppress residuals, deselect this item.

Plot. Plots are available for custom models only. (Plots are not available for saturated models, since all residuals are 0.) You can choose one or both of the following plots:

- ❑ **Residuals.** Produces a scatterplot matrix of standardized residuals against observed and expected cell counts.
- ❑ **Normal probability.** Displays normal and detrended normal probability plots of standardized residuals.

Display for Saturated Model. For saturated models, you can choose one or both of the following displays:

- ❑ **Parameter estimates.** Displays parameter estimates, standard errors, and confidence intervals. SPSS does not display parameter estimates if the table of expected frequencies has any zero cells (either structural or sampling) after delta is added.
- ❑ **Association table.** Displays tests of partial association.

Model Criteria. SPSS uses an iterative proportional-fitting algorithm to obtain parameter estimates. You can override one or more of the following estimation criteria:

Maximum iterations. By default, a maximum of 20 iterations is performed. To specify a different maximum, enter a positive integer.

Convergence. Iteration terminates if the change in expected frequency is less than or equal to the convergence value for at least one cell. The default convergence value is 10^{-3} times the largest observed cell frequency, or 0.25, whichever is larger. To override the default, select an alternate convergence value from the drop-down list.

Delta. The value added to all cell frequencies for saturated models. The default delta value is 0.5. To override the default delta, enter a value between 0 and 1.

Additional Features Available with Command Syntax

You can customize your model selection loglinear analysis if you paste your selections into a syntax window and edit the resulting HILOGLINEAR command syntax. (For information on syntax windows, see the SPSS Base system documentation.) Additional features include:

- The ability to specify cell weights in matrix form (using the CWEIGHT subcommand).
- The ability to generate analyses of several models with a single command (using the DESIGN subcommand).

See the Syntax Reference section of this manual for command syntax rules and for complete HILOGLINEAR command syntax.

6 General Loglinear Analysis

Loglinear models are statistical techniques used to examine relationships among categorical variables. For example, using loglinear models you can look at the relationships among job satisfaction, highest degree earned, and marital status. Or you can examine the relationships among family size, socioeconomic status, and religious affiliation. In a general loglinear model, the dependent variable is the number of cases in a cell of a multiway crosstabulation. The independent variables are the categorical variables. Continuous variables (covariates) can also be incorporated into the model.

Logit models, a special type of loglinear model in which one or more of the categorical variables are singled out as the dependent variable, are discussed in Chapter 7. Chapter 5 illustrates hierarchical loglinear model building. The general loglinear model facility described in this chapter can be used for both hierarchical and nonhierarchical loglinear models. It has many options not available in the Model Selection facility. Different types of parameter estimates are obtained from the two facilities, so the discussion of the loglinear model and corresponding parameter estimates is somewhat different in the two chapters.

Voting Behavior

As an example of what's involved in building a loglinear model, consider Figure 6.1, which is a crosstabulation of whether a person voted in the 1992 presidential election by whether or not they have a college degree. The data are a subset from the 1993 General Social Survey.

Figure 6.1 Crosstabulation of voting by degree

```
VOTE92  DID R VOTE IN 1992 ELECTION  by  DEGREE2  college/junior college degree

                     DEGREE2         Page 1 of 1
             Count  |
           Col Pct  |Yes       No
                    |                      Row
                    |     1.00|     2.00|  Total
VOTE92      --------+---------+---------+
                 1  |   372   |   659   |  1031
  VOTED             |  88.2   |  64.1   |  71.1
                    +---------+---------+
                 2  |    50   |   369   |   419
  DID NOT VOTE      |  11.8   |  35.9   |  28.9
                    +---------+---------+
            Column     422      1028      1450
             Total    29.1      70.9     100.0

Number of Missing Observations:  2
```

A loglinear model can be written for the natural log of the predicted number of cases, m_{ij}, in each of the four cells.

$$\ln(m_{11}) = \mu + \lambda^{\text{voter}} + \lambda^{\text{with degree}} + \lambda^{\text{voter with degree}}$$

$$\ln(m_{12}) = \mu + \lambda^{\text{voter}} + \lambda^{\text{without degree}} + \lambda^{\text{voter without degree}}$$

$$\ln(m_{21}) = \mu + \lambda^{\text{nonvoter}} + \lambda^{\text{with degree}} + \lambda^{\text{nonvoter with degree}}$$

$$\ln(m_{22}) = \mu + \lambda^{\text{nonvoter}} + \lambda^{\text{without degree}} + \lambda^{\text{nonvoter without degree}}$$

Equation 6.1

In the above equations the term μ is comparable to the intercept term in a regression model. It tells you the predicted number of cases in a cell if the values of all of the lambda (λ) parameters are 0. The lambda parameters represent the increments or decrements from the base value for particular combinations of values of the row and column variables.

Each individual category of the row and column variables has an associated lambda. There is a lambda for voters and a lambda for nonvoters. Similarly, there is a lambda for those with college degrees and for those without college degrees. There are also lambda parameters for all possible combinations of values of the row and column variables. They are called **interactions**. In this example, there are four voter-by-degree interactions:

$\lambda^{\text{voter without degree}}$

$\lambda^{\text{voter with degree}}$

$\lambda^{\text{nonvoter with degree}}$

$\lambda^{\text{nonvoter without degree}}$

For example, $\lambda^{\text{voter with degree}}$ represents how many more cases measured on the log scale (or fewer cases, if λ is negative) are expected in the cell than would be predicted from just the μ, λ^{voter}, and $\lambda^{\text{with degree}}$ parameters.

A model that contains lambda parameters for all of the categories of the independent variables as well as for all possible combinations of the categories of the independent variables is called a **saturated model**. A saturated model exactly reproduces the observed cell counts. The model in Equation 6.1 is a saturated model, since it contains terms for the values of voting and education as well as their interaction. Without the interaction terms, the model in Equation 6.1 would no longer be saturated. It would be called an **unsaturated model**.

Estimating the Parameters

Whenever you build a model that involves categorical variables, you must establish a frame of reference for them. For example, if you have four regions of the country in the model, you can't make an absolute statement about the effect of region 1. What you can do is compare region 1 to other regions. You can say sales in region 1 are higher than the average of sales across all regions. Or that sales in region 1 are higher than sales in region 4. Any comment about region 1 depends on a frame of reference. The same is true for the lambda parameters in a loglinear model. The lambda parameters must have a frame of reference. The General Loglinear Analysis procedure selects a frame of reference by setting some of the lambda parameters to 0. For example, if the lambda parameter for nonvoters is set to 0, then the lambda parameter for voters uses nonvoters as a frame of reference. Similarly, in the region example, if the lambda for region 4 is set to 0, then all other regions are compared to region 4. Usually, effects involving the last category of a variable are set to 0. If there are cells without any cases, additional parameters may also be set to 0. Parameters which are set to 0 are called **aliased** or **redundant parameters**.

Figure 6.2 shows how the 9 parameters in Equation 6.1 are numbered.

Figure 6.2 Parameter numbers

```
Correspondence Between Parameters and Terms of the Design

Parameter   Aliased  Term

        1            Constant
        2            [VOTE92 = 1]
        3        x   [VOTE92 = 2]
        4            [DEGREE2 = 1.00]
        5        x   [DEGREE2 = 2.00]
        6            [VOTE92 = 1]*[DEGREE2 = 1.00]
        7        x   [VOTE92 = 1]*[DEGREE2 = 2.00]
        8        x   [VOTE92 = 2]*[DEGREE2 = 1.00]
        9        x   [VOTE92 = 2]*[DEGREE2 = 2.00]

Note: 'x' indicates an aliased (or a redundant) parameter.
      These parameters are set to zero.
```

Parameters that are set to 0 are marked with an x. You can see that

- Parameter 2 corresponds to λ^{voter} (coded 1 in the data).
- Parameter 4 corresponds to $\lambda^{\text{with degree}}$ (coded 1 in the data).
- Parameter 6 corresponds to the interaction parameter $\lambda^{\text{voter with degree}}$.

All other lambda parameters, those involving the last category of a variable, are set to 0. The parameter estimates, their standard errors, and associated statistics are shown in Figure 6.3.

Figure 6.3 Parameter estimates

	Parameter	Estimate	SE	Z-value	Asymptotic 95% CI Lower	Upper
	1	5.9108	.0521	113.54	5.81	6.01
voter	2	.5799	.0650	8.92	.45	.71
	3	.0000	.	.	.	.
with degree	4	-1.9988	.1507	-13.26	-2.29	-1.70
	5	.0000	.	.	.	.
voter with degree	6	1.4269	.1641	8.70	1.11	1.75
	7	.0000	.	.	.	.
	8	.0000	.	.	.	.
	9	.0000	.	.	.	.

Calculating Expected Values

If you substitute the values of the parameter estimates in Figure 6.3 into the right side of Equation 6.1, you will obtain the natural logs of the observed cell counts. For example,

$$\begin{aligned}
\ln(372) &= \mu + \lambda^{\text{voter}} + \lambda^{\text{with degree}} + \lambda^{\text{voter with degree}} \\
&= 5.91 + 0.58 - 2.00 + 1.43 \\
\ln(659) &= \mu + \lambda^{\text{voter}} + \lambda^{\text{without degree}} + \lambda^{\text{voter without degree}} \\
&= 5.91 + 0.58 + 0 + 0 \\
\ln(50) &= \mu + \lambda^{\text{nonvoter}} + \lambda^{\text{with degree}} + \lambda^{\text{nonvoter with degree}} \\
&= 5.91 + 0 - 2.00 + 0 \\
\ln(369) &= \mu + \lambda^{\text{nonvoter}} + \lambda^{\text{without degree}} + \lambda^{\text{nonvoter without degree}} \\
&= 5.91 + 0 + 0 + 0
\end{aligned}$$

Equation 6.2

If the model is not saturated, the observed and expected cells will not be equal unless the model fits perfectly. A saturated model always fits perfectly because it has as many non-zero parameters as there are cells; thus, it can always reproduce the cell counts exactly.

Interpreting the Parameter Estimates

Each of the parameter estimates in a loglinear model can be expressed in terms of expected cell counts. If you set all of the parameters that involve nonvoters and no degrees to 0 in Equation 6.1 and then solve for the remaining parameters, you will obtain:

$$\begin{aligned}
\lambda^{\text{voter}} &= \ln\left(\frac{m_{12}}{m_{22}}\right) \\
\lambda^{\text{with degree}} &= \ln\left(\frac{m_{21}}{m_{22}}\right) \\
\lambda^{\text{voter with degree}} &= \ln\left(\frac{m_{11}m_{22}}{m_{12}m_{21}}\right)
\end{aligned}$$

Equation 6.3

In this example, λ^{voter} is the logarithm of the ratio of voters to nonvoters among people without degrees. The ratio of the number of voters to nonvoters is called the **odds** of being a voter. To obtain the odds for a particular outcome, you divide the number of cases with that outcome by the number of cases without that outcome. From Figure 6.1, the odds of having a college degree are 422/1,028. Similarly, the odds of being a voter are 1,031/419.

The odds for an event can also be thought of as the ratio of the probability that an outcome will happen to the probability that it will not happen. (When you calculate the probability of an outcome, you divide the number of cases with that outcome by the total number of cases. You do the same when calculating the probability that an outcome will not occur. Since the same denominator is used for both calculations, it cancels when you take the ratio of the two probabilities. That's why you're left with just the number of cases with the outcome divided by the number of cases without the outcome.)

- Odds close to 1 tell you that the two probabilities are almost equal. For example, an odds of 1 for voting means that a person is equally likely to vote as not to vote.
- An odds greater than 1 tells you that a person is more likely to vote than not to vote.
- An odds less than 1 tells you that a person is less likely to vote than not to vote.

Since in loglinear models the log of the frequencies is modeled, you will often encounter the log of the odds. It's called the **log odds**. An odds of 1 corresponds to a log odds of 0, because the logarithm of 1 is 0.

From the entries for parameter 2 in Figure 6.3, you see that the estimated log odds of being a voter for a person without a degree is 0.58. The ratio of the parameter estimate to its standard error is in the column labeled *Z-value*. For sufficiently large sample sizes, this ratio is normally distributed. The 95% confidence interval is from 0.45 to 0.71. Notice that the confidence interval does not include the value of 0. So you can reject the null hypothesis that the log odds are 0.

Since it is easier to think about odds than log odds, you may want to convert the log odds in Figure 6.3 to odds. You do that by finding e^{λ}. For example, a log odds of 0.58 corresponds to an odds of 1.79, since $e^{0.58}$ is 1.79. This means that people without degrees are almost 80% more likely to vote than not to vote. You can compute a confidence interval for the odds by finding e^{lower} and e^{upper}, where lower and upper are the values for the lower and upper bounds of the confidence interval for the log odds. The 95% confidence interval for the odds of a person without a degree voting are 1.57 ($e^{0.45}$) to 2.03 ($e^{0.71}$).

The lambda for degree (parameter 4) is the log odds that a person who doesn't vote has a degree. Since the value for lambda is negative, it means that for a nonvoter, the odds of having a degree are less than 1. The log odds of –2 translate to an odds of 0.13. That means that nonvoters are much less likely to have degrees than not to have them. For every 13 nonvoters who have a degree, 100 do not.

If you look at the formula for the interaction parameter, you see that it is the logarithm of the ratio of two odds: the odds that a person with a degree votes and the odds that a person without a degree votes. The ratio of two odds is called an **odds ratio**. Its logarithm is called the **log odds ratio**. From Figure 6.3 (parameter 6), the log odds ratio is 1.43, so the odds ratio is $e^{1.43}$, or 4.18. This means that the estimated odds of a person with a degree voting are more than 4 times the odds of a person without a degree voting. If there is no association between voting and degree, you expect this ratio to be close to 1. The 95% confidence interval for the odds ratio is 3.03 to 5.75, which does not include 1, so you can reject the null hypothesis that the odds ratio is 1. Degree and voting behavior do not appear to be independent.

A saturated model reproduces the observed counts perfectly. One of the primary reasons for fitting a saturated model to data is so that you can examine the parameter estimates and see which ones are not significantly different from 0. You may then want to build a model that excludes these parameters. For example, if the voting-with-degree interaction parameter in the previous example was not significant, you could build a model that contained only the main effects for voting and degree.

Fitting an Unsaturated Model

In the previous example, you looked at the relationship between voting and education. Let's see how you would fit a model that includes an additional variable, the sex of the respondent. Consider a model that contains terms for each of the three variables individually (the main effects) and interaction terms for degree and voting. The sex-by-degree, sex-by-vote, and sex-by-degree-by-vote interactions are not included in the model. If sex is independent of degree status and voting behavior, there is no need for these interactions. Figure 6.4 shows the SPSS shorthand for the model. The goodness-of-fit statistics for this model are shown in Figure 6.5.

Figure 6.4 Model and design information

```
Model: Poisson
Design: Constant + DEGREE2 + SEX + VOTE92 + VOTE92*DEGREE2
```

Figure 6.5 Goodness-of-fit statistics for an unsaturated model

Goodness-of-fit Statistics

	Chi-Square	DF	Sig.
Likelihood Ratio	9.8931	3	.0195
Pearson	9.9503	3	.0190

The Pearson chi-square is

$$\chi^2 = \sum \frac{(O-E)^2}{E}$$

Equation 6.4

where O is the observed count in a cell and E is the predicted count based on the model. The summation is taken over all cells in the table that do not have 0 for the expected count. The likelihood ratio chi-square is defined as

$$G^2 = 2\sum O \ln\left(\frac{O}{E}\right)$$

Equation 6.5

For large sample sizes, these statistics are equivalent. The advantage of the likelihood-ratio chi-square is that, like the total sum of squares in analysis of variance, it can be subdivided into interpretable parts that add up to the total (see Chapter 5 for illustrations). The degrees of freedom for both of the statistics is the same. It depends on the number of cells in the table, the number of parameters not set to 0, and the number of cells with expected counts of 0. If the model fits the data well, the observed significance levels for the goodness-of-fit statistics should be large.

From Figure 6.5, you see that the model does not fit the data well. The observed significance level is less than 0.05. One reason the model may not fit the data is that the distribution of college degrees may be different for males and females. That is, there may be a degree-by-sex interaction. You can test this by adding the interaction term for sex and degree to the model. The results from this model are shown in Figure 6.6.

Figure 6.6 Goodness of fit for a model with degree-by-sex term added

Goodness-of-fit Statistics

	Chi-Square	DF	Sig.
Likelihood Ratio	1.2210	2	.5431
Pearson	1.2162	2	.5444

You see that the observed significance level for the goodness-of-fit statistics is now larger, greater than 0.5. So you have no reason to believe that the model does not fit. It appears that the sex-by-vote and sex-by-degree-by-vote interactions are not needed to adequately describe the data. The voting behavior of males and females appears to be similar. The two-way interaction of voting and sex would have to be included in the model if men and women were not equally likely to vote. The three-way interaction, sex-by-degree-by-vote, would have to be included in the model if the relationship of degree to voting was different for males and females.

Whenever you fit a model to data, you want to select a model that has as few parameters as possible and yet fits the data well. The parameter correspondence for the model with the degree-by-sex interaction added is shown in Figure 6.7. If you look at the parameter estimates in Figure 6.8, you see that they are all significantly different from 0, suggesting that there are no additional terms that might be deleted from the model.

Figure 6.7 Parameter correspondence

Parameter	Aliased	Term
1		Constant
2		[DEGREE2 = 1.00]
3	x	[DEGREE2 = 2.00]
4		[SEX = 1]
5	x	[SEX = 2]
6		[VOTE92 = 1]
7	x	[VOTE92 = 2]
8		[DEGREE2 = 1.00]*[SEX = 1]
9	x	[DEGREE2 = 1.00]*[SEX = 2]
10	x	[DEGREE2 = 2.00]*[SEX = 1]
11	x	[DEGREE2 = 2.00]*[SEX = 2]
12		[VOTE92 = 1]*[DEGREE2 = 1.00]
13	x	[VOTE92 = 1]*[DEGREE2 = 2.00]
14	x	[VOTE92 = 2]*[DEGREE2 = 1.00]
15	x	[VOTE92 = 2]*[DEGREE2 = 2.00]

Note: 'x' indicates an aliased (or a redundant) parameter.
These parameters are set to zero.

Figure 6.8 Parameter estimates

				Asymptotic 95% CI	
Parameter	Estimate	SE	Z-value	Lower	Upper
1	5.3938	.0580	92.93	5.28	5.51
2	-2.1515	.1601	-13.44	-2.47	-1.84
3	.0000	.	.	.	.
4	-.3901	.0636	-6.14	-.51	-.27
5	.0000	.	.	.	.
6	.5799	.0650	8.92	.45	.71
7	.0000	.	.	.	.
8	.3427	.1163	2.95	.11	.57
9	.0000	.	.	.	.
10	.0000	.	.	.	.
11	.0000	.	.	.	.
12	1.4269	.1641	8.70	1.11	1.75
13	.0000	.	.	.	.
14	.0000	.	.	.	.
15	.0000	.	.	.	.

(To interpret the parameters in Figure 6.8, you must write out the model for each of the cells in the table, set to 0 the parameters that are assigned values of 0, and then solve for the remaining parameters.)

Examining Residuals

The overall goodness-of-fit test tells you if the model appears to fit the data. It doesn't tell you whether there are particular cells that the model fits poorly or whether there is a systematic lack of fit. To see how well the model fits the individual cells, you can examine several types of residuals. The simplest is the raw residual, which is just the difference between the observed and expected cell frequencies. The size of the raw residual depends not only on how well the model fits but also on the number of cases in a particular cell. For example, a raw residual of 5 may indicate poor fit if the observed number of cases in a cell is 4 but excellent fit if the observed number of cases in the cell is 12,000.

Just as in regression analysis, it's easier to interpret the residuals from a loglinear analysis if they are divided by an estimate of their standard deviation. A **standardized residual** is computed by dividing the raw residual by an estimate of the standard deviation of the observed count. If the model is correct, for a Poisson model the standard deviation of the observed count is the square root of the predicted count. (Poisson is the default distribution; see "Sampling Models" on p. 201 for further discussion.) For a multinomial model, the standard deviation is $\sqrt{E(1 - E/N)}$, where E is the predicted count and N is the total number of cases in the table. Standardized residuals are sometimes called Pearson residuals because for a Poisson model, if you square them and sum them over all cells, they equal the Pearson chi-square statistic. For large sample sizes, if the model fits, the distribution of standardized residuals is normal, with a mean of 0 and a standard deviation of less than 1.

Since for a complex model the standard deviation of the standardized residuals can be quite a bit less than 1, the adjusted residual is a better diagnostic aid. The **adjusted residual** is the standardized residual divided by an estimate of its standard error. For large samples, its distribution is normal, with a mean of 0 and a standard deviation of 1.

The **deviance residual** can also be used for examining departures from fit in a model. The deviance residual is the cell's contribution to the likelihood-ratio chi-square, maintaining the sign of the raw residual. The sum of the squared deviances for all cells is equal to the likelihood-ratio chi-square. For large sample sizes, the distribution of the deviances is normal, with a mean of 0 and a standard deviation of 1.

Figure 6.9 is a scatterplot matrix of the adjusted residuals and observed and expected counts for the voting behavior loglinear model. From the plots, you see that there doesn't appear to be a pattern when the residuals are plotted against the predicted values. Since there are only eight cells, it's difficult to detect departures from normality.

Figure 6.9 Scatterplot matrix of adjusted residuals

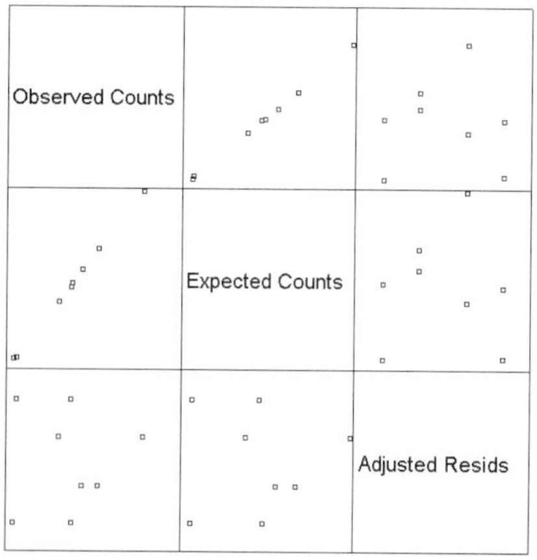

Models for Ordinal Data

In many situations, the categorical variables used in loglinear models are ordinal in nature. For example, income levels range from low to high, as does interest in a product or severity of a disease. Ordinal variables may result from grouping values of interval variables such as income or education, or they may arise when ordering (but not distance) between categories can be established. Happiness, interest, and opinions on various issues are measured on ordinal scales. Although "a lot" is more than "some," the actual distance between the two response categories cannot be determined. The additional information contained in the ordering of the categories can be incorporated into loglinear models, which may result in a more parsimonious representation of the data (Agresti, 1984).

Let's consider some common models for ordinal data, using data on job satisfaction and income from the 1984 General Social Survey. Income is grouped into four categories. Job satisfaction is measured on a four-point scale: very dissatisfied, a little dissatisfied, moderately satisfied, and very satisfied. Figure 6.10 is the crosstabulation of job satisfaction and income categories. You can see that as salary increases, so does job satisfaction, especially at the ends of the salary scale. Almost 10% of people earning less than $6,000 were very dissatisfied with their jobs, while only 4% of those earning $25,000 or more were very dissatisfied with theirs. Similarly, almost 54% of those earning over $25,000 were very satisfied with their jobs, while only 40% of those earning less than $6,000 were very satisfied. Although the chi-square test doesn't lead us to reject the hypothesis of independence, examination of the data suggests a relationship between the two variables.

Figure 6.10 Job satisfaction by income level

```
SATJOB  JOB OR HOUSEWORK  by  RINCOME  RESPONDENT'S INCOME

                  RINCOME                               Page 1 of 1
            Count |
          Col Pct |LESS 6000  6-15     15-25    25+
                  |                                        Row
                  |        1|       2|       3|       4| Total
SATJOB     -------+---------+--------+--------+--------+
                1 |    20   |   22   |   13   |    7   |    62
  VERY DIS        |   9.7   |  7.6   |  5.5   |  4.1   |   6.9
                  +---------+--------+--------+--------+
                2 |    24   |   38   |   28   |   18   |   108
  A LITTLE        |  11.7   | 13.1   | 11.9   | 10.5   |  12.0
                  +---------+--------+--------+--------+
                3 |    80   |  104   |   81   |   54   |   319
  MOD SAT         |  38.8   | 36.0   | 34.5   | 31.6   |  35.4
                  +---------+--------+--------+--------+
                4 |    82   |  125   |  113   |   92   |   412
  VER SAT         |  39.8   | 43.3   | 48.1   | 53.8   |  45.7
                  +---------+--------+--------+--------+
           Column     206      289      235      171      901
            Total    22.9     32.1     26.1     19.0    100.0

        Chi-Square                  Value          DF        Significance
--------------------         -----------        ----        ------------

Pearson                          11.98857          9              .21395
Likelihood Ratio                 12.03686          9              .21124
Mantel-Haenszel test for          9.54552          1              .00200
      linear association

Minimum Expected Frequency -    11.767

Number of Missing Observations:  572
```

A variety of loglinear models that use the ordering of the job satisfaction and income variables can be considered. Some of the models depend on the "scores" assigned to each category of response. Sometimes these are arbitrary, since the actual distances between the categories are unknown. In this example, scores from 1 to 4 are assigned to both the income and job satisfaction categories. Other scores, such as the midpoints of the salary categories, might also be considered.

Three types of models will be considered for these data. One is the **linear-by-linear association model**, which uses the ordering of both variables. Another is the **row-effects model**, which uses only the ordering of the column variable. The third is the **column-effects model**, which uses the ordering of the row variable.

The Linear-by-Linear Association Model

The linear-by-linear association model for two variables can be expressed as

Equation 6.6

$$\ln(m_{ij}) = \mu + \lambda_i^X + \lambda_j^Y + BU_iV_j$$

where the scores U_i and V_j are assigned to rows and columns. In this model, μ and the two lambda parameters are the usual loglinear terms for the constant and the main effects of income and job satisfaction. What differs is the inclusion of the term involving B. The coefficient B is essentially a regression coefficient that, for a particular cell, is multiplied by the scores assigned to that cell for income and job satisfaction. If the two variables are independent, the coefficient should be close to 0. (However, a coefficient of 0 does not necessarily imply independence, since the association between the two variables may be nonlinear.) If the coefficient is positive, more cases are expected to fall into cells with large scores or small scores for both variables than would be expected if the two variables were independent. If the coefficient is negative, an excess of cases is expected in cells that have small values for one variable and large values for the other.

Consider Figure 6.11, which contains parameter estimates for the linear-by-linear association model for job satisfaction and income.

Figure 6.11 Parameter correspondence and estimates

```
Correspondence Between Parameters and Terms of the Design

Parameter   Aliased  Term

        1            Constant
        2            [SATJOB = 1]
        3            [SATJOB = 2]
        4            [SATJOB = 3]
        5       x    [SATJOB = 4]
        6            [RINCOME = 1]
        7            [RINCOME = 2]
        8            [RINCOME = 3]
        9       x    [RINCOME = 4]
       10            B

Note: 'x' indicates an aliased (or a redundant) parameter.
      These parameters are set to zero.

Parameter Estimates

                                                Asymptotic 95% CI
Parameter   Estimate        SE     Z-value      Lower      Upper

        1     2.6987     .5501        4.91       1.62       3.78
        2    -1.1120     .2809       -3.96      -1.66       -.56
        3     -.8044     .2033       -3.96      -1.20       -.41
        4      .0181     .1172         .15       -.21        .25
        5      .0000     .             .          .          .
        6     1.2639     .3667        3.45        .55       1.98
        7     1.2534     .2597        4.83        .74       1.76
        8      .6870     .1592        4.31        .37       1.00
        9      .0000     .             .          .          .
       10      .1119     .0364        3.07        .04        .18
```

Parameter 10 is *B*, the regression coefficient. (The covariate is computed as the product of the variables *rincome* and *satjob*, using Compute from the Transform menu. It is entered into the model as a covariate, along with the factors *satjob* and *rincome*.) The coefficient is positive and large when compared to its standard error, indicating that there is a positive association between income and job satisfaction. That is, as income increases or decreases, so does job satisfaction. The goodness-of-fit statistics displayed in Figure 6.12, as well as the adjusted residual plots in Figure 6.13, indicate that the linear-by-linear interaction model fits the data very well. Inclusion of one additional parameter in the model has changed the observed significance level from 0.21 for the independence model (see Figure 6.10) to 0.97 for the linear-by-linear interaction model (see Figure 6.12).

Figure 6.12 Goodness of fit for the linear-by-linear association model

Goodness-of-fit Statistics

	Chi-Square	DF	Sig.
Likelihood Ratio	2.3859	8	.9668
Pearson	2.3296	8	.9692

Figure 6.13 Adjusted residual plots

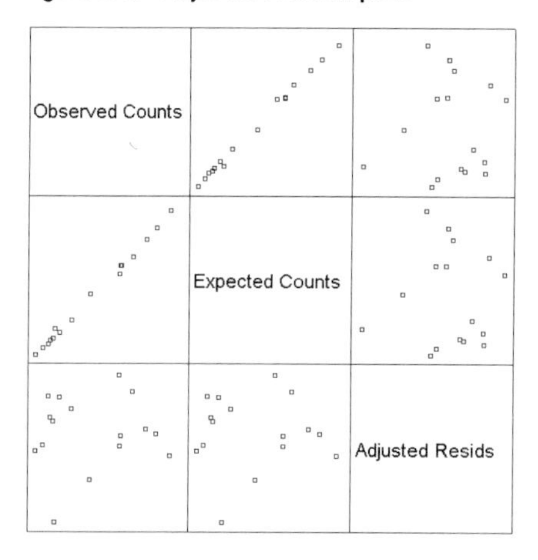

The regression coefficient *B* in the linear-by-linear association model can be interpreted in terms of an odds ratio. If two adjacent rows have scores which differ by + 1 , and two adjacent column have scores that differ by + 1, then

$$e^{B} = \frac{m_{ij}/m_{i(j+1)}}{m_{(i+1)j}/m_{(i+1)(j+1)}} \qquad \text{Equation 6.7}$$

The numerator of the above odds ratio is for row i the odds of being in column j instead of column $j + 1$. Similarly, the denominator is for row $i + 1$ the odds of being in column j instead of column $j + 1$. Let's consider an example. Figure 6.14 contains observed and expected cell counts for the linear-by-linear association model.

Figure 6.14 Observed and expected cell counts

```
Table Information

                          Observed                     Expected
Factor          Value       Count         %              Count         %

SATJOB  Very Dis
 RINCOME     < 6000         20.00 (   2.22)           19.35 (   2.15)
 RINCOME       6-15         22.00 (   2.44)           21.41 (   2.38)
 RINCOME      15-25         13.00 (   1.44)           13.59 (   1.51)
 RINCOME        25+          7.00 (    .78)            7.65 (    .85)

SATJOB  A Little
 RINCOME     < 6000         24.00 (   2.66)           29.43 (   3.27)
 RINCOME       6-15         38.00 (   4.22)           36.43 (   4.04)
 RINCOME      15-25         28.00 (   3.11)           25.86 (   2.87)
 RINCOME        25+         18.00 (   2.00)           16.28 (   1.81)

SATJOB    Mod Sat
 RINCOME     < 6000         80.00 (   8.88)           74.93 (   8.32)
 RINCOME       6-15        104.00 (  11.54)          103.73 (  11.51)
 RINCOME      15-25         81.00 (   8.99)           82.37 (   9.14)
 RINCOME        25+         54.00 (   5.99)           57.98 (   6.43)

SATJOB  Very Sat
 RINCOME     < 6000         82.00 (   9.10)           82.30 (   9.13)
 RINCOME       6-15        125.00 (  13.87)          127.43 (  14.14)
 RINCOME      15-25        113.00 (  12.54)          113.18 (  12.56)
 RINCOME        25+         92.00 (  10.21)           89.10 (   9.89)
```

Since we've assigned sequential scores from 1 to 4 to the rows and columns, adjacent rows and column differ by a score of 1. Consider two adjacent rows: those for the very dissatisfied and a little dissatisfied categories. Consider also two adjacent columns: income less than $6000 and income between $6,000 and $15,000. For very dissatisfied people, the predicted odds of having an income of less than $6000 compared to $6,000 to $15,000 is, from the expected counts in Figure 6.14, $19.35/21.41 = 0.904$. Similarly, for people who are a little dissatisfied, the same predicted odds are $29.43/36.43 = 0.808$.

That means that as job satisfaction increases, the odds of being in the lower income category decrease. The ratio of the two odds is 1.119. That's the value of $e^{0.112}$. The 95% confidence interval for the odds is from 1.041($e^{0.04}$) to 1.197 ($e^{0.18}$). Equation 6.7 can be generalized to the situation of arbitrary rows and columns. See Agresti (1990) for further discussion of this example.

Row- and Column-Effects Models

In a row-effects model, only the ordinal nature of the column variable is used. For each row, a separate slope based on the values of the column variables is estimated. The magnitude and sign of the coefficient indicates whether cases are more or less likely to fall into a column with a high or low score, as compared to the independence model.

Consider the row-effects model when job satisfaction is the row variable. The model contains the factor variables *satjob* and *rincome,* and a term for the interaction between the covariate *cov* (which is equal to *rincome* for a case) and *satjob.* The coefficients for each row are displayed in Figure 6.15 as parameters 10 through 13. The first coefficient, parameter 10, is negative, indicating that very dissatisfied people are less likely to be in high-income categories than the independence model would predict. The next two coefficients, parameters 11 and 12, are also negative, but smaller in value. The coefficient for the second job satisfaction category, parameter 11, is not statistically different from 0. The coefficient for the fourth job satisfaction category is 0 because it is the reference category. All of the other regression coefficients measure association compared to the last category.

Figure 6.15 Parameter correspondence and estimates for the row-effects model

Parameter	Aliased	Term
1		Constant
2		[SATJOB = 1]
3		[SATJOB = 2]
4		[SATJOB = 3]
5	x	[SATJOB = 4]
6		[RINCOME = 1]
7		[RINCOME = 2]
8		[RINCOME = 3]
9	x	[RINCOME = 4]
10		[SATJOB = 1]*COV
11		[SATJOB = 2]*COV
12		[SATJOB = 3]*COV
13	x	[SATJOB = 4]*COV

Note: 'x' indicates an aliased (or a redundant) parameter.
These parameters are set to zero.

Parameter Estimates

Parameter	Estimate	SE	Z-value	Asymptotic 95% CI Lower	Upper
1	4.5084	.0956	47.17	4.32	4.70
2	-.9896	.3270	-3.03	-1.63	-.35
3	-.9946	.2736	-3.64	-1.53	-.46
4	.1520	.1903	.80	-.22	.52
5	.0000	.	.	.	.
6	-.1195	.1472	-.81	-.41	.17
7	.3331	.1157	2.88	.11	.56
8	.2278	.1046	2.18	.02	.43
9	.0000	.	.	.	.
10	-.3908	.1372	-2.85	-.66	-.12
11	-.1408	.1046	-1.35	-.35	.06
12	-.1677	.0723	-2.32	-.31	-.03
13	.0000	.	.	.	.

Overall, the row-effects model fits quite well, as shown in the goodness-of-fit statistics in Figure 6.16. The observed significance level, 0.998, is quite large.

Figure 6.16 Goodness of fit for the row-effects model

Goodness-of-fit Statistics

	Chi-Square	DF	Sig.
Likelihood Ratio	.5256	6	.9975
Pearson	.5257	6	.9975

The column-effects model, which treats job satisfaction as an ordinal variable and ignores the ranking of the income categories, contains the factor variables *satjob* and *rincome* and a term for the interaction between the covariate *cov* (which is equal to *satjob* for a case) and *rincome.* The column-effects model fits the data reasonably well, as shown in Figure 6.17. The observed significance level is 0.91. In fact, all three models that incorporate the ordinal nature of the classification variables result in good fit. Of the three, the linear-by-linear model is the most parsimonious because when it is compared to the independence model, it estimates only one additional parameter. The row- and column-effects models are particularly useful when only one classification variable is ordinal or when both variables are ordinal but a linear trend across categories exists for only one.

Figure 6.17 Goodness of fit for the column-effects model

Goodness-of-fit Statistics

	Chi-Square	DF	Sig.
Likelihood Ratio	2.1372	6	.9067
Pearson	2.1105	6	.9093

Incomplete Tables

All models examined so far have been based on complete tables. That is, all cells of the crosstabulations can have non-zero observed frequencies. This is not necessarily the case. For example, if you are studying the association between types of surgery and sex of the patient, the cell corresponding to caesarean sections for males must be 0. This is termed a **fixed-zero** or **structural-zero cell**, since no cases can ever fall into it. Also, certain types of models "ignore" cells by treating them as if they were fixed zeros. Fixed-zero cells lead to incomplete tables, and special provisions must be made during analysis.

Cells in which the observed frequency is 0 but in which it is *possible* to have cases are sometimes called **random zeros** or **sampling zeros**. For example, in a cross-classification of occupation and ethnic origin, the cell corresponding to Lithuanian sword-swallowers would probably be a random zero.

If a table has many cells with small expected values (say, less than 5), the chi-square approximation for the goodness-of-fit statistics may be inadequate. In this case, pooling of categories should be considered.

There are many types of models for analyzing incomplete tables. In the next section, we will consider one of the simplest, the quasi-independence model, which considers the diagonal entries of a square table to be fixed zeros.

Testing Real against Ideal

What is the "ideal" number of children? The answer to the question is obviously influenced by various factors, including the size of the family in which one was raised. The 1982 General Social Survey asked respondents how many siblings they have as well as how many children should be in the ideal family. Figure 6.18 contains the crosstabulation of the number of children in the respondent's family and the number of children perceived as ideal.

Figure 6.18 Actual-by-ideal number of children

```
IDEAL  IDEAL NUMBER OF CHILDREN  by  REAL  ACTUAL NUMBER OF CHILDREN

                 REAL                                  Page 1 of 1
         Count  |
         Col Pct|0-1      2        3-4      5 +
                |                                     Row
                |    1.00|    2.00|    3.00|    4.00| Total
IDEAL   --------+--------+--------+--------+--------+
               1|     2  |     6  |    20  |    26  |    54
  0-1           |   2.7  |   3.0  |   4.5  |   3.9  |   3.9
                +--------+--------+--------+--------+
               2|    55  |   138  |   278  |   335  |   806
  2             |  74.3  |  68.7  |  62.3  |  49.7  |  57.8
                +--------+--------+--------+--------+
               3|    16  |    48  |   139  |   287  |   490
  3-4           |  21.6  |  23.9  |  31.2  |  42.6  |  35.1
                +--------+--------+--------+--------+
               4|     1  |     9  |     9  |    26  |    45
  5 +           |   1.4  |   4.5  |   2.0  |   3.9  |   3.2
                +--------+--------+--------+--------+
         Column      74       201      446      674     1395
          Total     5.3      14.4     32.0     48.3    100.0

      Chi-Square                    Value            DF              Significance
--------------------             -----------         ----            ------------

Pearson                           46.30172            9                 .00000
Likelihood Ratio                  47.48083            9                 .00000
Mantel-Haenszel test for          23.46950            1                 .00000
      linear association

Minimum Expected Frequency -     2.387
Cells with Expected Frequency < 5 -      2 OF     16 ( 12.5%)
```

To test the hypothesis that the actual (real) number of children and the ideal number are independent, the chi-square test of independence can be used. From Figure 6.18, the chi-square value is 46 with 9 degrees of freedom. The observed significance level is very small, indicating that it is unlikely that the numbers of real children and ideal children are independent.

Quasi-Independence

There are many reasons why the independence model may not fit the data. One possible explanation is that, in fact, the real and ideal sizes are fairly close to one another. However, if we ignore the diagonal entries of the table, we may find that the remaining cells are independent—in other words, that there is no tendency for children from small families to want large families or children from large families to want small families.

This hypothesis may be tested by ignoring the diagonal entries of the table and testing independence for the remaining cells using the test of **quasi-independence** (cell weighting is used to impose **structural zeros** on diagonal cells). Figure 6.19 contains the observed and expected cell frequencies for this model. Notice that all diagonal entries have values of 0 for observed and expected cell frequencies. However, the residuals and goodness-of-fit statistics indicate that the quasi-independence model does not fit well either.

Figure 6.19 Statistics for the quasi-independence model

```
Table Information

                       Observed                Expected
Factor    Value          Count        %          Count        %

REAL       0-1
 IDEAL      0-1            .00 (    .00)           .00 (    .00)
 IDEAL        2          55.00 (   5.05)         39.97 (   3.67)
 IDEAL      3-4          16.00 (   1.47)         30.06 (   2.76)
 IDEAL      5 +           1.00 (    .09)          1.97 (    .18)

REAL         2
 IDEAL      0-1           6.00 (    .55)          5.34 (    .49)
 IDEAL        2            .00 (    .00)           .00 (    .00)
 IDEAL      3-4          48.00 (   4.40)         54.11 (   4.96)
 IDEAL      5 +           9.00 (    .83)          3.55 (    .33)

REAL       3-4
 IDEAL      0-1          20.00 (   1.83)         20.30 (   1.86)
 IDEAL        2         278.00 (  25.50)        273.22 (  25.07)
 IDEAL      3-4            .00 (    .00)           .00 (    .00)
 IDEAL      5 +           9.00 (    .83)         13.48 (   1.24)

REAL       5 +
 IDEAL      0-1          26.00 (   2.39)         26.36 (   2.42)
 IDEAL        2         335.00 (  30.73)        354.81 (  32.55)
 IDEAL      3-4         287.00 (  26.33)        266.83 (  24.48)
 IDEAL      5 +            .00 (    .00)           .00 (    .00)


- - - - - - - - - - - - - - - - - - - - - - - - - - - - - - -

Goodness-of-fit Statistics

                    Chi-Square        DF         Sig.

Likelihood Ratio       24.6134         5        .0002
         Pearson       26.0583         5      9.E-05
```

Symmetry Models

If the diagonal terms of Figure 6.19 are ignored, the table can be viewed as consisting of two triangles. The lower left triangle, shown in Table 6.1, consists of cells in which the ideal number of children is larger than the actual number of children. The upper right triangle, shown in Table 6.2, consists of cells in which the ideal number is smaller than the actual number.

Table 6.1 Lower left triangle 1 (real < ideal)

		Real		
		0-1	**2**	**3-4**
	2	55		
Ideal	**3-4**	16	48	
	5+	1	9	9

Table 6.2 Upper right (rotated) triangle 2 (real > ideal)

		Ideal		
		0-1	**2**	**3-4**
	2	6		
Real	**3-4**	20	278	
	5+	26	335	287

Several hypotheses about the two triangles may be of interest. One is the symmetry hypothesis. That is, are the corresponding entries of the two triangles equal? In a loglinear model framework, symmetry implies that the main effects for the row and column variables, as well as their interaction, are the same for the two triangles. The loglinear representation of this model is

$$\ln(m_{ijk}) = \mu + \lambda_i^{\text{real}} + \lambda_j^{\text{ideal}} + \lambda_{ij}^{\text{real * ideal}} \quad \textbf{Equation 6.8}$$

In testing the symmetry hypothesis, the variables *triangle, ideal,* and *real* define the contingency table. The model contains *real, ideal,* and the *real*-by-*ideal* interaction. For this analysis, we use aggregate (frequency count) data, which are entered using the Data Editor. As shown in Figure 6.20, the variables *triangle, ideal, real, fre,* and *cweight* are recorded for each case (for the variables *ideal* and *real,* value labels are shown instead of data values). The variable *fre* contains the number of individuals in a particular cell of triangle 1 or triangle 2. Prior to analysis, cases are weighted by *fre,* using Weight Cases from the Data menu (see the SPSS Base system documentation for more information on case weighting). The *cweight* variable identifies the cells that are structural zeros. Cells that have observed and expected counts equal to 0 have the value of 0 for *cweight.* Note that you must enter these cells into the data file even if they have no cases. (The *cweight* variable is specified as a Cell Structure variable in the General Loglinear Analysis dialog box.)

Figure 6.20 Data triangles in the SPSS Data Editor

c:\advug61\genlog\nv5log3.sav

1:triangle 1

	triangle	ideal	real	fre	cweight
1	1.00	2.00	1.00	55.00	1.00
2	1.00	2.00	2.00	.00	.00
3	1.00	2.00	3.00	.00	.00
4	1.00	3.00	1.00	16.00	1.00
5	1.00	3.00	2.00	48.00	1.00
6	1.00	3.00	3.00	.00	.00
7	1.00	4.00	1.00	1.00	1.00
8	1.00	4.00	2.00	9.00	1.00
9	1.00	4.00	3.00	9.00	1.00
10	2.00	2.00	1.00	6.00	1.00

Figure 6.21 and Figure 6.22 contain the General Loglinear Analysis procedure output for the symmetry hypothesis. The expected values are simply the average of the observed frequencies for the two corresponding cells in the triangles. Thus, the expected values for corresponding cells are always equal. For example, the first expected frequency, 30.5, is the average of 55 and 6. Notice also that the second triangle has been "rotated" into lower-triangular form without changing the variable names, so that the variables are mislabeled in triangle 2. For example, the entry labeled *TRIANGLE 2, IDEAL 2, REAL 0-1* is really the entry for an ideal number of children of 0 or 1 and an actual number of 2.

Figure 6.21 Statistics for the symmetry model

Table Information

Factor	Value	Observed Count	%	Expected Count	%
TRIANGLE	1.00				
IDEAL	2				
REAL	0-1	55.00	(5.05)	30.50	(2.80)
REAL	2	.00	(.00)	.00	(.00)
REAL	3-4	.00	(.00)	.00	(.00)
IDEAL	3-4				
REAL	0-1	16.00	(1.47)	18.00	(1.65)
REAL	2	48.00	(4.40)	163.00	(14.95)
REAL	3-4	.00	(.00)	.00	(.00)
IDEAL	5+				
REAL	0-1	1.00	(.09)	13.50	(1.24)
REAL	2	9.00	(.83)	172.00	(15.78)
REAL	3-4	9.00	(.83)	148.00	(13.58)
TRIANGLE	2.00				
IDEAL	2				
REAL	0-1	6.00	(.55)	30.50	(2.80)
REAL	2	.00	(.00)	.00	(.00)
REAL	3-4	.00	(.00)	.00	(.00)
IDEAL	3-4				
REAL	0-1	20.00	(1.83)	18.00	(1.65)
REAL	2	278.00	(25.50)	163.00	(14.95)
REAL	3-4	.00	(.00)	.00	(.00)
IDEAL	5+				
REAL	0-1	26.00	(2.39)	13.50	(1.24)
REAL	2	335.00	(30.73)	172.00	(15.78)
REAL	3-4	287.00	(26.33)	148.00	(13.58)

Goodness-of-fit Statistics

	Chi-Square	DF	Sig.
Likelihood Ratio	977.4182	6	.0000
Pearson	795.2596	6	.0000

Figure 6.22 Residuals for the symmetry model

Table Information

Factor	Value	Resid.	Adj. Resid.	Dev. Resid.
TRIANGLE	1.00			
IDEAL	2			
REAL	0-1	24.50	6.27	3.98
REAL	2	.	.	.
REAL	3-4	.	.	.
IDEAL	3-4			
REAL	0-1	-2.00	-.67	-.48
REAL	2	-115.00	-12.74	-10.61
REAL	3-4	.	.	.
IDEAL	5+			
REAL	0-1	-12.50	-4.81	-4.45
REAL	2	-163.00	-17.58	-16.52
REAL	3-4	-139.00	-16.16	-15.09
TRIANGLE	2.00			
IDEAL	2			
REAL	0-1	-24.50	-6.27	-5.43
REAL	2	.	.	.
REAL	3-4	.	.	.
IDEAL	3-4			
REAL	0-1	2.00	.67	.46
REAL	2	115.00	12.74	8.18
REAL	3-4	.	.	.
IDEAL	5+			
REAL	0-1	12.50	4.81	3.01
REAL	2	163.00	17.58	10.98
REAL	3-4	139.00	16.16	10.11

The goodness-of-fit statistics, as well as the large residuals, indicate that the symmetry model fits poorly. This is not very surprising, since examination of Table 6.1 shows that the number of cases in each triangle is quite disparate.

The symmetry model provides a test of whether the probability of falling into cell (i,j) of triangle 1 is the same as the probability of falling into cell (i,j) of triangle 2. It does not take into account the fact that the overall observed probability of membership in triangle 2 is much greater than the probability of membership in triangle 1.

Since the triangle totals are so disparate, a more reasonable hypothesis to test is whether the probability of falling into corresponding cells in the two triangles is equal, adjusting for the observed totals in the two triangles. In other words, we test whether the probability of falling in cell (i,j) is the same for triangle 1 and triangle 2, assuming that the probability of membership in the two triangles is equal. The expected value for each cell is no longer the average of the observed frequencies for the two triangles but is a weighted average. The weights are the proportion of cases in each triangle. The expected value is the product of the expected probability and the sample size in the triangle.

The symmetry model that preserves triangle totals is represented by the following loglinear model:

$$\ln(m_{ijk}) = \mu + \lambda_i^{\text{real}} + \lambda_j^{\text{ideal}} + \lambda_k^{\text{triangle}} + \lambda_{ij}^{\text{real * ideal}}$$ **Equation 6.9**

This differs from the previous symmetry model in that the term $\lambda_k^{\text{triangle}}$, which preserves triangle totals, is included.

Figure 6.23 and Figure 6.24 show a symmetry model that preserves triangle totals. As in the previous symmetry analysis, the variables *triangle*, *ideal*, and *real* define the contingency table. The model contains *real*, *ideal*, *triangle*, and the *real*-by-*ideal* interaction. Notice that the expected number of cases in the first cell is 7.72. This is 5.59% of the total number of cases in triangle 1. Similarly, for triangle 2, the expected number of cases in the first cell is 53.28. This is 5.59% of the cases in triangle 2. Thus, the estimated probabilities are the same for the two triangles, although the actual numbers differ. The residuals and goodness-of-fit tests, however, suggest that a symmetry model preserving the observed totals in the two triangles does not fit the data well either.

Figure 6.23 Statistics for the symmetry model preserving observed totals

Table Information

Factor	Value	Observed Count	%	Expected Count	%
TRIANGLE	1.00				
IDEAL	2				
REAL	0-1	55.00	(5.05)	7.72	(.71)
REAL	2	.00	(.00)	.00	(.00)
REAL	3-4	.00	(.00)	.00	(.00)
IDEAL	3-4				
REAL	0-1	16.00	(1.47)	4.56	(.42)
REAL	2	48.00	(4.40)	41.27	(3.79)
REAL	3-4	.00	(.00)	.00	(.00)
IDEAL	5+				
REAL	0-1	1.00	(.09)	3.42	(.31)
REAL	2	9.00	(.83)	43.55	(4.00)
REAL	3-4	9.00	(.83)	37.48	(3.44)
TRIANGLE	2.00				
IDEAL	2				
REAL	0-1	6.00	(.55)	53.28	(4.89)
REAL	2	.00	(.00)	.00	(.00)
REAL	3-4	.00	(.00)	.00	(.00)
IDEAL	3-4				
REAL	0-1	20.00	(1.83)	31.44	(2.88)
REAL	2	278.00	(25.50)	284.73	(26.12)
REAL	3-4	.00	(.00)	.00	(.00)
IDEAL	5+				
REAL	0-1	26.00	(2.39)	23.58	(2.16)
REAL	2	335.00	(30.73)	300.45	(27.56)
REAL	3-4	287.00	(26.33)	258.52	(23.72)

\- -

Goodness-of-fit Statistics

	Chi-Square	DF	Sig.
Likelihood Ratio	294.5014	5	.0000
Pearson	423.6283	5	.0000

Figure 6.24 Residuals for the symmetry model preserving observed totals

Table Information

Factor	Value	Resid.	Adj. Resid.	Dev. Resid.
TRIANGLE	1.00			
IDEAL	2			
REAL	0-1	47.28	18.74	11.02
REAL	2	.	.	.
REAL	3-4	.	.	.
IDEAL	3-4			
REAL	0-1	11.44	5.83	4.16
REAL	2	6.73	1.34	1.02
REAL	3-4	.	.	.
IDEAL	5+			
REAL	0-1	-2.42	-1.42	-1.54
REAL	2	-34.55	-6.77	-6.38
REAL	3-4	-28.48	-5.83	-5.59
TRIANGLE	2.00			
IDEAL	2			
REAL	0-1	-47.28	-18.73	-8.27
REAL	2	.	.	.
REAL	3-4	.	.	.
IDEAL	3-4			
REAL	0-1	-11.44	-5.83	-2.19
REAL	2	-6.73	-1.34	-.40
REAL	3-4	.	.	.
IDEAL	5+			
REAL	0-1	2.42	1.42	.49
REAL	2	34.55	6.77	1.96
REAL	3-4	28.48	5.83	1.74

Adjusted Quasi-Symmetry

The previously described symmetry model preserves only totals in each triangle. It does not require that the row and column sums for the expected values equal the observed row and column sums. The original crosstabulation shows that the marginal distributions of real and ideal children are quite different. The **adjusted quasi-symmetry model** can be used to test whether the pattern of association in the two triangles is similar when row and column totals are preserved in each triangle.

The loglinear model for adjusted quasi-symmetry is

$$\ln(m_{ijk}) = \mu + \lambda_i^{\text{real}} + \lambda_j^{\text{ideal}} + \lambda_k^{\text{triangle}} + \lambda_{ij}^{\text{real * ideal}} + \lambda_{ik}^{\text{real * triangle}} + \lambda_{jk}^{\text{ideal * triangle}}$$

Equation 6.10

This model differs from the completely saturated model for the three variables only in that it does not contain the three-way interaction among number of real children, number of ideal children, and triangle number. Thus, the adjusted quasi-symmetry model tests whether the three-way interaction is significantly different from 0.

Figure 6.25 contains the goodness-of-fit statistics for the quasi-symmetry model. The small chi-square value suggests that there is no reason to believe that the model does not fit well. However, there is only one degree of freedom for the model, since many parameters have been estimated. The residuals for the adjusted quasi-symmetry model are shown in Figure 6.26.

Figure 6.25 Statistics for the adjusted quasi-symmetry model

```
Table Information

                         Observed                   Expected
Factor        Value        Count          %           Count          %

TRIANGLE  1.00
 IDEAL          2
  REAL        0-1          55.00 (   5.05)          55.00 (   5.05)
  REAL          2            .00 (    .00)            .00 (    .00)
  REAL        3-4            .00 (    .00)            .00 (    .00)
 IDEAL        3-4
  REAL        0-1          16.00 (   1.47)          14.76 (   1.35)
  REAL          2          48.00 (   4.40)          49.24 (   4.52)
  REAL        3-4            .00 (    .00)            .00 (    .00)
 IDEAL         5+
  REAL        0-1           1.00 (    .09)           2.24 (    .21)
  REAL          2           9.00 (    .83)           7.76 (    .71)
  REAL        3-4           9.00 (    .83)           9.00 (    .83)

TRIANGLE  2.00
 IDEAL          2
  REAL        0-1           6.00 (    .55)           6.00 (    .55)
  REAL          2            .00 (    .00)            .00 (    .00)
  REAL        3-4            .00 (    .00)            .00 (    .00)
 IDEAL        3-4
  REAL        0-1          20.00 (   1.83)          21.24 (   1.95)
  REAL          2         278.00 (  25.50)         276.76 (  25.39)
  REAL        3-4            .00 (    .00)            .00 (    .00)
 IDEAL         5+
  REAL        0-1          26.00 (   2.39)          24.76 (   2.27)
  REAL          2         335.00 (  30.73)         336.24 (  30.85)
  REAL        3-4         287.00 (  26.33)         287.00 (  26.33)
- - - - - - - - - - - - - - - - - - - - - - - - - - - - - - - - - - - - -
Goodness-of-fit Statistics

                       Chi-Square        DF          Sig.

Likelihood Ratio          1.3244          1         .2498
         Pearson          1.1570          1         .2821
```

Figure 6.26 Residuals for the adjusted quasi-symmetry model

Table Information

Factor	Value	Resid.	Adj. Resid.	Dev. Resid.
TRIANGLE	1.00			
IDEAL	2			
REAL	0-1	.00	.00	.00
REAL	2	.	.	.
REAL	3-4	.	.	.
IDEAL	3-4			
REAL	0-1	1.24	1.07	.32
REAL	2	-1.24	-1.08	-.18
REAL	3-4	.	.	.
IDEAL	5+			
REAL	0-1	-1.24	-1.08	-.93
REAL	2	1.24	1.07	.43
REAL	3-4	.00	.00	.00
TRIANGLE	2.00			
IDEAL	2			
REAL	0-1	.00	.00	.00
REAL	2	.	.	.
REAL	3-4	.	.	.
IDEAL	3-4			
REAL	0-1	-1.24	-1.08	-.27
REAL	2	1.24	1.08	.07
REAL	3-4	.	.	.
IDEAL	5+			
REAL	0-1	1.24	1.07	.25
REAL	2	-1.24	-1.08	-.07
REAL	3-4	.00	.00	.00

An Ordinal Model for Real versus Ideal

Although the various symmetry models provide information about the relationship between the two triangles of a square table, we might wish to develop a more general model for the association between number of siblings in a family and one's view on the ideal number of children. The ordinal models considered in "Models for Ordinal Data" on p. 181 through "Row- and Column-Effects Models" on p. 185 might be a good place to start.

Figure 6.18 reveals that two children seems to be the most popular number. However, as the number of real children in a family increases, so does the tendency toward a larger ideal family size. For example, only 21.6% of only children consider three to four children to be ideal. Almost 43% of those from families of five or more children consider three or four children to be optimal. Thus, we might consider a row-effects model that incorporates the ordinal nature of the real number of children. For this analysis, *ideal* and *real* define the contingency table. The variable *cov*, which is equal to *real* for a case, is a covariate. The model contains the factors *ideal* and *real*, and the *ideal*-by-*cov* interaction. Figure 6.27 shows the observed and expected frequencies for the row-effects model. Figure 6.28 shows the residuals for the row-effects model. The large observed significance level and small residuals indicate that the model fits reasonably well.

Figure 6.27 Statistics for the row-effects ordinal model

Table Information

Factor	Value	Observed Count	%	Expected Count	%
IDEAL	0-1				
REAL	0-1	2.00	(.14)	2.26	(.16)
REAL	2	6.00	(.43)	7.02	(.50)
REAL	3-4	20.00	(1.43)	17.18	(1.23)
REAL	5 +	26.00	(1.86)	27.54	(1.97)
IDEAL	2				
REAL	0-1	55.00	(3.94)	56.98	(4.08)
REAL	2	138.00	(9.89)	139.41	(9.99)
REAL	3-4	278.00	(19.93)	269.22	(19.30)
REAL	5 +	335.00	(24.01)	340.38	(24.40)
IDEAL	3-4				
REAL	0-1	16.00	(1.15)	13.08	(.94)
REAL	2	48.00	(3.44)	49.11	(3.52)
REAL	3-4	139.00	(9.96)	145.52	(10.43)
REAL	5 +	287.00	(20.57)	282.28	(20.24)
IDEAL	5 +				
REAL	0-1	1.00	(.07)	1.67	(.12)
REAL	2	9.00	(.65)	5.46	(.39)
REAL	3-4	9.00	(.65)	14.08	(1.01)
REAL	5 +	26.00	(1.86)	23.79	(1.71)

Figure 6.28 Residuals for the row-effects ordinal model

Table Information

Factor	Value	Resid.	Adj. Resid.	Dev. Resid.
IDEAL	0-1			
REAL	0-1	-.26	-.22	-.18
REAL	2	-1.02	-.52	-.39
REAL	3-4	2.82	.87	.66
REAL	5 +	-1.54	-.80	-.30
IDEAL	2			
REAL	0-1	-1.98	-.67	-.26
REAL	2	-1.41	-.29	-.12
REAL	3-4	8.78	1.07	.53
REAL	5 +	-5.38	-1.12	-.29
IDEAL	3-4			
REAL	0-1	2.92	1.08	.78
REAL	2	-1.11	-.24	-.16
REAL	3-4	-6.52	-.83	-.54
REAL	5 +	4.72	1.03	.28
IDEAL	5 +			
REAL	0-1	-.67	-.64	-.56
REAL	2	3.54	2.05	1.39
REAL	3-4	-5.08	-1.72	-1.45
REAL	5 +	2.21	1.28	.45

- -

Goodness-of-fit Statistics

	Chi-Square	DF	Sig.
Likelihood Ratio	6.7118	6	.3483
Pearson	6.8354	6	.3363

Modeling Rates

The loglinear models we've considered so far predicted the expected number of cases in a cell based on values of categorical independent variables and covariates. The model fit was

$$\text{predicted count} = e^{\mu + \Sigma\lambda^i} \qquad \textbf{Equation 6.11}$$

where the summation is over the lambda parameters for that cell.

Loglinear models can also be used to model rates. For example, you can model the incidence of lung cancer based on duration and intensity of smoking. Or you can model the unemployment rate based on sex, education, and age. In both of these situations, the number of events for each combination of values of the independent variables is assumed to follow a Poisson distribution (see "Sampling Models" on p. 201).

In the rate model, you have two values for each cell of the table: the number of events and the number at risk. For example, if you follow 100,000 heavy smokers for one year and observe 2 cases of lung cancer, the cell "heavy smokers" has two values: 100,000 cases at risk and 2 lung cancer cases. The model for the rate in a particular cell is

$$\frac{\text{predicted count}}{\text{number at risk}} = e^{\mu + \Sigma\lambda^i} \qquad \textbf{Equation 6.12}$$

where the summation is over the lambda parameters for that cell. Taking the logarithm of both sides results in

$$\ln(\text{predicted count}) - \ln(\text{number at risk}) = \mu + \Sigma\lambda^i \qquad \textbf{Equation 6.13}$$

The term *ln(number at risk)* is called an **offset term**. (In the General Loglinear procedure, the number at risk is specified as a cell structure variable.)

As an example, consider data presented by Frome (1983) and discussed by Kleinbaum et al. (1988). The number of lung cancer deaths was recorded in 63 subgroups of cases. The subgroups were formed based on 9 age groups and 7 intensity-of-smoking groups. (These are described in Figure 6.29, the variable information from the General Loglinear Analysis procedure.)

Figure 6.29 Variable information for lung cancer data

```
Variable Information

Factor       Levels     Value

AGE              9           age
                      1.00 35-39 years
                      2.00 40-44 years
                      3.00 45-49 years
                      4.00 50-54 years
                      5.00 55-59 years
                      6.00 60-64 years
                      7.00 65-69 years
                      8.00 70-74 years
                      9.00 75-79 years

CIGS             7           number of cigs per day
                      1.00 0
                      2.00 1-9
                      3.00 10-14
                      4.00 15-19
                      5.00 20-24
                      6.00 25-34
                      7.00 35+
```

For each of the subgroups you have two values: the number of lung cancer deaths and the person years at risk. The model we'll fit is

$$m_{ij} = N_{ij}e^{\mu + \lambda_i^{\text{age}} + \lambda_j^{\text{smoke}}}$$ **Equation 6.14**

where N_{ij} is the number at risk.

This can be written as the loglinear model

$$\ln(m_{ij}) - \ln(N_{ij}) = \mu + \lambda_i^{\text{age}} + \lambda_j^{\text{smoke}}$$ **Equation 6.15**

This model assumes that the effects of smoking and age are independent. That means that the age effect is the same for all categories of smoking. Similarly, the effect of the intensity of the smoking is the same for all categories of age.

The goodness-of-fit statistics for this model are shown in Figure 6.30.

Figure 6.30 Goodness-of-fit statistics for lung cancer example

```
Goodness-of-fit Statistics

                   Chi-Square        DF        Sig.

Likelihood Ratio      51.4709        48       .3395
         Pearson      65.6613        48       .0459
```

You should use the likelihood-ratio chi-square value, since many of the expected counts are small in value, and the Pearson chi-square is not appropriate in such a situation. Based on the likelihood-ratio chi-square, it appears that the model fits reasonably well. The normal probability plot of the deviance residuals is shown in Figure 6.31. No serious departures from normality are apparent.

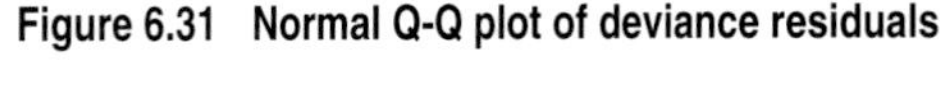

Figure 6.31 Normal Q-Q plot of deviance residuals

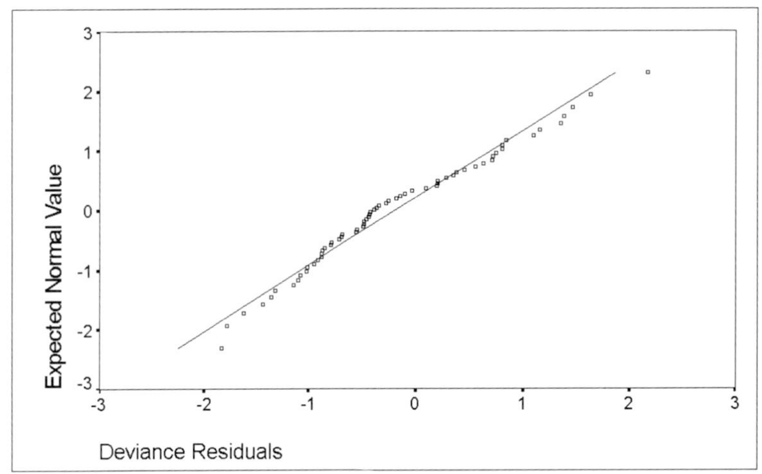

Interpreting Parameter Estimates

Figure 6.32 contains the parameter estimates for the model.

Figure 6.32 Parameter estimates for lung cancer example

	Parameter	Estimate	SE	Z-value	Asymptotic 95% CI Lower	Asymptotic 95% CI Upper
	1	-3.5592	.2772	-12.84	-4.10	-3.02
Age	2	-5.4134	1.0239	-5.29	-7.42	-3.41
Age	3	-4.4664	.6178	-7.23	-5.68	-3.26
Age	4	-3.7118	.4640	-8.00	-4.62	-2.80
Age	5	-2.2104	.2981	-7.42	-2.79	-1.63
Age	6	-2.1711	.3107	-6.99	-2.78	-1.56
Age	7	-1.2045	.2741	-4.39	-1.74	-.67
Age	8	-.9657	.2861	-3.38	-1.53	-.40
Age	9	-.5085	.2966	-1.71	-1.09	.07
Age	10	.0000	.	.	.	.
Smoking	11	-3.6059	.6048	-5.96	-4.79	-2.42
Smoking	12	-2.3859	.4472	-5.34	-3.26	-1.51
Smoking	13	-1.5068	.3220	-4.68	-2.14	-.88
Smoking	14	-1.2970	.3135	-4.14	-1.91	-.68
Smoking	15	-.7051	.2286	-3.08	-1.15	-.26
Smoking	16	-.4897	.2247	-2.18	-.93	-.05
Smoking	17	.0000	.	.	.	.

Parameter 1 is the constant. Parameters 2 through 10 are for the nine age categories. Parameters 11 through 17 are for the seven categories of smoking. Notice that the parameters for the last categories of smoking and age are set to 0. That means that they serve as the frame of reference for the other parameter estimates. The parameter estimate for the first smoking category (parameter 11) is for all age groups the log of the ratio of the predicted death rates from lung cancer for the nonsmokers (group 1) and the heaviest smokers (group 7). If you calculate $e^{\lambda} = e^{-3.61} = 0.027$, the interpretation is even simpler. For all age groups, the ratio of the predicted death rate from lung cancer for the nonsmokers to the heaviest smokers is 0.027. That means that the predicted death rate from lung cancer in nonsmokers is 2.7% of the death rate in the heaviest smokers. Since age and smoking intensity are independent in the model, the effect of smoking intensity is the same for all age groups. For the second to the last category (25–34 cigarettes per day), the parameter estimate is –0.49. Since $e^{-0.49}$ is 0.61, the predicted death rate from lung cancer for this group is 61% of the predicted death rate from lung cancer for the last group, smokers who smoke 35 or more cigarettes per day.

The parameter estimates for the age categories are interpreted the same way as for the smoking categories. For example, the predicted death rate from lung cancer in the first age group (35–39 years) is only 0.4% of the predicted death rate in the last category (ages 75–79), since $e^{-5.41} = 0.004$. Again, since the effects of age and smoking intensity are independent in the model, the effect of age is the same for all categories of smoking. If you look at parameter 9, you will see that its 95% confidence interval includes 0. That means that the predicted death rate for this group is not significantly different from that of the last group.

Sampling Models

The data in a multiway crosstabulation can originate in many different ways. For example, let's say you are interested in determining whether there is a relationship between type of treatment and cancer recurrence within one year of diagnosis in patients who are considered to be cancer free after their initial course of treatment. You can gather data in several different ways. For example, you might select the records of 1000 eligible patients from a cancer registry and determine what the initial treatment was and whether they had a recurrence within one year. In this case, your sample size of 1000 patients is fixed. Or you may decide to look at the records of all eligible cancer patients diagnosed in a particular two-year period. In this case, the sample size is not fixed, since you do not know how many eligible patients you will find in this two-year period. Another way you might obtain data is to select 500 eligible patients with recurrences and 500 patients without recurrences and see which of the treatments each patient received. Again, the sample size is fixed not only for the total number of patients but also for the number of patients in the two recurrence groups. Regardless of how you conducted the study, the results can be displayed in a crosstabulation of treatment and recurrence status. How you conduct the study determines, in part, which sampling model is appropriate for the data.

The two distributions most often used to describe the distribution of counts in a crosstabulation are the Poisson distribution and the multinomial distribution. The **Poisson distribution** is useful for modeling rare events such as suicides, deaths, or the number of raisins in a tablespoon of cereal. A Poisson sampling model for a crosstabulation arises when the total sample size is not fixed and the number of cases in each cell of the table is independent of the others and has a Poisson distribution. The counts arising from the study in which you obtained records for cases diagnosed in a two-year period may have a Poisson distribution, since the total sample size was not fixed.

The **multinomial distribution** is a generalization of the binomial distribution to more than two events. Under a multinomial sampling model, each cell of the crosstabulation table has a probability that indicates how likely an observation is to fall into it. The sum of the probabilities across all cells is 1. The total sample size in a multinomial model is fixed; thus, the cell counts are not independent, since they must sum to the total. In a multinomial sampling model, you know how many cases will be included in the study before you start. If you fix the row or column totals, such as when you selected 500 cases with recurrence and 500 cases without recurrence, the counts for each row or column have a multinomial distribution, and the distribution for the entire table is called the product multinomial distribution.

Multinomial sampling is the type more commonly encountered in analysis of categorical data. Fortunately, the parameter estimates are usually the same for the different sampling models. When you specify multinomial sampling in the General Loglinear Analysis procedure, the constant in the model is treated as known. That's why standard errors and confidence intervals are not calculated for it. For further discussion of sampling models as well as discussion of loglinear models, see Agresti (1990).

How to Obtain a General Loglinear Analysis

The General Loglinear Analysis procedure estimates parameters of hierarchical and nonhierarchical loglinear models using the Newton-Raphson method.

The minimum specifications are:

- One or more factor variables that define the tabulation.
- A model specification—Poisson or multinomial.

To obtain a general loglinear analysis, from the menus choose:

Statistics
  Loglinear ▶
    General...

This opens the General Loglinear Analysis dialog box, as shown in Figure 6.33.

Figure 6.33 General Loglinear Analysis dialog box

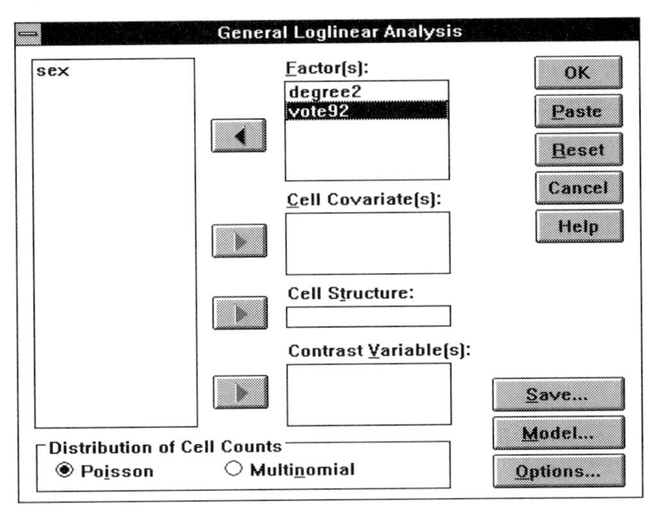

The numeric variables in your data file appear on the source list.

Factor(s). Select up to 10 categorical variables that define the cells of a table.

Cell Covariate(s). Optionally, you can select one or more continuous cell covariates. When a covariate is in the model, SPSS applies the mean covariate value for cases in a cell to that cell. To analyze an equiprobability model, select a variable that is actually a constant of 1. You can select a maximum of 200 covariates.

Cell Structure. By default, SPSS uses a weight of 1 for each cell. To provide your own cell weights, select a variable whose values are the weights. SPSS weights a cell by the average weight for cases in that cell. You can use this feature to define fixed (structural) zeros for incomplete tables. Do not use this facility to weight aggregate data; instead, choose Weight Cases on the Data menu (see the SPSS Base system documentation for more information on case weighting). This facility can also be used to include an offset term in the model.

Contrast Variable(s). Specify one or more continuous variables to be used as contrast variables. Contrast variables are used to compute generalized log-odds ratios (GLOR). The values of the contrast variable are the coefficients for the linear combination of the logs of the expected cell counts.

Distribution of Cell Counts. Choose one of the following distribution alternatives:

- **Poisson.** Choose this distribution if the total sample size is not fixed and the cells are independent. This is the default.
- **Multinomial.** Choose this distribution if the total sample size is fixed and the cells are not statistically independent.

Model Specification

By default, SPSS analyzes a saturated model. To analyze an unsaturated loglinear model, click on Model... in the General Loglinear Analysis dialog box. This opens the General Loglinear Analysis Model dialog box, as shown in Figure 6.34.

Figure 6.34 General Loglinear Analysis Model dialog box

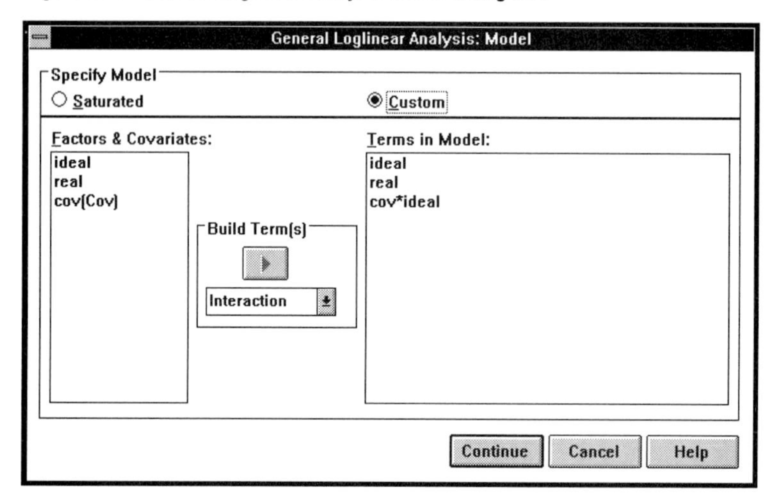

Specify Model. You can choose one of the following models:

- **Saturated.** The saturated model contains all main effects and interactions involving factor variables. This is the default. Covariates are not included in the model. To analyze a model containing covariates, define a custom model and include the covariate in the model.
- **Custom.** Select this item to define an unsaturated model. You must specify the terms to include in the model.

  If you remove a variable in the main dialog box, any terms in the model that contain that variable will also be deleted.

Terms in Model. To add a term to a custom model, select one or more factors or covariates, or a combination. If you don't want to create the highest-order interaction for the selected variables, select an alternate method for building a term from the Build Term(s) drop-down list. To add more terms to the model, repeat this process. You cannot use a term more than once in the model.

For example, to define a model containing main effects for the factors *ideal* and *real* and for the covariate-by-factor interaction *cov*ideal*, first highlight *ideal* and *real*, select Main effects from the Build Term(s) drop-down list, and click on [▸]. Next, highlight *cov* and *ideal*, select Interaction (the default) from the Build Term(s) drop-down list, and click on [▸].

Macintosh: Use ⌘-click to select multiple factors or covariates, or a combination.

- **Build Term(s).** You can build main effects or interactions. If you request an interaction of a higher order than the number of variables, SPSS creates a term for the highest-order interaction possible for the variables. If only one variable is selected, a main-effects term is added to the model. You can choose one of the following alternatives:

  Interaction. Creates the highest-level interaction term for the variables. This is the default for a selected group of variables.

  Main effects. Creates a main-effects term for each variable.

  All 2-way. Creates all possible two-way interactions for the variables.

  All 3-way. Creates all possible three-way interactions for the variables.

  All 4-way. Creates all possible four-way interactions for the variables.

  All 5-way. Creates all possible five-way interactions for the variables.

Options

To obtain optional statistics or plots, or to control model criteria, click on Options... in the General Loglinear Analysis dialog box. This opens the General Loglinear Analysis Options dialog box, as shown in Figure 6.35.

Figure 6.35 General Loglinear Analysis Options dialog box

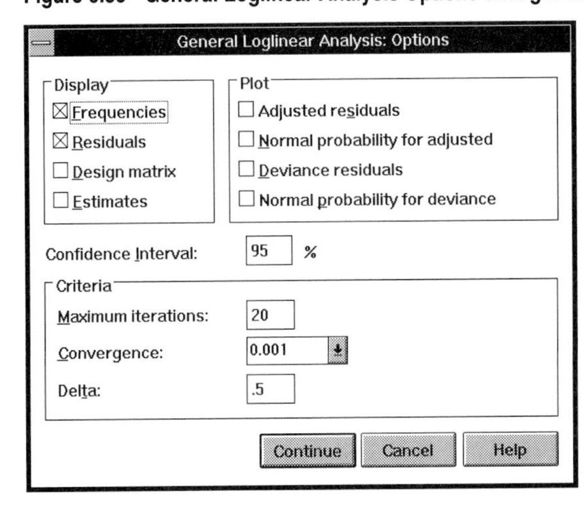

Display. SPSS displays model information and goodness-of-fit statistics. You can also choose one or more of the following displays:

- ❑ **Frequencies.** Observed and expected cell frequencies. Displayed by default. To suppress frequencies, deselect this item.
- ❑ **Residuals.** Raw, adjusted, and deviance residuals. Displayed by default. To suppress residuals, deselect this item.
- ❑ **Design matrix.** Design matrix of the model, showing the basis matrix corresponding to terms used in the analysis.
- ❑ **Estimates.** The parameter estimates of the model. The parameter estimates refer to the original categories.

Plot. For custom models, you can obtain one or more of the following plots:

- ❑ **Adjusted residuals.** Produces a scatterplot matrix of adjusted residuals against observed and expected cell counts.
- ❑ **Normal probability for adjusted.** Displays normal and detrended normal plots of adjusted residuals.
- ❑ **Deviance residuals.** Produces a scatterplot matrix of deviance residuals against observed and expected cell counts.
- ❑ **Normal probability for deviance.** Displays normal and detrended normal plots of deviance residuals.

Confidence Interval. By default, the confidence interval is 95%. If you want another confidence interval, enter a value between 50 and 99.99.

Criteria. The Newton-Raphson method is used to obtain maximum-likelihood parameter estimates. You can control one or more of the following algorithm criteria:

Maximum iterations. By default, a maximum of 20 iterations is performed. To specify a different maximum, enter a positive integer.

Convergence. By default, the convergence criterion is 0.001. To override the default, select an alternate convergence value from the drop-down list.

Delta. Constant added to all cells for initial approximations. Delta remains in the cells only for saturated models. The default value is 0.5. To override the default delta, enter a value between 0 and 1.

Saving Residuals or Predicted Values

To save residuals or predicted values as new variables, click on Save... in the General Loglinear Analysis dialog box. This opens the General Loglinear Analysis Save dialog box, as shown in Figure 6.36.

Figure 6.36 General Loglinear Analysis Save dialog box

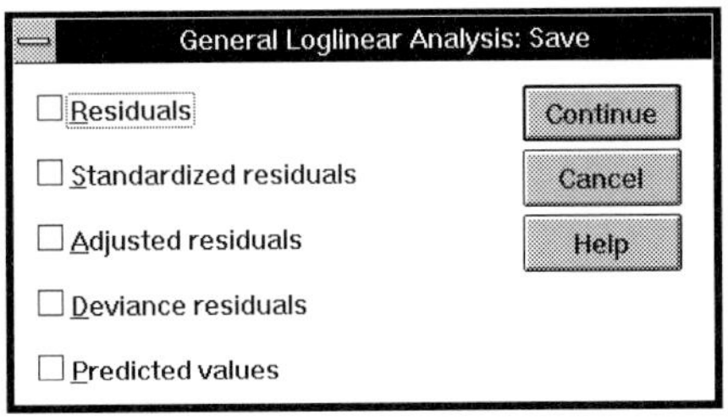

General Loglinear Analysis: Save

- ☐ Residuals
- ☐ Standardized residuals
- ☐ Adjusted residuals
- ☐ Deviance residuals
- ☐ Predicted values

Continue

Cancel

Help

The saved values refer to the aggregated data (to cells in the contingency table), even if the data are recorded in individual observations in the Data Editor. If you save residuals or predicted values for unaggregated data, the saved value for a cell in the contingency table is entered in the Data Editor for each case in that cell. To make sense of the saved values, you should aggregate the data to obtain the cell counts.

You can choose one or more of the following alternatives:

- ❑ **Residuals.** The difference between the observed and the expected count.
- ❑ **Standardized residuals.** The residual divided by the standard error of the observed count.
- ❑ **Adjusted residuals.** The residual divided by its estimated standard error.
- ❑ **Deviance residuals.** The signed square root of the individual contribution to the likelihood-ratio chi-square statistic.
- ❑ **Predicted values.** Estimated number of observations in a cell estimated from the sample under a specified model.

Additional Features Available with Command Syntax

You can customize your general loglinear analysis if you paste your selections into a syntax window and edit the resulting GENLOG command syntax. (For information on syntax windows, see the SPSS Base system documentation.) Additional features include:

- The default threshold value for redundancy checking can be changed by using the keyword EPS in the CRITERIA subcommand. The default value is 0.0000001, or 10^{-8}.
- Generalized residuals (using the GRESID subcommand).
- Standardized residuals (using the PRINT subcommand).

See the Syntax Reference section of this manual for command syntax rules and for complete GENLOG command syntax.

7 Logit Loglinear Analysis

The loglinear models described in the previous chapters are used to examine the interrelationships among categorical variables. For example, to see whether there is a relationship between happiness and marital status, you fit a loglinear model that includes terms for the various values of the two variables, as well as their interactions. The dependent variable in the model is the number of cases in the various cells of the table. All of the other variables are considered independent variables.

Often, however, you can single out one of the factor variables as a dependent variable. For example, you want to know if education level and sex are related to job satisfaction. Or, you want to know if a person's willingness to buy your product depends on age, income, and place of residence. In the first example, job satisfaction is the dependent variable, since you want to know if the other variables change the likelihood that a person is satisfied with his or her job. Similarly, in the second example, willingness to buy the product is the dependent variable.

A special class of loglinear models, called **logit models**, is used to model the relationship between one or more dependent categorical variables and a set of independent categorical variables (as well as covariates). Sometimes they are called **multinomial logit models**, since for each combination of values of the independent variable you assume that there is a multinomial distribution of the dependent variable, and that the cell counts across combinations are independent. As an example, let's look at a logit model in which the dependent variable is dichotomous. We'll examine factors that may be related to a person's belief that the U.S. will fight in a world war within the next 10 years. The data are a subset of cases from the 1993 General Social Survey.

Crosstabulation

Figure 7.1 is a crosstabulation of responses to the question "Do you expect the United States to fight in another world war within the next 10 years?" by the respondent's sex. The dependent variable, the answer to the question about expectation of war, is named *uswary* and has only two values—yes and no.

Figure 7.1 Crosstabulation of uswary and sex

```
                  SEX              Page 1 of 1
             Count
             Col Pct |MALE      FEMALE
                     |                      Row
                     |       1 |       2 |  Total
USWARY      ---------+---------+---------+
                   1 |    175  |    299  |   474
  YES                |   39.3  |   56.3  |  48.6
                     +---------+---------+
                   2 |    270  |    232  |   502
  NO                 |   60.7  |   43.7  |  51.4
                     +---------+---------+
             Column       445       531      976
              Total      45.6      54.4    100.0

Number of Missing Observations:  524
```

If you fit a loglinear model to the data, you have an equation that relates each of the expected frequencies to a set of parameters. For example, the saturated model for the first cell, the count of males who answer yes, is

$$\ln(m_{11}) = \mu + \lambda^{\text{yes}} + \lambda^{\text{male}} + \lambda^{\text{male yes}}$$ **Equation 7.1**

Similarly, the model for males who answer no is

$$\ln(m_{21}) = \mu + \lambda^{\text{no}} + \lambda^{\text{male}} + \lambda^{\text{male no}}$$ **Equation 7.2**

If you consider one of the variables as the dependent variable, instead of modeling the counts of cases for each cell, you can model the *ratios of the counts of the dependent variable* for each of the combinations of values of the independent variables. For example, if the respondent's sex is the independent variable and the answer to the question about expectation of war is the dependent variable, you model the ratio of males who answer yes to males who answer no, instead of simply modeling the cell count for males who answer yes and the cell count for males who answer no. Similarly, you model the ratio of the number of females who answer yes to the number of females who answer no. So, instead of having two separate models for the male cells, you have the single model

$$\ln\left(\frac{m_{11}}{m_{21}}\right) = \lambda^{\text{yes}} + \lambda^{\text{male yes}}$$ **Equation 7.3**

You recognize the ratio of the count of males who answer yes to the count of males who answer no as the odds of a male answering yes. (Remember that **odds** are the ratio of the probability that an event will occur to the probability that the event will not occur. See Chapter 6 for a detailed discussion of odds.) The log of the odds is called a **logit**, which is why these models are called logit models. From the data shown in Figure 7.1, the observed odds that a male answers yes are 175/270 (0.648), which is the ratio of the number of men who answer yes to the number of men who answer no. The natural log of the odds is –0.43.

You can write the similar model for females who answer yes as

$$\ln\left(\frac{m_{12}}{m_{22}}\right) = \lambda^{\text{yes}} + \lambda^{\text{female yes}}$$

Equation 7.4

In this case, the dependent variable is the log of the odds for females answering yes. From the data shown in Figure 7.1, the observed odds that a female answers yes are 299/232 (1.29). The log of the odds is 0.25.

Missing Parameters

You may have noticed that the saturated logit models don't have as many parameters as the saturated models for the corresponding individual cells. For example, Equation 7.1 and Equation 7.2 have parameters for the constant, for the answers yes and no, for the sex male, and for the male-by-answer interactions. The logit model in Equation 7.3 has parameters only for the answer yes and the male-by-yes interaction. The logit model has fewer parameters because the constant and all of the parameters that involve only the independent variables have canceled, since they appear in both the numerator and the denominator. Let's see how this works.

Using the fact that the log of a ratio is the log of the numerator minus the log of the denominator, the logit model in Equation 7.3 can be rewritten as

$$\ln\left(\frac{m_{11}}{m_{21}}\right) = \ln(m_{11}) - \ln(m_{21})$$

Equation 7.5

You can get the logit model by subtracting Equation 7.2 from Equation 7.1. This results in

$$\ln\left(\frac{m_{11}}{m_{21}}\right) = \lambda^{\text{yes}} + \lambda^{\text{male yes}} - \lambda^{\text{no}} - \lambda^{\text{male no}}$$

Equation 7.6

In Chapter 6, you saw that some of the parameters of a loglinear model are set to 0. SPSS prints a table telling you which of the parameters are set to 0. From the data shown in Figure 7.2, you see that in the saturated loglinear model only the constant and the parameters involving male and yes have not been set to 0.

Figure 7.2 Parameter correspondence table

```
Correspondence Between Parameters and Terms of the Design

Parameter   Aliased  Term

        1            Constant
        2            [SEX = 1]
        3       x    [SEX = 2]
        4            [USWARY = 1]
        5       x    [USWARY = 2]
        6            [SEX = 1]*[USWARY = 1]
        7       x    [SEX = 1]*[USWARY = 2]
        8       x    [SEX = 2]*[USWARY = 1]
        9       x    [SEX = 2]*[USWARY = 2]

Note: 'x' indicates an aliased (or a redundant) parameter.
      These parameters are set to zero.
```

Substituting 0 for the appropriate parameters in Equation 7.6 results in

$$\ln\left(\frac{m_{11}}{m_{21}}\right) = \lambda^{\text{yes}} + \lambda^{\text{male yes}} \qquad \textbf{Equation 7.7}$$

Whenever you fit a logit model, only parameters that include the dependent variable are in the model. Parameters that involve only independent variables are not included in the model. Every logit model with two values for the dependent variable can be converted to a loglinear model by including the appropriate terms that involve the independent variables. The parameter estimates for the logit model are the same as the parameter estimates for corresponding terms in the loglinear representation of the model. So, if you run a saturated loglinear model for the data shown in Figure 7.1, the parameter estimates for λ^{yes} and $\lambda^{\text{yes male}}$ would be identical to those from the corresponding logit model. All of the goodness-of-fit statistics are the same as well.

Parameter Estimates

Let's look at the output from the saturated logit model for the data shown in Figure 7.1. The parameter estimates are shown in Figure 7.3, and their explanation is shown in Figure 7.4.

Figure 7.3 Parameter estimates

```
Parameter Estimates

 Constant    Estimate

        1      5.5984
        2      5.4467

Note: Constants are not parameters under multinomial assumption.
      Therefore, standard errors are not calculated.

                                                      Asymptotic 95% CI
Parameter    Estimate          SE      Z-value         Lower        Upper

        3       .2537       .0875         2.90           .08          .43
        4       .0000         .             .             .            .
        5     -.6873        .1307        -5.26          -.94         -.43
        6       .0000         .             .             .            .
        7       .0000         .             .             .            .
        8       .0000         .             .             .            .
```

Figure 7.4 Parameter correspondence table

```
Correspondence Between Parameters and Terms of the Design

Parameter    Aliased   Term

        1              Constant for [SEX = 1]
        2              Constant for [SEX = 2]
        3              [USWARY = 1]
        4         x    [USWARY = 2]
        5              [USWARY = 1]*[SEX = 1]
        6         x    [USWARY = 1]*[SEX = 2]
        7         x    [USWARY = 2]*[SEX = 1]
        8         x    [USWARY = 2]*[SEX = 2]

Note: 'x' indicates an aliased (or a redundant) parameter.
      These parameters are set to zero.
```

As expected, you see only the parameter estimates for effects involving the dependent variable, *uswary*. Of these, only the estimates for λ^{yes} and $\lambda^{\text{yes male}}$ have not been set to 0. In addition to the parameter estimates, you see that there are constants for males and females. You can ignore these constants. (For each combination of values of the independent variables, the constants are the sum of the parameter estimates for terms that canceled when you went from the loglinear model to the corresponding logit model.)

The parameter estimates for the logit model are interpreted in the same way as the parameter estimates for the loglinear model described in Chapter 6. The parameter of interest is $\lambda^{\text{male yes}}$, which is parameter 5 in the output. In this example, since the model is saturated, you can obtain the parameter estimate directly from the observed values. First find the log of the ratio of two odds—the odds that a male answers yes and the odds that a female answers yes. The odds that a male answers yes are 0.648. The odds that a female answers yes are 1.29. The ratio of these two odds, called the **odds ratio**, is 0.502. This tells you that the odds of a yes response for a male are about half of the odds of a yes response for a female.

The log of the odds ratio, called the **log odds ratio**, is –0.69, the value for parameter 5. The log odds ratio tells you the difference in the log odds of a male answering yes and the log odds of a female answering yes. The asymptotic 95% confidence interval for the log odds ratio is from –0.94 to –0.43 (see Figure 7.3). This translates to a 95% confidence interval for the odds ratio of 0.39 to 0.65. (If sex and expectation of war are independent, the expected value of the odds ratio is 1, corresponding to a log odds ratio of 0.)

An Unsaturated Logit Model

In the previous example, you fit a saturated logit model with only one independent variable. You saw that the sex of the respondent is related to the belief that the U.S. will fight in a world war within the next 10 years. Of course, sex is only one of many possible predictors. Variables such as age, education level, and political affiliation may also be related to the response. Let's add one additional variable, highest degree received, to the model. The variable has four categories: less than a high school degree, a high school degree or junior college degree, a bachelor's degree, and a graduate degree. The saturated logit model includes parameters for the response, the response-by-sex interaction, the response-by-degree interaction, and the response-by-sex-by-degree interaction. We will fit a logit model that omits the three-way interaction among response, sex, and degree.

First look at Figure 7.5, which shows the goodness-of-fit statistics for the logit model with only two-way interactions.

Figure 7.5 Goodness-of-fit statistics

	Chi-Square	DF	Sig.
Likelihood Ratio	1.3768	3	.7110
Pearson	1.3653	3	.7137

The observed significance level of 0.7 for the chi-square statistics indicates that the model appears to fit reasonably well. The plots of the residuals do not suggest any glaring deficiencies, although the fit is somewhat poorer for cases with graduate degrees.

Figure 7.6 shows the observed and expected cell counts.

Figure 7.6 Observed and expected counts

```
Table Information

                                   Observed                 Expected
Factor                    Value       Count        %           Count        %

SEX                        MALE
 DEGREE4 Less than high schoo
  USWARY                    YES       49.00 ( 59.04)           47.82 ( 57.62)
  USWARY                     NO       34.00 ( 40.96)           35.18 ( 42.38)
 DEGREE4 High school or junio
  USWARY                    YES      103.00 ( 43.46)          101.49 ( 42.82)
  USWARY                     NO      134.00 ( 56.54)          135.51 ( 57.18)
 DEGREE4    Bachelor's degree
  USWARY                    YES       16.00 ( 20.25)           16.80 ( 21.27)
  USWARY                     NO       63.00 ( 79.75)           62.20 ( 78.73)
 DEGREE4      Graduate degree
  USWARY                    YES        7.00 ( 15.22)            8.89 ( 19.32)
  USWARY                     NO       39.00 ( 84.78)           37.11 ( 80.68)

SEX                      FEMALE
 DEGREE4 Less than high schoo
  USWARY                    YES       66.00 ( 70.97)           67.18 ( 72.24)
  USWARY                     NO       27.00 ( 29.03)           25.82 ( 27.76)
 DEGREE4 High school or junio
  USWARY                    YES      197.00 ( 58.46)          198.51 ( 58.91)
  USWARY                     NO      140.00 ( 41.54)          138.49 ( 41.09)
 DEGREE4    Bachelor's degree
  USWARY                    YES       25.00 ( 35.21)           24.20 ( 34.08)
  USWARY                     NO       46.00 ( 64.79)           46.80 ( 65.92)
 DEGREE4      Graduate degree
  USWARY                    YES       11.00 ( 37.93)            9.11 ( 31.43)
  USWARY                     NO       18.00 ( 62.07)           19.89 ( 68.57)
```

Notice that the percentages in the table are no longer based on the total sample size. Instead, for each combination of values of the independent variables, the percentages sum to 100 over the categories of the dependent variable. In Figure 7.6, you see that 15.2% of males with graduate degrees answer yes regarding the belief that the U.S. will fight in a world war within the next 10 years. The model predicts that 19.3% will answer yes. Of the women in the data set with graduate degrees, 38% answer yes. The model predicts that 31.4% will answer yes.

Measures of Dispersion and Association

When you build a logit model, you can analyze the **dispersion**, or spread, in the dependent variable. Two statistics that are used to measure the spread of a nominal variable are Shannon's entropy measure:

$$H = -\Sigma p_j \log p_j \qquad \textbf{Equation 7.8}$$

and Gini's concentration measure:

$$C = 1 - \Sigma p_j^2$$ **Equation 7.9**

Using either of these measures, you can subdivide the total dispersion of the dependent variable into the dispersion explained by the model and the residual (or unexplained) dispersion. Figure 7.7 shows the analysis-of-dispersion table for the current model. If the model is correct, two times the entropy for the model has an asymptotic chi-square distribution, with the same degrees of freedom as the model. For the concentration measure, the ratio of the model concentration divided by its degrees of freedom to the residual concentration divided by its degrees of freedom has an *F* distribution with model and residual degrees of freedom.

Figure 7.7 Analysis-of-dispersion table

Analysis of Dispersion

Source of Dispersion	Entropy	Concentration	DF
Due to Model	46.8378	44.8740	4
Due to Residual	628.6068	442.2522	970
Total	675.4446	487.1262	974

Measures of Association

```
      Entropy =  .0693
Concentration =  .0921
```

From the analysis-of-dispersion table, you can calculate statistics similar to R^2 in regression. They indicate the proportion of the total dispersion in the dependent variable that is attributable to the model (Magidson, 1981). In Figure 7.7, when dispersion is measured by the entropy criterion, the ratio of the dispersion explained by the model to the total dispersion is 0.069. When measured by the concentration criterion, it is 0.092. These values can be interpreted as measures of association. For a two-way table, the concentration measure is the square of Kendall's tau-*b*, while the entropy measure is the same as the uncertainty coefficient with response as the dependent variable. Although it is tempting to interpret the magnitude of these measures similarly to R^2 in regression, this may be misleading, since the coefficients may be small even when the variables are strongly related (Haberman, 1982). The coefficients are best interpreted in light of experience.

Parameter Estimates

The parameter estimates are shown in Figure 7.8, and their explanation is shown in Figure 7.9.

Figure 7.8 Parameter estimates

```
Parameter Estimates

 Constant   Estimate

        1     3.5604
        2     4.9090
        3     4.1303
        4     3.6140
        5     3.2512
        6     4.9308
        7     3.8459
        8     2.9900

Note: Constants are not parameters under multinomial assumption.
      Therefore, standard errors are not calculated.

                                                 Asymptotic 95% CI
Parameter   Estimate          SE     Z-value      Lower        Upper

        9     -.7803       .2825       -2.76      -1.33         -.23
       10      .0000           .           .          .            .
       11     1.7365       .3166        5.49       1.12         2.36
       12     1.1403       .2863        3.98        .58         1.70
       13      .1206       .3300         .37       -.53          .77
       14      .0000           .           .          .            .
       15      .0000           .           .          .            .
       16      .0000           .           .          .            .
       17      .0000           .           .          .            .
       18      .0000           .           .          .            .
       19     -.6491       .1358       -4.78       -.92         -.38
       20      .0000           .           .          .            .
       21      .0000           .           .          .            .
       22      .0000           .           .          .            .
```

Figure 7.9 Parameter correspondence table

```
Correspondence Between Parameters and Terms of the Design

Parameter   Aliased  Term

        1            Constant for [SEX = 1]*[DEGREE4 = .00]
        2            Constant for [SEX = 1]*[DEGREE4 = 1.00]
        3            Constant for [SEX = 1]*[DEGREE4 = 2.00]
        4            Constant for [SEX = 1]*[DEGREE4 = 3.00]
        5            Constant for [SEX = 2]*[DEGREE4 = .00]
        6            Constant for [SEX = 2]*[DEGREE4 = 1.00]
        7            Constant for [SEX = 2]*[DEGREE4 = 2.00]
        8            Constant for [SEX = 2]*[DEGREE4 = 3.00]
        9            [USWARY = 1]
       10       x    [USWARY = 2]
       11            [USWARY = 1]*[DEGREE4 = .00]
       12            [USWARY = 1]*[DEGREE4 = 1.00]
       13            [USWARY = 1]*[DEGREE4 = 2.00]
       14       x    [USWARY = 1]*[DEGREE4 = 3.00]
       15       x    [USWARY = 2]*[DEGREE4 = .00]
       16       x    [USWARY = 2]*[DEGREE4 = 1.00]
       17       x    [USWARY = 2]*[DEGREE4 = 2.00]
       18       x    [USWARY = 2]*[DEGREE4 = 3.00]
       19            [USWARY = 1]*[SEX = 1]
       20       x    [USWARY = 1]*[SEX = 2]
       21       x    [USWARY = 2]*[SEX = 1]
       22       x    [USWARY = 2]*[SEX = 2]

Note: 'x' indicates an aliased (or a redundant) parameter.
      These parameters are set to zero.
```

Parameters 11–14 are for degree by response, and parameter 19 is for sex by response. The parameters involving females and graduate degrees are set to 0, so they serve as the basis for comparison. Parameter 11 tells you that the log of the predicted odds ratio of responding yes for a person without a high school degree compared to the odds of responding yes for a person with a graduate degree is 1.74. This translates to an odds ratio of $e^{1.74}$, or 5.7. Similarly, parameter 12 tells you that the predicted odds ratio of responding yes for a person with a high school degree compared to a person with a graduate degree is $e^{1.14}$, or 3.1. Parameter 13 tells you that the predicted odds ratio of responding yes for a person with a bachelor's degree compared to a person with a graduate degree is $e^{0.12}$, or 1.13. Parameter 19, the parameter for sex, tells you that the predicted odds ratio for a male responding yes compared to a female responding yes is 0.52. Notice that the 95% confidence intervals for all of the parameters, except the parameter for bachelor's degree, do *not* include the value 0, which means that you can reject the null hypotheses that each of these log odds ratios is 0 in the population. Since the 95% confidence interval of the log odds ratio for bachelor's degree does include 0, you can't reject the hypothesis that the log odds ratio is 0 when people with bachelor's degrees are compared to people with graduate degrees.

Polychotomous Logit Model

In the previous examples, the dependent variable had two categories, which is the simplest example of a logit model. Logit models can also be used for categorical dependent variables that have more than two values. Such models are called **polychotomous logit models**. As an example, let's consider the relationship between marital status and sex and age. The categories of the marital status variable are never married, widowed, divorced or separated, and married. The age categories are 18–29, 30–45, 46–65, and 65+. Once again, the data are a subset from the 1993 General Social Survey.

When the dependent variable has more than two categories, several different types of logit models can be used. We'll look at one of the simplest types, the **baseline category logit model**. When your dependent variable had two values, you modeled the log of the odds of the first value of the dependent variable for the different combinations of values of the independent variables. When your dependent variable has several values, you no longer have a unique category to serve as the reference. Instead, you can select any one of the categories of the dependent variable to serve as the baseline, or comparison, category.

For example, if you select the married category as the comparison category and fit a logit model without the three-way interaction, the model for the never-married males in age group 1 is

$$\ln\left(\frac{m_{111}}{m_{211}}\right) = \lambda^{\text{never married}} + \lambda^{\text{never-married male}} + \lambda^{\text{never-married age1}}$$ **Equation 7.10**

The predicted count for never-married males in age group 1 is m_{111}, and the predicted count for married males in age group 1 is m_{211}. (The values of all of the parameters for the reference cell are 0, so they do not appear in the model.) For each combination of values of sex, age, and marital status, the dependent variable is the log of the ratio of the probability of that marital category compared to the probability of being married.

Goodness of Fit of the Model

The goodness-of-fit statistics for fitting a polychotomous logit model without the three-way interaction of sex, age, and marital status are shown in Figure 7.10.

Figure 7.10 Goodness-of-fit statistics

	Chi-Square	DF	Sig.
Likelihood Ratio	8.5593	9	.4789
Pearson	9.0183	9	.4356

The observed significance level is reasonably large, indicating that the model appears to fit the data. Figure 7.11, the plot of the adjusted residuals, does not suggest any problems.

Figure 7.11 Plot of adjusted residuals

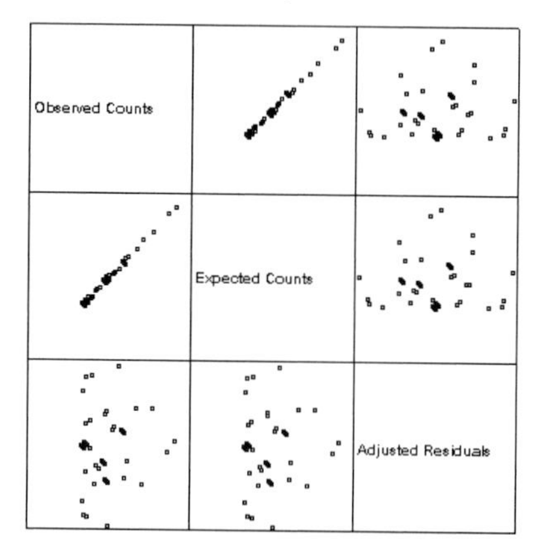

The normal probability plot of the adjusted residuals, shown in Figure 7.12, appears consistent with normality.

Figure 7.12 Normal probability plot of adjusted residuals

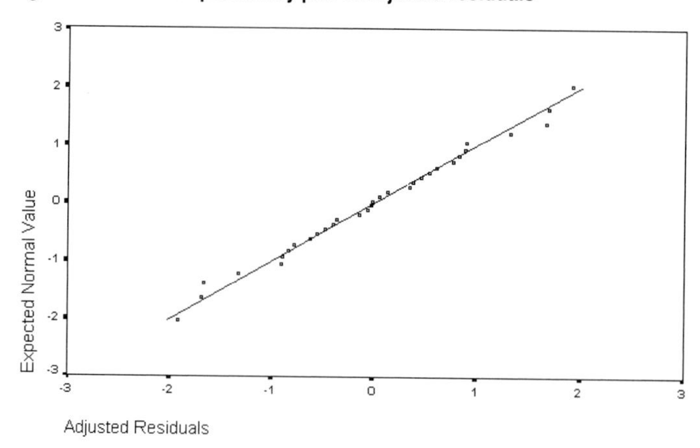

Interpreting Parameter Estimates

To see how the independent variables relate to the dependent variable, you must look at the parameter estimates. Figure 7.13 is an excerpt from the table that identifies the parameters. Only parameters that are not set to 0 are included in the table.

Figure 7.13 Correspondence table of parameters not set to 0

```
Correspondence Between Parameters and Terms of the Design

Parameter   Aliased  Term

        1            Constant for [SEX = 1]*[AGE4 = 1.00]
        2            Constant for [SEX = 1]*[AGE4 = 2.00]
        3            Constant for [SEX = 1]*[AGE4 = 3.00]
        4            Constant for [SEX = 1]*[AGE4 = 4.00]
        5            Constant for [SEX = 2]*[AGE4 = 1.00]
        6            Constant for [SEX = 2]*[AGE4 = 2.00]
        7            Constant for [SEX = 2]*[AGE4 = 3.00]
        8            Constant for [SEX = 2]*[AGE4 = 4.00]
        9            [MARITAL4 = 1.00]
       10            [MARITAL4 = 2.00]
       11            [MARITAL4 = 3.00]

       13            [MARITAL4 = 1.00]*[AGE4 = 1.00]
       14            [MARITAL4 = 1.00]*[AGE4 = 2.00]
       15            [MARITAL4 = 1.00]*[AGE4 = 3.00]

       17            [MARITAL4 = 2.00]*[AGE4 = 1.00]
       18            [MARITAL4 = 2.00]*[AGE4 = 2.00]
       19            [MARITAL4 = 2.00]*[AGE4 = 3.00]

       21            [MARITAL4 = 3.00]*[AGE4 = 1.00]
       22            [MARITAL4 = 3.00]*[AGE4 = 2.00]
       23            [MARITAL4 = 3.00]*[AGE4 = 3.00]

       29            [MARITAL4 = 1.00]*[SEX = 1]

       31            [MARITAL4 = 2.00]*[SEX = 1]

       33            [MARITAL4 = 3.00]*[SEX = 1]
```

The actual parameter estimates are shown in Figure 7.14. Again, for readability, parameters that are set to 0 are excluded from the table.

Figure 7.14 Estimates of parameters not set to 0

Parameter Estimates

Constant	Estimate
1	3.7194
2	5.0851
3	4.7573
4	4.1549
5	4.0388
6	5.1557
7	4.8565
8	3.9751

Note: Constants are not parameters under multinomial assumption.
Therefore, standard errors are not calculated.

Parameter	Estimate	SE	Z-value	Asymptotic 95% CI Lower	Upper
9	-2.6733	.3578	-7.47	-3.37	-1.97
10	.6330	.1571	4.03	.33	.94
11	-1.4275	.2465	-5.79	-1.91	-.94
13	3.0788	.3698	8.33	2.35	3.80
14	1.2625	.3660	3.45	.55	1.98
15	.3727	.4012	.93	-.41	1.16
17	-12.5252	28.7039	-.44	-68.78	43.73
18	-5.2710	.7238	-7.28	-6.69	-3.85
19	-2.0228	.2267	-8.92	-2.47	-1.58
21	.1034	.3395	.30	-.56	.77
22	.6859	.2608	2.63	.17	1.20
23	.6402	.2694	2.38	.11	1.17
29	.1907	.1549	1.23	-.11	.49
31	-1.5807	.2385	-6.63	-2.05	-1.11
33	-.6205	.1519	-4.08	-.92	-.32

To make it easier to interpret, the parameter estimates (e^{λ}) are summarized in Table 7.1.

Table 7.1 Parameter estimate (e^{λ}) summary

	Marital Status			
Age	**Never married**	**Widowed**	**Divorced/Separated**	**Married**
18–29	3.08 (21.76)	–12.52 (<0.001)	0.10 (1.11)	0 (1)
30–45	1.26 (3.52)	–5.27 (0.005)	0.69 (1.99)	0 (1)
46–65	0.37 (1.45)	–2.02 (0.13)	0.64 (1.90)	0 (1)
65+	0 (1)	0 (1)	0 (1)	0 (1)
Sex				
Male	0.19 (1.21)	–1.58 (0.20)	–0.62 (0.54)	0 (1)
Female	0 (1)	0 (1)	0 (1)	0 (1)

The parameter estimate lambda (λ) for age 18–29 and never married is 3.08. The value of e^{λ} is 21.76, which tells you that, based on the model, you are almost 22 times more likely to be never married than married at age 18–29 compared to being never married to married at age 65+. Since the model is not saturated, your estimate of the odds ratio is not exactly the same as that observed in the data. Let's see how much the estimated and observed odds ratios differ. To calculate the observed odds ratio, look at Figure 7.15, which is the crosstabulation of marital status by age for all of the cases.

Figure 7.15 Crosstabulation of marital status by age for all cases

AGE4 Page 1 of 1

MARITAL4 / Count	18-29 years 1.00	30-45 years 2.00	46-65 years 3.00	65+ years 4.00	Row Total
1.00 Never married	160	90	27	9	286 19.1
2.00 Widowed		2	38	125	165 11.0
3.00 Divorced or sepa	21	124	87	21	253 16.9
4.00 Married	98	335	245	117	795 53.0
Column Total	279 18.6	551 36.8	397 26.5	272 18.1	1499 100.0

Number of Missing Observations: 1

You'll need only the first and last rows for your computations. The odds of a person being never married to married at age 18–29 are 160/98. The odds of a person being never married to married at age 65+ are 9/117. The ratio of these two odds is 21.22. The estimated odds ratio of 21.76 is quite close to the observed ratio.

In Table 7.1, you see that the likelihood of being never married to married decreases with age. However, in Figure 7.14 you see that the parameter estimate for age 46–65 is not significantly different from 0. This means that by this age group, the odds of being never married to married are not significantly different from those at age 65+.

Now let's look at the widowed row. The parameter estimate for age 18–29 is very large and negative. That's because the odds of being widowed to being married at age 18–29 are very small compared to those at age 65+. Notice in Figure 7.14 that the standard error of the parameter estimate is quite large and that the parameter estimate is not significantly different from 0. This is because there are no young widowed cases in the sample, so the parameter cannot be estimated well. The correct conclusion is not that the value of the parameter is close to 0, but that it can't be estimated well from the data. For age group 30–45 compared to age group 65+, the estimated odds ratio of being widowed to married is 0.005. This means that the odds of being widowed to married are almost 200 times as large for people at age 65+ when compared to age group 30–45. Notice that this parameter estimate has a much smaller standard error than the previous parameter estimate and *is* significantly different from 0.

Now let's look at how a person's sex relates to marital status. The odds ratio of 1.2 for never-married males tells you that for a male, the estimated odds of being never married to married are 1.2 times as large as the same odds for a female. In Figure 7.14, you see that the parameter estimate is not significantly different from 0, so you can't conclude that the observed odds ratio of 1.2 is significantly different from 1, the odds ratio under the null hypothesis. The estimated odds ratio for widowed males is 0.20. This means that the odds for a male of being widowed to married are one-fifth as large as the same odds for a female. Similarly, the estimated odds ratio for divorced or separated males tells you that the odds that a male is divorced or separated compared to married are a little more than half of the odds that a female is divorced or separated compared to married. Thus, a female is almost two times as likely as a male to be divorced or separated compared to married.

Other Logit Models

In the previous example, the reference, or baseline, category for comparisons was one of the values of the dependent variable. (SPSS always chooses the last category of the dependent variable as the baseline category. By recoding the values of your dependent variable, you can select any category to be the baseline.) There are other ways to choose the comparisons. For example, when the dependent variable is ordinal, you can fit an **adjacent-categories logit model**. In an adjacent-categories logit model, each category of the dependent variable is compared to the next category. In a **continuation-ratio logit model**, each category of the dependent variable is compared to the sum of the categories that follow. SPSS will not fit these models directly; however, it is possible to fit the models in SPSS by estimating a series of logit models. See Agresti (1990) for details.

The models in this chapter include a single dependent variable. However, you can specify more than one dependent variable in a logit model. The interaction terms let you examine the relationships among the dependent variables as well as their interactions with the independent variables. The interpretation of these interaction terms is the same as for the identical terms in a general loglinear model.

How to Obtain a Logit Loglinear Analysis

The Logit Loglinear Analysis procedure estimates parameters of logit loglinear models using the Newton-Raphson algorithm.

The minimum specification is:

- For saturated logit models, one categorical dependent variable.

To obtain a logit loglinear analysis, from the menus choose:

Statistics
 Loglinear ▸
  Logit...

This opens the Logit Loglinear Analysis dialog box, as shown in Figure 7.16.

Figure 7.16 Logit Loglinear Analysis dialog box

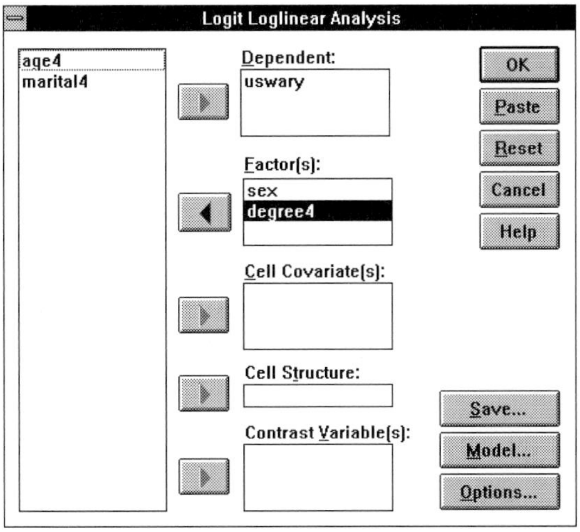

The numeric variables in your data file appear on the source list.

Dependent. Select one or more categorical variables. The total number of dependent and factor variables must be less than or equal to 10.

Factor(s). Optionally, you can select one or more categorical factor variables. The total number of dependent and factor variables must be less than or equal to 10.

Cell Covariate(s). Optionally, you can select one or more continuous cell covariates. When a covariate is in the model, SPSS applies the mean covariate value for cases in a cell to that cell. To analyze an equiprobability model, select a variable that is actually a constant of 1. You can select a maximum of 200 covariates.

Cell Structure. By default, SPSS uses a weight of 1 for all cells. To provide your own cell weights, select a variable whose values are the weights. SPSS weights each cell by the average weight for cases in the cell. You can use this feature to define fixed (structural) zeros for incomplete tables. Do not use this facility to weight aggregate data; instead, choose Weight Cases on the Data menu (see the SPSS Base system documentation for more information on case weighting).

Contrast Variable(s). Specify one or more continuous variables to be used as contrast variables. Contrast variables are used to compute generalized log-odds ratios (GLOR). The values of the contrast variable are the coefficients for the linear combination of the logs of the expected cell counts.

Model Specification

By default, SPSS analyzes a saturated logit model. To analyze an unsaturated logit model, click on Model... in the Logit Loglinear Analysis dialog box. This opens the Logit Loglinear Analysis Model dialog box, as shown in Figure 7.17.

Figure 7.17 Logit Loglinear Analysis Model dialog box

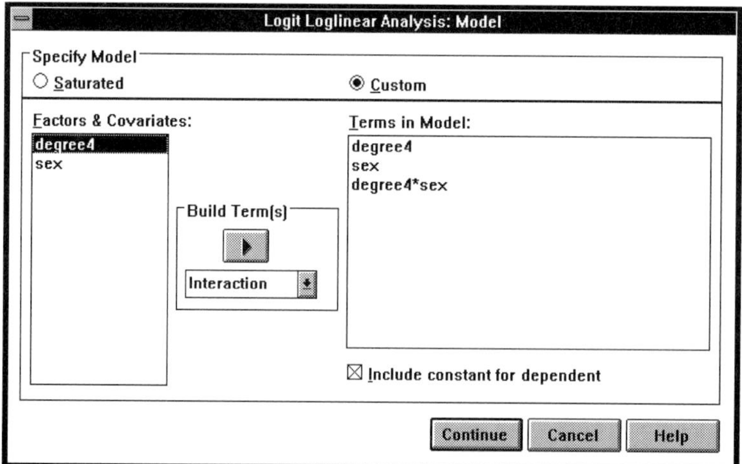

Specify Model. You can choose one of the following models:

❍ **Saturated.** Analyzes a saturated logit model. This is the default. Covariates are not included in the model. To analyze a model containing covariates, define a custom model and include the covariate in the model.

❍ **Custom.** Defines an unsaturated logit model. You must specify which factors, if any, to include in the model (you do not need to specify any factors if the model includes a main effect for the dependent variable; see below).

For custom models, if you remove a variable in the main dialog box, any terms in the model that contain that variable will also be deleted.

Terms in Model. To create a term for the model, select one or more factors or covariates. If you don't want to create the highest-order interaction for the variables, select an alternate method for building a term from the Build Term(s) drop-down list. To add more terms to the model, repeat this process. You cannot use a term more than once in the model.

For example, to define a model containing the main effects of *sex* and *degree4* on the dependent variable, first highlight *sex* and *degree4*. Next, select Main effects from the Build Term(s) drop-down list, and click on ▶. These terms appear in the output as *sex**[dependent variable] and *degree4**[dependent variable]. To add a term for the effect of the *sex*-by-*degree4* interaction on the dependent variable, highlight *sex* and *degree4*, select Interaction (the default) from the Build Term(s) drop-down list, and click on ▶. This term appears in the output as *sex*degree4**[dependent variable].

Macintosh: Use ⌘-click to select multiple factors or covariates, or a combination.

🠗 **Build Term(s).** You can build main effects or interactions for the selected variables. If you request an interaction that has a higher order than the number of variables, SPSS creates a term for the highest-order interaction possible for the selected variables. If only one variable is selected, the main effect of the selected variable is added to the model. Choose one of the following alternatives:

Interaction. Creates the highest-level interaction term for the variables. This is the default for a selected group of variables.

Main effects. Creates a main-effects term for each variable.

All 2-way. Creates all possible two-way interactions for the variables.

All 3-way. Creates all possible three-way interactions for the variables.

All 4-way. Creates all possible four-way interactions for the variables.

All 5-way. Creates all possible five-way interactions for the variables.

The following option is also available for custom models only:

❑ **Include constant for dependent.** Includes a constant for the dependent variable in a custom model. This is the default. To exclude the dependent variable from a custom model, deselect this item. If the dependent variable is excluded from a custom model, the model must contain at least one factor.

How Terms Are Used in the Analysis

Terms are added to the design by taking all possible combinations of the dependent terms and matching each combination with each term in the model list. If the Include constant option is selected, there is also a unit term (1) added to the model list.

For example, suppose variables *D1* and *D2* are the dependent variables. A dependent terms list is created by the Logit Loglinear Analysis procedure (*D1*, *D2*, *D1*D2*). If the terms in model list contains *M1* and *M2* and a constant is included, the model list contains *1*, *M1*, and *M2*. The resultant design includes combinations of each model term with each dependent term, as shown in the following list:

D1, D2, D1*D2,
M1*D1, M1*D2, M1*D1*D2,
M2*D1, M2*D2, M2*D1*D2

Options

To obtain optional statistics or plots, or to control model criteria, click on Options... in the Logit Loglinear Analysis dialog box. This opens the Logit Loglinear Analysis Options dialog box, as shown in Figure 7.18.

Figure 7.18 Logit Loglinear Analysis Options dialog box

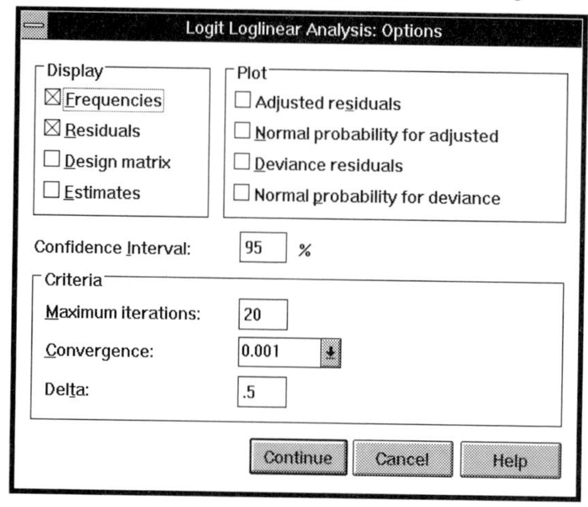

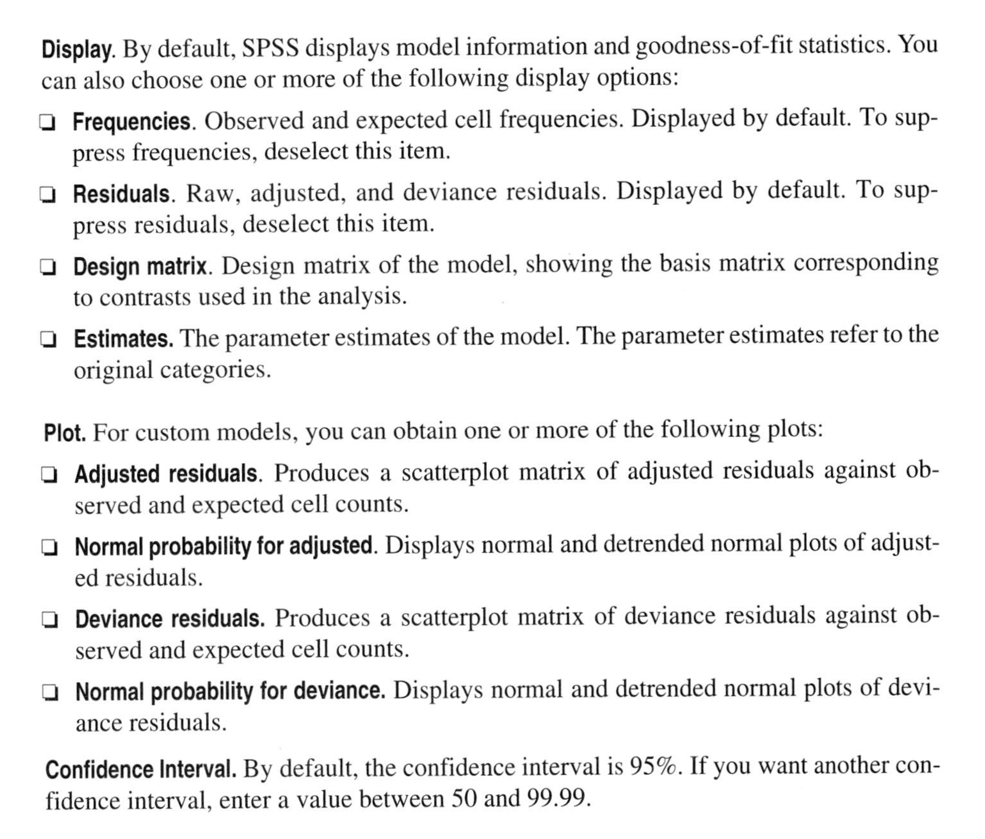

Display. By default, SPSS displays model information and goodness-of-fit statistics. You can also choose one or more of the following display options:

- ❑ **Frequencies**. Observed and expected cell frequencies. Displayed by default. To suppress frequencies, deselect this item.
- ❑ **Residuals**. Raw, adjusted, and deviance residuals. Displayed by default. To suppress residuals, deselect this item.
- ❑ **Design matrix**. Design matrix of the model, showing the basis matrix corresponding to contrasts used in the analysis.
- ❑ **Estimates.** The parameter estimates of the model. The parameter estimates refer to the original categories.

Plot. For custom models, you can obtain one or more of the following plots:

- ❑ **Adjusted residuals**. Produces a scatterplot matrix of adjusted residuals against observed and expected cell counts.
- ❑ **Normal probability for adjusted**. Displays normal and detrended normal plots of adjusted residuals.
- ❑ **Deviance residuals.** Produces a scatterplot matrix of deviance residuals against observed and expected cell counts.
- ❑ **Normal probability for deviance.** Displays normal and detrended normal plots of deviance residuals.

Confidence Interval. By default, the confidence interval is 95%. If you want another confidence interval, enter a value between 50 and 99.99.

Criteria. The Newton-Raphson method is used to obtain maximum-likelihood parameter estimates. You can control one or more of the following algorithm criteria:

> **Maximum iterations**. By default, a maximum of 20 iterations is performed. To specify a different maximum, enter a positive integer.
>
> **Convergence**. By default, the convergence criterion is 0.001. To override the default, select an alternate convergence value from the drop-down list.
>
> **Delta**. Constant added to all cells for initial approximations. Delta remains in the cells only for saturated models. The default value is 0.5. To override the default delta, enter a value between 0 and 1.

Saving Residuals or Predicted Values

To save residuals or predicted values as new variables, click on Save... in the Logit Loglinear Analysis dialog box. This opens the Logit Loglinear Analysis Save dialog box, as shown in Figure 7.19.

Figure 7.19 Logit Loglinear Analysis Save dialog box

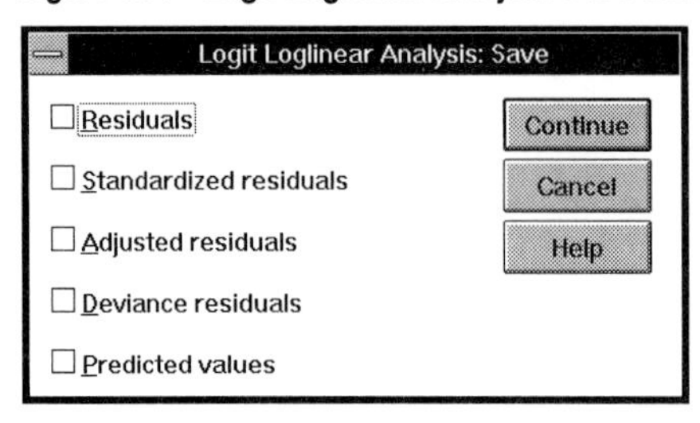

The saved values refer to the aggregated data (to cells in the contingency table), even if the data are recorded in individual observations in the Data Editor. If you save residuals or predicted values for unaggregated data, the saved value for a cell in the contingency table is entered in the Data Editor for each case in that cell. To make sense of the saved values, you should aggregate the data to obtain the cell counts.

You can choose one or more of the following alternatives:

❑ **Residuals.** The difference between the observed and the expected count.

❑ **Standardized residuals**. The residual divided by the standard error of the observed count.

❑ **Adjusted residuals.** The residual divided by its estimated standard error.

❑ **Deviance residuals.** The signed square root of the individual contribution to the likelihood-ratio chi-square statistic.

❑ **Predicted values.** Estimated number of observations in a cell estimated from the sample under a specified model.

Additional Features Available with Command Syntax

You can customize your logit loglinear analysis if you paste your selections into a syntax window and edit the resulting GENLOG command syntax. (For information on syntax windows, see the SPSS Base system documentation.) Additional features include:

- The default threshold value for redundancy checking can be changed by using the keyword EPS in the CRITERIA subcommand. The default value is 0.0000001, or 10^{-8}.
- Generalized residuals (using the GRESID subcommand).
- Standardized residuals (using the PRINT subcommand).

See the Syntax Reference section of this manual for command syntax rules and for complete GENLOG command syntax.

8 Nonlinear Regression

Many real-world relationships are approximated with linear models, especially in the absence of theoretical models that can serve as guides. We would be unwise to model the relationship between speed of a vehicle and stopping time with a linear model, since the laws of physics dictate otherwise. However, nothing deters us from modeling salary as a linear function of variables such as age, education, and experience. In general, we choose the simplest model that fits an observed relationship. Another reason that explains our affinity to linear models is the accompanying simplicity of statistical estimation and hypothesis testing. Algorithms for estimating parameters of linear models are straightforward; direct solutions are available; iteration is not required. There are, however, situations in which it is necessary to fit nonlinear models. Before considering the steps involved in nonlinear model estimation, let's consider what makes a model nonlinear.

What Is a Nonlinear Model?

There is often confusion about the characteristics of a nonlinear model. Consider the following equation:

$$Y = B_0 + B_1 X_1^2 \qquad \text{Equation 8.1}$$

Is this a linear or nonlinear model? The equation is certainly not that of a straight line—it is the equation for a parabola. However, the word *linear,* in this context, does not refer to whether the equation is that of a straight line or a curve. It refers to the functional form of the equation. That is, can the dependent variable be expressed as a linear combination of parameter values times values of the independent variables? The parameters must be linear. The independent variables can be transformed in any fashion. They can be raised to various powers, logged, and so on. The transformation cannot involve the parameters in any way, however.

The previous model is a linear model, since it is nonlinear in only the independent variable X. It is linear in the parameters B_0 and B_1. In fact, we can write the model as

$$Y = B_0 + B_1 X' \qquad \text{Equation 8.2}$$

where X' is the square of X_1. The parameters in the model can be estimated using the usual linear model techniques.

Transforming Nonlinear Models

Consider the model

$$Y = e^{B_0 + B_1 X_1 + B_2 X_2 + E} \qquad \text{Equation 8.3}$$

The model, as it stands, is not of the form

$$Y = B_0 + B_1 Z_1 + B_2 Z_2 + \ldots + B_p Z_p + E \qquad \text{Equation 8.4}$$

where the B's are the parameters and the Z's are functions of the independent variables, so it is a nonlinear model. However, if we take natural logs of both sides of Equation 8.3, we get the model

$$\ln(Y) = B_0 + B_1 X_1 + B_2 X_2 + E \qquad \text{Equation 8.5}$$

The transformed equation is linear in the parameters, and we can use the usual techniques for estimating them. Models that initially appear to be nonlinear but that can be transformed to a linear form are sometimes called **intrinsically linear models**. It is a good idea to examine what appears to be a nonlinear model to see if it can be transformed to a linear one. Transformation to linearity makes estimation much easier.

Another example of a transformable nonlinear model is

$$Y = e^B X + E \qquad \text{Equation 8.6}$$

The transformation $B' = e^B$ results in the model

$$Y = B'X + E \qquad \text{Equation 8.7}$$

We can use the usual methods to estimate B' and then take its natural log to get the values of B.

Error Terms in Transformed Models

In both linear and nonlinear models, we assume that the error term is additive. When we transform a model to linearity, we must make sure that the transformed error term satisfies the requisite assumptions. For example, if our original model is

$$Y = e^{BX} + E \qquad \text{Equation 8.8}$$

taking natural logs does not result in a model that has an additive error term. To have an additive error term in the transformed model, our original model would have had to be

$$Y = e^{BX + E} = e^{BX}e^{E}$$ **Equation 8.9**

Intrinsically Nonlinear Models

A model such as

$$Y = B_0 + e^{B_1X_1} + e^{B_2X_2} + e^{B_3X_3} + E$$ **Equation 8.10**

is **intrinsically nonlinear**. We can't apply a transformation to linearize it. We must estimate the parameters using nonlinear regression. In nonlinear regression, just as in linear regression, we choose values for the parameters so that the sum of squared residuals is a minimum. There is not, however, a closed solution. We must solve for the values iteratively. There are several algorithms for the estimation of nonlinear models (see Fox, 1984; Draper & Smith, 1981).

Fitting the Logistic Population Growth Model

As an example of fitting a nonlinear equation, we will consider a model for population growth. Population growth is often modeled using a logistic population growth model of the form

$$Y_i = \frac{C}{1 + e^{A + BT_i}} + E_i$$ **Equation 8.11**

where Y_i is the population size at time T_i. Although the model often fits the observed data reasonably well, the assumptions of independent error and constant variance may be violated, since with time-series data errors are not independent and the size of the error may be dependent on the magnitude of the population. Since the logistic population growth model is not transformable to a linear model, we will have to use nonlinear regression to estimate the parameters.

Figure 8.1 contains a listing of decennial populations (in millions) of the United States from 1790 to 1960, as found in Fox (1984). Figure 8.2 is a plot of the same data. For the nonlinear regression, we will use the variable *decade*, which represents the number of decades since 1790, as the independent variable. This should prevent possible computational difficulties arising from large data values (see "Computational Problems" on p. 240).

Figure 8.1 Decennial population of the United States

POP	YEAR	DECADE
3.895	1790	0
5.267	1800	1
7.182	1810	2
9.566	1820	3
12.834	1830	4
16.985	1840	5
23.069	1850	6
31.278	1860	7
38.416	1870	8
49.924	1880	9
62.692	1890	10
75.734	1900	11
91.812	1910	12
109.806	1920	13
122.775	1930	14
131.669	1940	15
150.697	1950	16
178.464	1960	17

Figure 8.2 Plot of decennial population of the United States

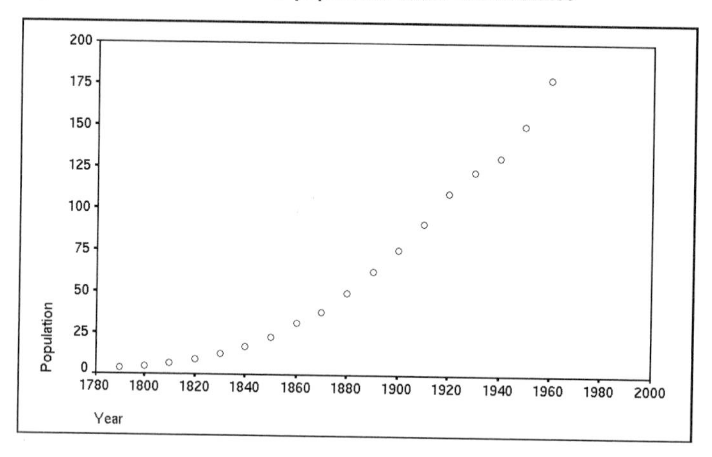

In order to start the nonlinear estimation algorithm, we must have initial values for the parameters. Unfortunately, the results of nonlinear estimation often depend on having good starting values for the parameters. There are several ways for obtaining starting values (see “Estimating Starting Values” on p. 238 through “Use Properties of the Nonlinear Model” on p. 239).

For this example, we can obtain starting values by making some simple assumptions. In the logistic growth model, the parameter *C* represents the asymptote. We’ll arbitrarily choose an asymptote that is not too far from the largest observed value. Let’s take an asymptote of 200, since the largest observed value for the population is 178.

Using the value of 200 for C, we can estimate a value for A based on the observed population at time 0:

$$3.895 = \frac{200}{1 + e^{A}}$$ **Equation 8.12**

So,

$$A = \ln\left(\frac{200}{3.895} - 1\right) = 3.9$$ **Equation 8.13**

To estimate a value for B, we can use the population at time 1, and our estimates of C and A. This gives us

$$5.267 = \frac{200}{1 + e^{B + 3.9}}$$ **Equation 8.14**

from which we derive

$$B = \ln\left(\frac{200}{5.27} - 1\right) - 3.9 = -0.29$$ **Equation 8.15**

We use these values as initial values in the nonlinear regression routine.

Estimating the Parameters

Figure 8.3 shows the residual sums of squares and parameter estimates at each iteration. At step 1, the parameter estimates are the starting values that we have supplied. At the major iterations, which are identified with integer numbers, the derivatives are evaluated and the direction of the search determined. At the minor iterations, the distance is established. As the note at the end of the table indicates, iteration stops when the relative change in residual sums of squares between iterations is less than or equal to the convergence criterion.

Figure 8.3 Parameter estimates for nonlinear regression

Iteration	Residual SS	A	B	C
1	969.6898219	3.90000000	-.30000000	200.000000
1.1	240.3756627	3.87148504	-.27852485	237.513990
2	240.3756627	3.87148504	-.27852485	237.513990
2.1	186.5020615	3.89003377	-.27910189	243.721558
3	186.5020615	3.89003377	-.27910189	243.721558
3.1	186.4972404	3.88880287	-.27886478	243.975460
4	186.4972404	3.88880287	-.27886478	243.975460
4.1	186.4972278	3.88885123	-.27886164	243.985980
5	186.4972278	3.88885123	-.27886164	243.985980
5.1	186.4972277	3.88884856	-.27886059	243.987296

```
Run stopped after 10 model evaluations and 5 derivative evaluations.
Iterations have been stopped because the relative reduction between successive
residual sums of squares is at most SSCON = 1.000E-08
```

Summary statistics for the nonlinear regression are shown in Figure 8.4. For a nonlinear model, the tests used for linear models are not appropriate. In this situation, the residual mean square is not an unbiased estimate of the error variance, even if the model is correct. For practical purposes, we can still compare the residual variance with an estimate of the total variance, but the usual *F* statistic cannot be used for testing hypotheses.

The entry in Figure 8.4 labeled *Uncorrected Total* is the sum of the squared values of the dependent variable. The entry labeled *(Corrected Total)* is the sum of squared deviations around the mean. The *Regression* sum of squares is the sum of the squared predicted values. The entry labeled *R squared* is the coefficient of determination. It may be interpreted as the proportion of the total variation of the dependent variable around its mean that is explained by the fitted model. For nonlinear models, its value can be negative if the selected model fits worse than the mean. (For a discussion of this statistic, see Kvalseth, 1985.) It appears from the R^2 value of 0.9965 that the model fits the observed values well. Figure 8.5 is a plot of the observed and predicted values for the model.

Figure 8.4 Summary statistics for nonlinear regression

Nonlinear Regression Summary Statistics   Dependent Variable POP

Source	DF	Sum of Squares	Mean Square
Regression	3	123053.53112	41017.84371
Residual	15	186.49723	12.43315
Uncorrected Total	18	123240.02834	
(Corrected Total)	17	53293.92477	

R squared = 1 - Residual SS / Corrected SS =   .99650

Figure 8.5 Observed and predicted values for nonlinear model

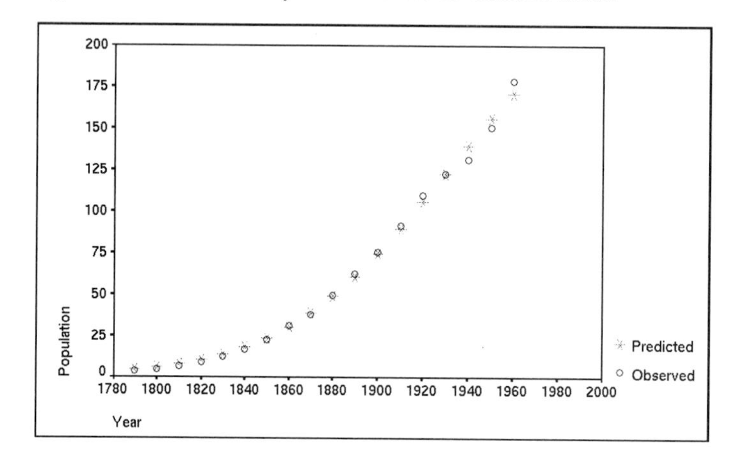

Approximate Confidence Intervals for the Parameters

In the case of nonlinear regression, it is not possible to obtain exact confidence intervals for each of the parameters. Instead, we must rely on **asymptotic** (large sample) approximations. Figure 8.6 shows the estimated parameters, standard errors, and asymptotic 95% confidence intervals. The asymptotic correlation matrix of the parameter estimates is shown in Figure 8.7. If there are very large positive or negative values for the correlation coefficients, it is possible that the model is **overparameterized**. That is, a model with fewer parameters may fit the observed data as well. This does not necessarily mean that the model is inappropriate; it may mean that the amount of data is not sufficient to estimate all of the parameters.

Figure 8.6 Estimated parameters and confidence intervals

Parameter	Estimate	Asymptotic Std. Error	Asymptotic 95 % Confidence Interval Lower	Upper
A	3.888848562	.093704407	3.689122346	4.088574778
B	-.278860588	.015593951	-.312098308	-.245622868
C	243.98729636	17.967399750	205.69069033	282.28390239

Figure 8.7 Asymptotic correlation matrix of parameter estimates

Asymptotic Correlation Matrix of the Parameter Estimates

	A	B	C
A	1.0000	-.7244	-.3762
B	-.7244	1.0000	.9042
C	-.3762	.9042	1.0000

Examining the Residuals

The SPSS Nonlinear Regression procedure allows you to save predicted values and residuals that can be used for exploring the goodness of fit of the model. Figure 8.8 is a plot of residuals against the observed year values. You will note that the errors appear to be correlated and that the variance of the residuals increases with time.

Figure 8.8 Plot of residuals against observed values

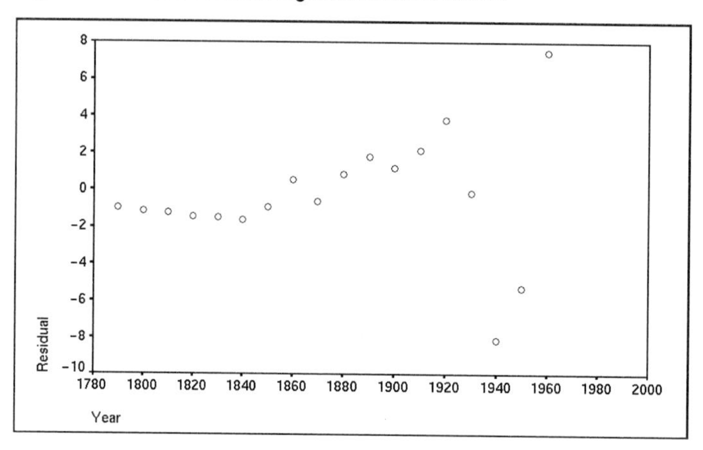

To compute asymptotic standard errors of the predicted values and statistics used for outlier detection and influential case analysis, you can use the SPSS Linear Regression procedure, specifying the residuals from the SPSS Nonlinear Regression procedure as the dependent variable and the derivatives (see "Saving New Variables" on p. 245) as the independent variables.

Estimating Starting Values

As previously indicated, you must specify initial values for all parameters. Good initial values are important and may provide a better solution in fewer iterations. In addition, computational difficulties can sometimes be avoided by a good choice of initial values. Poor initial values can result in nonconvergence, a local rather than global solution, or a physically impossible solution.

There are a number of ways to determine initial values for nonlinear models. Milliken (1987) and Draper and Smith (1981) describe several approaches, which are summarized in the following sections. Generally, a combination of techniques will be most useful. If you don't have starting values, don't just set them all to 0. Use values in the neighborhood of what you expect to see.

If you ignore the error term, sometimes a linear form of the model can be derived. Linear regression can then be used to obtain initial values. For example, consider the model:

$$Y = e^{A + BX} + E$$

Equation 8.16

If we ignore the error term and take the natural log of both sides of the equation, we obtain the model

$$\ln(Y) = A + BX \quad \text{Equation 8.17}$$

We can use linear regression to estimate A and B and specify these values as starting values in nonlinear regression.

Use Properties of the Nonlinear Model

Sometimes we know the values of the dependent variable for certain combinations of parameter values. For example, if in the model

$$Y = e^{A + BX} \quad \text{Equation 8.18}$$

we know that when X is 0, Y is 2, we would select the natural log of 2 as a starting value for A. Examination of an equation at its maximum, minimum, and when all the independent variables approach 0 or infinity may help in selection of initial values.

Solve a System of Equations

By taking as many data points as you have parameters, you can solve a simultaneous system of equations. For example, in the previous model, we could solve the equations

$$\ln(Y_1) = A + BX_1$$
$$\ln(Y_2) = A + BX_2 \quad \text{Equation 8.19}$$

Using subtraction,

$$\ln(Y_1) - \ln(Y_2) = BX_1 - BX_2 \quad \text{Equation 8.20}$$

we can solve for the values of the parameters

$$B = \frac{\ln(Y_1) - \ln(Y_2)}{X_1 - X_2} \quad \text{Equation 8.21}$$

and

$$A = \ln(Y_1) - BX_1 \quad \text{Equation 8.22}$$

Computational Problems

Computationally, nonlinear regression problems can be difficult to solve. Models that require exponentiation or powers of large data values may cause underflows or overflows. (An **overflow** is caused by a number that is too large for the computer to handle, while an **underflow** is caused by a number that is too small for the computer to handle.) Sometimes the program may continue and produce a reasonable solution, especially if only a few data points caused the problem. If this is not the case, you must eliminate the cause of the problem. If your data values are large—for example, years—you can subtract the smallest year from all of the values. That's what was done with the population example. Instead of using the actual years, we used decades since 1790 (to compute the number of decades, we subtracted the smallest year from each year value and divided the result by 10). You must, however, consider the effect of rescaling on the parameter values. Many nonlinear models are not scale invariant. You can also consider rescaling the parameter values.

If the program fails to arrive at a solution—that is, if it doesn't converge—you might consider choosing different starting values. You can also change the criterion used for convergence.

If none of these strategies works, you can use the sequential quadratic programming algorithm to try to solve problems that are causing difficulties in the Nonlinear Regression procedure, which uses a Levenberg-Marquardt algorithm by default. For a particular problem, one algorithm may perform better than the other.

Additional Nonlinear Regression Options

Three additional options are available for nonlinear models when the sequential quadratic programming algorithm is used. You can supply linear and nonlinear constraints for the values of the parameter estimates, and you can specify your own loss function. (By default, the loss function that is minimized is the sum of the squared residuals.) In addition, standard errors for the parameter estimates as well as asymptotic confidence intervals can be obtained with **bootstrapping**, in which repeated random samples are selected from the data and the model is estimated from each one.

How to Obtain a Nonlinear Regression Analysis

The Nonlinear Regression procedure provides two methods for estimating parameters of nonlinear models. The default Levenberg-Marquardt algorithm analyzes unconstrained models. The sequential quadratic programming algorithm is more general: it allows you to specify constraints on parameter estimates, provide your own loss function, and obtain bootstrap estimates of standard errors. For either method, you can save predicted values, residuals, and derivatives.

The minimum specifications are:

- One dependent variable.
- The formula for predicted values of the model you want to fit. The expression can use numeric independent variables, SPSS functions, and defined parameters.
- Model parameters and their starting values.

To obtain a nonlinear regression analysis, from the menus choose:

Statistics
  Regression ▸
    Nonlinear...

This opens the Nonlinear Regression dialog box, as shown in Figure 8.9.

Figure 8.9 Nonlinear Regression dialog box

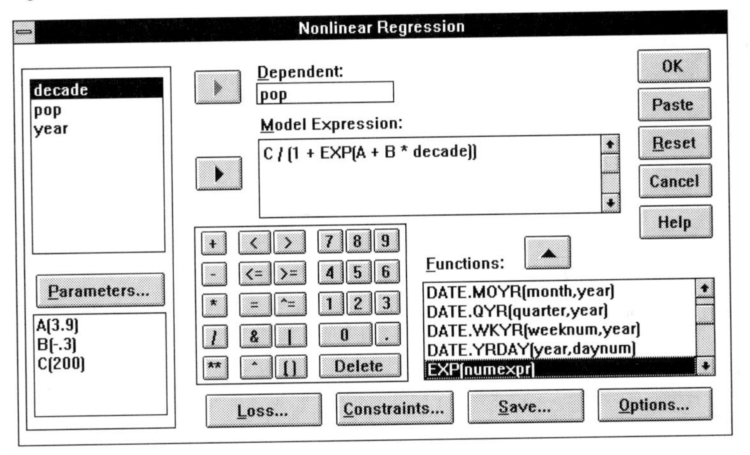

The numeric and short string variables in your working data file appear on the source list.

Dependent. Select one numeric variable from the source list to be the dependent variable.

Model Expression. Specify the formula for the predicted values of the dependent variable. The expression should use at least one independent variable. You can type and edit the expression in the text box or use the calculator pad and the lists of source variables, parameters, and functions to paste elements into the model equation (you must define parameters before you can select them from the list of parameters; see "Defining Model Parameters," below).

For example, for the logistic growth model described in "Fitting the Logistic Population Growth Model" on p. 233, *pop* was selected as the dependent variable and the model was specified as $'C/(1+\text{EXP}('A+'B*\text{decade}))$. Note that the model expression does not include "[dependent variable] =", since this is implicit in the model expression.

See the SPSS Base system documentation for a complete description of calculator-pad operators. For a description of SPSS functions, see the *SPSS Base System Syntax Reference Guide*.

Segmented Models

In a segmented model, the formula for a predicted value changes for different values of one or more variables. An example of a segmented model is one in which the predicted value is equal to B when the variable *decade* is less than 1 and the predicted value is $'B+\text{decade}$ for larger values of *decade*. You can define a segmented model using boolean logic, in which an expression evaluates to 1 when it is true and 0 when it is false. For example, the expression $(\text{decade}<1)$ is equal to 1 for cases in which *decade* is less than 1, and it is equal to 0 for larger values of *decade*. Thus, $(\text{decade}<1)*'B+(\text{decade}\geq 1)*('B+\text{decade})$ is the appropriate specification of the segmented *decade* model.

Defining Model Parameters

You must provide starting values for each model parameter. Click on Parameters... in the Nonlinear Regression dialog box. This opens the Nonlinear Regression Parameters dialog box, as shown in Figure 8.10.

Figure 8.10 Nonlinear Regression Parameters dialog box

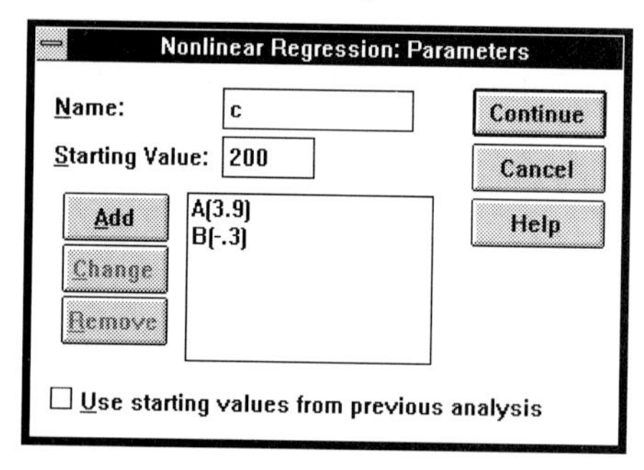

To define a starting value for a parameter, name the parameter and enter a starting value; then click on Add to add the parameter to the list of defined parameters. To change a parameter, highlight it on the list, modify the name or the starting value, and click on Change. To remove a parameter, highlight it on the list of parameters and click on Remove.

Name. Specify a valid name for a new SPSS variable. The name must begin with a letter and can be up to eight characters in length (for complete variable naming rules, see the SPSS Base system documentation). The parameter name cannot duplicate the name of an existing variable or another parameter. Parameters are used only in the current SPSS session; they cannot be saved.

Starting Value. Enter a starting value for the parameter.

The following option is also available:

❑ **Use starting values from previous analysis**. Uses final parameter values from the most recent nonlinear analysis as starting values in the current model. The final parameter values from the previous analysis override the manually entered values that you have specified.

This option is available only if you have run a previous nonlinear analysis in the current SPSS session, using the same working data file. (If you reset the nonlinear regression settings in the Nonlinear Regression dialog box, this option is disabled.) The previous model must contain all of the parameters used in the current model. Otherwise, if the previous model did not contain all of the parameters, the analysis terminates.

Defining a Loss Function

By default, SPSS chooses parameter estimates that minimize the sum of squared residuals for the model. To define an alternate minimization criterion (loss function), click on Loss... in the Nonlinear Regression dialog box. This opens the Nonlinear Regression Loss Function dialog box, as shown in Figure 8.11.

Figure 8.11 Nonlinear Regression Loss Function dialog box

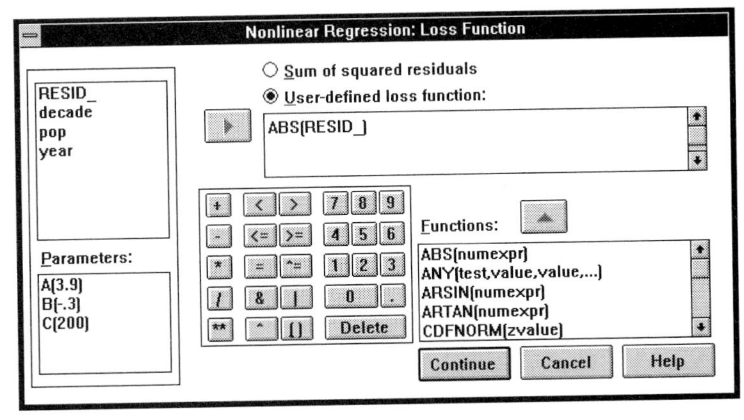

You can choose one of the following alternatives:

❍ **Sum of squared residuals.** Parameter estimates are selected according to the least-squares criterion. This is the default loss function.

❍ **User-defined loss function.** Select this to define your own loss function. You can type and edit the expression in the text box or use the calculator pad and the lists of source variables (including the residual variable *RESID_*), parameters, and functions to paste elements into the expression. For example, to minimize the sum of the absolute value of the residuals, define the loss function as ABS(RESID_).

Specifying Parameter Constraints

To constrain model parameters, click on Constraints... in the Nonlinear Regression dialog box. This opens the Nonlinear Regression Parameter Constraints dialog box, as shown in Figure 8.12.

Figure 8.12 Nonlinear Regression Parameter Constraints dialog box

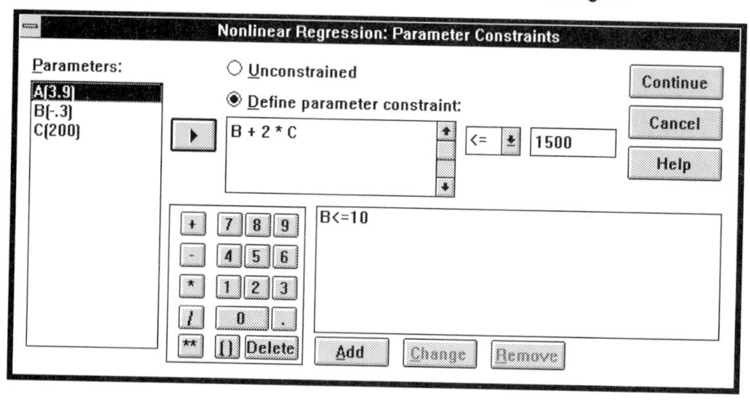

You can choose one of the following alternatives:

❍ **Unconstrained.** Model parameters are not constrained. This is the default.

❍ **Define parameter constraint.** Select this to define one or more expressions that impose bounds on parameters in the model. To constrain a single parameter, select it from the list of parameters. Then choose a relational operator and enter a numeric value. For example, to constrain the parameter *B* to be less than or equal to 10, select *B* for the first text box, choose <= (the default) from the drop-down list, and enter 10 in the second text box. Click on Add to add the constraint to the list of defined parameter constraints.

To constrain a combination of parameters, such as 'B + 2'C, build the parameter combination in the first text box using parameters, constants, and arithmetic operators. Then choose a relational operator and enter a numeric value. Click on Add to add the constraint to the list of parameter constraints.

To define additional constraints, repeat this process. To remove a constraint, highlight it on the list of constraints and click on Remove. To change a constraint, highlight it on the list, modify the parameter(s) or the relation, and click on Change.

Saving New Variables

To save diagnostic variables, click on Save... in the Nonlinear Regression dialog box. This opens the Nonlinear Regression Save New Variables dialog box, as shown in Figure 8.13.

Figure 8.13 Nonlinear Regression Save New Variables dialog box

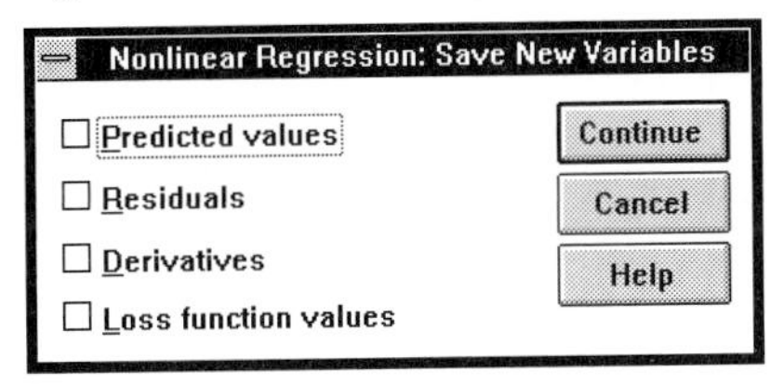

You can choose one or more of the following measures:

- **Predicted values**. The variable name *pred_* is assigned to the predicted values.
- **Residuals**. The variable name *resid* is assigned to the residuals.
- **Derivatives**. One derivative is saved for each model parameter. Derivative names are created by prefixing *'d.'* to the first six characters of parameter names.
- **Loss function values**. This option is available if you specify your own loss function. The variable name *loss_* is assigned to the values of the loss function.

If these variable names already exist in the data file, a sequential number is appended to the end of the new variable name. For example, if the data file contains a variable called *pred_*, the new variable for predicted values is named *pred_1*.

Options

To obtain bootstrap estimates of standard errors or to control the parameter estimation method, click on Options... in the Nonlinear Regression dialog box. This opens the Nonlinear Regression Options dialog box, as shown in Figure 8.14.

Figure 8.14 Nonlinear Regression Options dialog box

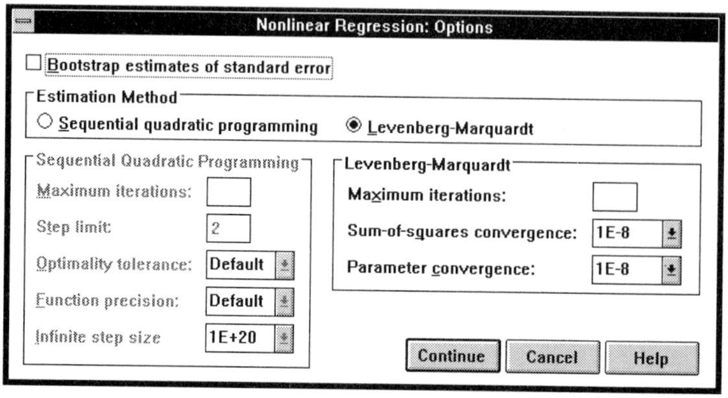

Nonlinear Regression: Options

Bootstrap estimates of standard error

Estimation Method
Sequential quadratic programming
Levenberg-Marquardt

Sequential Quadratic Programming
Maximum iterations:
Step limit: 2
Optimality tolerance: Default
Function precision: Default
Infinite step size 1E+20

Levenberg-Marquardt
Maximum iterations:
Sum-of-squares convergence: 1E-8
Parameter convergence: 1E-8

Continue Cancel Help

❑ **Bootstrap estimates of standard error.** Estimates standard errors of parameters by sampling with replacement from the working data file. The nonlinear equation is estimated for $10 \times p \times (p + 1) / 2$ samples, where p is the number of defined parameters. Standard errors are computed as the standard deviation of bootstrap estimates. Also displays confidence intervals for parameter estimates and the correlation matrix of parameter estimates. This option is most useful when a user-specified loss function is defined, since, by default, standard errors are displayed only when the least-squares criterion loss function is used.

Estimation Method. SPSS has two iterative methods for computing parameter estimates. For either method, derivatives are estimated by the program. You can choose one of the following alternatives:

❍ **Sequential quadratic programming.** This method is available for constrained and unconstrained models. Sequential quadratic programming is used automatically if you specify a constrained model, a user-defined loss function, or bootstrapping.

❍ **Levenberg-Marquardt.** This is the default algorithm for unconstrained models. The Levenberg-Marquardt method is not available if you specify a constrained model, a user-defined loss function, or bootstrapping.

Sequential Quadratic Programming. If you choose the sequential quadratic programming method, you can override one or more of the following algorithm settings:

Maximum iterations. Maximum number of major iterations. The default maximum is the larger of 50 or $(3(p + l) + 10n)$, where p is the number of parameters, l is the number of linear parameter constraints, and n is the number of nonlinear constraints. To specify a different maximum, enter a positive integer.

Step limit. Maximum permissible change in the length of the parameter vector. By default, the parameter vector can change by no more than a factor of 2. To override this step limit, enter a value greater than 0.

Optimality tolerance. Approximate accuracy of the loss function. For example, if optimality tolerance is 10^{-6}, the loss function is accurate to approximately six significant digits. The default value for optimality tolerance is your computer's epsilon raised to the power of 0.8 (epsilon is the smallest number that, when added to 1, results in a sum greater than 1). To override this value, select an alternate tolerance value from the drop-down list.

Function precision. The accuracy with which the loss function can be measured. Function precision acts as a relative precision when the function is large and an absolute precision when the function is small. The default value is your computer's epsilon raised to the power of 0.9 (epsilon is the smallest number that, when added to 1, results in a sum greater than 1). To override this value, select an alternate precision value from the drop-down list or select disable to disable this setting.

Infinite step size. The magnitude of the change in parameters that is defined as infinite. If the change in the parameters across successive iterations is greater than the infinite step size, the problem is considered unbounded and estimation stops. The default value is $1\text{'E} + 20$. To override this value, select an alternate infinite step size from the drop-down list.

Levenberg-Marquardt. If you choose the Levenberg-Marquardt method, you can override one or more of the following algorithm settings:

Maximum iterations. Maximum number of iterations. The default is 100 iterations per parameter. To override this value, enter a positive integer value for the maximum total number of iterations for the model.

Sum-of-squares convergence. Convergence criterion for the model sum of squares. Iteration stops if successive iterations fail to reduce the sum of squares by this proportion or more. The default is $1\text{'E} - 8$. To override this value, select an alternate convergence value from the drop-down list or select disable to disable this criterion.

Parameter convergence. Convergence criterion for parameter estimates. Iteration stops if successive iterations fail to change any of the parameter values by this proportion or more. The default is $1\text{'E} - 8$. To override this value, select an alternate convergence value from the drop-down list or select disable to disable this criterion.

Additional Features Available with Command Syntax

You can customize your nonlinear regression analysis if you paste your selections into a syntax window and edit the resulting NLR, CNLR, and MODEL PROGRAM command syntax. (For information on syntax windows, see the SPSS Base system documentation.) Additional features include:

- The ability to specify derivatives (using the DERIVATIVES subcommand).
- The ability to read starting values from a user-specified file and to write final values to a file (using the FILE and OUTFILE subcommands).
- The ability to specify multiple-statement model and loss functions (using the MODEL PROGRAM command).
- The ability to specify the number of samples for bootstrapping (using the BOOTSTRAP subcommand).
- Additional parameter estimation criteria (using the CRITERIA subcommand).

See the Syntax Reference section of this manual for command syntax rules and for complete NLR and CNLR command syntax.

9 Probit Analysis

How much insecticide does it take to kill a pest? How low does a sale price have to be to induce a consumer to buy a product? In both of these situations, we are concerned with evaluating the potency of a stimulus. In the first example, the stimulus is the amount of insecticide; in the second, it is the sale price of an object. The response we are interested in is all-or-none. An insect is either dead or alive; a sale is made or not. Since all insects and shoppers do not respond in the same way—that is, they have different tolerances for insecticides and sale prices—the problem must be formulated in terms of the proportion responding at each level of the stimulus.

Different mathematical models can be used to express the relationship between the proportion responding and the "dose" of one or more stimuli. In this chapter, we will consider two commonly used models: the probit response model and the logit response model. We will assume that we have one or more stimuli of interest and that each stimulus can have several doses. We expose different groups of individuals to the desired combinations of stimuli. For each combination, we record the number of individuals exposed and the number who respond.

Probit and Logit Response Models

In probit and logit models, instead of regressing the actual proportion responding on the values of the stimuli, we transform the proportion responding using either a logit or probit transformation. For a probit transformation, we replace each of the observed proportions with the value of the standard normal curve below which the observed proportion of the area is found.

For example, if half (0.5) of the subjects respond at a particular dose, the corresponding probit value is 0, since half of the area in a standard normal curve falls below a Z score of 0. If the observed proportion is 0.95, the corresponding probit value is 1.64.

If the logit transformation is used, the observed proportion P is replaced by

$$\ln\left(\frac{P}{1-P}\right)$$

Equation 9.1

This quantity is called a **logit**. If the observed proportion is 0.5, the logit-transformed value is 0, the same as the probit-transformed value. Similarly, if the observed proportion is 0.95, the logit-transformed value is 1.47. This differs somewhat from the corresponding probit value of 1.64. (In most situations, analyses based on logits and probits give very similar results.)

The regression model for the transformed response can be written as

$$\text{Transformed}P_i = A + BX_i$$ **Equation 9.2**

where P_i is the observed proportion responding at dose X_i. (Usually, the log of the dose is used instead of the actual dose.) If there is more than one stimulus variable, terms are added to the model for each of the stimuli. The SPSS Probit Analysis procedure obtains maximum-likelihood estimates of the regression coefficients.

An Example

Finney (1971) presents data showing the effect of a series of doses of rotenone (an insecticide) when sprayed on *Macrosiphoniella sanborni*. Table 9.1 contains the concentration, the number of insects tested at each dose, the proportion dying, and the probit transformation of each of the observed proportions.

Table 9.1 Effects of rotenone

Dose	Number observed	Number dead	Proportion dead	Probit
10.2	50	44	0.88	1.18
7.7	49	42	0.86	1.08
5.1	46	24	0.52	0.05
3.8	48	16	0.33	-0.44
2.6	50	6	0.12	-1.18

Figure 9.1 is a plot of the observed probits against the logs of the concentrations. (On the menus, you specify *died* as the response frequency variable, *total* as the observation frequency variable, and *dose* as the covariate. You also ask for a log transformation of *dose*.) You can see that the relationship between the two variables is linear. If the relationship did not appear to be linear, the concentrations would have to be transformed in some other way in order to achieve linearity. If a suitable transformation could not be found, fitting a straight line would not be a reasonable strategy for modeling the data.

Figure 9.1 Plot of observed probits against logs of concentrations

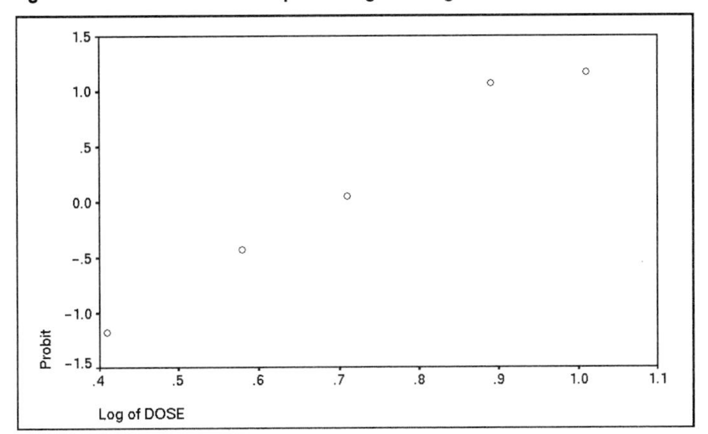

The parameter estimates and standard errors for this example are shown in Figure 9.2.

Figure 9.2 Parameter estimates and standard errors

```
Parameter estimates converged after 10 iterations.
Optimal solution found.

Parameter Estimates (PROBIT model:  (PROBIT(p)) = Intercept + BX):

          Regression Coeff.  Standard Error     Coeff./S.E.

 DOSE              4.16914           .47306         8.81306


                  Intercept  Standard Error  Intercept/S.E.

                  -2.85940           .34717        -8.23640

 Pearson  Goodness-of-Fit  Chi Square =      1.621     DF = 3    P =  .655

 Since Goodness-of-Fit Chi square is NOT significant, no heterogeneity
 factor is used in the calculation of confidence limits.
```

The regression equation is

$$\text{Probit}\,(P_i) \;=\; -2.86 + 4.17\,(\log_{10}(\text{dose}_i))$$ **Equation 9.3**

To see how well this model fits, consider Figure 9.3, which contains observed and expected frequencies, residuals, and the predicted probability of a response for each of the log concentrations.

Figure 9.3 Statistics for each concentration

Observed and Expected Frequencies

DOSE	Number of Subjects	Observed Responses	Expected Responses	Residual	Prob
1.01	50.0	44.0	45.586	-1.586	.91172
.89	49.0	42.0	39.330	2.670	.80265
.71	46.0	24.0	24.845	-.845	.54010
.58	48.0	16.0	15.816	.184	.32950
.41	50.0	6.0	6.253	-.253	.12506

You can see that the model appears to fit the data reasonably well. A goodness-of-fit test for the model, based on the residuals, is shown in Figure 9.2. The chi-square goodness-of-fit test is calculated as

$$\chi^2 = \sum \frac{(\text{residual}_i)^2}{n_i \hat{P}_i (1 - \hat{P}_i)}$$

Equation 9.4

where n_i is the number of subjects exposed to dose i, and $\hat{P}_i$ is the predicted proportion responding at dose i. The degrees of freedom are equal to the number of doses minus the number of estimated parameters. In this example, we have five doses and two estimated parameters, so there are three degrees of freedom for the chi-square statistic. Since the observed significance level for the chi-square statistic is large—0.655—there is no reason to doubt the model.

When the significance level of the chi-square statistic is small, several explanations are possible. It may be that the relationship between the concentration and the probit is not linear. Or it may be that the relationship is linear but the spread of the observed points around the regression line is unequal. That is, the data are heterogeneous. If this is the case, a correction must be applied to the estimated variances for each concentration group (see "Confidence Intervals for Expected Dosages," below).

Confidence Intervals for Expected Dosages

Often you want to know what the concentration of an agent must be in order to achieve a certain proportion of response. For example, you may want to know what the concentration would have to be in order to kill half of the insects. This is known as the **median lethal dose**. It can be obtained from the previous regression equation by solving for the concentration that corresponds to a probit value of 0. For this example,

$$\log_{10}(\text{median lethal dose}) = 2.86/4.17$$
$$\text{median lethal dose} = 4.85$$

Equation 9.5

Confidence intervals can be constructed for the median lethal dose as well as for the dose required to achieve any response. The SPSS Probit Analysis procedure calculates 95% intervals for the concentrations required to achieve various levels of response. The values for this example are shown in Figure 9.4.

Figure 9.4 Confidence intervals

Confidence Limits for Effective DOSE

Prob	DOSE	95% Confidence Limits Lower	Upper
.01	1.34232	.90152	1.73955
.02	1.56042	1.09195	1.97144
.03	1.71682	1.23282	2.13489
.04	1.84473	1.35041	2.26709
.05	1.95577	1.45411	2.38094
.06	2.05553	1.54847	2.48260
.07	2.14718	1.63607	2.57552
.08	2.23270	1.71858	2.66189
.09	2.31344	1.79709	2.74314
.10	2.39033	1.87239	2.82033
.15	2.73686	2.21737	3.16638
.20	3.04775	2.53300	3.47603
.25	3.34246	2.83556	3.77074
.30	3.63134	3.13342	4.06251
.35	3.92126	3.43173	4.36004
.40	4.21775	3.73415	4.67105
.45	4.52592	4.04368	5.00346
.50	4.85119	4.36322	5.36609
.55	5.19983	4.69624	5.76942
.60	5.57976	5.04753	6.22651
.65	6.00164	5.42413	6.75456
.70	6.48081	5.83681	7.37806
.75	7.04092	6.30250	8.13488
.80	7.72177	6.84956	9.08968
.85	8.59893	7.53102	10.36751
.90	9.84550	8.46614	12.26163
.91	10.17274	8.70644	12.77237
.92	10.54059	8.97434	13.35269
.93	10.96042	9.27744	14.02276
.94	11.44911	9.62693	14.81270
.95	12.03312	10.04028	15.77021
.96	12.75742	10.54699	16.97720
.97	13.70788	11.20286	18.59208
.98	15.08186	12.13489	20.98491
.99	17.53233	13.75688	25.40958

The column labeled *Prob* is the proportion responding. The column labeled *DOSE* is the estimated dosage required to achieve this proportion. The 95% confidence limits for the dose are shown in the next two columns. If the chi-square goodness-of-fit test has a significance level less than 0.15 (the program default), a heterogeneity correction is automatically included in the computation of the intervals (Finney, 1971).

Comparing Several Groups

In the previous example, only one stimulus at several doses was studied. If you want to compare several different stimuli, each measured at several doses, additional statistics may be useful. Consider the inclusion of two additional insecticides in the previously described problem. Besides rotenone at five concentrations, we also have five concentrations of deguelin and four concentrations of a mixture of the two. Figure 9.5 shows a partial listing of these data, as entered in the Data Editor. As in the previous example, variable *dose* contains the insecticide concentration, *total* contains the total number of cases, and *died* contains the number of deaths. Factor variable *agent* is coded 1 (rotenone), 2 (deguelin), or 3 (mixture).

Figure 9.5 Data for rotenone and deguelin

Newdata

	dose	agent	total	died
1	2.57	1.00	50.00	6.00
2	3.80	1.00	48.00	16.00
3	5.13	1.00	46.00	24.00
4	7.76	1.00	49.00	42.00
5	10.23	1.00	50.00	44.00
6	10.00	2.00	48.00	18.00
7	20.42	2.00	48.00	34.00
8	30.20	2.00	49.00	47.00
9	40.74	2.00	50.00	47.00
10	50.12	2.00	48.00	48.00

Figure 9.6 is a plot of the observed probits against the logs of the concentrations for each of the three groups separately.

Figure 9.6 Plot of observed probits against logs of concentrations

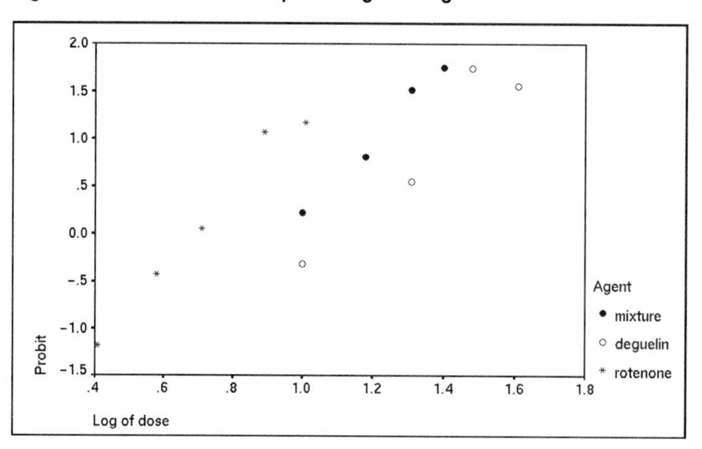

You can see that there appears to be a linear relationship between the two variables for all three groups. One of the questions of interest is whether all three lines are parallel. If so, it would make sense to estimate a common slope for them. Figure 9.7 contains the estimate of the common slope, separate intercept estimates for each of the groups, and a test of parallelism.

Figure 9.7 Intercept estimates and test of parallelism

```
            Regression Coeff.  Standard Error      Coeff./S.E.

DOSE                 3.90635           .30691         12.72803

                    Intercept  Standard Error  Intercept/S.E.  AGENT

                     -2.67343          .23577       -11.33913  rotenone
                     -4.36573          .40722       -10.72071  deguelin
                     -3.71153          .37491        -9.89977  mixture

Pearson  Goodness-of-Fit  Chi Square =       7.471     DF = 24    P =  .999
         PARALLELISM TEST CHI SQUARE =       1.162     DF = 2    P =  .559

Since Goodness-of-Fit Chi square is NOT significant, no heterogeneity
factor is used in the calculation of confidence limits.
```

The observed significance level for the test of parallelism is large—0.559—so there is no reason to reject the hypothesis that all three lines are parallel. Thus, the equation for rotenone is estimated to be

$$\text{Probit}(P_i) = -2.67 + 3.91(\log_{10}(\text{dose}_i)) \quad \textbf{Equation 9.6}$$

The equation for deguelin is

$$\text{Probit}(P_i) = -4.37 + 3.91(\log_{10}(\text{dose}_i)) \quad \textbf{Equation 9.7}$$

and the equation for the mixture is

$$\text{Probit}(P_i) = -3.71 + 3.91(\log_{10}(\text{dose}_i)) \quad \textbf{Equation 9.8}$$

Comparing Relative Potencies of the Agents

The relative potency of two stimuli is defined as the ratio of two doses that are equally effective. For example, the relative median potency is the ratio of two doses that achieve a response rate of 50%. In the case of parallel regression lines, there is a constant relative potency at all levels of response. For example, consider Figure 9.8, which shows some of the doses needed to achieve a particular response for each of the three agents.

Figure 9.8 Expected doses

AGENT 1 rotenone

Prob	DOSE	95% Confidence Limits Lower	Upper
.25	3.24875	2.82553	3.65426
.30	3.54926	3.11581	3.97227
.35	3.85249	3.40758	4.29636
.40	4.16414	3.70545	4.63356
.45	4.48965	4.01370	4.99083
.50	4.83482	4.33680	5.37586
.55	5.20654	4.68002	5.79788
.60	5.61352	5.05003	6.26880
.65	6.06764	5.45592	6.80487
.70	6.58603	5.91086	7.42974
.75	7.19524	6.43524	8.18030

AGENT 2 deguelin

Prob	DOSE	95% Confidence Limits Lower	Upper
.25	8.80913	7.24085	10.32755
.30	9.62399	8.00041	11.20430
.35	10.44620	8.76976	12.09054
.40	11.29127	9.56202	13.00451
.45	12.17389	10.38960	13.96387
.50	13.10986	11.26577	14.98799
.55	14.11778	12.20605	16.10010
.60	15.22135	13.23000	17.33026
.65	16.45271	14.36393	18.71981
.70	17.85833	15.64554	20.32922
.75	19.51024	17.13277	22.25327

AGENT 3 mixture

Prob	DOSE	95% Confidence Limits Lower	Upper
.25	5.99049	4.82754	7.12052
.30	6.54462	5.33737	7.72006
.35	7.10375	5.85480	8.32478
.40	7.67843	6.38880	8.94696
.45	8.27864	6.94795	9.59841
.50	8.91512	7.54150	10.29194
.55	9.60054	8.18034	11.04290
.60	10.35100	8.87822	11.87107
.65	11.18837	9.65364	12.80367
.70	12.14424	10.53309	13.88054
.75	13.26759	11.55720	15.16421

For rotenone, the expected dosage to kill half of the insects is 4.83; for deguelin, it is 13.11; and for the mixture, it is 8.91. The relative median potency for rotenone compared to deguelin is $4.83/13.11$, or 0.37; for rotenone compared to the mixture, it is 0.54; and for deguelin compared to the mixture, it is 1.47. These relative median potencies and their confidence intervals are shown in Figure 9.9.

Figure 9.9 Relative potencies and their confidence intervals

```
Estimates of Relative Median Potency

                              95% Confidence Limits
 AGENT          Estimate        Lower           Upper

 1 VS.  2          .3688        .23353          .52071
 1 VS.  3          .5423        .38085          .71248
 2 VS.  3         1.4705       1.20619         1.85007
```

If a confidence interval does not include the value of 1, we have reason to suspect the hypothesis that the two agents are equally potent.

Estimating the Natural Response Rate

In some situations, the response of interest is expected to occur even if the stimulus is not present. For example, if the organism of interest has a very short life span, you would expect to observe deaths even without the agent. In such situations, you must adjust the observed proportions to reflect deaths due to the agent alone.

If the natural response rate is known, it can be entered into the SPSS Probit Analysis procedure. It can also be estimated from the data, provided that data for a dose of 0 are entered together with the other doses. If the natural response rate is estimated from the data, an additional degree of freedom must be subtracted from the chi-square goodness-of-fit degrees of freedom.

More than One Stimulus Variable

If several stimuli are evaluated simultaneously, an additional term is added to the regression model for each stimulus. Regression coefficients and standard errors are displayed for each stimulus. In the case of several stimuli, relative potencies and confidence intervals for the doses needed to achieve a particular response cannot be calculated in the usual fashion, since you need to consider various combinations of the levels of the stimuli.

How to Obtain a Probit Analysis

The SPSS Probit Analysis procedure estimates the relationship between one or more stimuli and the proportion of cases having a particular response. The response proportions are transformed using either the probit or logit transformation. Optionally, you can compare subgroups, transform predictors, and control the natural response rate.

The Probit Analysis procedure is designed for use with aggregated data, in which each row contains the number of responses and the number of total observations for a group of cases with a specific combination of levels of the predictor variables and a specific level of a factor variable, if there is one. SPSS computes the proportion of cases responding from the number of responses and the number of total observations.

To obtain the required summary counts when you have case-by-case data, see the Aggregate Data procedure in the SPSS Base system documentation. If you want to fit a logistic regression model when you do not have groups of subjects at each set of stimulus levels, you can use the Logistic Regression procedure, which contains many additional diagnostics.

The minimum specifications are:

- A response frequency variable.
- An observation frequency variable.
- One or more stimulus variables (covariates).

To obtain a probit analysis, from the menus choose:

Statistics
  Regression ▶
    Probit...

This opens the Probit Analysis dialog box, as shown in Figure 9.10.

Figure 9.10 Probit Analysis dialog box

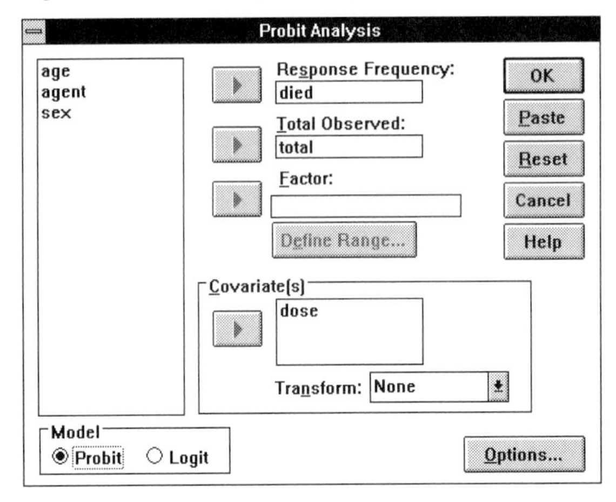

The numeric variables in your data file appear on the source list.

Response Frequency. Select a response frequency variable. Each case should contain the number of individuals that responded when exposed to a particular stimulus level (or combination of stimulus levels, if you have several stimuli).

Total Observed. Select an observation frequency variable. Each case should contain the total number of individuals exposed to a particular stimulus level or combination of stimulus levels.

Factor. By default, the model is estimated for all cases in the working data file as if they were from a single sample. To estimate separate intercepts for subgroups, select a factor (grouping) variable. You must indicate which factor levels you want to use in the analysis (see "Defining Factor Range," below).

Covariate(s). Select one or more predictor (stimulus) variables. A term is included in the model for each covariate.

- **Transform.** By default, covariates are not transformed. Choose one of the following alternatives from the drop-down list:

  None. Do not transform the covariates. This is the default.

  Log base 10. Apply the $\log_{10}$ transformation to each covariate for the analysis.

  Natural log. Apply the natural log transformation to each covariate for the analysis.

Model. Model selection controls how response proportions are transformed. You can choose one of the following alternatives:

- **Probit.** Applies the probit transformation to the response proportions. This is the default.
- **Logit.** Applies the logit (log odds) transformation to the response proportions.

Defining Factor Range

If you have a factor variable, you must specify which factor levels you want to use in the analysis. Highlight the factor and click on Define Range... in the Probit Analysis dialog box. This opens the Probit Analysis Define Range dialog box, as shown in Figure 9.11.

Figure 9.11 Probit Analysis Define Range dialog box

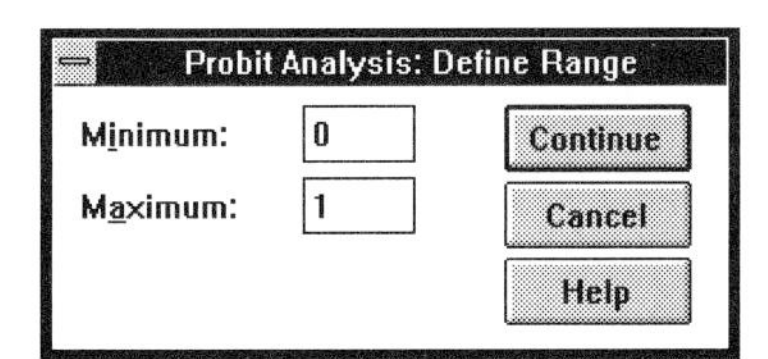

Specify minimum and maximum integer categories to use in the analysis. The first value must be less than the second value. Each non-empty integer category in the range is used as a factor level (non-integer values in the range are truncated, so cases with the same integer portion are combined into the same factor level). Cases with values outside the range are excluded from the analysis.

Options

To obtain a test of parallelism, to specify the natural response rate, or to control parameter estimation criteria, click on Options... in the Probit Analysis dialog box. This opens the Probit Analysis Options dialog box, as shown in Figure 9.12.

Figure 9.12 Probit Analysis Options dialog box

Probit Analysis: Options
Statistics
☒ Frequencies
☒ Relative median potency
☐ Parallelism test
☒ Fiducial confidence intervals
Significance level for use of heterogeneity factor: .15
Natural Response Rate
◉ None ○ Calculate from data ○ Value:
Criteria
Maximum iterations: 20
Step limit: .1
Optimality tolerance: Default
Continue Cancel Help

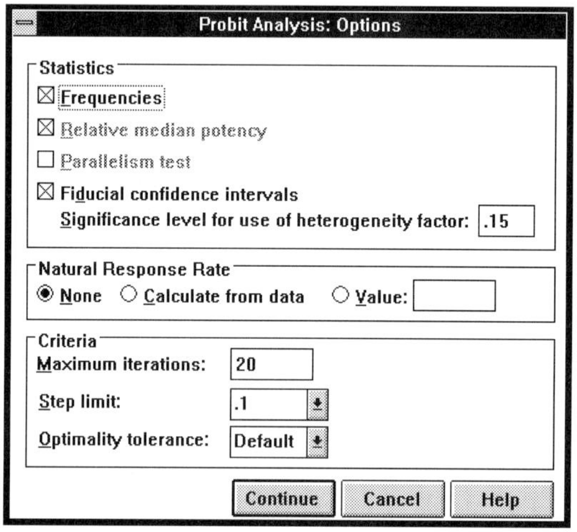

Statistics. For all models, SPSS displays regression coefficients, their standard errors, and a chi-square goodness-of-fit test. For single-covariate models, a plot is displayed showing the transformed proportions against values of the covariate. A covariance-correlation matrix of regression coefficients is displayed for models containing two or more covariates. You can choose one or more of the following additional statistics:

- **Frequencies**. Displays observed and expected frequencies and residuals for each case. Frequencies and residuals are displayed by default.
- **Relative median potency**. Displays the ratio of median potencies for each pair of factor levels. Also shows 95% confidence limits for each relative median potency. Relative median potencies are displayed by default when you have a factor variable but are not available if you do not have a factor variable or if you have more than one covariate.

❑ **Parallelism test**. Tests the homogeneity of regression slopes for factor levels. The parallelism test is available only if you have a factor variable.

❑ **Fiducial confidence intervals**. Displays 95% confidence limits for various response proportions. If you have a factor variable, SPSS displays confidence intervals for each factor level. Fiducial confidence intervals are shown by default for models having a single covariate, but are not available for models containing more than one covariate.

Significance level for use of heterogeneity factor. By default, SPSS uses a heterogeneity correction in its calculation of confidence intervals if the probability associated with the goodness-of-fit test is less than 0.15. You can enter a probability to override the default cutoff probability.

Natural Response Rate. In some situations, the response is expected to occur even when a stimulus is absent. You can select one of the following alternatives:

❍ **None**. Does not estimate a natural response rate.

❍ **Calculate from data**. Estimates the natural response rate from the sample data. If you select this item, your data should contain a case representing the control level, for which the value of the covariate(s) is 0. SPSS estimates the natural response rate using the proportion of responses for the control level as an initial value. If you have not selected a log transformation, the control group is also included in the analysis.

❍ **Value**. Sets the natural response rate in the model (select this item when you know the natural response rate in advance). Enter the natural response proportion (the proportion must be less than 1). For example, if the response occurs 10% of the time when the stimulus is 0, enter 0.10.

Criteria. SPSS uses the sequential quadratic programming method to estimate model parameters. You can override one or all of the following estimation criteria:

Maximum iterations. Iteration terminates if the maximum number of iterations is reached. The default maximum number of iterations is 20. To override the default maximum, enter a positive integer.

Step limit. Maximum permissible change in the length of the parameter vector. By default, the parameter vector can change by no more than a factor of 0.1. To override this step limit, select a value from the drop-down list.

Optimality tolerance. Approximate accuracy of the loss function. For example, if optimality tolerance is 10^{-6}, the loss function should be accurate to approximately six significant digits. The default value for optimality tolerance is your computer's epsilon raised to the power of 0.8 (epsilon is the smallest number that, when added to 1, results in a sum greater than 1). To override this value, select an optimality tolerance value from the drop-down list.

Additional Features Available with Command Syntax

You can customize your probit analysis if you paste your selections into a syntax window and edit the resulting PROBIT syntax. (For information on syntax windows, see the SPSS Base system documentation.) Additional features include:

- The ability to obtain both probit and logit models in a single step (using the MODEL subcommand).
- User-specified bases for log transformation of the covariates (using the LOG subcommand).

See the Syntax Reference section of this manual for command syntax rules and for complete PROBIT command syntax.

10 Life Tables

How long do marriages last? How long do people work for a company? How long do patients with a particular cancer live? To answer these questions, you must evaluate the interval between two events—marriage and divorce, hiring and departure, diagnosis and death. Solution of the problem is complicated by the fact that the event of interest (divorce, termination, or death) may not occur for all people during the period in which they are observed, and the actual period of observation may not be the same for all people. That is, not everyone gets divorced or quits a job, and not everyone gets married or starts a job on the same day.

These complicating factors eliminate the possibility of doing something simple, such as calculating the average time between the two events. In this chapter, we will consider special statistical techniques for looking at the interval between two events when the second event does not necessarily happen to everyone and when people are observed for different periods of time.

The Follow-up Life Table

A statistical technique useful for these types of data is called a follow-up **life table**. (The technique was first applied to the analysis of survival data, from which the term life table originates.) The basic idea of the life table is to subdivide the period of observation after a starting point, such as beginning work at a company, into smaller time intervals—say, single years. For each interval, all people who have been observed at least that long are used to calculate the probability of a **terminal event**, such as leaving the company, occurring in that interval. The probabilities estimated from each of the intervals are then used to estimate the overall probability of the event occurring at different time points. All available data are used for the computations.

A Personnel Example

As an example of life table techniques, consider the following problem. As personnel director of a small company, you are asked to prepare a report on the longevity of employees in your company. You have available in the corporate database information on the date that employees started employment and the last date they worked. You know that it is wrong to look at just the average time on the job for people who left. It doesn't tell you anything about the length of employment for people who are still employed. For example, if your only departures were 10 people who left during their first year with the company, an average employment time based only on them would be highly misleading. You need a way to use information from both the people who left and those who are still with the company.

The employment times for people who are still with the company are known as **censored observations**, since you don't know exactly how long these people will work for the company. You do know, however, that their employment time will be at least as long as the time they have already been at the company. People who have already left the company are **uncensored**, since you know their employment times exactly.

Organizing the Data

As a first step in analyzing the data, you must construct a summary table like that shown in Table 10.1. Each row of the table corresponds to a time interval of one year. (You can choose any interval length you want. For a rapid-turnover company, you might want to consider monthly intervals; for a company with minimal turnover, you might want to consider intervals of several years.) The first interval corresponds to a time period of less than one year, the second interval corresponds to a time period of one year or more but less than two, and so on. The starting point of each interval is shown in the column labeled *Start of interval*.

For each interval, you count the number of people who left within that interval (*Left*). Similarly, you count the number of people who have worked that long and are still working for the company. For them, this is the latest information available. For each interval, you can also count the number of people who were observed in each of the intervals. That is, you can count how many people worked at least one year, at least two years, and so on.

Table 10.1 Summary table

Start of interval (years)	Left	Current employees working this long	Employees working at least this long	At risk
0	2	2	100	99
1	1	2	96	95
2	7	16	93	85
3	6	15	70	62
4	5	12	49	43
5	5	10	32	27
6	4	9	17	12.5
7	1	1	4	3.5
8	0	2	2	1.0
TOTAL	31	69		

From Table 10.1, for example, you see that you have information for 100 people, of whom 31 have left the company and 69 are still with the company. You see that 2 people left the company during their first year, and 2 current employees have been employed for less than one year. Since 4 people have been observed for one year or less, all of the rest, 96 people, have worked for one year or more (this is the entry in the second to the last column for the interval that starts at 1).

Calculating Probabilities

Based on the data shown in Table 10.1, you can calculate some useful summary statistics to describe the longevity of the employees. These statistics will make maximum use of the available information by estimating a series of probabilities, each based on as much data as possible.

The first probability you want to estimate is the probability that an employee will quit during the first year of employment. You have observations on 100 employees, 2 of whom quit during the first year. Initially, you may think that the estimate of the probability of leaving in the first year should be 2 out of 100. However, such a calculation does not take into account the fact that you have on staff 2 employees who have not yet completed their first year. They haven't been observed for the entire interval. You can assume, for simplicity, that they have been observed, on average, for half of the length of the interval. Thus, each is considered as contributing only half of an observation. So, for the first interval, instead of having observations for 100 people, we have observations for 99 ($100 - (0.5 \times 2)$). The column labeled *At risk* in Table 10.1 shows the number of observations for each interval when the number is adjusted for the current employees who are assumed to be observed for half of the interval.

The probability of an employee leaving during the first year, using the number in the *At risk* column as the denominator, is $2/99$, or 0.0202. The probability of staying until the end of the first interval is 1 minus the probability of leaving. For the first interval, it is 0.9798.

The next probability you must estimate is that of an employee leaving during the second year, assuming that the employee did not leave during the first year. All employees who have been employed for at least one year contribute information to this calculation. From Table 10.1, you see that 1 employee out of 95 at risk left during the second year. The probability of leaving during the second year, given that an employee made it through the first, is then $1/95$, which is 0.0105. The probability of staying to the end of the second year, if the employee made it to the beginning of the second year, is 0.9895.

From these two probabilities (the probability of making it through the first year, and the probability of making it through the second year given that an employee has made it through the first), you can estimate the probability that the employee will make it to the end of the second year of employment. The formula is

$$\text{P (second)} = \text{P (first)} \times \text{P (second given first)} \qquad \textbf{Equation 10.1}$$

For this example, the probability of surviving to the end of the second year is 0.9798×0.9895, or 0.9695. This is the cumulative probability of surviving to the end of the second interval.

The Life Table

Figure 10.1 is the life table computed by the SPSS Life Tables procedure for the personnel data (for this analysis, *length* is the survival time variable, and yearly intervals are requested). The contents of the columns of the life table are as follows:

Interval Start Time. The beginning value for each interval. Each interval extends from its start time up to the start time of the next interval.

Number Entering This Interval. The number of cases that have survived to the beginning of the current interval.

Number Withdrawn during This Interval. The number of cases entering the interval for which follow-up ends somewhere in the interval. These are censored cases; that is, these are cases for which the event of interest has not occurred at the time of last contact.

Number Exposed to Risk. This is calculated as the number of cases entering the interval minus one half of those withdrawn during the interval.

Number of Terminal Events. The number of cases for which the event of interest occurs within the interval.

Proportion of Terminal Events. An estimate of the probability of the event of interest occurring in an interval for a case that has made it to the beginning of that interval. It is computed as the number of terminal events divided by the number exposed to risk.

Proportion Surviving. The proportion surviving is 1 minus the proportion of terminal events.

Cumulative Proportion Surviving at End. This is an estimate of the probability of surviving to the end of an interval. It is computed as the product of the proportion surviving this interval and the proportion surviving all previous intervals.

Probability Density. The probability density is an estimate of the probability per unit time of experiencing an event in the interval.

Hazard Rate. The hazard rate is an estimate of the probability per unit time that a case that has survived to the beginning of an interval will experience an event in that interval.

Standard Error of the Cumulative Proportion Surviving. This is an estimate of the variability of the estimate of the cumulative proportion surviving.

Standard Error of the Probability Density. This is an estimate of the variability of the estimated probability density.

Standard Error of the Hazard Rate. This is an estimate of the variability of the estimated hazard rate.

Figure 10.1 Life table

Life Table
Survival Variable LENGTH

Intrvl Start Time	Number Entrng this Intrvl	Number Wdrawn During Intrvl	Number Exposd to Risk	Number of Termnl Events	Propn Termi-nating	Propn Sur-viving	Cumul Propn Surv at End	Proba-bility Densty	Hazard Rate	Se of Cumul Sur-viving	Se of Proba-bility Densty	Se of Hazard Rate
.0	100.0	2.0	99.0	2.0	.0202	.9798	.9798	.0202	.0204	.0141	.0141	.0144
1.0	96.0	2.0	95.0	1.0	.0105	.9895	.9695	.0103	.0106	.0173	.0103	.0106
2.0	93.0	16.0	85.0	7.0	.0824	.9176	.8896	.0798	.0859	.0330	.0289	.0324
3.0	70.0	15.0	62.5	6.0	.0960	.9040	.8042	.0854	.1008	.0446	.0333	.0411
4.0	49.0	12.0	43.0	5.0	.1163	.8837	.7107	.0935	.1235	.0557	.0397	.0551
5.0	32.0	10.0	27.0	5.0	.1852	.8148	.5791	.1316	.2041	.0699	.0541	.0908
6.0	17.0	9.0	12.5	4.0	.3200	.6800	.3938	.1853	.3810	.0900	.0796	.1870
7.0	4.0	1.0	3.5	1.0	.2857	.7143	.2813	.1125	.3333	.1148	.0985	.3287
8.0	2.0	2.0	1.0	.0	.0000	1.0000	.2813	.0000	.0000	.1148	.0000	.0000

The median survival time for these data is 6.43

Median Survival Time

An estimate of the median survival time is displayed below the life table. The median survival time is the time point at which the value of the cumulative survival function is 0.5. That is, it is the time point by which half of the cases are expected to experience the event. Linear interpolation is used to calculate this value. If the cumulative proportion surviving at the end of the last interval is greater than 0.5, the start time of the last interval is identified with a plus sign (+) to indicate that the median time exceeds the start value of the last interval.

From Figure 10.1, you see that the median survival time for the personnel data is 6.43 years. This means that half of the people have left the company after 6.43 years. At six years, almost 58% of the employees remain. At seven years, only 39% remain.

Assumptions Needed to Use the Life Table

The basic assumption underlying life table calculations is that survival experience does not change during the course of the study. For example, if the employment possibilities or work conditions change during the period of the study, it makes no sense to combine all of the cases into a single life table. To use a life table, you must assume that a person hired today will behave the same way as a person who was hired five years ago. You must also assume that observations that are censored do not differ from those that are not censored. These are critical assumptions that determine whether life table analysis is an appropriate technique.

Lost to Follow-up

In the personnel example, we had information available for all employees in the company. We knew who left and when. This is not always the case. If you are studying the survival experience of patients who have undergone a particular therapeutic procedure, you may have two different types of censored observations. You will have patients whose length of survival is not known because they are still alive. You may also have patients with whom you have lost contact and for whom you know only that they were alive at some date in the past. If patients with whom you have lost contact differ from patients who remain in contact, the results of a life table analysis will be misleading.

Consider the situation in which patients with whom you have lost contact are healthier than patients who remain in contact. By assuming that patients who are lost to follow-up behave the same way as patients who are not, the life table will underestimate the survival experience of the group. Similarly, if sicker patients lose contact, the life table will overestimate the proportion surviving at various time points. It cannot be emphasized too strongly that no statistical procedure can atone for problems associated with incomplete follow-up.

Plotting Survival Functions

The survival functions displayed in the life table can also be plotted. This allows you to better examine the functions. It also allows you to compare the functions for several groups. For example, Figure 10.2 is a plot of the cumulative percentage surviving when the employees from Figure 10.1 are subdivided into two groups: clerical and professional (you specify *type* as the factor variable). You see that clerical employees have shorter employment times than professional employees.

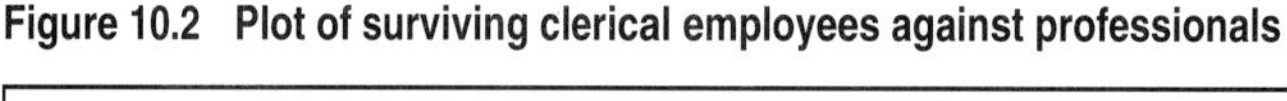

Figure 10.2 Plot of surviving clerical employees against professionals

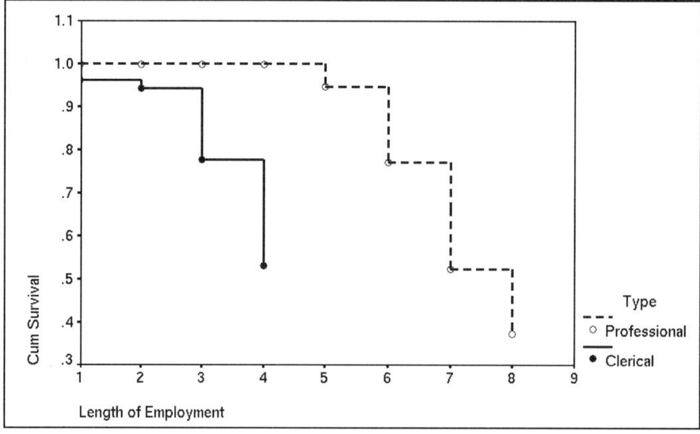

Comparing Survival Functions

If two or more groups in your study can be considered as samples from some larger population, you may want to test the null hypothesis that the survival distributions are the same for the subgroups. The SPSS Life Tables procedure uses the Wilcoxon (Gehan) test (Lee, 1992).

As shown in Figure 10.3, the number of censored and uncensored cases, as well as an average score, are displayed for each group. The average score is calculated by comparing each case to all others and incrementing the score for a case by 1 if the case has a longer survival time than another case and decrementing it by 1 if the case has a shorter survival time.

In Figure 10.3, the observed significance level for the test that all groups come from the same distribution is less than 0.00005, leading you to reject the null hypothesis that the groups do not differ.

Figure 10.3 Comparing subgroups

```
Comparison of survival experience using the Wilcoxon (Gehan) statistic
   Survival Variable  LENGTH
          grouped by  TYPE

  Overall comparison    statistic       26.786  D.F.      1   Prob.   .0000

  Group  label                 Total N   Uncen     Cen  Pct Cen  Mean Score

      1  Clerical                   54      19      35    64.81    -15.4444
      2  Professional               46      12      34    73.91     18.1304
```

How to Obtain Life Tables

The SPSS Life Tables procedure produces nonparametric life tables and related statistics for examining the length of time between two events. You can obtain survival and hazard plots and compare subgroups. For the life tables, the Life Tables procedure groups cases into intervals that you specify. To obtain life tables that estimate survival at the event times, use the Kaplan-Meier procedure (see Chapter 11).

The minimum specifications are:

- A survival time variable.
- A survival status variable whose codes indicate whether an event has occurred.
- Status variable codes that identify uncensored cases.
- Time intervals to display in the life table.

To obtain life tables, from the menus choose:

Statistics
 Survival ▸
  Life Tables...

This opens the Life Tables dialog box, as shown in Figure 10.4.

Figure 10.4 Life Tables dialog box

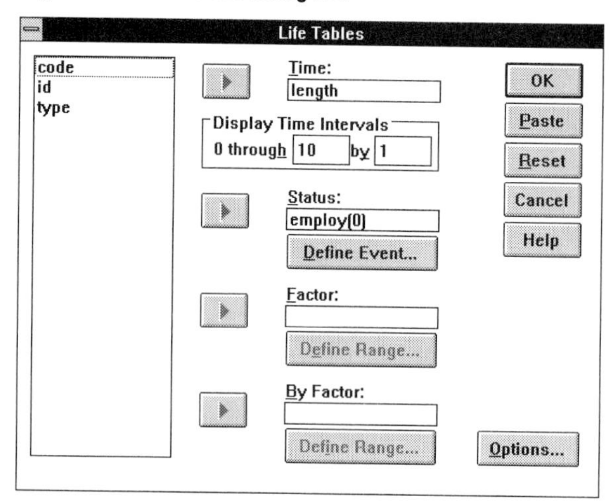

The numeric variables in your data file appear on the source list.

Time. Select a survival time variable whose values indicate how long cases survived (uncensored cases) or were tracked (censored cases). The time variable can be measured in any unit of time. SPSS excludes cases with negative survival times from the analysis.

Display Time Intervals. SPSS uses time 0 as the starting point of the first time interval displayed in life tables. Enter a starting value for the last interval you want to display. For example, if survival times are recorded in years, enter 10 to analyze a time period of 0 through 10 years. Also enter a value for the length of time intervals. For example, if the survival times are recorded in years, enter 1 to display yearly intervals; enter 10 to display decennial intervals.

Status. Select a survival status variable whose values indicate whether an event occurred for a case. You must also indicate which status codes identify uncensored cases (see "Define Event for Status Variable" on p. 271).

Factor. By default, SPSS produces a life table for all cases as if they were from a single sample. To obtain life tables for subgroups, select a first-order factor variable. SPSS generates a separate life table for each category of the variable. You must indicate which levels of your first-order factor variable to use in the analysis (see "Define Range for Factor Variable" on p. 271).

By Factor. Optionally, you can select a second-order factor variable. SPSS produces one life table for each combination of values of first- and second-order factor variables. For example, if the first-order factor has two categories and the second-order factor has four categories, eight life tables are produced. You must indicate which levels of your second-order factor variable to use in the analysis (see "Define Range for Factor Variable," below). The variable is treated as a first-order factor if you do not also select a variable for Factor.

Define Event for Status Variable

To specify which values of the status variable identify events, highlight the status variable and click on Define Event... in the Life Tables dialog box. This opens the Life Tables Define Event for Status Variable dialog box, as shown in Figure 10.5.

Figure 10.5 Life Tables Define Event for Status Variable dialog box

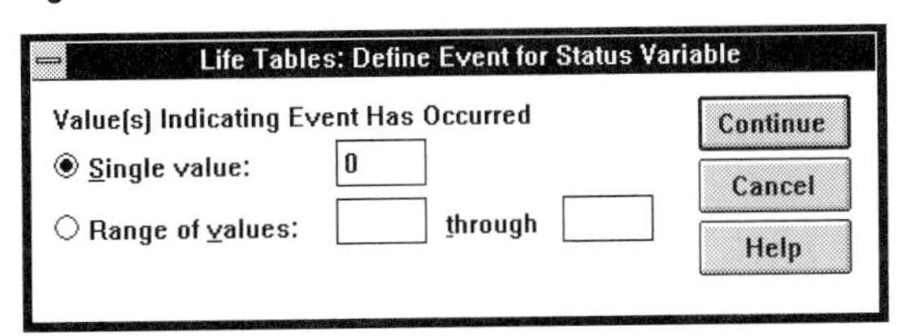

A single value or a range of values can indicate that an event has occurred. Cases with other values are treated as censored cases. Choose one of the following alternatives:

❍ **Single value**. This is the default. Enter the value.

❍ **Range of values**. Enter the lowest and highest values in the range.

Define Range for Factor Variable

For each factor variable, you must indicate which factor levels you want to use in the analysis. Highlight the factor and click on Define Range... in the Life Tables dialog box. This opens the Life Tables Define Range for Factor Variable dialog box, as shown in Figure 10.6.

Figure 10.6 Life Tables Define Range for Factor Variable dialog box

Life Tables: Define Range for Factor Variable

Minimum:

Maximum:

Continue

Cancel

Help

Enter integer values corresponding to the lowest and highest levels to use. Each integer category in the inclusive range is used as a factor level. Non-integer factor values are truncated for the analysis, so cases having the same integer portion are combined into the same factor level. Cases with factor levels outside the range are excluded from the analysis.

Options

To obtain survival plots, compare subgroups, control the handling of missing values, or suppress life tables, click on **Options...** in the Life Tables dialog box. This opens the Life Tables Options dialog box, as shown in Figure 10.7.

Figure 10.7 Life Tables Options dialog box

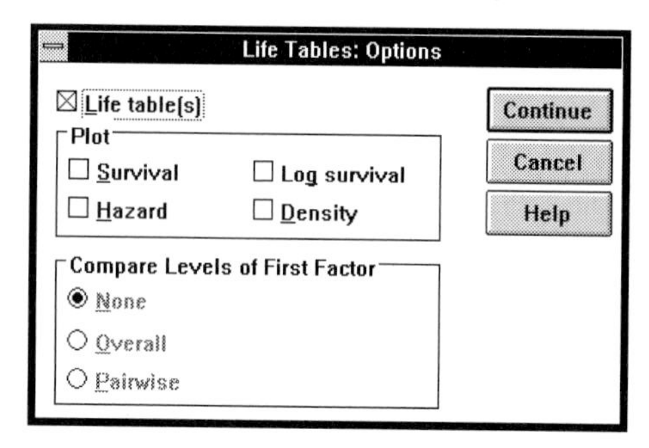

At least one life table, plot, or factor-level comparison must be selected.

Plot. For plots, SPSS treats all cases in the working data file as if they were from a single sample unless you have a factor variable. If you have a first-order factor, plots show each subgroup. If you also have a second-order factor, one plot of first-order factor levels is displayed for each level of the second-order factor. You can choose one or more of the following plots:

- **Survival.** Displays the cumulative survival function on a linear scale.
- **Hazard.** Displays the hazard function.
- **Log survival.** Displays the cumulative survival function on a logarithmic scale.
- **Density.** Displays the density function.

Compare Levels of First Factor. The Wilcoxon (Gehan) test compares the survival distributions of first-order factor levels (see "Comparing Survival Functions" on p. 269). A first-order factor is required to obtain this test. If you also have a second-order factor, a separate test is performed for each level of the second-order factor. You can choose one of the following alternatives:

❍ **None.** No subgroup comparison is performed. This is the default.

❍ **Overall.** Compares all levels of the first-order factor simultaneously.

❍ **Pairwise.** Compares each pair of first-order factor levels. For example, if the factor variable has three categories, tests compare levels 1 versus 2, 2 versus 3, and 1 versus 3. An overall comparison is also displayed.

The following display option is also available:

❑ **Life table(s).** By default, SPSS displays life tables with its output. If you want only plots or group comparisons, deselect this item.

Additional Features Available with Command Syntax

You can customize your life tables if you paste your selections into a syntax window and edit the resulting SURVIVAL command syntax. (For information on syntax windows, see the SPSS Base system documentation.) Additional features include:

- The ability to generate analyses of multiple survival time variables in a single step (using the TABLES and STATUS subcommands).
- Time intervals of varying lengths (using the INTERVALS subcommand).
- Subgroup comparison options for large samples and for aggregated data (using the CALCULATE subcommand).
- The ability to write survival statistics to a text file for use in subsequent analyses (using the WRITE subcommand).

See the Syntax Reference section of this manual for command syntax rules and for complete SURVIVAL command syntax.

11 Kaplan-Meier Survival Analysis

In Chapter 10, you saw how actuarial life tables are used to analyze the time interval between two events when the second event does not necessarily happen to everyone during the course of the study. For example, you can construct a life table to examine the length of time between diagnosis of a disease and death. Although death is an event from which no one escapes, it need not occur during the time period in which patients are enrolled in a study. Cases for which the event does not occur during the period of observation are called **censored cases**. Special statistical techniques are needed for analysis of data that contain censored observations. Since these techniques are often used to analyze data in which the event of interest is death, they are known as survival time or failure time techniques.

In this chapter, we will describe the distribution of times to an event using a method proposed by Kaplan and Meier (1958). This method is very closely related to the actuarial method described in Chapter 10. In the next chapter, the time to the occurrence of an event is modeled using Cox regression models, also known as **proportional hazards models**. Cox regression models are similar to usual regression models in that you predict a dependent variable (length of time to the occurrence of an event) as a function of a set of independent variables. However, unlike ordinary regression models, Cox regression models can be used when there are censored observations.

Kaplan-Meier Estimators

Before considering how Kaplan-Meier estimates are calculated, recall how the actuarial estimates are obtained. When you construct a life table using the actuarial method, you subdivide the period of time under study into intervals, such as years or months. For each of the intervals, based on the number of people under observation in that interval and the number of events in that interval, you estimate the probability of an event occurring during the interval. Estimates from the individual intervals are combined to estimate cumulative probabilities. Censored individuals contribute information to all intervals during which they are observed. For example, a censored individual who is known to be event-free for four years contributes information to the estimation of rates for the first four years. A very important assumption of both the actuarial and Kaplan-

Meier methods is that censored cases do not differ from those that remain under observation. See Chapter 10 for a more detailed discussion.

To compute Kaplan-Meier estimates of the probability of being event-free at various time points, you don't need to establish intervals at which the various probabilities are evaluated. Instead, you estimate the probability of an event each time an event is observed.

Let's consider an example. Table 11.1 contains 10 hypothetical survival times, measured in months, for patients diagnosed with a particular disease. Survival status codes shown in the table differentiate censored and uncensored cases. (The SPSS Kaplan-Meier procedure accepts short string and numeric survival status codes.)

Table 11.1 Hypothetical survival times

Time	Status
2	event
5	event
12	censored
35	event
40	censored
43	event
49	event
64	event
69	censored
72	event

From Table 11.1, you see that the observed survival times range from 2 months to 72 months. Three of the times are censored. That is, three patients with survival times of 12 months, 40 months, and 69 months were alive at last contact. If you were constructing an actuarial life table, you'd have to decide at what points you want to estimate the survival curve. For example, you might want to calculate yearly survival rates by grouping the data into 12-month intervals.

To calculate Kaplan-Meier estimates of the survival curve, you evaluate the survival curve at each of the time points at which an event occurs. That is, you evaluate it at each of the uncensored time points. The computations are simple. Consider Table 11.2.

The first column of Table 11.2 contains the sorted observed survival times. The next column indicates whether the time corresponds to an event or to a censored observation. The third column tells you the total number of cases that are alive before the current time. The next column tells you the number of cases alive after the current time. To estimate the conditional probability of surviving at each of the observed event times, given that one is alive after the last event, you divide the number of cases alive just after that time by the number of cases alive just prior to that time. For example, the probability of surviving at two months is estimated to be 0.90, since of 10 cases alive up to two months, only 9 survive past two months.

Table 11.2 Calculating a Kaplan-Meier curve

Time	Status	Number Alive Prior	Number Remaining Alive	Proportion Alive	Cumulative Survival Proportion
2	event	10	9	9/10	9/10=0.90
5	event	9	8	8/9	0.9x(8/9)=0.80
12	censored	8	7		
35	event	7	6	6/7	0.8x(6/7)=0.69
40	censored	6	5		
43	event	5	4	4/5	0.69x(4/5)=0.55
49	event	4	3	3/4	0.55x(3/4)=0.41
64	event	3	2	2/3	0.41x(2/3)=0.27
69	censored	2	1		
72	event	1	0	0/1	0.27x(0/1)=0

The second death is observed at five months. The probability of surviving past five months, assuming that you were alive past two months, is 8 out of 9, since nine cases are alive just prior to five months and only eight cases are alive after five months. To calculate the *cumulative* probability of surviving five months, you must multiply the probability of surviving up to five months by the probability of surviving five months, given that you were alive prior to five months. In general, to calculate cumulative survival probabilities for a particular event time, you must multiply all of the individual survival probabilities up to and including that time. The cumulative survival probability estimates the proportion of all cases that are still alive at a particular time point. Notice that probabilities are not estimated at the censored times. That is because the proportion surviving does not change at these points. Only the number of cases still under observation is decreased.

The SPSS Kaplan-Meier Table

Figure 11.1 contains output from the SPSS Kaplan-Meier procedure for the previous example. The format is very similar to that of Table 11.2. The first two columns are identical. The third column shows the cumulative survival. The fourth column contains estimates of the standard error of the cumulative survival. These can be used to construct confidence intervals for the survival curve.

Figure 11.1 Kaplan-Meier table

Survival Analysis for TIME

Time	Status	Cumulative Survival	Standard Error	Cumulative Events	Number Remaining
2.00	Event	.9000	.0949	1	9
5.00	Event	.8000	.1265	2	8
12.00	Censored			2	7
35.00	Event	.6857	.1515	3	6
40.00	Censored			3	5
43.00	Event	.5486	.1724	4	4
49.00	Event	.4114	.1756	5	3
64.00	Event	.2743	.1620	6	2
69.00	Censored			6	1
72.00	Event	.0000	.0000	7	0

Number of Cases: 10 Censored: 3 (30.00%) Events: 7

	Survival Time	Standard Error	95% Confidence Interval
Mean:	45.84	8.73	(28.74, 62.95)
Median:	49.00	8.96	(31.43, 66.57)

The standard error of the cumulative proportion surviving at time k is

$$se(t_k) = S(t_k)\sqrt{\sum_{i=1}^{k}\frac{d_i}{n_i(n_i-d_i)}}$$ **Equation 11.1**

where $S(t_k)$ is the cumulative survival probability, d_i is the number of events at time t_i, and n_i is the number of cases alive prior to time t_i.

The column labeled *Cumulative Events* is a count of all events that have occurred up to and including the current time. The *Number Remaining* column contains the number of cases alive after the current time.

The total number of cases, the number and percentage of censored cases, and the number of events are printed below the table. Also printed are estimates of the mean and median survival times, their standard errors, and 95% confidence intervals. The median survival time is the first observed time at which the cumulative survival is 50% or less.

The mean survival time is *not* the average of the observed survival times, since it does not make sense to compute the usual arithmetic average if all of the cases are not dead. Special techniques are used to estimate mean survival time when there are censored observations. If the largest observed survival time is for a censored observation, the estimate of the mean survival time is said to be restricted to the largest observed survival time.

Plotting the Cumulative Survival

A plot of the cumulative survival function for the sample data is shown in Figure 11.2. You will notice that the survival function has steps. That is, it is only at each of the observed event times that the estimate of the survival function changes. It is constant for all times between two adjacent event times. A "+" indicates a censored case.

Figure 11.2 Cumulative survival function

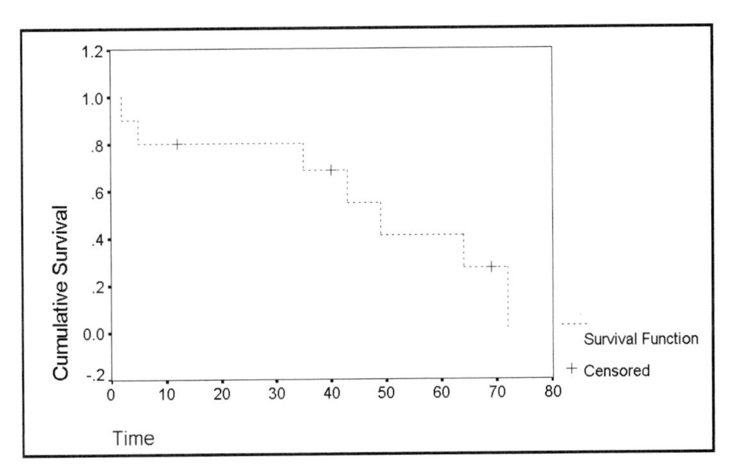

Comparing Survival Functions

In many situations, you want to do more than just estimate a cumulative survival function for a single group. You want to compare survival functions across several groups. Consider the following data from a Veteran's Administration Lung Cancer Trial presented by Kalbfleisch and Prentice (1980). Survival times, as well as other information, were obtained for a sample of 137 men with advanced inoperable lung cancer.

One of the hypotheses of interest is whether the different histological types of lung cancer affect survival. The patients can be subdivided into four groups based on the histology of their tumors: squamous cell (group 1), small cell (group 2), adenocarcinoma (group 3), or large cell (group 4). Using the SPSS Kaplan-Meier procedure, we can estimate four survival functions, one for each of the histology categories. (You specify *histol* as the factor variable.)

Figure 11.3 is a summary table of the number of cases and events in each of the histology groups. You see that there are a total of nine censored cases. Eleven percent of the cases in the squamous histology group are censored, 6% in the small-cell group are censored, and almost 4% in both the adenocarcinoma and large-cell groups are censored.

Figure 11.3 Summary table for histology factor

Survival Analysis for TIME

		Total	Number Events	Number Censored	Percent Censored
HISTOL	squamous	35	31	4	11.43
HISTOL	small cell	48	45	3	6.25
HISTOL	adeno	27	26	1	3.70
HISTOL	large cell	27	26	1	3.70
Overall		137	128	9	6.57

The plot of the estimated survival functions for each of the histology groups is shown in Figure 11.4. From the plot, it appears that the four groups differ in survival. Patients with adenocarcinoma seem to survive the shortest time, while patients with squamous-cell histology appear to have the best prognosis.

Figure 11.4 Survival functions for histology groups

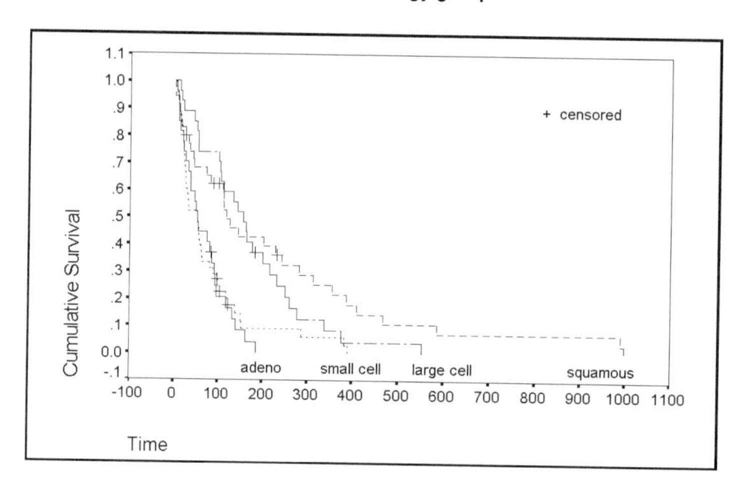

Several statistical tests are available for evaluating the null hypothesis that, in the population, two or more survival functions are equal. The SPSS Kaplan-Meier procedure lets you choose among the **log rank test**, the **Breslow test** (also known as the **generalized Wilcoxon test**), and the **Tarone-Ware test**. All three tests are based on computing the weighted difference between the observed and expected number of deaths at each of the time points. A component of the statistic can be written as

$$U = \sum_{i=1}^{k} w_i (0_i - E_i)$$

Equation 11.2

where w_i is the weight for time point i, and k is the number of distinct time points. For the log rank test, all of the weights are equal to 1; for the Breslow test, the weights are the number at risk at each time point; and for the Tarone-Ware test, the weights are the square root of the number at risk. From the formula, you see that the log rank test weights all deaths equally. The Breslow test weights early deaths more than later deaths because the number at risk always decreases with time.

The log rank test is more powerful than the Breslow test for detecting differences if the mortality rate in one group is a multiple of that in another group. If this is not the case, the log rank test may not be as powerful as the Breslow test. If the percentage of censored cases is large, Prentice and Marek (1979) have shown that the Breslow test has very low power, since the early deaths dominate the statistic.

Figure 11.5 shows the log rank test, as well as the other tests, for testing the null hypothesis that the survival functions are the same for the four histologic groups. Since the observed significance level is small ($p < 0.00005$), you can reject the null hypothesis. Although, based on Figure 11.5, you can reject the overall hypothesis that the four groups have the same survival function, you can't tell which of the groups are significantly different from each other. To do that, you must compare all pairs of groups. Figure 11.6 contains the output for all pairwise comparisons.

Figure 11.5 Overall test statistics for histology factor

Test Statistics for Equality of Survival Distributions for HISTOL

	Statistic	df	Significance
Log Rank	25.40	3	.0000
Breslow	19.43	3	.0002
Tarone-Ware	22.57	3	.0000

Figure 11.6 Pairwise log rank tests for histology factor

Log Rank Statistic and (Significance)

Factor	1.00	2.00	3.00
2.00	11.57 (.0007)		
3.00	12.05 (.0005)	.10 (.7557)	
4.00	.82 (.3644)	9.37 (.0022)	17.67 (.0000)

From Figure 11.6, you can see that all comparisons are significant except squamous versus large cell (group 1 versus group 4), and small cell versus adenocarcinoma (group 2 versus group 3).

Making all pairwise comparisons for survival data entails the same problems as have been previously discussed in one-way analysis of variance. As the number of comparisons increases, the probability of calling a difference significant by chance increases as well. You can protect yourself from rejecting the null hypothesis too often by applying a Bonferroni correction. You multiply each of the observed significance levels by the number of comparisons being made.

Stratified Comparisons of Survival Functions

The goal of the Veteran's Administration Lung Trial was to study the efficacy of a new treatment in prolonging survival time of inoperable lung cancer patients. All of the patients were assigned to either the standard treatment or the new treatment. In analyzing the data, your first impulse might be to estimate two survival functions, one for the standard treatment and one for the new treatment, and then to compare them. The problem with this approach is that you have already seen that the type of histology affects the survival prognosis. What you must do is to compare the two treatments within each of the four histology categories. (In the SPSS Kaplan-Meier procedure, the histology categories are identified as **strata**, while the treatment is the factor of interest.)

Figure 11.7 shows the summary table for the data when the cases are classified into both histology groups and treatment categories. From this table, you see that the number of cases receiving the two treatments is not the same in each of the histology strata. For example, of the 48 cases with small-cell carcinoma, 18 were assigned to the new treatment and 30 to the standard treatment.

Figure 11.7 Summary table for histology group and treatment method

Survival Analysis for TIME

		Total	Number Events	Number Censored	Percent Censored
HISTOL	squamous	35	31	4	11.43
TRTMENT	new	20	18	2	10.00
TRTMENT	standard	15	13	2	13.33
HISTOL	small cell	48	45	3	6.25
TRTMENT	new	18	17	1	5.56
TRTMENT	standard	30	28	2	6.67
HISTOL	adeno	27	26	1	3.70
TRTMENT	new	18	17	1	5.56
TRTMENT	standard	9	9	0	.00
HISTOL	large cell	27	26	1	3.70
TRTMENT	new	12	12	0	.00
TRTMENT	standard	15	14	1	6.67
Overall		137	128	9	6.57

The survival functions for the two types of treatments for each of the four histology types are shown in Figure 11.8. Each of the separate plots is for a histology category. Within each histology category, you see survival functions for each of the two treatments. From these plots, the two treatments don't appear to be very different.

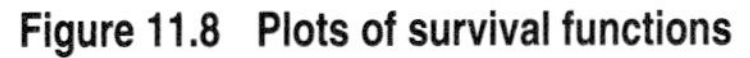

Figure 11.8 Plots of survival functions

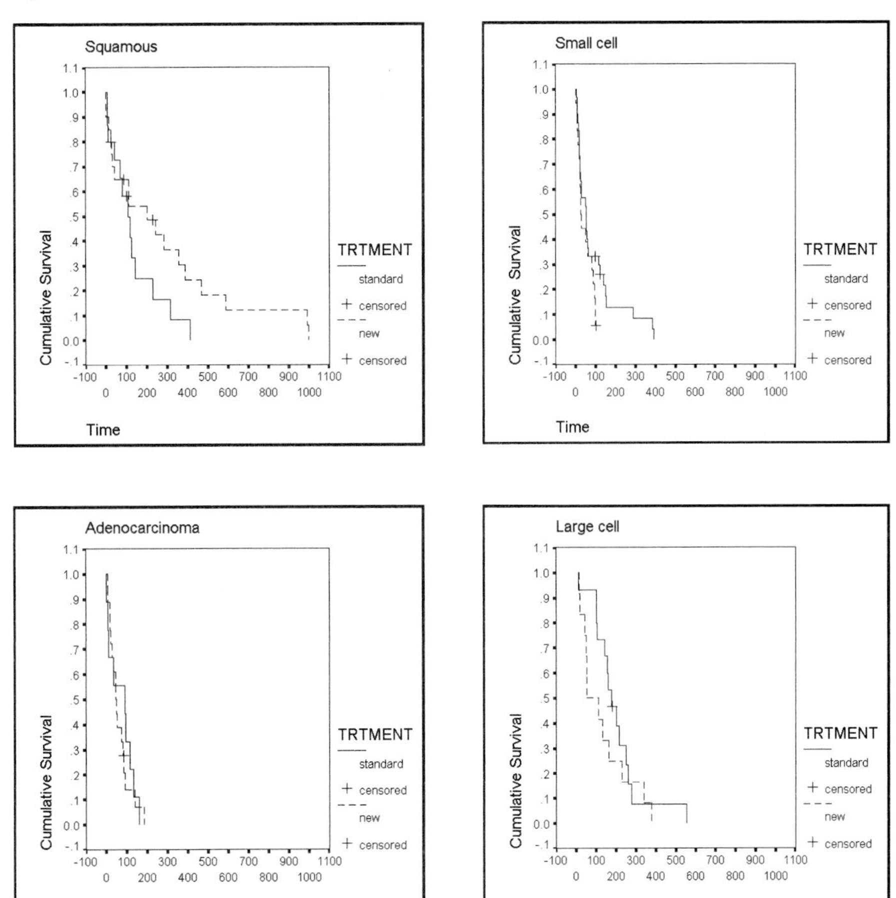

You can test the null hypothesis that the two survival functions are the same, pooling the results across the histology category. All three of the previously described tests can be used when the cases are subdivided into strata.

Figure 11.9 shows the results from the log rank test. The large observed significance level ($p = 0.4022$) tells you that there is not enough evidence to reject the null hypothesis that the survival functions for the two treatments are identical.

Figure 11.9 Test statistics for treatment method

```
Test Statistics for Equality of Survival Distributions for TRTMENT
Adjusted for HISTOL

              Statistic        df       Significance

Log Rank            .70        1            .4022
Breslow            1.04        1            .3070
Tarone-Ware        1.02        1            .3119
```

How to Obtain a Kaplan-Meier Survival Analysis

The SPSS Kaplan-Meier procedure estimates survival curves using the product-limit method. Like the Life Tables procedure, the Kaplan-Meier procedure produces nonparametric survival tables, and both procedures can compare subgroups and analyze samples containing right-censored cases. However, while the Life Tables procedure groups survival times into arbitrary intervals, the Kaplan-Meier procedure estimates survival at the event times. To use continuous predictors (covariates) to predict survival times, the Cox Regression procedure is available (see Chapter 12).

The minimum specifications are:

- A survival time variable.
- A survival status variable whose values indicate whether an event has occurred.
- Status variable codes that identify uncensored cases.

To obtain a Kaplan-Meier survival analysis, from the menus choose:

Statistics
 Survival ▸
  Kaplan-Meier...

This opens the Kaplan-Meier dialog box, as shown in Figure 11.10.

Figure 11.10 Kaplan-Meier dialog box

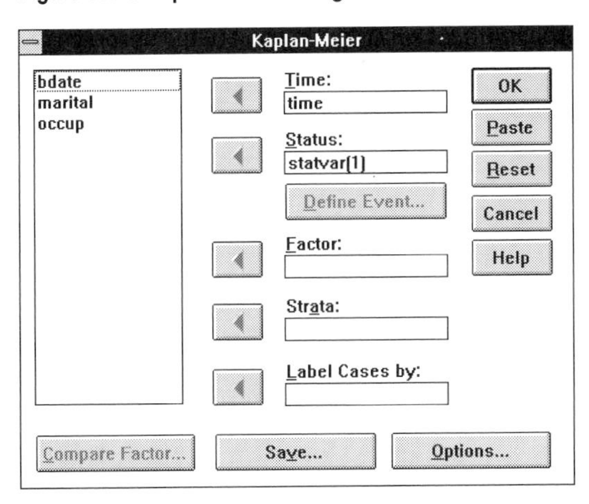

The variables in your working data file appear on the source list.

Time. Select a numeric survival time variable indicating how long cases survived (uncensored cases) or were tracked (censored cases). The time variable can be measured in any unit of time. SPSS excludes cases having negative survival times from the analysis.

Status. Select a numeric or short string survival status variable whose values indicate whether an event occurred for a case. You must also specify which status codes identify uncensored cases (see "Define Event for Status Variable" on p. 286).

Factor. By default, the analysis treats all cases in the working data file as if they were from a single sample. To compute separate survival curves for subgroups, select a short string or numeric factor variable whose categories divide your sample into two or more groups.

Strata. You can control for a categorical stratification variable in your analysis. Select a short string or numeric variable whose values form strata within levels of the factor variable. Analysis is performed within each stratum for each factor level.

Label Cases by. By default, cases are identified in the survival table by their survival times. You can use the first 20 characters of a variable's values to label cases in the survival table. Select a case-identification variable. If value labels are defined for the case-identification variable, SPSS uses the value labels to identify cases in the table.

Define Event for Status Variable

To specify which values of the status variable identify events, highlight the status variable and click on Define Event... in the Kaplan-Meier dialog box. This opens the Kaplan-Meier Define Event for Status Variable dialog box, as shown in Figure 11.11.

Figure 11.11 Kaplan-Meier Define Event for Status Variable dialog box

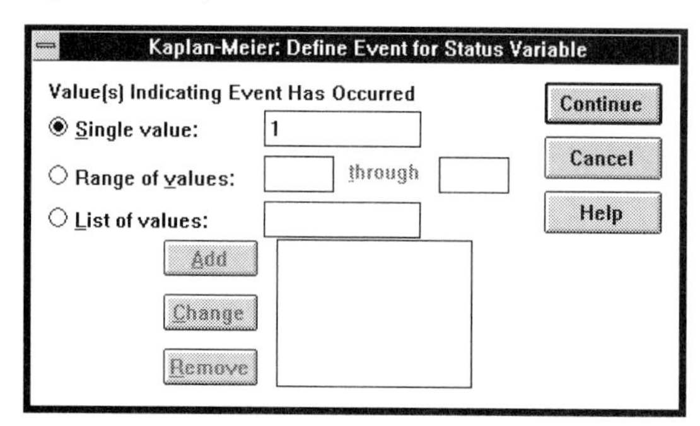

A single value or several values can indicate that an event has occurred for a case. Cases with other nonmissing values for the status variable are treated as censored cases. Choose one of the following alternatives:

- **Single value**. This is the default. Enter a value.
- **Range of values**. This item is available only if your status variable is numeric. Enter the lowest and highest values in the range. The first value must be less than the second value.
- **List of values**. Select this item to define a list of values. Enter a value and click on Add to add the value to the list. To add more values to the list, repeat this process. To change a value, highlight it on the list, modify the value, and click on Change. To remove a value, highlight it on the list and click on Remove.

Comparing Factor Levels

If you have a factor variable, you can obtain tests comparing survival functions for different factor levels. Click on Compare Factor... in the Kaplan-Meier dialog box. This opens the Kaplan-Meier Compare Factor Levels dialog box, as shown in Figure 11.12.

Figure 11.12 Kaplan-Meier Compare Factor Levels dialog box

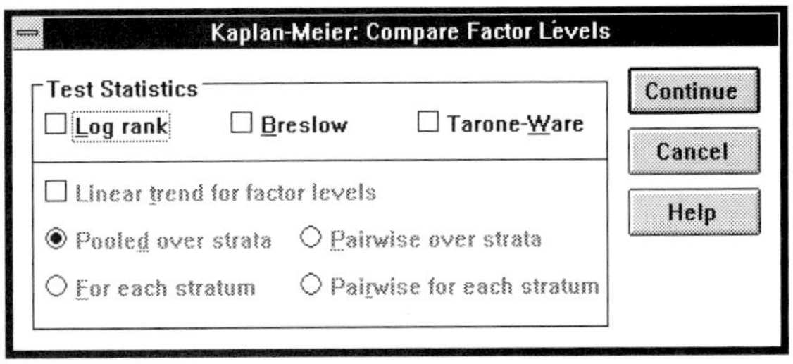

Test Statistics. You can choose one or more of the following chi-square tests of equality of survival functions for factor levels:

- **Log rank**. Displays the log rank (Mantel-Cox) test. This test weights all time points equally.
- **Breslow**. Displays the Breslow (generalized Wilcoxon) test. This test weights time points by the number of cases at risk.
- **Tarone-Ware**. Displays the Tarone-Ware test. This test weights time points by the square root of the number of cases at risk.

If you request a test comparing all factor levels (see below), the following option is available:

- **Linear trend for factor levels**. Uses trend information in overall tests of equality of survival functions. This test is appropriate if factor levels have a natural ordering (for example, when factor codes represent doses applied to different groups). SPSS assumes that the factor levels are equally spaced.

For tests comparing factor levels, you can choose tests that compare all factor levels simultaneously or tests between pairs of factors. You can pool results across strata or obtain separate tests for each stratum. Choose one of the following alternatives:

- **Pooled over strata**. Compares all factor levels in a single test. This is the default test of equality of survival curves.
- **For each stratum**. Performs a separate test of equality of all factor levels for each stratum. If you do not have a stratification variable, the tests are not performed.
- **Pairwise over strata**. Compares each distinct pair of factor levels. Pairwise trend tests are not available.
- **Pairwise for each stratum**. Compares each distinct pair of factor levels for each stratum. Pairwise trend tests are not available. If you do not have a stratification variable, the tests are not performed.

Saving New Variables

To save survival variables, click on Save... in the Kaplan-Meier dialog box. This opens the Kaplan-Meier Save New Variables dialog box, as shown in Figure 11.13.

Figure 11.13 Kaplan-Meier Save New Variables dialog box

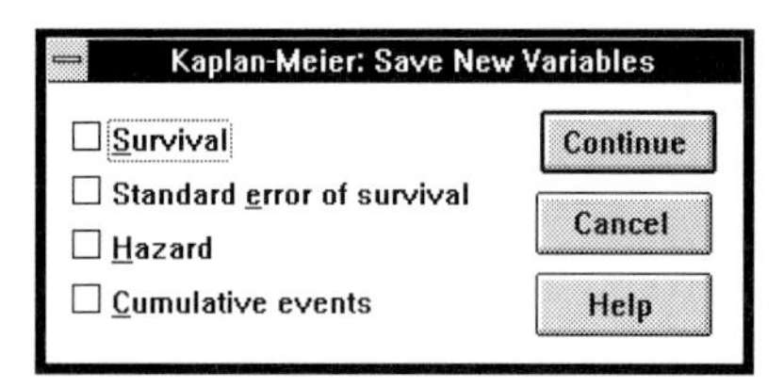

You can save one or more of the following variables:

- **Survival**. Cumulative survival probability estimate. The default variable name is the prefix *sur_* with a sequential number appended to it. For example, if *sur_1* already exists, SPSS assigns the variable name *sur_2*.
- **Standard error of survival**. Standard error of the cumulative survival estimate. The default variable name is the prefix *se_* with a sequential number appended to it. For example, if *se_1* already exists, SPSS assigns the variable name *se_2*.
- **Hazard**. Cumulative hazard function estimate. The default variable name is the prefix *haz_* with a sequential number appended to it. For example, if *haz_1* already exists, SPSS assigns the variable name *haz_2*.
- **Cumulative events**. Cumulative frequency of events when cases are sorted by their survival times and status codes. The default variable name is the prefix *cum_* with a sequential number appended to it. For example, if *cum_1* already exists, SPSS assigns the variable name *cum_2*.

Options

To obtain optional statistics and plots, click on Options... in the Kaplan-Meier dialog box. This opens the Kaplan-Meier Options dialog box, as shown in Figure 11.14.

Figure 11.14 Kaplan-Meier Options dialog box

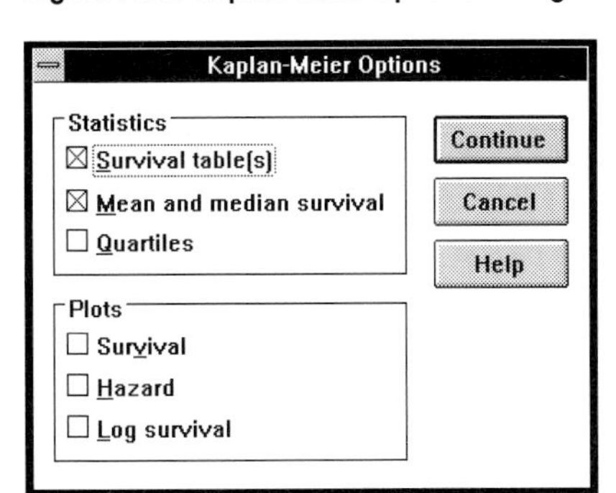

Statistics. Statistics are displayed for each combination of factor level and stratum. You can choose one or more of the following statistical displays:

❑ **Survival table(s).** Shows the product-limit survival estimate, its standard error, the cumulative frequency of events, and the number of cases at risk. Survival tables are displayed by default.

❑ **Mean and median survival.** Shows mean and median survival times with their standard errors and confidence intervals. Mean and median survival times are displayed by default.

❑ **Quartiles.** Displays 25th, 50th, and 75th percentiles and their standard errors for the survival time variable.

Plots. If you have a stratification variable, a separate plot is produced for each stratum. Each plot shows a separate curve for each factor level. You can choose one or more of the following plots:

❑ **Survival.** Plots the cumulative survival distribution on a linear scale.

❑ **Hazard.** Plots the cumulative hazard function on a linear scale.

❑ **Log survival.** Plots the cumulative survival distribution on a logarithmic scale.

Additional Features Available with Command Syntax

You can customize your Kaplan-Meier survival analysis if you paste your selections into a syntax window and edit the resulting KM command syntax. (For information on syntax windows, see the SPSS Base system documentation.) Additional features include:

- The ability to identify cases as lost to follow-up (using the STATUS subcommand).
- Unequal spacing for linear trend tests (using the TREND subcommand).

- User-specified percentiles for the survival time variable (using the PERCENTILES subcommand).

See the Syntax Reference section of this manual for command syntax rules and for complete KM command syntax.

12 Cox Regression

The time interval between two events almost certainly depends on a variety of factors. The duration of a marriage is based on economic, social, and emotional considerations. Similarly, the survival time after diagnosis of a disease depends on predictor variables such as the severity of the disease, the treatment method, and the general condition of the patient.

In Chapter 10 and Chapter 11, you saw how the time interval between two events can be examined using actuarial and Kaplan-Meier estimators of the survival distribution. When using these methods to assess the influence of predictor variables on the survival times, you had to calculate separate survival distributions for each of the categories of a predictor variable. Values of continuous predictor variables, such as age, had to be grouped into categories.

Analyses requiring that cases be subdivided into groups based on the values of the predictor variables are often unsatisfactory, since the number of cases in a group rapidly diminishes with increasing numbers of predictor variables. That's why techniques such as multiple linear regression are invaluable for studying the relationship between a dependent variable and a set of independent variables. However, multiple linear regression, as described in the SPSS Base system documentation, cannot be used for analysis of time-to-event data, since there is no way to handle censored observations. (Remember, censored cases are those for which the event has not yet occurred. They cannot be ignored but must be incorporated into any analysis.) In this chapter, you'll see how a special type of regression model, the **Cox regression model**, can be used to analyze data that contain censored observations.

The Cox Regression Model

The regression model proposed by Cox can be written in several different ways. Let's consider the simple case with only one predictor variable, say *age*. (In the Cox model, the independent or predictor variables are usually called **covariates**.) When the **cumulative survival function**—that is, the proportion of cases "surviving" at a particular point in time—is the dependent variable, the model is

$$S(t) = [S_0(t)]^p$$ **Equation 12.1**

where $p = e^{(B \times age)}$.

From the model, you see that for any value of the age variable, the proportion surviving at time *t* depends on two quantities. The first, designated as $S_0(t)$ in the equation, is called the **baseline survival function**. The baseline survival function does not depend on age; it depends only on time. The baseline survival function is similar to the constant term in multiple regression in that it is the reference value that is increased or decreased depending on the values of the independent variables and their relationship with the dependent variable. The second part of the equation, the term designated *p* in the model, depends not on time but on the value of the covariate *age* and on the value of the regression coefficient *B*.

To better understand the components of the Cox regression model, let's consider a data set presented by Bartolucci and Fraser (1977). The data track the survival experience of 60 patients with Hodgkin's disease who were receiving standard treatment for the disease. Age, sex, stage of the disease, histology, survival time in months, and censoring status are recorded for each patient.

To examine the effect of age upon survival, we can calculate a Cox regression model with survival time as the dependent variable and the age of the patient at diagnosis as the covariate. For computational details on the method of partial likelihood, the technique used for estimating the model parameters, see Kalbfleisch and Prentice (1980) or Blossfeld et al. (1989).

Figure 12.1 contains statistics for the *age* variable when it is entered into the Cox regression model. From the figure, you see that the coefficient for *age* is 0.03, with an observed significance level of 0.009. Based on the observed significance level, you can reject the null hypothesis that the population value of the coefficient is 0.

Figure 12.1 Statistics for age

```
------------------- Variables in the Equation --------------------

Variable           B       S.E.      Wald  df      Sig        R     Exp(B)

AGE            .0301      .0115    6.8034   1    .0091    .1474     1.0306
```

Now let's consider the meaning of the regression coefficient. Using Equation 12.1, the baseline survival function, and the regression coefficient, you can estimate the cumulative survival function for persons of different ages. (In the usual regression model, you generate one predicted value for each case. In the Cox regression model, you generate an entire survival curve for each value of the independent variable.) Consider Figure 12.2. The first column, *TIME*, contains the survival times at which the first ten events occur. The next column, *BASE*, contains the baseline survival function at these times.

Figure 12.2 Cumulative survival functions

TIME	BASE	AGE20
.37	.9954	.9916
.53	.9906	.9830
.70	.9857	.9741
1.17	.9808	.9653
2.37	.9759	.9564
3.30	.9708	.9474
3.50	.9656	.9381
4.00	.9603	.9286
5.60	.9548	.9191
5.97	.9493	.9094

To calculate the estimated survival curve for 20-year-olds, you use the equation

$$S(t|age_{20}) = [S_0(t)]^{1.8258} \quad \textbf{Equation 12.2}$$

The exponent is $e^{(0.0301 \times 20)}$, in which 0.0301 is the regression coefficient for *age*, and 20 is the age group of interest. For example, the four-month survival for a 20-year-old is estimated to be $0.9603^{1.8258}$, which is 0.929. In Figure 12.2, the cumulative survival function for 20-year-olds is shown in the column labeled *AGE20*. (You can save cumulative survival and hazard functions for a value of a covariate using COX REGRESSION command syntax. For this example, age 20 is specified as the covariate value on the PATTERN subcommand, and the OUTFILE subcommand is used to save the survival functions. See the COXREG command in the Syntax Reference section of this manual for more information.) You can estimate the survival curve for persons of any age just by changing the value of age in the previous equation. All of the curves generated in this fashion are just different powers of the baseline survival function.

If the value of the exponent to which the baseline survival curve is raised is greater than 1, the resulting survival times are smaller than those of the baseline. If the exponent to which the baseline survival curve is raised is less than 1, the resulting survival times are greater than those of the baseline. This means that variables with positive coefficients are associated with decreased survival times, while variables with negative coefficients are associated with increased survival times.

The Hazard Function

The cumulative survival function is probably the most intuitive way of characterizing survival times. However, there are several other closely related functions that are also used. The **hazard function**, or death rate at time t, tells you how likely to experience an event a case is, given that it has survived to that time. The hazard function is not a probability but a death rate per unit of time, so it need not be less than 1.

Since the hazard function can be derived from the survival function, the Cox regression model defined in Equation 12.1 can be written in terms of the hazard function as

$$h(t) = [h_0(t)]e^{(BX)}$$ **Equation 12.3**

Equation 12.3 is very similar to Equation 12.1. The hazard function, like the survival function, is factored into two component pieces. The baseline hazard, $h_0(t)$, depends only on time, while $e^{(BX)}$ depends only on the values of the covariates and the regression coefficients. Instead of raising the cumulative survival function to the $e^{(BX)}$ power, the baseline hazard is multiplied by $e^{(BX)}$. Since the Cox model for the hazard function results in a simpler equation than that for the survival function, the Cox model is usually expressed in hazard form and is called the **Cox proportional hazards model**. The reason for the name "proportional hazards" is the fact that for two cases, the ratio of their hazards will be a constant for all time points.

To see this, let's plot the cumulative hazard functions for the Hodgkin's disease data for two cases: one 40 years of age, the other 20 years of age. The results are shown in Figure 12.3. You see that the difference between the two cumulative hazard rates is not constant. In fact, the longer the time, the larger the difference. What is constant, however, is the ratio of the two hazards—hence the name, proportional hazards regression.

Figure 12.3 Cumulative hazard functions

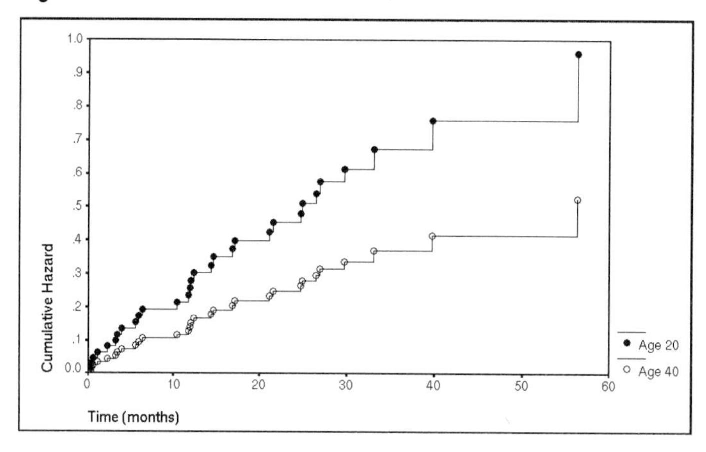

The assumption that hazards are proportional restricts the applicability of the Cox regression model. There are situations in which such an assumption is not realistic. For example, if you are comparing surgical and medical treatments of a condition, it is quite likely that for early time points, the hazards associated with surgery exceed those of the medical treatment. However, at later time points, the two hazard curves may be comparable. Thus, the ratio of the hazard functions is not a constant over time. Instead, it is large for early times and close to 1 for later times. The assumption of proportional hazards for two or more groups implies that the cumulative survival curves and the hazard curves for groups do not cross. (Later on in this chapter, you'll see how to test the assumption of proportional hazards.)

Multiple Covariates

Although we've considered a model with only one covariate—age—the Cox regression model can be used for any number of predictor variables. The general form of the model is

$$h(t) = [h_0(t)]\, e^{(B_1X_1 + B_2X_2 + \ldots + B_pX_p)}$$ **Equation 12.4**

where X_1 to X_p are the covariates. The covariates can be either continuous variables, such as age, blood pressure, and temperature, or categorical variables, such as stage of a disease and histology. If you have categorical covariates, you must transform their values by creating a new set of variables that correspond in some way to the original categories. The number of new variables required to represent a categorical variable is one less than the number of categories. A separate coefficient is estimated for each of the new variables. The interpretation of the coefficients for the coded variables depends on the coding scheme selected. (See the section on categorical variables in Chapter 1 for a more detailed discussion of coding schemes and interpretation of coefficients for categorical variables. The Cox Regression procedure will automatically code variables that are identified as categorical.) You can also include terms in the model that are interactions of individual variables. For example, if you think that the effect of the stage of the disease is different for men and women, you can include a stage-by-sex term in the model.

The Model with Three Covariates

In the Hodgkin's disease example considered earlier, age was found to be significantly associated with the hazard of dying. Let's see how the inclusion of two additional variables, *sex* (coded 1 for males and 0 for females) and *stage* (coded 0 for less advanced and 1 for more advanced), changes the model.

Figure 12.4 Statistics for the three-variable model

----------------------------- Variables in the Equation ---------------------------

Variable	B	S.E.	Wald	df	Sig	R	Exp(B)	95% CI for Exp(B) Lower	Upper
AGE	.0355	.0123	8.3975	1	.0038	.1701	1.0361	1.0116	1.0613
SEX	-.1042	.4044	.0664	1	.7966	.0000	.9010	.4079	1.9905
STAGE	.9655	.4060	5.6565	1	.0174	.1286	2.6262	1.1851	5.8196

From the column labeled *B* in Figure 12.4, you see that the estimated model can be written as

$$h(t) = [h_0(t)]\, e^{(0.0355 \times age - 0.1042 \times sex + 0.9655 \times stage)} \qquad \textbf{Equation 12.5}$$

The standard errors of the coefficients are shown in the column labeled *S.E.* The square of the ratio of the coefficient to its standard error is in the column labeled *Wald.* For large sample sizes, under the null hypothesis that the coefficient is 0, the Wald statistic has a chi-square distribution. The degrees of freedom associated with the statistic are 1, unless the test is for a categorical variable, in which case the degrees of freedom for the overall test of the variable are equal to the number of categories of the variable minus 1. From the column labeled *Sig*, you see that the coefficients for *age* and *stage* are significantly different from 0, while the coefficient for *sex* is not. The next column, labeled *R*, is an attempt to assess the partial correlation of each independent variable with the dependent variable. *R* can range from –1 to +1. *R* is 0 if the Wald statistic is less than twice the degrees of freedom for the variable. Otherwise, *R* is defined as

$$R = \pm\sqrt{\frac{\text{Wald statistic} - (2 \times \text{df})}{-2 \times \text{log likelihood for initial model}}} \qquad \textbf{Equation 12.6}$$

where *df* is the degrees of freedom for the coefficient. The sign of the coefficient is attached to *R*.

As in linear regression and logistic regression models, the value of the coefficients and their significance levels depends not only on the strength of the association of the individual variables with the dependent variable but also on the other independent variables in the model. If some of your independent variables are highly correlated, the resulting coefficients may be misleading. That is, variables may have wrong signs and the significance levels may not correctly reflect the strength of the association with the dependent variable. In such situations, you should include only one of the set of highly correlated variables in the model.

From the column labeled *Exp(B)* in Figure 12.4, you can tell the percentage change in the hazard rate for a unit increase in the covariate. For example, a one-year increase in age increases the hazard rate by 3.61%, since the ratio of the hazard rates for cases one year apart in age is 1.0361.

For a dichotomous variable, such as *stage* or *sex*, when two sequential numbers are used for coding and the larger of the two indicates presence of the characteristic, e^B is the ratio of the estimated hazard for a case with the characteristic to that for a case without the characteristic. This is often called the **relative risk** associated with the variable. If the relative risk is 1, the variable does not influence survival. If the relative risk is less than 1, a positive value for the variable is associated with increased survival, since the hazard rate is decreased.

From Figure 12.4, the relative risk for the advanced stage of the disease is 2.63. This means that the estimated risk of dying is 2.63 times greater for a person with advanced disease, compared to a person without advanced disease, adjusting for the other factors in the model. You can calculate confidence intervals for e^B by finding the confidence interval for β, the population value, and then raising the lower limit and the upper limit to the power of *e*. The 95% confidence intervals for e^B are given in the last two columns of Figure 12.4. If the 95% confidence interval does not include the value of 1, you can reject the null hypothesis that the variable is not related to survival.

Testing the Model

Whenever you build a statistical model, you want an overall test of the hypothesis that all parameters are 0. In the Cox regression model, as in logistic regression, several somewhat different statistics can be used. For large sample sizes, they all have a chi-square distribution. Since the estimation method used in the Cox Regression procedure depends on maximizing the partial likelihood function, comparing changes in the values of –2 times the log likelihood (*–2LL*) for different models is the basis for a variety of tests.

The likelihood-ratio test for the hypothesis that all parameters are 0 is obtained by comparing *–2LL* for a model in which all coefficients are 0 (the initial model) with *–2LL* for a model that contains the coefficients of interest.

Figure 12.5 Likelihood statistics

```
Beginning Block Number 0.  Initial Log Likelihood Function

-2 Log Likelihood      221.222

Beginning Block Number 1.  Method:  Enter

Variable(s) Entered at Step Number 1..
    AGE
    SEX          sex
    STAGE        stage of disease

Coefficients converged after 4 iterations.

-2 Log Likelihood      208.062

                     Chi-Square    df     Sig
Overall (score)         13.176      3   .0043
Change (-2LL) from
 Previous Block         13.159      3   .0043
 Previous Step          13.159      3   .0043
```

From Figure 12.5, you see that *–2LL* for the initial model is 221.222. For the model that contains age, sex, and stage of disease, *–2LL* is 208.062. The difference between these two numbers is 13.159. This is the entry labeled *Change (–2LL) from Previous Block*. The degrees of freedom are the difference between the number of parameters in the two models. In this example, the degrees of freedom are 3, since we are comparing a model with no parameters to one with three parameters. The observed significance level is small—0.0043—so the null hypothesis that all of the coefficients are 0 can be rejected.

Another test of the null hypothesis that all parameters are 0 can be based on the **score statistic**. This test is also sometimes called the **global chi-square** or the **overall chi-**

square. (See, for example, Blossfeld et al., 1989.) From Figure 12.5, you see that the value for the overall chi-square is very similar to that of the change in the likelihood statistic. Once again, you reject the null hypothesis that the parameters are 0.

Selecting Predictor Variables

Often, you have many variables that may or may not be predictors of survival. One of your goals is to identify variables related to survival and build a model that excludes variables that do not appear to be good predictors. The usual methods for variable selection can be used to build models in Cox Regression. All of the problems associated with variable-selection algorithms remain. None of the algorithms results in a "best" model in any true sense. As always, the model will fit the sample from which it is estimated better than it will fit the population from which the sample is selected. Another sample from the same population may result in a different model. It is a good idea to examine several possible models and choose among them on the basis of interpretability, parsimony, and ease of variable acquisition.

The Cox Regression procedure provides several methods for model selection. You can enter all variables into a model with a single step using the **forced-entry method**. You can also use forward and backward variable selection for model building. The score statistic is always used for entering variables into a model. One of three statistics can be used for variable removal: the likelihood-ratio statistic based on the maximum partial likelihood estimates, the likelihood-ratio statistic based on conditional parameter estimates (Lawless & Singhal, 1978), and the Wald statistic. The likelihood-ratio statistic based on the maximum partial likelihood estimates is probably the best criterion for variable removal; however, it is computationally intensive because it requires repeated model reestimations. The change in log likelihood is computed by deleting each variable in turn from the model and calculating the change in the log likelihood. The statistic based on conditional estimates does not require reestimation of the model and in many situations tends to give similar results to the statistic based on the maximum partial likelihood estimates.

Except for the actual statistics used for variable entry and removal, the variable-selection algorithms are identical to those described in Chapter 1. As in logistic regression, all variables that are used to represent the same categorical variable are entered or removed from the model in the same step.

In **forward selection**, variables are considered one at a time for entry into the model. After a variable is added to a model, all variables already in the model are examined for removal. The algorithm stops when no more variables meet entry or removal criteria, or when the resulting model is identical to a previous one (which means the algorithm is cycling). In **backward selection**, all variables are first entered into the model in a single step. Then the variables are examined for removal. Once no more variables meet removal criteria, variables are again considered for entry. The algorithm stops when no more variables meet entry or removal criteria.

An Example of Forward Selection

To see what the output looks like when a model is built using forward selection, let's return to the Hodgkin's disease data set. The independent variables we will consider for entry into the model are *age*, *sex*, *stage*, *histology* (coded 1 for nodular sclerosis, 2 for mixed cellular, and 3 for other) and the two-way interactions among *sex*, *stage*, and *histology* (*histology*, *stage*, and *sex* are declared as categorical covariates). Figure 12.6 shows the output used to select the first variable for inclusion into the model.

Figure 12.6 Statistics for initial model

```
Dependent Variable: TIME survival time in months

          Events Censored

              30         30 (50%)

Beginning Block Number 0. Initial Log Likelihood Function

-2 Log Likelihood       221.222

---------- Variables not in the Equation -----------
Residual Chi Square = 52.91 with 10 df    Sig =.0000

Variable              Score  df     Sig        R

AGE                  7.1036   1  .0077    .1519
SEX                   .0303   1  .8618    .0000
HIST                12.7973   2  .0017    .1994
 HIST(1)             6.2787   1  .0122    .1391
 HIST(2)             2.6526   1  .1034    .0543
STAGE                4.1426   1  .0418    .0984
HIST*STAGE            .8294   2  .6605    .0000
 HIST(1)*STAGE        .5820   1  .4455    .0000
 HIST(2)*STAGE        .3366   1  .5618    .0000
HIST*SEX             3.8242   2  .1478    .0000
 HIST(1)*SEX         1.0055   1  .3160    .0000
 HIST(2)*SEX         2.7206   1  .0991    .0571
STAGE*SEX             .1691   1  .6809    .0000
```

For each of the candidate variables, you see the score statistic and its observed significance level. The *R* statistic is computed as previously described, with the score statistic replacing the Wald statistic. (For categorical variables and interactions with more than one degree of freedom, an overall score statistic for the variable as well as the score statistics for the individual components are printed. The overall score statistic determines whether the variable is entered.) The smallest observed significance level is for the *histology* variable. Since it is less than 0.05, the default for variable entry, *histology* will be the first variable to enter the model.

The residual chi-square, printed before the table of score statistics, tests the null hypothesis that the coefficients for all variables not in the model are 0. If the observed significance level for the residual chi-square is small, you can reject the hypothesis that all of the coefficients are 0. In this example, the observed significance level is less than 0.00005, so it appears that the variables not in the model are related to survival.

Figure 12.7 shows the output when the categorical variable *histology* is entered into the model. Both of the statistics that test whether all coefficients for variables in the model are 0 (the overall chi-square and the likelihood-ratio statistic) have small observed significance levels, so you can reject the hypothesis that the population coefficients are 0. From the estimated deviation coefficients for the *histology* variable, you can see that the first two histology types (nodular sclerosis and mixed cellular) are associated with decreased death rates. This means that the third category of "other," which is not shown, is associated with increased rates. (Its coefficient is 1.006, since with deviation contrasts, the sum of all of the coefficients is 0.)

Figure 12.7 Statistics for model containing variable histology

```
Variable(s) Entered at Step Number 1..
    HIST      histology

Log likelihood converged after 3 iterations.

-2 Log Likelihood        213.181

                      Chi-Square      df     Sig
Overall (score)          12.797        2   .0017
Change (-2LL) from
 Previous Block           8.041        2   .0179
 Previous Step            8.041        2   .0179


----------------------- Variables in the Equation -----------------------

Variable              B        S.E.       Wald  df      Sig        R     Exp(B)

HIST                                   10.6520   2    .0049    .1734
 HIST(1)          -.5911      .2511     5.5402   1    .0186   -.1265      .5537
 HIST(2)          -.4150      .2725     2.3196   1    .1278   -.0380      .6604
```

To decide which variable enters the model next, look at the *Variables not in the Equation* table in Figure 12.8. The smallest observed significance level is for the variable *age*, so it is entered into the model containing *histology*. The $-2LL$ for the model containing *age* and *histology* is 207.517, as shown in Figure 12.9. Based on changes in $-2LL$, you can test the hypothesis that the regression coefficients for both *age* and *histology* are 0 and that the coefficient for *age* is 0 when the variable is added to a model that already contains *histology*.

Figure 12.8 Statistics for variables not in the equation

```
---------- Variables not in the Equation ----------
Residual Chi Square = 25.35 with 8 df    Sig = .0014

Variable              Score  df     Sig         R

AGE                  5.7623   1   .0164     .1304
SEX                   .5876   1   .4434     .0000
STAGE                4.0764   1   .0435     .0969
HIST*STAGE            .6603   2   .7188     .0000
 HIST(1)*STAGE        .5254   1   .4686     .0000
 HIST(2)*STAGE        .3495   1   .5544     .0000
HIST*SEX             6.1921   2   .0452     .0995
 HIST(1)*SEX         1.7432   1   .1867     .0000
 HIST(2)*SEX         5.7828   1   .0162     .1308
STAGE*SEX             .3967   1   .5288     .0000
```

For the model with *age* alone, *–2LL* is 213.181, while for the initial model, it is 221.222. Based on these numbers, the change in *–2LL* when *age* and *histology* are entered into the model is 221.222 – 207.517 , or 13.705. This is the chi-square value labeled *Change (–2LL) from Previous Block* in Figure 12.9. Since the observed significance level is small, you can reject the null hypothesis that the coefficients for the variables added to the model in the current block are 0.

Figure 12.9 Statistics for model containing variables histology and age

```
Variable(s) Entered at Step Number 2..
    AGE

Coefficients converged after 5 iterations.

-2 Log Likelihood        207.517

                    Chi-Square     df      Sig
Overall (score)         19.130      3   .0003
Change (-2LL) from
 Previous Block         13.705      3   .0033
 Previous Step           5.664      1   .0173
```

If you build a model by sequentially specifying different steps, such as by entering *age* into a model containing no variables and then performing forward stepwise selection using the remaining variables as candidates for entry and removal, each of the methods corresponds to a block. The current model is always compared to the final model from the previous block. If you do not have multiple blocks, the previous block is the model with no variables.

Another test for the same hypothesis is provided by the overall (score) chi-square in Figure 12.9. The conclusion based on this statistic, as you would expect, is the same.

To test the hypothesis that the coefficient for the variable entered or removed at a step is 0, you calculate the difference in *–2LL* between the two models. In this example, the change in *–2LL* when *age* is entered into the model containing *histology* is 213.181 – 207.517 , or 5.664. This is labeled *Change (–2LL) from Previous Step* in Figure 12.9. Since the observed significance level is small, you can reject the null hypothesis that the coefficient for *age* is 0 when *age* is added to the model containing histology.

After a variable is added to the model, all variables in the equation are examined for removal. Figure 12.10 contains removal statistics for the two variables currently in the model. As previously explained, you can base the decision to remove a variable on three different statistics: changes in likelihood based on the maximum partial likelihood estimates, changes in likelihood based on the conditional likelihood estimates, and the Wald statistic. In Figure 12.10, the changes in likelihood are based on the conditional parameter estimates. Since both of the variables have observed significance levels less than 0.1 (the default for variable removal), neither of them can be removed, and the algorithm continues screening the variables for entry.

Figure 12.10 Removal statistics for variables age and histology

```
------Model if Term Removed-------
Based on conditional coefficients

Term              Loss
Removed       Chi-square  df    Sig

AGE              5.6751    1  .0172
HIST             6.8714    2  .0322
```

The next variable to be entered is stage of the disease. Statistics for the variables in the equation are shown in Figure 12.11. Again, the coefficient for *stage* is negative, indicating that the early stage has a better prognosis than the later stage. Since *stage* is a categorical variable with only two values (early and late), it has one degree of freedom, and only one coefficient is printed. The printed coefficient corresponds to deviation parameter estimates (the default), so it tells you the effect of early stage compared to the average effect of both stages.

Figure 12.11 Statistics for model containing variables age, histology, and stage

```
---------------------- Variables in the Equation -----------------------

Variable                 B        S.E.      Wald  df      Sig        R      Exp(B)

AGE                  .0301       .0125    5.8311   1    .0157    .1316      1.0305
HIST                                      6.3091   2    .0427    .1022
 HIST(1)            -.3481       .2727    1.6295   1    .2018    .0000       .7060
 HIST(2)            -.4833       .2722    3.1528   1    .0758   -.0722       .6167
STAGE               -.4237       .2052    4.2617   1    .0390   -.1011       .6546
```

From the statistics for variables not in the equation shown in Figure 12.12, you see that none of the variables meets entry criteria. Since components of categorical variables are all entered and removed as a single variable, only the significance for the overall variable is considered for entry and removal. (If you want the component variables of a categorical variable to be entered and removed individually, you must use the transformation language outside of the Cox Regression procedure to create the component variables; then you can use these variables as you would any others in the model.) From the statistics labeled *Model if Term Removed*, you see that none of the variables is eligible for removal because the observed significance level is less than 0.1 for all of them. Forward model selection stops at this point because no more variables are eligible for entry or removal.

Figure 12.12 Statistics for variables not in the equation

```
---------- Variables not in the Equation ----------
Residual Chi Square = 12.13 with 6 df    Sig = .0592

Variable            Score  df    Sig        R

SEX                 .2889   1  .5909    .0000
HIST*STAGE         2.0054   2  .3669    .0000
 HIST(1)*STAGE      .6948   1  .4045    .0000
 HIST(2)*STAGE     1.3945   1  .2376    .0000
HIST*SEX           5.0668   2  .0794    .0694
 HIST(1)*SEX       2.5326   1  .1115    .0491
 HIST(2)*SEX       4.1600   1  .0414    .0988
STAGE*SEX           .0186   1  .8916    .0000

------Model if Term Removed-------
Based on conditional coefficients

Term               Loss
Removed        Chi-square  df    Sig

AGE               6.1419   1  .0132
HIST              5.2752   2  .0715
STAGE             4.6393   1  .0312

No more variables can be added or deleted.
```

Plotting the Estimated Functions

You can examine the estimated cumulative survival and cumulative hazard functions by plotting them for different covariate values (patterns). For example, Figure 12.13 is a plot of the cumulative survival function for 40-year-old patients with nodular sclerosis and advanced stage of the disease. Figure 12.14 is also a plot of the cumulative survival function for 40-year-olds with nodular sclerosis. Two separate curves are shown—one for early-stage patients and one for late-stage patients.

Figure 12.13 Survival function for 40-year-olds with nodular sclerosis and advanced stage

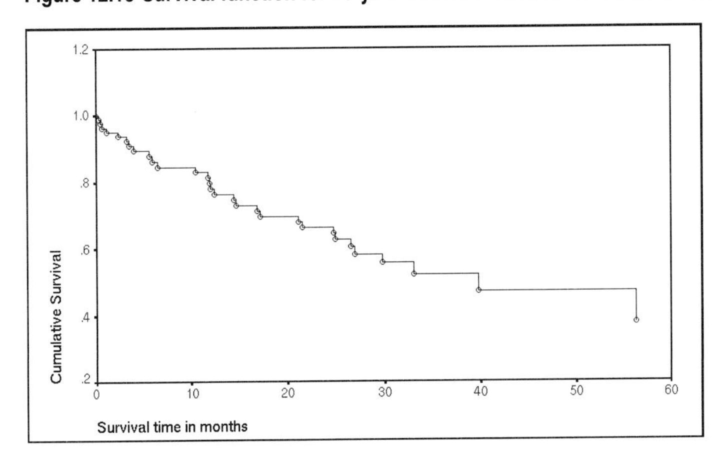

Figure 12.14 Survival functions for early and late stages

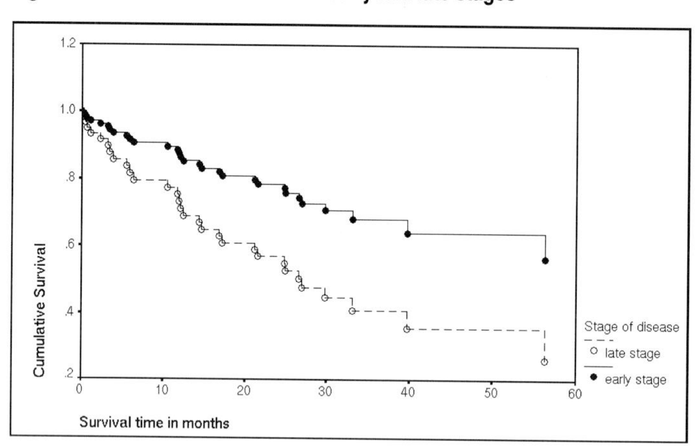

Checking the Proportional Hazards Assumption

As discussed in "The Hazard Function" on p. 293, one of the assumptions of the Cox regression model is that for any two cases, the ratio of the estimated hazard across time is a constant. For example, if you have two patients with the same age and histology but different stages of the disease, the ratio of their estimated hazard rates across all time points is e^{B}, where B is the regression coefficient for the case with the stage coded as 1. This is not an assumption to be made lightly. It is quite possible that the hazard functions of early- and late-stage patients are not related by a constant multiplier and that their ratio depends on time.

It is possible to modify the Cox regression model to incorporate separate baseline hazard functions for each group. (The groups for which separate models are estimated are usually called **strata**.) For example, if you have reason to believe that strata have different baseline hazard functions, you can fit the model

$$h_i(t) = h_{0,i}(t)\, e^{(BX)}$$ **Equation 12.7**

where $h_i(t)$ is the hazard function for stratum i. In this model, separate baseline hazard functions $h_{0,i}(t)$ are estimated for each of the strata. The effects of the covariates are still assumed to be the same in all of the groups.

Let's apply this model to the Hodgkin's disease data, using stage of disease as the strata variable. The SPSS Cox Regression procedure can estimate two separate baseline hazard functions, one for each of the stages. Only one set of coefficients for the predictor variables is estimated because we are assuming that the influence of the covariates is the same in all of the strata. We may want to see whether the baseline hazard functions are proportional in the strata. If they are, the variable used to form the strata can be used as a covariate in the model, and a common baseline hazard can be estimated for all of the groups.

A useful plot for assessing whether the baseline hazard functions are proportional is called the **log-minus-log (LML) survival plot**. If the baseline hazard functions are proportional, the lines for the individual strata should be parallel. Figure 12.15 is the LML plot for the Hodgkin's disease data when *stage* of disease is the stratification variable, and *age* and *histology* are covariates. You see that the two lines are more or less parallel and that it may be unnecessary to stratify the data based on *stage*. In the next section, you'll see how a formal test of the proportional hazards assumption can be constructed using a time-dependent covariate.

Figure 12.15 Log-minus-log plot

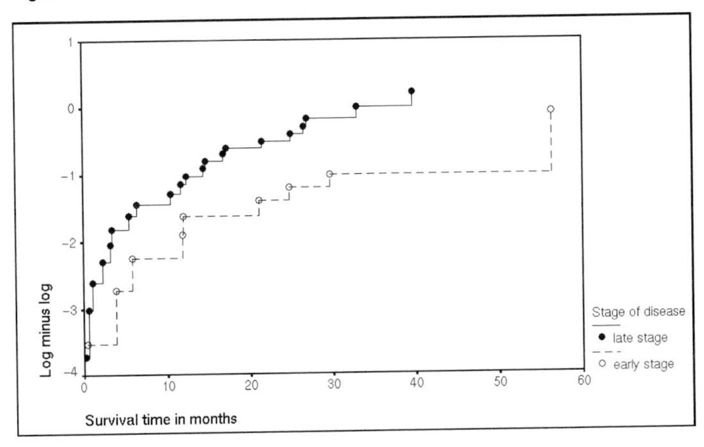

Time-Dependent Covariates

To see how you can analyze data when hazards are not proportional, consider the following example presented by Stablein et al. (1981). They report survival times for two groups of patients with locally advanced, non-resectable gastric carcinoma. One group of patients was treated using a combination of chemotherapy and radiation, while the other group was treated with radiation alone. The question of interest is whether the two treatments are equally effective in prolonging survival.

Before you start to build a model, it is always a good idea to look at plots that describe the data. Figure 12.16 contains plots of the cumulative survival function for the two treatment groups. Notice that the two survival curves cross. This is an indication that the proportional hazards assumption may not be appropriate for these data. The log-minus-log plot in Figure 12.17 also suggests that the hazards in the two groups are not proportional. The points for the two treatment groups do not appear to fall on parallel lines.

Figure 12.16 Cumulative survival functions for treatment groups

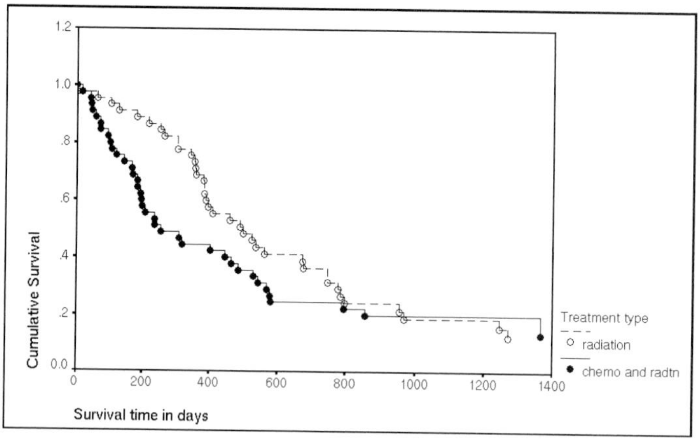

Figure 12.17 Log-minus-log plot of hazard functions for treatment groups

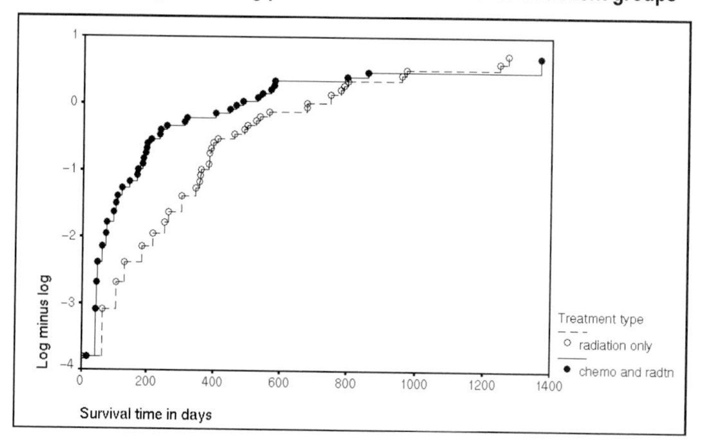

Since the main question of interest in this analysis is whether survival differs for the two treatments, we need a model that incorporates nonproportional hazards over time and allows us to estimate the treatment effect. A simplification of the model suggested by Stablein et al. is

$$h(t) = h_0(t)\, e^{B_1 * \text{treat} + B_2 * \text{treat} * \text{t\_cov\_}}$$ **Equation 12.8**

This model differs from those we have previously considered because time is included as a predictor variable. The above model contains two predictor variables: treatment (*treat*), coded as 0 for chemotherapy and radiation and 1 for radiation alone, and treatment multiplied by time (*treat*t_cov_*). This type of model does not force the ratio of the hazard rates for the two groups to be constant over time. Instead, the ratio can vary over time.

Figure 12.18 contains the output from fitting the model with the treatment-by-time interaction. From the indicator variable coefficients and their significance levels, you see that there is a treatment effect and that it is not constant over time. (If it were constant over time, you would not reject the null hypothesis that the treatment-by-time coefficient is 0.) The chemotherapy treatment is better for early time points, but its superiority decreases with time, since the coefficient for treatment by time is positive. Whenever you want to test that hazards are proportional for different strata, you can incorporate a time-by-stratification-variable interaction. If you can reject the null hypothesis that the coefficient for this term is 0, the hazards are not proportional. Any predictor variable whose values change with time is known as a **time-dependent covariate**. Such covariates are easily incorporated into Cox regression models (*t_cov_* , which is a time-dependent covariate in this example, is computed as survival time divided by 30 to convert survival in days to survival in months). However, the actual computations take much longer because the values of the covariates must be generated at each time point. Examples of time-dependent covariates are physiological measurements or lab results that are repeated throughout the course of a study and the treatment a patient is receiving, if each patient receives several treatments during a study. (You have to define time-dependent covariates in a special way in the Cox Regression procedure. For details, see "How to Obtain a Time-Dependent Cox Regression Analysis" on p. 320.)

Figure 12.18 Statistics for the time-dependent Cox model

```
Variable(s) Entered at Step Number 1..
    TREAT       treatment type
    TREAT*T_COV_

Coefficients converged after 4 iterations.

-2 Log Likelihood       557.527

                   Chi-Square    df     Sig
Overall (score)         8.887     2   .0118
Change (-2LL) from
 Previous Block         9.263     2   .0097
 Previous Step          9.263     2   .0097

----------------------- Variables in the Equation ----------------------

Variable              B         S.E.       Wald  df      Sig        R     Exp(B)

TREAT           -1.2089        .4309     7.8710   1    .0050   -.1018      .2985
TREAT*T_COV_      .0715        .0278     6.5966   1    .0102    .0901     1.0741
```

Graphical Displays

Graphical examination of the data is an important part of any statistical analysis. In the analysis of survival data with censored observations, displays are necessarily more complex because you must consider an additional component, whether or not a data point is censored. One way to incorporate the censoring pattern in a display is to choose a different plotting symbol for censored and uncensored points. The failure times are more important, so the uncensored cases should be more prominent. Gentleman and Crowley (1991) recommend that, instead of selecting different symbols for the two types of cases, they be distinguished by color or grey scale. This makes it easier to extract the important information.

As an example, consider Figure 12.19, which is a plot of survival time against age, with censored cases represented by open circles and uncensored cases by closed circles. You see that for the very early ages, there appears to be somewhat more censoring. You also see that there appears to be a relationship between age and survival time.

Figure 12.19 Plot of survival time against age

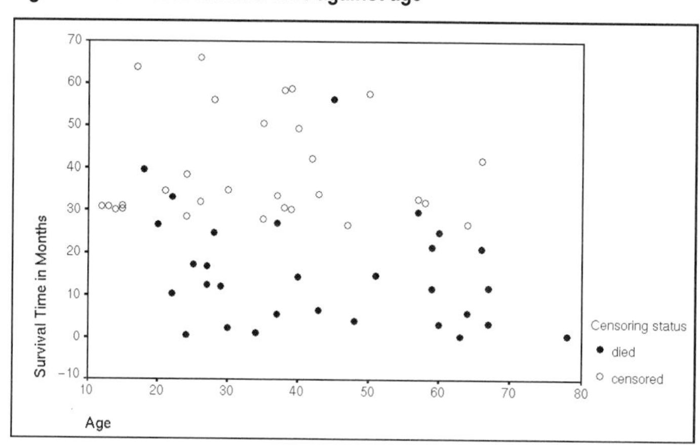

Looking for Influential Cases

You do not want the results of any statistical analysis to depend heavily on the values for an individual case. That's why it is always important to look for **influential cases** in your data. (A case is influential if there are substantial changes in the values of the coefficients when the model is computed with the case and without the case.) A statistic that estimates the change in regression coefficients with and without a case is called **DfBeta**. For each case, you can compute a DfBeta for each variable in the model. Cases with outlying values for DfBeta should be examined. An easy way of examining the DfBetas is to plot for each variable in the model the DfBetas against the case-identification number.

Figure 12.20 is a plot of the estimated changes in the age coefficient when each of the cases in turn is excluded from the model that contains *age*, *histology*, and *stage*. You will note that as expected, most of the values fall in a horizontal band around 0. There is one value that is far removed from the rest. It is for case number 16. If you exclude case 16 from the estimation of the model, the coefficient for age changes from 0.0301 to 0.0406, resulting in a DfBeta for age for case 16 of 0.0113. (The DfBetas calculated in the Cox Regression procedure are an approximation, so their values will not be identical to those obtained from actually removing the case from the analysis.) If you look at the data, you will see that case 16 is a 60-year-old who lived for a long time. (The average age for all of the patients is 38.73). This case decreases the value of the coefficient for age, since, overall, as age increases, so does the probability of dying. Whenever you find influential observations in your data, you should check the values for the case to make sure they are not the result of coding or data-entry errors. If the recorded values are correct, when reporting the results of your analyses, you should include discussion of the impact of the influential case on the results of the analyses. See Belsey et al. (1980) for further discussion.

Figure 12.20 Plot of DfBeta against case-identification number

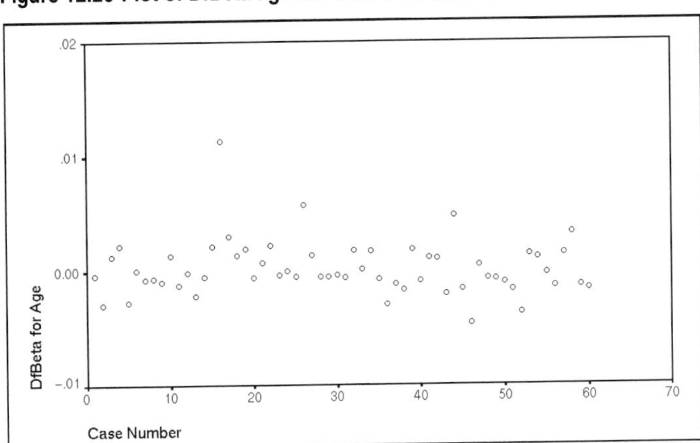

Examination of Residuals

There are several types of residuals that have been found to be useful in detecting violations of the proportional hazards model. The **partial residual** has been suggested by Schoenfeld (1982). It is computed for each covariate in the model for uncensored cases. For each case, the partial residual for a variable is the difference between the observed value of the covariate and its conditional expectation based on the cases still under observation when the case fails. These residuals can be plotted against time to test the proportional hazards assumption. Crowley and Storer (1983) discuss some of the problems associated with plotting the partial residuals against the values of the covariates. Figure 12.21 is a plot of the partial residuals for age against time. You notice that the values appear to be randomly distributed in a band around 0.

Figure 12.21 Plot of partial residuals against survival time

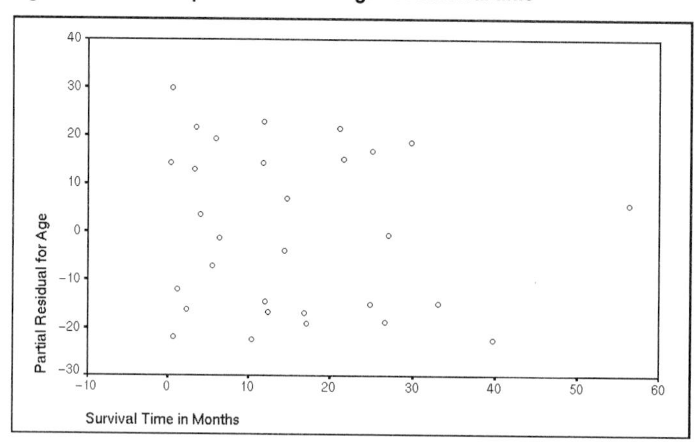

Another type of residual used in proportional hazards models is the **martingale residual** (Therneau et al., 1990). Although it is not directly saved in the SPSS Cox Regression procedure, it can be computed easily by saving the **Cox-Snell residual** (see Cox & Snell, 1968), which is just the estimated cumulative hazard, and using the transformation language. For a censored case, the martingale residual is the negative of the Cox-Snell residual. For an uncensored case, it is 1 minus the Cox-Snell residual. The martingale residuals can be plotted against the values of the independent variables or the linear predictor score ($X'Beta$), which is the sum of the products of the coefficients and the mean-corrected variables for a case.

Figure 12.22 is a plot of the martingale residuals against the values of age. Most of the points fall in a horizontal band around 0. However, there is one outlying case with a martingale residual close to –3. As you would expect, this is case 16 again. The plot of the martingale residuals against the linear predictor scores, Figure 12.23, again shows case 16 as an outlier.

Figure 12.22 Plot of martingale residuals against age

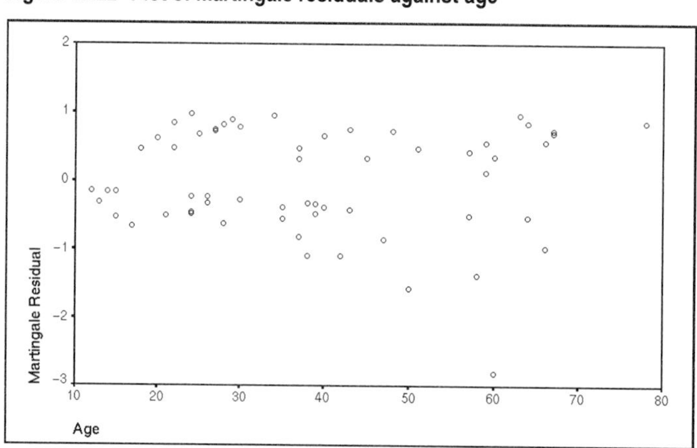

Figure 12.23 Plot of martingale residuals against linear predictor score

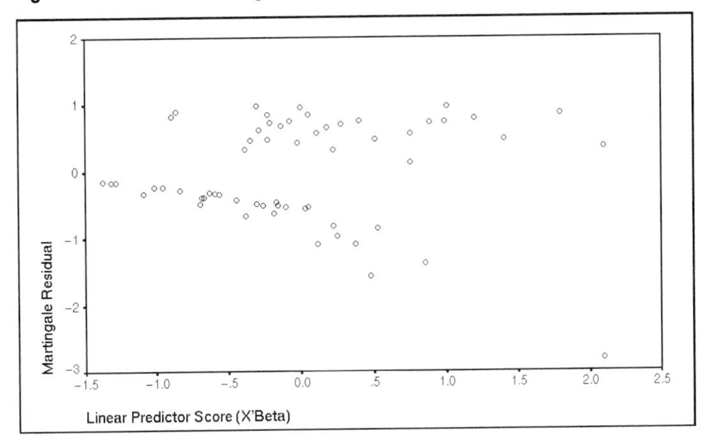

SPSS Cox Regression Procedures

SPSS provides two procedures for proportional hazards regression. The Cox Regression procedure analyzes models with **time-constant covariates** (covariates that do not change as a function of time). The Time-Dependent Cox Regression procedure analyzes models containing one time-dependent covariate, one or more time-constant covariates, or both. Both procedures use the Newton-Raphson method to estimate model parameters. However, you should use the Cox Regression procedure if your model contains only time-constant covariates, since the procedure allows you to save diagnostic variables and obtain plots that are not available in the Time-Dependent Cox Regression procedure.

How to Obtain a Time-Constant Cox Regression Analysis

The Cox Regression procedure estimates proportional hazard regression models for time-constant covariates. You can enter variables into the model using forced entry or one of six stepwise methods. You can also obtain subgroup analyses, save diagnostic variables, and plot hazard and survival functions.

The minimum specifications are:

- A survival time variable.
- A survival status variable whose values indicate whether an event occurred.
- Survival status codes that identify uncensored cases.

To obtain a Cox regression analysis with time-constant covariates, from the menus choose:

Statistics
  Survival ▸
    Cox Regression...

This opens the Cox Regression dialog box, as shown in Figure 12.24.

Figure 12.24 Cox Regression dialog box

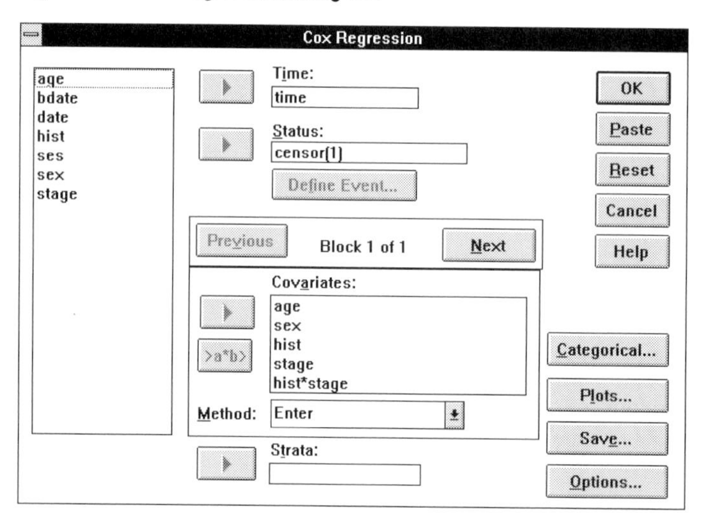

The numeric and short string variables in your working data file appear on the source list.

Time. Select a numeric survival time variable indicating how long cases survived (uncensored cases) or were tracked (censored cases). The time variable can be measured in any unit of time. SPSS excludes cases with nonpositive survival times from the analysis.

Status. Select a numeric or short string variable whose values indicate whether an event has occurred for a case. You must also specify which status codes identify uncensored cases (see "Define Event for Status Variable" on p. 314).

Strata. By default, hazard and survival functions are computed for all cases in the working data file as if they were from a single sample. To obtain separate functions for subgroups, select a short string or numeric variable that divides the sample into two or more strata. A separate baseline hazard function is computed for each stratum.

Covariates. Optionally, you can select a block of one or more numeric or short string predictor variables. To create interaction terms, highlight two or more variables on the source list and click on a*b.

You must indicate which, if any, numeric covariates you want to treat as categorical (see "Defining Categorical Predictors" on p. 314). Each term in the block must be unique; however, a term can contain components of another term. For example, while *hist*stage* can appear only once in a block, the same block can contain *hist, stage,* and *hist*stage*age*.

Macintosh: Use ⌘-click to select multiple covariates.

- **Method**. Method selection controls the entry of the block of covariates into the model. For stepwise methods, the score statistic is used for entering variables; removal testing is based on the likelihood-ratio statistic or the Wald statistic. You can choose one of the following alternatives:

  Enter. Forced-entry method. All variables in the block are entered in a single step. This is the default.

  Forward: Conditional. Forward stepwise selection. Removal testing is based on the probability of the likelihood-ratio statistic based on conditional parameter estimates.

  Forward: LR. Forward stepwise selection. Removal testing is based on the probability of the likelihood-ratio statistic based on the maximum partial likelihood estimates.

  Forward: Wald. Forward stepwise selection. Removal testing is based on the probability of the Wald statistic.

  Backward: Conditional. Backward stepwise selection. Removal testing is based on the probability of the likelihood-ratio statistic based on conditional parameter estimates.

  Backward: LR. Backward stepwise selection. Removal testing is based on the probability of the likelihood-ratio statistic based on the maximum partial likelihood estimates.

  Backward: Wald. Backward stepwise selection. Removal testing is based on the probability of the Wald statistic.

Covariates are entered into, or removed from, a single model. However, you can use different entry methods for different sets of covariates. For example, you can enter one block using forced entry and another block using forward selection. After entering the first block of covariates into the model, click on Next to add a second block. If you do not want the default (forced entry), select an alternate method. To move back and forth between blocks of covariates, use Previous and Next.

Define Event for Status Variable

To specify which values of the status variable identify events, highlight the status variable and click on Define Event... in the Cox Regression dialog box. This opens the Cox Regression Define Event for Status Variable dialog box, as shown in Figure 12.25.

Figure 12.25 Cox Regression Define Event for Status Variable dialog box

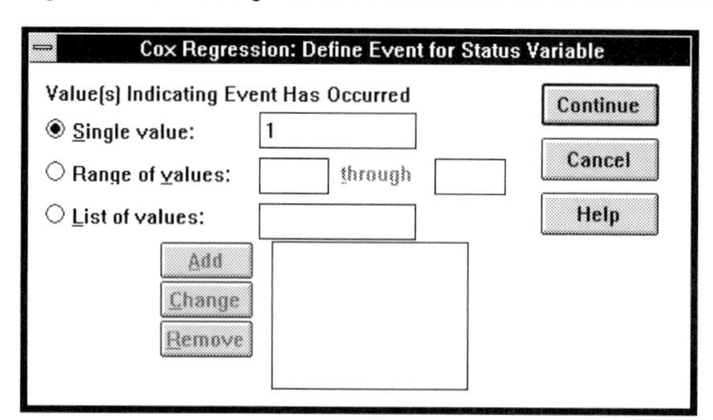

A single value or several values can indicate that an event has occurred. Cases having other nonmissing values of the status variable are treated as censored cases. Choose one of the following alternatives:

- **Single value**. This is the default. Enter a value.
- **Range of values**. This option is available only if your status variable is numeric. Enter the lowest and highest values in the range. The first value must be less than the second value.
- **List of values**. Select this to define a list of values. Enter a value and click on Add to add the value to the list. To add more values to the list, repeat this process. To change a value, highlight it on the list, modify the value, and click on Change. To remove a value, highlight it on the list and click on Remove.

Defining Categorical Predictors

By default, SPSS treats string covariates as categorical and numeric covariates as continuous. To treat one or more numeric covariates as categorical, click on Categorical... in the Cox Regression dialog box. This opens the Cox Regression Define Categorical Covariates dialog box, as shown in Figure 12.26.

Figure 12.26 Cox Regression Define Categorical Covariates dialog box

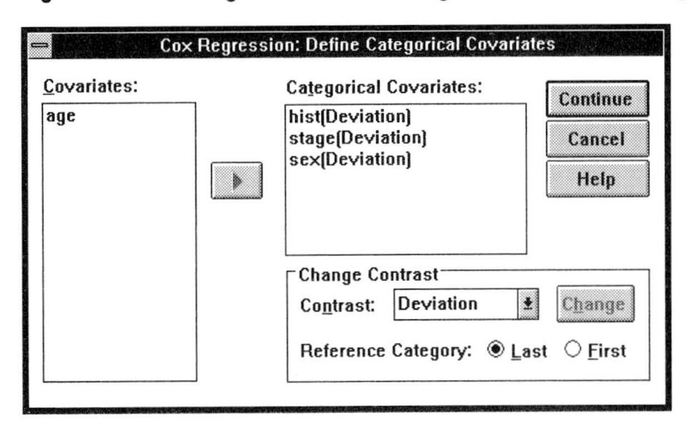

Categorical Covariates. Select variables you want to treat as categorical from the list of numeric covariates. SPSS transforms categorical variables and interaction terms containing categorical variables into contrast variables, which are entered or removed from the model as a set. String covariates are always treated as categorical (you can remove string variables from the list of categorical covariates only by removing them from the covariates list in the Cox Regression dialog box).

Change Contrast. By default, each categorical covariate is transformed into a set of deviation contrasts. To get a different type of contrast, highlight one or more covariates, select a contrast type, and click on Change. Optionally, you can change the default reference category.

- **Contrast.** You can choose one of the following types of contrasts:

  Deviation. Each category of the predictor variable except the reference category is compared to the overall effect. This is the default.

  Simple. Each category of the predictor variable (except the reference category) is compared to the reference category.

  Difference. Also known as reverse Helmert contrasts. Each category of the predictor variable except the first category is compared to the average effect of previous categories.

  Helmert. Each category of the predictor variable except the last category is compared to the mean effect of subsequent categories.

  Repeated. Each category of the predictor variable except the first category is compared to the category that precedes it.

Polynomial. The predictor variable is transformed into linear, quadratic, and cubic components, and so forth (depending on the number of categories). Categories are assumed to be equally spaced.

Indicator. Contrasts indicate the presence or absence of category membership. The reference category is represented in the contrast matrix as a row of zeros.

Reference Category. For deviation, simple, and indicator contrasts, you can override the default reference category. Choose one of the following alternatives:

❍ **Last.** Use the last category of the variable as the reference category. This is the default.

❍ **First.** Use the first category as the reference category.

Plots

To obtain optional plots, click on Plots... in the Cox Regression dialog box. This opens the Cox Regression Plots dialog box, as shown in Figure 12.27.

Figure 12.27 Cox Regression Plots dialog box

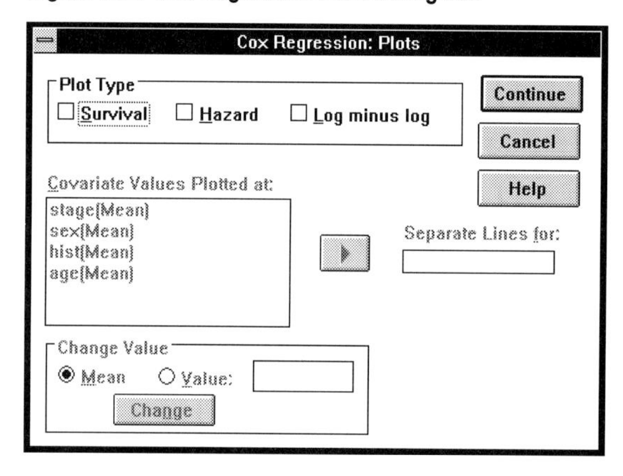

Plot Type. You can choose one or more of the following plots:

❏ **Survival.** Plots the cumulative survival function.

❏ **Hazard.** Plots the cumulative hazard function.

❏ **Log minus log.** For each stratum, plots the cumulative survival function after the $ln\,(-ln)$ transformation is applied to the function. Log-minus-log plots provide a graphical test of the proportional hazards assumption.

Covariate Values Plotted at. By default, functions plotted by SPSS are evaluated at the means of covariates and contrast variables in the model. To evaluate the functions at a value other than the means, highlight one or more covariates, and after you change the value, click on Change. To change the value for other covariates in the model, repeat this process.

Change Value. You can choose one of the following alternatives:

- ❍ **Mean.** Evaluates the hazard and survival functions at the mean values of the covariate(s). This is the default.
- ❍ **Value.** Evaluates the hazard and survival functions at a user-specified value of the covariate(s). Enter a value.

Separate Lines for. To plot separate functions for subgroups, select a categorical covariate which is in the model that splits the data into two or more groups. If you have a stratification variable, one plot is shown for each stratum. Plots of separate functions are available only for covariates that have been defined as categorical.

Saving New Variables

To save survival or diagnostic variables, click on Save... in the Cox Regression dialog box. This opens the Cox Regression Save New Variables dialog box, as shown in Figure 12.28.

Figure 12.28 Cox Regression Save New Variables dialog box

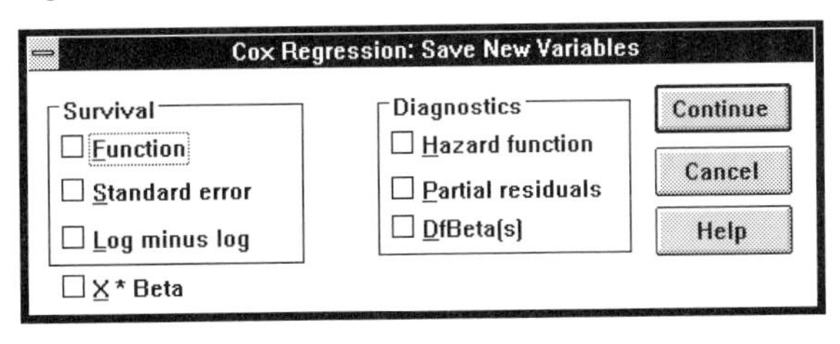

SPSS displays a table in the output showing the names it assigns to any new variables.

Survival. You can choose one or more of the following survival variables:

- ❑ **Function.** Cumulative survival estimate.
- ❑ **Standard error.** Standard error of the survival estimate.
- ❑ **Log minus log.** Cumulative survival estimate after the $ln(-ln)$ transformation is applied to the estimate.

Diagnostics. You can choose one or more of the following diagnostic variables:

❑ **Hazard function**. Cumulative hazard function estimate (also called the Cox-Snell residual).

❑ **Partial residuals**. You can plot partial residuals against survival time to test the proportional hazards assumption. One variable is saved for each covariate in the final model. Partial residuals are available only for models containing at least one covariate.

❑ **DfBeta(s)**. Estimated change in a coefficient if a case is removed. One variable is saved for each covariate in the final model. DfBeta(s) are available only for models containing at least one covariate.

You can also save the following measure:

❑ **X*Beta**. Linear predictor score. The sum of mean-corrected covariate values multiplied by their corresponding parameter estimates.

Options

To display optional statistics, or to control criteria for parameter estimation or stepwise model building, click on Options... in the Cox Regression dialog box. This opens the Cox Regression Options dialog box, as shown in Figure 12.29.

Figure 12.29 Cox Regression Options dialog box

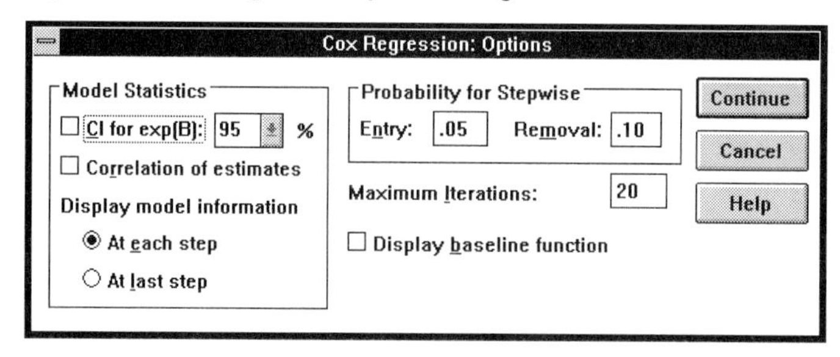

Model Statistics. You can choose one or both of the following displays:

❑ **CI for exp(B)**. Displays confidence intervals for exponentiated parameter estimates. The default confidence level for the intervals is 95%. To use a different confidence level, select a value from the drop-down list.

❑ **Correlation of estimates**. Displays the correlation matrix of regression coefficients.

Display model information. For the current model, SPSS displays $-2LL$, the likelihood-ratio statistic, and the overall chi-square. For variables in the model, parameter estimates, their standard errors, and the Wald statistic are shown. For variables not in the model, the score statistics and residual chi-square are displayed. You can choose one of the following alternatives:

- ❍ **At each step**. Displays full model information for each entry step. This is the default.
- ❍ **At last step**. Displays full model information for the final model of an entry block; summarizes intermediate steps.

Probability for Stepwise. You can override one or both of the following criteria for stepwise model building:

Entry. By default, a variable is entered into the model if the probability of its score statistic is 0.05 or less. To override this value, enter an entry probability between 0 and 1.

Removal. By default, a variable is removed from the model if the probability of its removal statistic (Wald, likelihood-ratio, or conditional likelihood-ratio statistic) is 0.10 or greater. To override this value, enter a removal probability between 0 and 1.

Maximum Iterations. SPSS computes parameter estimates using the iterative Newton-Raphson method. Iteration terminates if the relative change in the parameter estimates is less than 1E–4, if the percentage change in the log likelihood is less than 1–E5, or if the maximum number of iterations (20) is reached. To use a different maximum number of iterations, enter a positive integer.

The following option is also available:

- ❑ **Display baseline function**. Displays a table showing the baseline hazard function and the survival and hazard functions evaluated at the covariate means. If you have a stratification variable, a separate table is shown for each stratum.

Additional Features Available with Command Syntax

You can customize your proportional hazards regression analysis if you paste your selections to an input window and edit the resulting COX REGRESSION command syntax. (For information on syntax windows, see the SPSS Base system documentation.) Additional features include:

- The ability to identify cases as lost to follow-up (using the STATUS subcommand).
- Additional contrast options, such as user-defined contrasts and alternative reference categories (using the CONTRASTS subcommand).
- Additional parameter estimation criteria (using the CRITERIA subcommand).
- User-specified names for new variables (using the SAVE subcommand).
- The ability to save coefficients and survival statistics for strata to an SPSS data file (using the OUTFILE subcommand).

See the Syntax Reference section of this manual for command syntax rules and for complete COX REGRESSION command syntax.

How to Obtain a Time-Dependent Cox Regression Analysis

The Time-Dependent Cox Regression procedure analyzes proportional hazards models containing a covariate whose values change during the course of a study. You can enter variables into the model using forced entry or one of six stepwise methods. Subgroup analyses are also available. Before you specify the proportional hazards model you want to analyze, you must define the expression for computing the time-dependent covariate.

The minimum specifications are:

- The expression for computing the time-dependent covariate.
- A survival time variable.
- A survival status variable whose values indicate whether an event occurred.
- Survival status codes that identify uncensored cases.

To obtain a time-dependent Cox regression analysis, from the menus choose:

Statistics
 Survival ▸
  Cox w/ Time-Dep. Cov...

This opens the Compute Time-Dependent Covariate dialog box, as shown in Figure 12.30.

Figure 12.30 Compute Time-Dependent Covariate dialog box

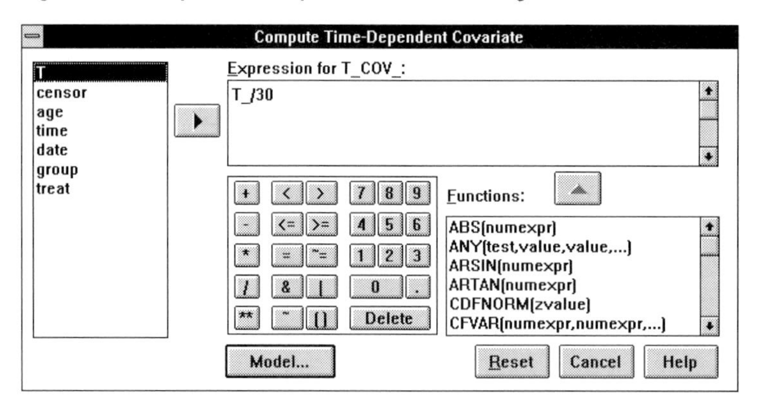

The numeric and short string variables in your working data file appear on the source list. To compute a time-dependent covariate, your data file cannot contain a variable named *T_COV_* or *T_*. If you have a variable that uses one of these names, use the Define Variable procedure on the Data menu to rename the variable before using the Time-Dependent Cox Regression procedure.

Expression for T_COV_. Specify the formula for the time-dependent covariate (*T_COV_*). You can type and edit the expression in the text box or use the calculator pad and the lists of variables and functions to paste elements into the expression. The expression should take as one of its arguments the SPSS variable *T_*, which represents the observed survival time for an uncensored case.

For example, to compute the time-dependent covariate used in "Time-Dependent Covariates" on p. 305, specify the expression as T_/30 . To compute a time-dependent covariate that is equal to the natural log of survival time, specify the expression as LN(T_) . Note that "T_COV_=" does not appear in the model expression since it is implicit in the model expression. The time-dependent covariate is used only in the current SPSS session; it cannot be saved.

See the SPSS Base system documentation for a complete description of calculator pad operators. For a description of SPSS functions, see the *SPSS Base System Syntax Reference Guide*.

Segmented Time-Dependent Covariates

The formula for computing a segmented time-dependent covariate changes for different time values. An example of a segmented time-dependent covariate is one that is equal to variable *treat* when *T_* is less than 1 and is equal to variable *age* for larger values of *T_*. You can define a segmented time-dependent covariate using boolean logic, in which an expression evaluates to 1 when it is true and 0 when it is false. For example, the expression $(T\_ < 1)$ is equal to 1 for cases in which *T_* is less than 1, and it is equal to 0 for larger values of *T_*. Thus, $(T\_ < 1) * treat + (T\_ \geq 1) * age$ is the appropriate specification of the segmented covariate that depends on *treat* and *age*.

Defining the Time-Dependent Cox Regression Model

To define the Cox regression model, click on Model... in the Compute Time-Dependent Covariate dialog box. This opens the Time-Dependent Cox Regression dialog box, as shown in Figure 12.31.

Figure 12.31 Time-Dependent Cox Regression dialog box

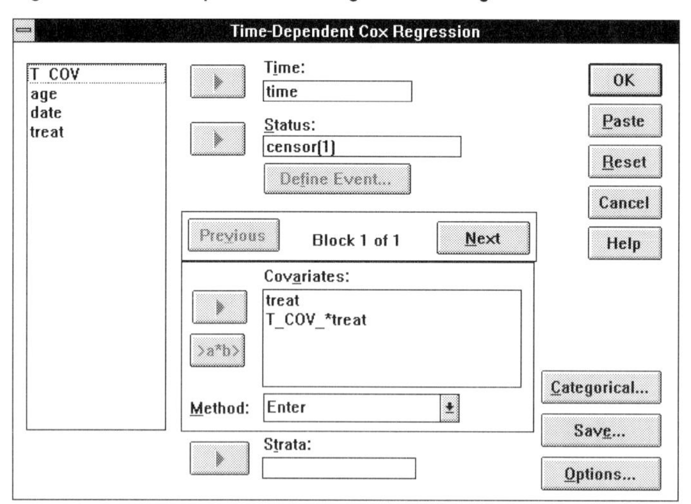

The numeric and short string variables in your working data file appear on the source list.

Time. Select a numeric survival time variable indicating how long cases survived (uncensored cases) or were tracked (censored cases). The time variable can be measured in any unit of time. SPSS excludes cases with non-positive times from the analysis. You cannot use the time-dependent covariate as the survival time variable.

Status. Select a numeric or short string variable whose values indicate whether an event has occurred for a case. You must also specify which status codes identify uncensored cases (see "Define Event for Status Variable" on p. 323). You cannot use the time-dependent covariate as the status variable.

Strata. By default, hazard and survival functions are computed for all cases in the working data file as if they were from a single sample. To obtain separate functions for subgroups, select a short string or numeric variable that divides the sample into two or more strata. A separate baseline hazard function is computed for each stratum. You cannot use the time-dependent covariate as the stratification variable.

Covariates. Optionally, you can select a block of one or more numeric or short string predictor variables. One of these variables can be the time-dependent covariate. To create interaction terms, highlight two or more variables on the source list and click on a*b.

You must indicate which, if any, numeric covariates you want to treat as categorical (see "Defining Categorical Predictors" on p. 314). Each term in the block must be unique; however, a term can contain components of another term. For example, while *sex* and *age* can appear only once in a block, the same block can contain *sex*, *age*, and *sex*age*stage*.

Macintosh: Use ⌘-click to select multiple covariates.

🠗 **Method**. Method selection controls the entry of the block of covariates into the model. For stepwise methods, the score statistic is used for entering variables; removal testing is based on the likelihood-ratio statistic or the Wald statistic. You can choose one of the following alternatives:

Enter. Forced-entry method. All variables in the block are entered in a single step. This is the default.

Forward: Conditional. Forward stepwise selection. Removal testing is based on the probability of the likelihood-ratio statistic based on conditional parameter estimates.

Forward: LR. Forward stepwise selection. Removal testing is based on the probability of the likelihood-ratio statistic based on the maximum partial likelihood estimates.

Forward: Wald. Forward stepwise selection. Removal testing is based on the probability of the Wald statistic.

Backward: Conditional. Backward stepwise selection. Removal testing is based on the probability of the likelihood-ratio statistic based on conditional parameter estimates.

Backward: LR. Backward stepwise selection. Removal testing is based on the probability of the likelihood-ratio statistic based on the maximum partial likelihood estimates.

Backward: Wald. Backward stepwise selection. Removal testing is based on the probability of the Wald statistic.

Covariates are entered into, or removed from, a single model. However, you can use different entry methods for different sets of covariates. For example, you can enter one block using forced entry and another block using forward selection. After entering the first block of covariates into the model, click on Next to add a second block. If you do not want the default (forced entry), select an alternate method.

Define Event for Status Variable

To specify which values of the status variable identify events, highlight the status variable and click on Define Event.... This opens the Time-Dependent Cox Regression Define Event for Status Variable dialog box, as shown in Figure 12.32.

Figure 12.32 Time-Dependent Cox Regression Define Event for Status Variable dialog box

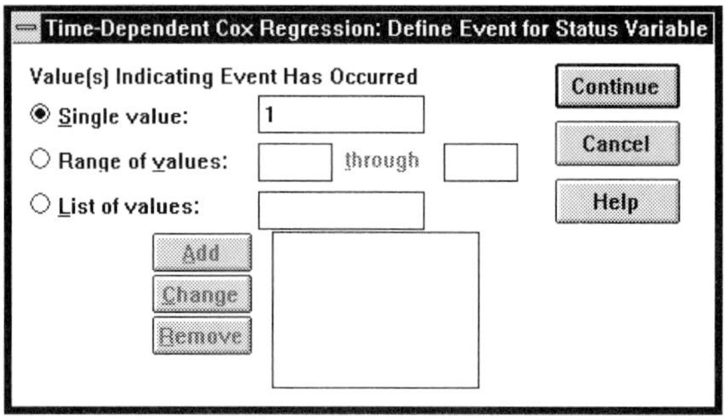

A single value or several values can indicate that an event has occurred. Cases having other nonmissing values of the status variable are treated as censored cases. Choose one of the following alternatives:

- **Single value**. This is the default. Enter a value.
- **Range of values**. This option is available only if your status variable is numeric. Enter the lowest and highest values in the range. The first value must be less than the second value.
- **List of values**. Select this item to define a list of values. Enter a value and click on Add to add the value to the list. To add more values to the list, repeat this process. To change a value, highlight it on the list, modify the value, and click on Change. To remove a value, highlight it on the list and click on Remove.

Defining Categorical Predictors

By default, SPSS treats string covariates as categorical and numeric covariates as continuous. To treat one or more numeric covariates as categorical, click on Categorical... in the Time-Dependent Cox Regression dialog box. This opens the Time-Dependent Cox Regression Define Categorical Covariates dialog box, as shown in Figure 12.33.

Figure 12.33 Time-Dependent Cox Regression Define Categorical Covariates dialog box

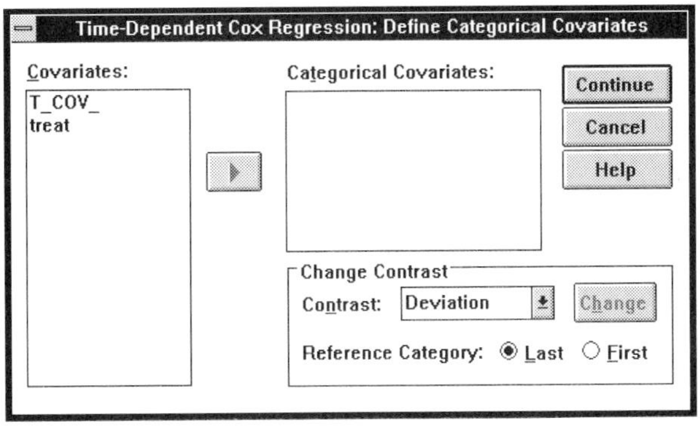

Categorical Covariates. Select variables you want to treat as categorical from the list of numeric covariates. SPSS transforms categorical variables and interaction terms containing categorical variables into contrast variables, which are entered or removed from the model as a set. String covariates are always treated as categorical (you can remove string variables from the list of categorical covariates only by removing them from the covariate list in the Cox Regression dialog box).

Change Contrast. By default, each categorical covariate is transformed into a set of deviation contrasts. To obtain a different type of contrast, highlight one or more covariates, select a contrast type, and click on Change. Optionally, you can change the default reference category.

- **Contrast.** You can choose one of the following alternatives:

  Deviation. Each category of the predictor variable except the reference category is compared to the overall effect. This is the default.

  Simple. Each category of the predictor variable (except the reference category) is compared to the reference category.

  Difference. Also known as reverse Helmert contrasts. Each category of the predictor variable except the first category is compared to the average effect of previous categories.

  Helmert. Each category of the predictor variable except the last category is compared to the mean effect of subsequent categories.

  Repeated. Each category of the predictor variable except the first category is compared to the category that precedes it.

Polynomial. The predictor variable is transformed into linear, quadratic, and cubic components, and so forth (depending on the number of categories). Categories are assumed to be equally spaced.

Indicator. Contrasts indicate the presence or absence of category membership. The reference category is represented in the contrast matrix as a row of zeros.

Reference Category. For deviation, simple, and indicator contrasts, you can override the default reference category. Choose one of the following alternatives:

❍ **Last**. Use the last category of the variable as the reference category. This is the default.

❍ **First**. Use the first category as the reference category.

Saving the Change in Regression Coefficients

To save the change in regression coefficients when a case is omitted (DfBeta), click on Save... in the Time-Dependent Cox Regression dialog box. This opens the Time-Dependent Cox Regression Save New Variables dialog box, as shown in Figure 12.34.

Figure 12.34 Time-Dependent Cox Regression Save New Variables dialog box

Time-Dependent Cox Regression: Save New Variables
Survival
☐ Function
☐ Standard error
☐ Log minus log
☐ X * Beta
Diagnostics
☐ Hazard function
☐ Partial residuals
☐ DfBeta(s)
Continue
Cancel
Help

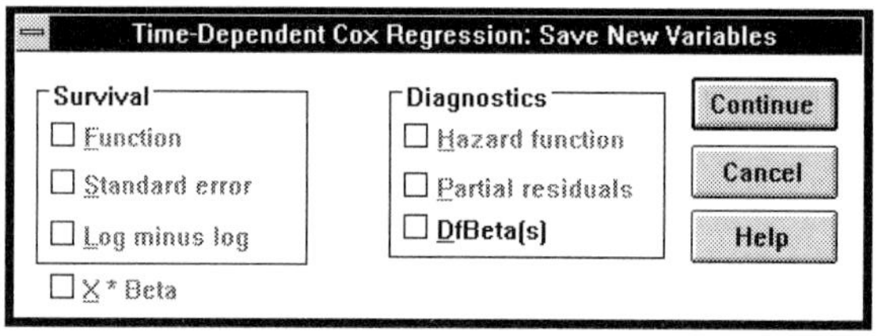

SPSS displays a table in the output showing the name it assigns to the new variable.

Diagnostics. You can choose the following diagnostic measure:

❑ **DfBeta(s)**. Estimated change in a regression coefficient if a case is removed. One variable is saved for each covariate in the final model.

Options

To display optional statistics or to control criteria for parameter estimation or stepwise model building, click on Options... in the Time-Dependent Cox Regression dialog box. This opens the Time-Dependent Cox Regression Options dialog box, as shown in Figure 12.35.

Figure 12.35 Time-Dependent Cox Regression Options dialog box

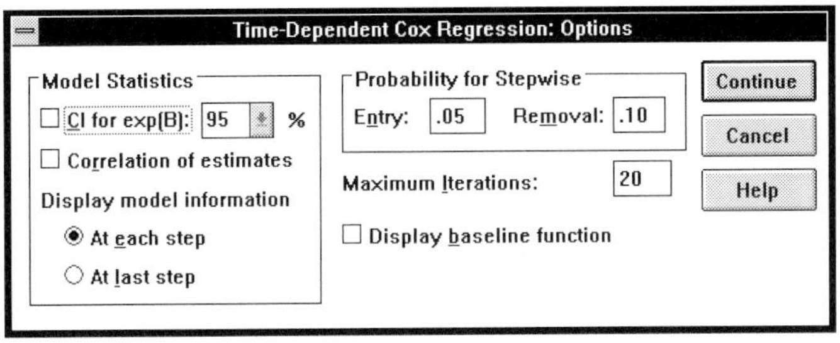

Time-Dependent Cox Regression: Options

Model Statistics
CI for exp(B): 95 %
Correlation of estimates
Display model information
At each step
At last step

Probability for Stepwise
Entry: .05 Removal: .10

Maximum Iterations: 20

Display baseline function

Continue
Cancel
Help

Model Statistics. You can choose one or both of the following displays:

- **CI for exp(B).** Displays confidence intervals for exponentiated parameter estimates. The default confidence level for the intervals is 95%. To use a different confidence level, select a value from the drop-down list.
- **Correlation of estimates.** Displays the correlation matrix of regression coefficients.

Display model information. SPSS displays $-2LL$, the likelihood-ratio statistic, and the overall chi-square. For variables in the model, parameter estimates, their standard errors, and the Wald statistic are shown. For variables not in the model, the score statistic and residual chi-square are displayed. You can choose one of the following alternatives:

- **At each step.** Displays full model information for each entry step. This is the default.
- **At last step.** Displays full model information for the final model of an entry block; summarizes intermediate steps.

Probability for Stepwise. You can override one or both of the following criteria for stepwise model building:

Entry. By default, a variable is entered into the model if the probability of its score statistic is 0.05 or less. To override this value, enter an entry probability between 0 and 1.

Removal. By default, a variable is removed from the model if the probability of its removal statistic (Wald, likelihood-ratio, or conditional likelihood-ratio statistic) is 0.10 or greater. To override this value, enter a removal probability between 0 and 1.

Maximum Iterations. SPSS computes parameter estimates using the iterative Newton-Raphson method. Iteration terminates if the relative change in the parameter estimates is less than 1E–4, if the percentage change in the log likelihood is less than 1–E5, or if the maximum number of iterations (20) is reached. To use a different maximum number of iterations, enter a positive integer.

The following option is not available for time-dependent models:

- ❑ **Display baseline function.** Displays a table showing the baseline hazard function and the survival and hazard functions evaluated at the covariate means. If you have a stratification variable, a separate table is shown for each stratum.

Additional Features Available with Command Syntax

You can customize your time-dependent proportional hazards regression analysis if you paste your selections into an output window and edit the resulting COX REGRESSION and TIME PROGRAM command syntax. (For information on syntax windows, see the SPSS Base system documentation.) Additional features include:

- The ability to obtain analyses of models having several time-dependent covariates (using the TIME PROGRAM command).
- The ability to use more than one statement to define a time-dependent covariate.
- The ability to identify cases as lost to follow-up (using the STATUS subcommand).
- Additional contrast options, such as user-defined contrasts and alternative reference categories (using the CONTRASTS subcommand).
- Additional parameter estimation criteria (using the CRITERIA subcommand).
- User-specified names for new variables (using the SAVE subcommand).
- The ability to save coefficients and survival statistics to an SPSS data file (using the OUTFILE subcommand).

See the Syntax Reference section of this manual for command syntax rules and for complete COX REGRESSION and TIME PROGRAM command syntax.

13 Writing Your Own Program: The Matrix Language

Although SPSS provides a large number of commonly used routines, you may on occasion want to perform an analysis that is not currently available in SPSS. The matrix language provides you with tools that enable you to write your own programs. You can perform all of the standard operations of matrix arithmetic plus over 50 matrix functions, including inversion, eigenvalue extraction, and the cumulative distribution functions for the normal, t, F, and chi-square distributions. Looping and conditional statements let you specify the flow of control in the program, and you can read from or write to raw data files, SPSS data files, and SPSS matrix-format data files.

In this chapter, we will consider two statistical techniques and how they can be implemented in SPSS using the matrix language. The first technique is nearest-neighbor discriminant analysis; the second, a repeated measures analysis of variance for multivariate categorical data.

Nearest-Neighbor Discriminant Analysis

In all forms of discriminant analysis, we are concerned with the allocation of cases to known groups. Methods differ in the rules they use. In the simplest form of nearest-neighbor discriminant analysis, we assign a case to the same group as the case nearest to it. That is, we assign a case to the same group as its nearest neighbor. (See *SPSS Professional Statistics* for a discussion of the use of the nearest-neighbor rule in cluster analysis.)

The computational algorithm involves finding the distances between all pairs of cases and then finding for each case the case nearest to it. The group to which the nearest neighbor belongs is the group to which the case is then assigned. If a case has more than one nearest neighbor and the nearest neighbors are in different groups, classification is not done.

The nearest-neighbor rule can be modified to examine k nearest neighbors instead of just one. A case is classified into that group to which the majority of the k nearest neighbors belongs. Again, if there is no group to which the majority of nearest neighbors belongs, classification is not done. See Hand (1981) for further discussion.

The Matrix Job

A matrix program to carry out nearest-neighbor discriminant analysis appears in Figure 13.1.

Figure 13.1 Matrix program for nearest-neighbor discriminant analysis

```
*         NEAREST-NEIGHBOR DISCRIMINANT ANALYSIS
*         ====================================
*  Lines preceded with the comment "*Debug*" display intermediate
*     results so that you can monitor the progress of the algorithm.
*     These lines can be removed without affecting the solution.

*  Assume an SPSS data file containing a square distance matrix
*     and an SPSS data file containing the data, with the grouping
*     variable named GROUP.  These could be raw data files;
*     if so you'd use READ rather than GET to read them.
*  Distance matrix can be created by the PROXIMITIES command, e.g.
*  PROXIMITIES V1 TO V10 /MATRIX OUT('distance matrix').

*  Known groups are coded from 1 to NGROUPS, specified at the
*     beginning of the MATRIX program.  If NGROUPS is more than
*     10, the labeling vector GRPNAMES must be extended.
*  Number of neighbors to be examined is LOOKAT, specified at the
*     beginning of the MATRIX program.

MATRIX.
PRINT /TITLE "Nearest-Neighbor Discriminant Analysis" /SPACE=NEWPAGE.
*Number of groups.
COMPUTE NGROUPS = 3.
*Number of neighbors to examine.
COMPUTE LOOKAT  = 5.

GET DISTANCE /FILE='SPSS distance matrix'.
GET GROUPS   /FILE='SPSS data file' /VARIABLES=GROUP /MISSING=0.

*  A case isn't considered one of its own neighbors, so change
*     the distance between each case and itself to a big number.
COMPUTE BIGNUMBR = MMAX(DISTANCE) + 1.
CALL SETDIAG(DISTANCE, BIGNUMBR).

*Number of cases.
COMPUTE NCASES   = NROW(DISTANCE).
*For display purposes.
COMPUTE IDNUMS   = MAKE(NCASES,1,0).
*Predicted group.
COMPUTE PRED     = MAKE(NCASES,1,NGROUPS+1).
*Classification table.
COMPUTE CLASTABL = MAKE(NGROUPS+1,NGROUPS+1,0).
*  PRED is initialized to the "unknown group" code NGROUPS+1, and
*  CLASTABL leaves room for cases with this code.

LOOP CASE = 1 TO NCASES.
*Debug*.
+  PRINT CASE /TITLE 'Analysis of case number:'/SPACE=1.
+  COMPUTE IDNUMS(CASE) = CASE.

*  Assign the 'missing' group code where group code out of range.
+  DO IF (GROUPS(CASE) < 1 OR GROUPS(CASE) > NGROUPS).
+     COMPUTE GROUPS(CASE) = NGROUPS + 1.
+  END IF.
```

```
*  Find the nearest neighbors. FOUND = number found so far.
+  COMPUTE NEIGHBRS = MAKE(1,NCASES,0).
+  COMPUTE FOUND = 0.
+  LOOP.
+     COMPUTE TARGET = MMIN(DISTANCE(CASE,:)).
*     TARGET = smallest remaining distance in this row.

*     Identify all cases with distance = TARGET.
+     LOOP THISCASE = 1 TO NCASES.
+        DO IF (DISTANCE(CASE,THISCASE) = TARGET).
+           COMPUTE FOUND = FOUND + 1.
+           COMPUTE NEIGHBRS(FOUND) = THISCASE.

*           Now set this distance to a big number, so we
*           don't use it as the TARGET next time around.
+           COMPUTE DISTANCE(CASE,THISCASE) = BIGNUMBR.
+        END IF.
+     END LOOP.
+  END LOOP IF (FOUND >= LOOKAT).
*  Stop looking for neighbors when we've found enough of them;
*     otherwise go back and get the next smallest target.
*  Note that we may have found more than we were looking for,
*     because of tied distances.

+  COMPUTE GRPCOUNT=MAKE(1,NGROUPS,0).
*  GRPCOUNT = How many neighbors in each group.

*Debug*.
+  COMPUTE NBRGRP = MAKE(1,FOUND,0).

*  Count how many neighbors were found in each group.
+  LOOP THISNBR = 1 TO FOUND.
+     COMPUTE NEARGRP = GROUPS(NEIGHBRS(THISNBR)).
*     NEARGRP = Group membership of this neighboring case.
+     DO IF (NEARGRP > 0 AND NEARGRP <= NGROUPS).
+        COMPUTE GRPCOUNT(NEARGRP) = GRPCOUNT(NEARGRP) + 1.
+     END IF.
*Debug*.
+     COMPUTE NBRGRP(THISNBR) = NEARGRP.
+     END LOOP.
*Debug*.
+  PRINT NEIGHBRS(1:FOUND)/TITLE 'ID numbers of neighbors'.
*Debug*.
+  PRINT NBRGRP /TITLE 'Groups of neighbors'.

*  Determine which group had the highest count.
+  COMPUTE GRPRANK = RNKORDER(GRPCOUNT).
*    GRPRANK contains ranks of the corresponding elements of GRPCOUNT.
+  LOOP THISGRP=1 TO NGROUPS.
*        Search for a rank of NGROUPS, the highest-ranking group count.
*        We search for the highest possible rank rather than the
*        maximum group count because this algorithm doesn't predict
*        a group in the case of ties.  (When several groups tie for the
*        highest count, the RNKORDER function assigns an averaged rank.)
+     DO IF (GRPRANK(THISGRP) = NGROUPS).
+        COMPUTE PRED(CASE)=THISGRP.
*No need to continue searching.
+        BREAK.
+     END IF.
+  END LOOP.
*Debug*.
+  PRINT PRED(CASE) /TITLE 'Predicted group'.
+  COMPUTE CLASTABL(GROUPS(CASE),PRED(CASE)) =
           CLASTABL(GROUPS(CASE),PRED(CASE)) + 1.
END LOOP.

* Add row and column totals to classification table.
COMPUTE CLASTABL = {CLASTABL,       RSUM(CLASTABL);
                    CSUM(CLASTABL), MSUM(CLASTABL)}.
```

```
* Assign row & column labels for classification table.
* If more than 10 valid groups, you must extend the GRPNAMES vector.
COMPUTE GRPNAMES = {"1","2","3","4","5","6","7","8","9","10","11"," "}.
COMPUTE GRPNAMES(NGROUPS+2) = "Total".

PRINT (NGROUPS+1) /TITLE "Group code representing UNKNOWN"
   /SPACE=NEWPAGE.
PRINT CLASTABL
   /TITLE "Classification table (rows=actual,columns=predicted)"
   /SPACE=2 /RNAMES=GRPNAMES /CNAMES=GRPNAMES.
PRINT {IDNUMS,GROUPS,PRED} /TITLE "    Results for each case"
   /SPACE=2 /CLABELS="ID","Actual","Pred".

* Free some memory to make room for the data file.
RELEASE DISTANCE, GROUPS, NEIGHBRS, GRPCOUNT, GRPRANK,
    IDNUMS, CLASTABL, GRPNAMES.
*Debug*.
RELEASE NBRGRP.

* Add predicted group to data file, save as the active system file.
GET DATA /FILE='SPSS data file' /NAMES=NAMEVEC.
*Append a column to data matrix.
COMPUTE DATA={DATA,PRED}.
*Append element to names vector.
COMPUTE NAMEVEC={NAMEVEC,"P_GROUP"}.
SAVE DATA /OUTFILE=* /NAMES=NAMEVEC.
END MATRIX.
```

After some initial comments, the matrix program begins by defining the number of groups in the data (*ngroups*, which is set equal to 3 in Figure 13.1) and the number of nearest neighbors to examine (*lookat*, set equal to 5 in Figure 13.1). It then reads the distance matrix (obtained from the PROXIMITIES command) into a variable named *distance* and the group memberships (where known) of the cases into a variable named *groups*. Since group membership is only a single variable in the SPSS file *data*, the variable *groups* is read as a **column vector**—that is, a matrix with only one column.

After initializing some variables using the NROW (number of rows) and MAKE (create a constant matrix) functions that are provided with the matrix language, the program enters a loop over all the cases in the file. Inside the loop, predicted group membership is assigned according to the nearest-neighbor algorithm in the following steps:

1. Assigns missing or unknown group codes to a special value. Since the valid group codes range from 1 to the value of the variable *ngroups*, or 3, the code *ngroups*+1, or 4, is used for any case with a group that is out of this range.

2. Identifies the nearest neighbors. The MAKE function creates a vector, *neighbrs*, to hold their ID numbers. Then the program enters a loop that continues until the number of neighbors identified is at least as large as the value of the variable *lookat*, which was set to 5 for this example. (More than five neighbors may be identified if several cases are at the same distance as the fifth-nearest neighbor.)

3. Finds the distance of the nearest neighbor with the MMIN function, which returns the smallest value in a matrix expression. Here it operates on the expression *distance (case,:)*, which is the row of the *distance* matrix that is indexed by the value of *case*.

4. Identifies, in a third loop, all cases in that row for which distance from the current case equals the minimum distance found in the previous step. After each case is identified, its distance is set to a large number so that it won't get in the way of finding other neighbors.

5. Once enough neighbors have been found, counts how many are in each of the valid groups. (The variable *nbrgrp* is used to keep a list of the group membership of each neighbor. This list isn't really needed in the algorithm, but it is useful to print it when you are developing the program and want to make sure that it's doing the right thing. Lines that are used for debugging in this way are preceded with the comment **Debug** in Figure 13.1.)

6. Determines which group had the highest count among the nearest neighbors. This step is based on the RNKORDER function, which creates a vector containing the ranks (from lowest to highest) of the group counts. Where two group counts are the same, the algorithm doesn't assign a predicted group, so only a group with the highest possible rank (equal to *ngroups*) is wanted. (If several groups tie for the highest value, RNKORDER gives them an averaged rank that will be less than *ngroups*.)

7. Once the predicted group membership is determined, adds the case to the classification results table, a crosstabulation of actual and predicted group memberships. The variable *clastabl* is a square matrix with a row for each actual membership code and a column for each predicted membership code. Before *clastabl* is printed, the RSUM, CSUM, and MSUM functions are used to add row, column, and overall totals to it. (The expression inside the braces on the COMPUTE statement constructs a new 5×5 matrix out of the 4×4 matrix *clastabl*, the 4×1 matrix created by the RSUM function, the 1×4 matrix created by the CSUM function, and the 1×1 matrix, or **scalar**, created by the MSUM function.)

8. Finally, reads in the original data matrix, adds a new column to it representing the predicted-group variable, and saves it into the working data file for further analysis.

A portion of the output from this matrix program on a sample data file of 11 cases is shown in Figure 13.2. Only the classification results table and casewise listing are shown. The debugging output giving intermediate results is omitted.

Figure 13.2 Results of nearest-neighbor discriminant analysis

```
Classification table (rows=actual,columns=predicted)
          1     2     3     4 Total
    1     2     1     0     0     3
    2     0     4     0     0     4
    3     1     0     0     1     2
    4     1     0     0     1     2
Total     4     5     0     2    11

     Results for each case
       ID Actual    Pred
        1      1       1
        2      2       2
        3      1       1
        4      4       4
        5      3       4
        6      1       2
        7      4       1
        8      2       2
        9      3       1
       10      2       2
       11      2       2
```

Repeated Measures Analysis of Categorical Data

Koch et al. (1977) present examples of using weighted least-squares methods for the analysis of repeated measures categorical data. Consider the data in Table 13.1, a tabulation of the responses of 46 patients who were each administered three drugs and were then categorized as having a favorable or unfavorable response to each of them.

Table 13.1 Responses to three drugs

Drug			Number with pattern
1	2	3	
F	F	F	6
F	F	U	16
F	U	F	2
F	U	U	4
U	F	F	2
U	F	U	4
U	U	F	6
U	U	U	6

Several hypotheses may be of interest. We will consider only the hypothesis of first-order marginal symmetry (homogeneity), which states that the probability of a favorable response is the same for each drug. To test the hypothesis, you must perform the following calculations:

1. Compute the marginal totals necessary for hypothesis testing. This involves calculating the proportion of individuals who responded favorably to each drug. These marginal proportions are the dependent variable in this analysis.

   To compute the marginal proportions, we can calculate the product $F = A'*P$, where

   $$A' = \begin{bmatrix} 1 & 1 & 1 & 1 & 0 & 0 & 0 & 0 \\ 1 & 1 & 0 & 0 & 1 & 1 & 0 & 0 \\ 1 & 0 & 1 & 0 & 1 & 0 & 1 & 0 \end{bmatrix}$$

   and where P, in transposed form (denoted P'), is

   $$P' = \begin{bmatrix} 0.13 & 0.35 & 0.04 & 0.09 & 0.04 & 0.09 & 0.13 & 0.13 \end{bmatrix}$$

   A value of 1 in the first row of A' indicates a cell in which response to drug 1 was favorable; a value of 1 in the second row indicates a cell in which response to drug 2 was favorable; and, similarly, a value of 1 in the third row indicates a cell in which response to drug 3 was favorable. The vector P is just the observed proportions in each of the eight cells. The resulting vector F contains the observed proportions of favorable response for each drug.

2. Compute the variance-covariance matrix of the observed proportions using

   $$VP = (DP - P*P')/N$$

   where N is the number of cases in the study. DP is a diagonal matrix with the observed proportions on the diagonal.

3. Compute the variance-covariance matrix of the marginal totals using

   $$VF = A'*VP*A$$

4. Compute the variance-covariance matrix of the parameter estimates using

   $$VB = (X'*(VF)^{-1}*X)^{-1}$$

   where X is the design matrix and the superscript -1 denotes the inverse of a matrix. We will use the following design matrix:

   $$X = \begin{bmatrix} 1 & 1 & 0 \\ 1 & 0 & 1 \\ 1 & -1 & -1 \end{bmatrix}$$

5. Compute the parameter estimates using

$$B = VB*X'*(VF)^{-1}*F$$

6. Compute the test of the hypothesis that all marginal proportions are equal using

$$QCHI = (L*B)'*(L*VB*L')^{-1}*L*B$$

where

$$L = \begin{bmatrix} 0 & 1 & 0 \\ 0 & 0 & 1 \end{bmatrix}$$

Determine the significance level of *QCHI* using the chi-square distribution with two degrees of freedom.

7. Compute chi-square statistics and their significance levels for each of the parameters. To do this, divide each parameter estimate by its standard error and square the result. This ratio is approximately chi-square distributed with a single degree of freedom.

The Matrix Job

The matrix job to perform the previous computations and display the results is shown in Figure 13.3.

Figure 13.3 Matrix program for categorical repeated measures analysis

```
*  REPEATED MEASURES OF CATEGORICAL VARIABLES VIA WLS.
*
* Data are from Koch, Landis, and Lehnen (1977), "A General
*   Methodology for the Analysis of Experiments with Repeated
*   Measurement of Categorical Data," Biometrics, vol.33, pp.133-158.

* Each subject is given three drugs.  The effectiveness of each
*   is rated as favorable (1) or unfavorable (0).

*  Variables:
*  A is the response configuration matrix.
*  P is the proportion of cases in each response configuration.
*  F is the marginal total, and is the dependent variable.
*  VF is the asymptotic covariance matrix of F.
*  X is the design matrix.
*  B is the vector of parameter estimates.
*  VB is the asymptotic covariance matrix of B.
*  QCHI is overall chi-square test, and QSIG is its significance.
*  BCHI is parameter chi-square tests, and BSIG their significance.

DATA LIST FREE/ D1 D2 D3 COUNT.
BEGIN DATA.
1 1 1 6
1 1 0 16
1 0 1 2
1 0 0 4
0 1 1 2
0 1 0 4
0 0 1 6
0 0 0 6
END DATA.

MATRIX.
GET A     /VARIABLES=D1 D2 D3.
GET COUNT /VARIABLE=COUNT.

COMPUTE N = CSUM(COUNT).
COMPUTE P = COUNT/N.
*1. Marginal totals.
COMPUTE F = T(A)*P.
*2. Covariance matrix of proportions.
COMPUTE VP = (MDIAG(P) - P*T(P))/N.
*3. Covariance matrix of marginals.
COMPUTE VF = T(A)*VP*A.
COMPUTE X={1, 1, 0;
           1, 0, 1;
           1,-1,-1}.
*4. Covariance matrix of estimates.
COMPUTE VB = INV(T(X)*INV(VF)*X).
*5. Parameter estimates.
COMPUTE B = VB*T(X)*INV(VF)*F.

COMPUTE L = {0, 1, 0;
             0, 0, 1}.
*6. Test homogeneity.
COMPUTE QCHI = T(L*B)*INV(L*VB*T(L))*(L*B).
COMPUTE QSIG =1 - CHICDF(QCHI,2).
*7. Test coefficients.
COMPUTE BCHI = (B &* B) &/ DIAG(VB).
COMPUTE BSIG =1 - CHICDF(BCHI,1).

PRINT F /FORMAT=F12.6 /TITLE='Marginal totals' /SPACE=NEWPAGE.
PRINT VF /FORMAT=F14.8
 /TITLE='Asymptotic covariance of marginal totals'.
PRINT {QCHI,QSIG} /FORMAT=F12.6 /TITLE='Homogeneity test'
 /CLABELS='Chi Sq.' 'Prob.'.
PRINT {B,BCHI,BSIG} /FORMAT=F12.6 /TITLE='Parameters'
 /CLABELS='Estimates' 'Chi Sq.' 'Prob.'.
PRINT VB /FORMAT=F14.8
 /TITLE='Asymptotic covariance of parameter estimates'.
END MATRIX.
```

From the output in Figure 13.4, you can see that the chi-square test for marginal homogeneity results in a chi-squared value of 6.58 with two degrees of freedom and an observed significance level of 0.0372. Thus, we can reject the null hypothesis of marginal homogeneity. Although, except for the constant, none of the other parameters is significantly different from 0, it appears that the third drug may be different from the first two.

Figure 13.4 Output from categorical repeated measures program

```
Marginal totals
      .608696
      .608696
      .347826

Asymptotic covariance of marginal totals
      .00517794       .00234240      -.00082190
      .00234240       .00517794      -.00082190
     -.00082190      -.00082190       .00493137

Homogeneity test
      Chi Sq.         Prob.
     6.584493       .037170

Parameters
     Estimate       Chi Sq.          Prob.
      .521739    146.837438        .000000
      .086957      2.946619        .086057
      .086957      2.946619        .086057

Asymptotic covariance of parameter estimates
      .00185383       .00037899       .00037899
      .00037899       .00256614      -.00026940
      .00037899      -.00026940       .00256614
```

Using the Matrix Language

The matrix language allows you to create your own statistical procedures using matrix operations. Matrix programs can perform mathematical calculations, display results, and read and write ordinary and matrix-format SPSS data files. Since MATRIX uses the compact language of matrix algebra, matrix programs are usually much shorter than those created with standard programming languages.

Matrix programs are created using command syntax. Every matrix job begins with the MATRIX command and ends with END MATRIX. These commands enclose statements that perform matrix operations. (Matrix command syntax is described in the Syntax Reference section of this manual.)

Like regular SPSS command syntax, you can create and run matrix programs in a syntax window.

Using Matrix Commands in a Syntax Window

To edit and run matrix language commands in a syntax window, follow the steps described below:

1. From the menus choose:

   File
     New ▸
       SPSS Syntax

   This opens a syntax window.

2. Enter your matrix language commands in the syntax window. The matrix language commands in Figure 13.5 display standardized (Z) scores for variables in the SPSS data file *v5mtrx3.sav* (standardized scores are also available from the Descriptives procedure).

Figure 13.5 Matrix commands in a syntax window

!Syntax1

Run Syntax

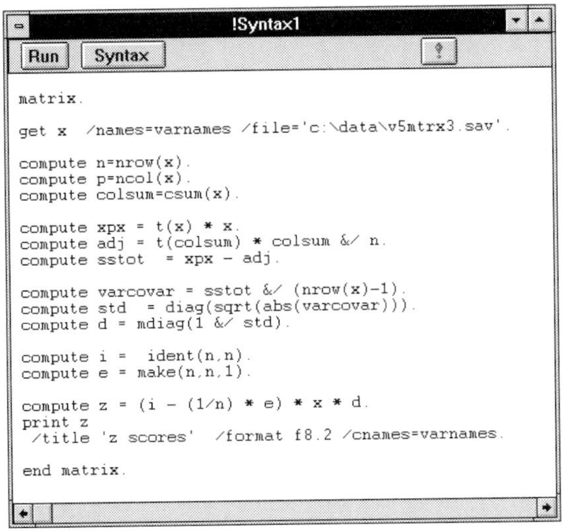

```
matrix.

get x  /names=varnames /file='c:\data\v5mtrx3.sav'.

compute n=nrow(x).
compute p=ncol(x).
compute colsum=csum(x).

compute xpx = t(x) * x.
compute adj = t(colsum) * colsum &/ n.
compute sstot  = xpx - adj.

compute varcovar = sstot &/ (nrow(x)-1).
compute std  = diag(sqrt(abs(varcovar))).
compute d = mdiag(1 &/ std).

compute i =  ident(n,n).
compute e = make(n,n,1).

compute z = (i - (1/n) * e) * x * d.
print z
 /title 'z scores'  /format f8.2 /cnames=varnames.

end matrix.
```

3. Highlight the commands you want to run and click on the Run pushbutton. Output is displayed in an output window. If your command syntax generates any error messages, you can edit the syntax and rerun it from the syntax window.

4. Remember to execute the END MATRIX command before attempting to use non-matrix commands.

For more information about using SPSS command syntax, including general SPSS syntax rules, see the introduction to the Syntax Reference section of this manual.

SPSS

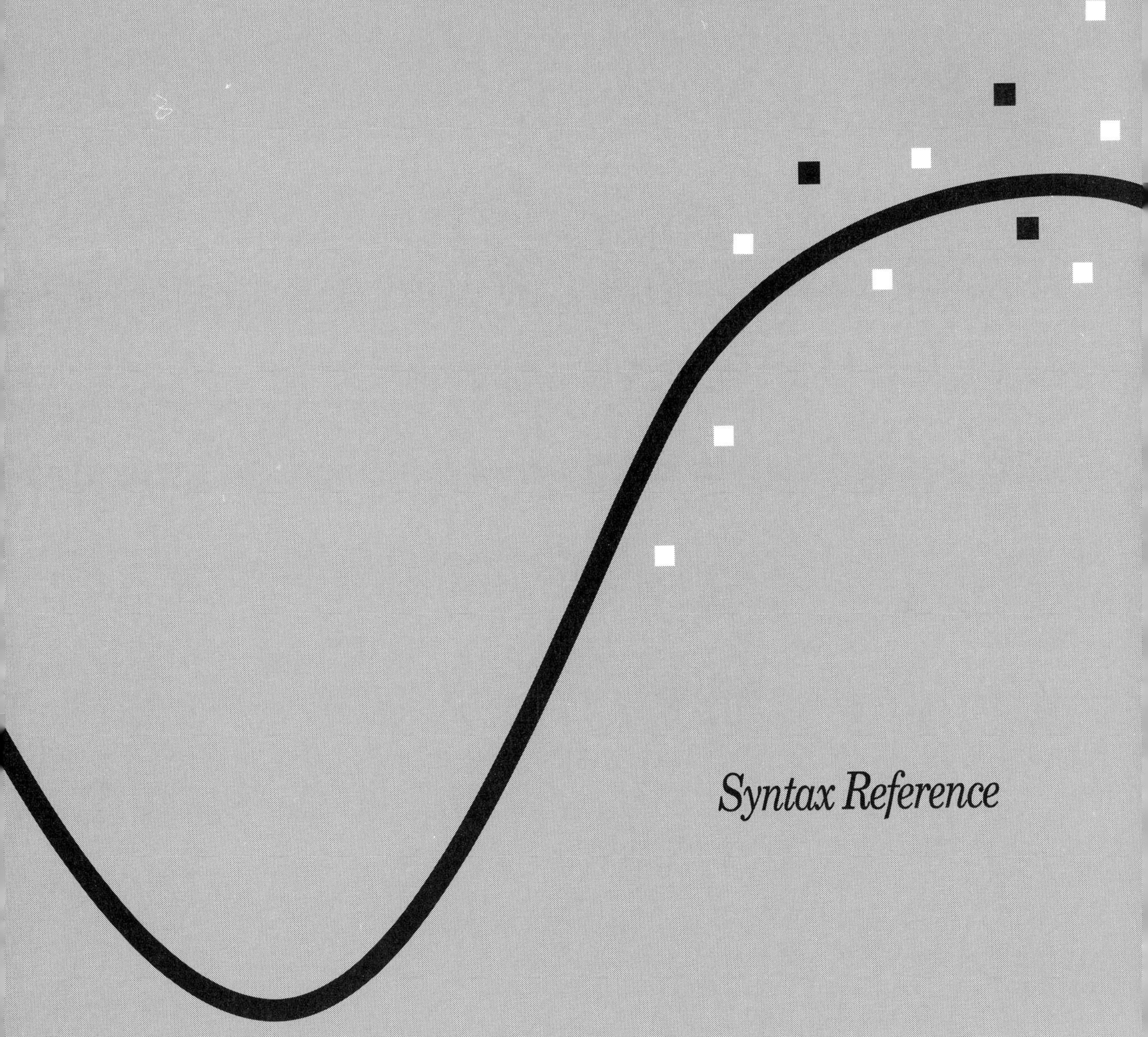

Introduction

This syntax reference guide describes the SPSS command language underlying SPSS Advanced Statistics. Most of the features of these commands are implemented in the dialog boxes and can be used directly from the dialog boxes. Or you can paste the syntax to a syntax window and edit it or build a command file, which you can save and reuse. The features that are available only in command syntax are summarized following the discussion of the dialog box interface in the corresponding chapter on each statistical procedure.

This introduction to the Syntax Reference section gives basic rules for specifying command syntax and shows how to edit and run syntax in a syntax window. For more information about SPSS command syntax, see the *SPSS Base System Syntax Reference Guide*. For more information about running commands in SPSS, see the SPSS Base system documentation.

A Few Useful Terms

All terms in the SPSS command language fall into one or more of the following categories:

- **Keyword**. A word already defined by SPSS to identify a command, subcommand, or specification. Most keywords are, or resemble, common English words.
- **Command**. A specific instruction that controls the execution of SPSS.
- **Subcommand**. Additional instructions for SPSS commands. A command can contain more than one subcommand, each with its own specifications.
- **Specifications**. Instructions added to a command or subcommand. Specifications may include subcommands, keywords, numbers, arithmetic operators, variable names, special delimiters, and so forth.

Each command begins with a command keyword (which may contain more than one word). The command keyword is followed by at least one blank space and then any additional specifications. Each command ends with a command terminator, which is a period. For example:

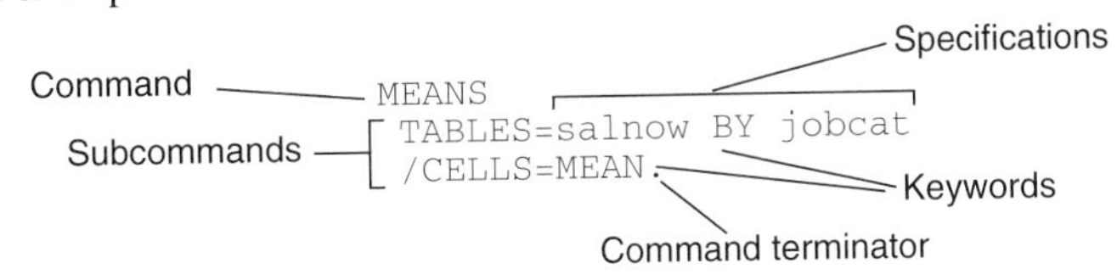

Syntax Diagrams

Each SPSS command described in this manual includes a syntax diagram that shows all the subcommands, keywords, and specifications allowed for that command. These syntax diagrams are also available in the online help system for easy reference when entering commands in a syntax window. By remembering the following rules, you can use the syntax diagram as a quick reference for any command:

- Elements shown in all capital letters are keywords defined by SPSS to identify commands, subcommands, functions, operators, and other specifications.
- Elements in lower case describe specifications you supply.
- Elements in boldface type are defaults. A default indicated with ** is in effect when the keyword is not specified. (Boldface is not used in the online help system syntax diagrams.)
- Parentheses, apostrophes, and quotation marks are required where indicated.
- Elements enclosed in square brackets ([ ]) are optional.
- Braces ({ }) indicate a choice among elements. You can specify any one of the elements enclosed within the aligned braces.
- Ellipses indicate that an element can be repeated.
- Most abbreviations are obvious; for example, varname stands for variable name and varlist stands for a list of variables.
- The command terminator is not shown in the syntax diagrams.

Syntax Rules

Keep in mind the following simple rules when writing and editing commands in a syntax window:

- Each command must begin on a new line and end with a period.
- Subcommands are separated by slashes. The slash before the first subcommand in a command is optional in most commands.
- SPSS keywords are not case-sensitive, and three-letter abbreviations can be used for most keywords.
- Variable names must be spelled out in full.
- You can use as many lines as you want to specify a single command. However, text included within apostrophes or quotation marks must be contained on a single line.
- You can add space or break lines at almost any point where a single blank is allowed, such as around slashes, parentheses, arithmetic operators, or between variable names.
- Each line of syntax cannot exceed 80 characters.
- The period must be used as the decimal indicator.

For example,

```
FREQUENCIES
 VARIABLES=JOBCAT SEXRACE
 /PERCENTILES=25 50 75
 /BARCHART.
```

and

```
freq var=jobcat sexrace /percent=25 50 75 /bar.
```

are both acceptable alternatives that generate the same results. The second example uses three-letter abbreviations and lower case, and the command is on one line.

INCLUDE Files

If your SPSS commands are contained in a command file that is specified on the SPSS INCLUDE command, the syntax rules are slightly different:

- Each command must begin in the first column of a new line.
- Continuation lines within a command must be indented at least one space.
- The period at the end of the command is optional.

If you generate command syntax by pasting dialog box choices into a syntax window, the format of the commands is suitable for both INCLUDE files and commands run in a syntax window.

Using Command Syntax in a Syntax Window

To edit and run command syntax in a syntax window:

1. From the menus choose:

   File
   New ▸
   SPSS Syntax

   This opens a syntax window.

2. Enter the commands in the syntax window. For example, the MANOVA specifications shown in Figure 1 produce a repeated measures analysis of variance.

Figure 1 MANOVA commands in a syntax window

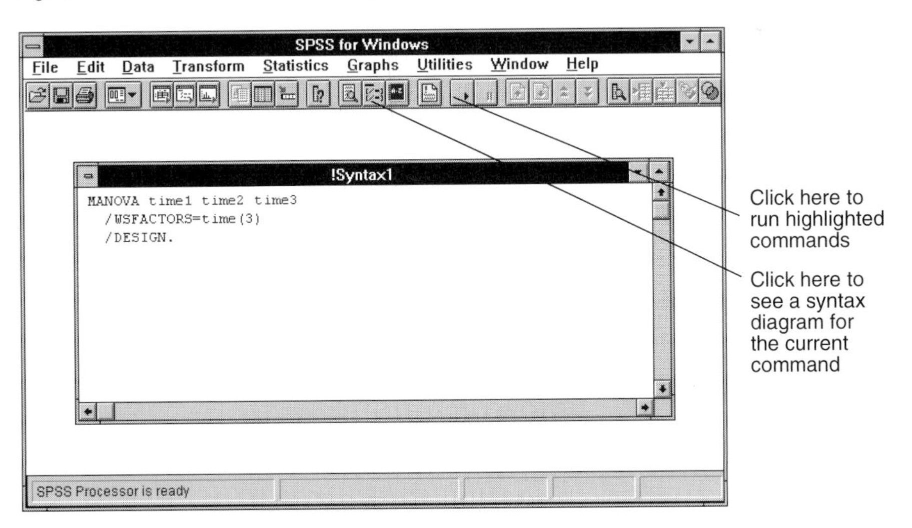

3. Highlight the commands you want to run and click on the Run Syntax tool (▸). Output is displayed in an output window. If your command syntax generates any error messages, you can edit the syntax and rerun it from the syntax window.

Running an Existing Command File

To run an existing command file:

1. From the menus choose:

   File
   Open ▸
   Syntax

2. Select the command file you want from the Open Syntax File dialog box (see the SPSS Base system documentation for information on how to locate and open syntax files).
3. Highlight the commands you want to run and click on the Run Syntax tool (▸).

Online Syntax Help

Syntax diagrams for the SPSS command language are available to assist you when you work with command syntax in a syntax window.

1. In a syntax window, type the name of the command you want to use. For example, to obtain the syntax chart for the PROBIT command, type `probit`. (SPSS commands are not case-sensitive.)
2. With the cursor on the same line, click on the Syntax Help tool (). This opens the Syntax Help window containing the syntax diagram for the command, as shown in Figure 2.

Figure 2 Syntax Help window for the PROBIT command

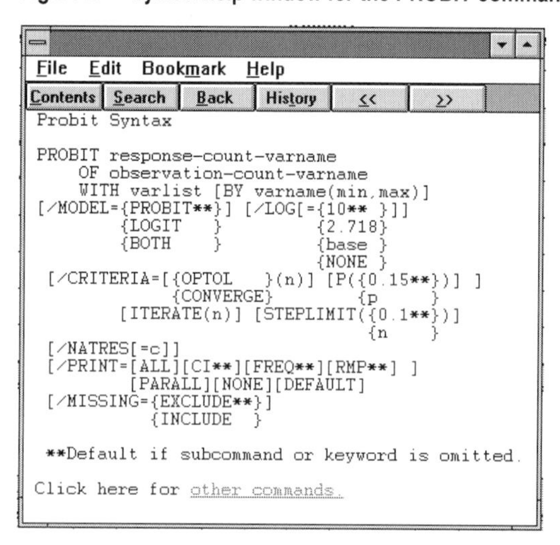

To see the syntax diagram as you work in a syntax window, reduce the main SPSS application window so that the Help window fits beside it or below it.

Copying Syntax from the Help Window

You can also copy some or all of the syntax diagram to a syntax window.

1. From the Help window menu choose:

   Edit
     Copy...

   This opens the Copy dialog box, as shown in Figure 3.

Figure 3 Copy dialog box

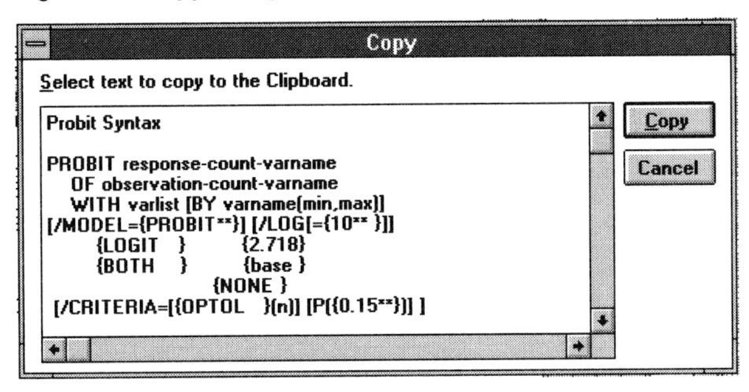

2. Highlight the text you want to copy and then click on the Copy pushbutton. From there you can paste it into a syntax window. Delete all the brackets, braces, and asterisks used to denote optional and alternative specifications and defaults, and replace the lowercase generic user specifications with your own.

COXREG

```
[TIME PROGRAM]*
[commands to compute time dependent covariates]

[CLEAR TIME PROGRAM]

COXREG [VARIABLES =] survival varname [WITH varlist]
    / STATUS = varname [EVENT] (vallist) [LOST (vallist)]
   [/STRATA  = varname]
   [/CATEGORICAL = varname]
   [/CONTRAST (varname) = {DEVIATION (refcat)}]
                          {SIMPLE (refcat)   }
                          {DIFFERENCE        }
                          {HELMERT           }
                          {REPEATED          }
                          {POLYNOMIAL(metric)}
                          {SPECIAL (matrix)  }
                          {INDICATOR (refcat)}

   [/METHOD = {ENTER**          }   [{varlist}]]
              {BSTEP [{COND}]}       {ALL    }
                      {LR  }
                      {WALD}
              {FSTEP [{COND}]}
                      {LR  }
                      {WALD}

   [/MISSING = {EXCLUDE**}]
               {INCLUDE  }

   [/PRINT = [{DEFAULT**}]  [CI ({95})]]
              {SUMMARY  }         {n }
              {BASELINE }
              {CORR     }
              {ALL      }

   [/CRITERIA = [{BCON}({1E-4**})]   [LCON({1E-5**})]
                 {PCON} { n     }         { n     }
                [ITERATE({20**})]
                         { n  }
                [PIN({0.05**})]        [POUT({0.1**})]]
                     { n     }               { n    }

   [/PLOT = [NONE**] [SURVIVAL] [HAZARD] [LML]]
   [/PATTERN = [varname(value)...] [BY varname]]
   [/OUTFILE = [COEFF(file)] [TABLE(file)]]
   [/SAVE = tempvar [(newvarname)],tempvar ...]
   [/EXTERNAL]
```

* TIME PROGRAM is required to generate time-dependent covariates.

**Default if subcommand or keyword is omitted.

Temporary variables created by COXREG are:

SURVIVAL
SE
HAZARD
RESID
LML
DFBETA
PRESID
XBETA

Example:

```
TIME PROGRAM.
COMPUTE Z=AGE + T_.

COXREG SURVIVAL WITH Z
  /STATUS SURVSTA EVENT(1).
```

Overview

COXREG applies Cox proportional hazards regression to analysis of survival times—that is, the length of time before the occurrence of an event. COXREG supports continuous and categorical independent variables (covariates), which can be time-dependent. Unlike SURVIVAL and KM, which compare only distinct subgroups of cases, COXREG provides an easy way of considering differences in subgroups as well as analyzing effects of a set of covariates.

Options

Processing of Independent Variables. You can specify which of the independent variables are categorical with the CATEGORICAL subcommand and control treatment of these variables with the CONTRAST subcommand. You can select one of seven methods for entering independent variables into the model using the METHOD subcommand. You can also indicate interaction terms using the keyword BY between variable names on either the VARIABLES subcommand or the METHOD subcommand.

Specifying Termination and Model-building Criteria. You can specify the criteria for termination of iteration and control variable entry and removal with the CRITERIA subcommand.

Adding New Variables to Working Data File. You can use the SAVE subcommand to save the cumulative survival, standard error, cumulative hazard, log-minus-log-of-survival function, residuals, XBeta, and, wherever available, partial residuals and DfBeta.

Output. You can print optional output using the PRINT subcommand, suppress or request plots with the PLOT subcommand, and, with the OUTFILE subcommand, write SPSS data files containing coefficients from the final model or a survival table. When only time-constant covariates are used, you can use the PATTERN subcommand to specify a pattern of covariate values in addition to the covariate means to use for the plots and the survival table.

Basic Specification

- The minimum specification on COXREG is a dependent variable with the STATUS subcommand.
- To analyze the influence of time-constant covariates on the survival times, the minimum specification requires either the WITH keyword followed by at least one covariate (independent variable) on the VARIABLES subcommand or a METHOD subcommand with at least one independent variable.

- To analyze the influence of time-dependent covariates on the survival times, the TIME PROGRAM command and transformation language are required to define the functions for the time-dependent covariate(s).

Subcommand Order

- The VARIABLES subcommand must be specified first; the subcommand keyword is optional.
- Remaining subcommands can be named in any order.

Syntax Rules

- Only one dependent variable can be specified for each COXREG command.
- Any number of covariates (independent variables) can be specified. The dependent variable cannot appear on the covariate list.
- The covariate list is required if any of the METHOD subcommands are used without a variable list or if the METHOD subcommand is not used.
- Only one status variable can be specified on the STATUS subcommand. If multiple STATUS subcommands are specified, only the last specification is in effect.
- You can use the BY keyword to specify interaction between covariates.

Operations

- The TIME PROGRAM computes the values for time-dependent covariates.
- COXREG replaces covariates specified on CATEGORICAL with sets of contrast variables. In stepwise analyses, the set of contrast variables associated with one categorical variable is entered or removed from the model as a block.
- Covariates are screened to detect and eliminate redundancies.
- COXREG deletes all cases that have negative values for the dependent variable.

Limitations

- Only one dependent variable is allowed.
- Maximum 100 covariates in a single interaction term.
- Maximum 35 levels for a BY variable on PATTERN.

Example

```
TIME PROGRAM.
COMPUTE Z=AGE + T_.

COXREG SURVIVAL WITH Z
  /STATUS SURVSTA EVENT (1).
```

- TIME PROGRAM defines the time-dependent covariate *Z* as the current age. *Z* is then specified as a covariate.
- The dependent variable *SURVIVAL* contains the length of time to the terminal event or to censoring.
- A value of 1 on the variable *SURVSTA* indicates an event.

TIME PROGRAM Command

TIME PROGRAM is required to define time-dependent covariates. These are covariates whose values change during the course of the study.

- TIME PROGRAM and the transformations that define the time-dependent covariate(s) must precede the COXREG command.
- A time-dependent covariate is a function of the current time, which is represented by the special variable *T_*.
- The working data file must not have a variable named *T_*. If it does, rename the variable before you run the COXREG command. Otherwise, you will trigger an error.
- *T_* cannot be specified as a covariate. Any other variable in the TIME PROGRAM can be specified on the covariate list.
- For every time-dependent covariate, values are generated for each valid case for all uncensored times in the same stratum that occur before the observed time. If no STRATA subcommand is specified, all cases are considered to belong to one stratum.
- If any function defined by the time program results in a missing value for a case that has no missing values for any other variable used in the procedure, COXREG terminates with an error.

CLEAR TIME PROGRAM Command

CLEAR TIME PROGRAM deletes all time-dependent covariates created in the previous time program. It is primarily used in interactive mode to remove temporary variables associated with the time program so that you can redefine time-dependent covariates for the Cox Regression procedure. It is not necessary to use this command if you have already executed COXREG. All temporary variables created by the time program are automatically deleted.

VARIABLES Subcommand

VARIABLES identifies the dependent variable and the covariates to be included in the analysis.

- The minimum specification is the dependent variable. The subcommand keyword is optional.
- You must specify the keyword WITH and a list of all covariates if no METHOD subcommand is specified or if a METHOD subcommand is specified without naming the variables to be used.
- If the covariate list is not specified on VARIABLES but one or more METHOD subcommands are used, the covariate list is assumed to be the union of the sets of variables listed on all the METHOD subcommands.

- You can specify an interaction of two or more covariates using the keyword BY. For example, `A B BY C D` specifies the three terms *A*, *B*C*, and *D*.
- The keyword TO can be used to specify a list of covariates. The implied variable order is the same as in the working data file.

STATUS Subcommand

To determine whether the event has occurred for a particular observation, COXREG checks the value of a status variable. STATUS lists the status variable and the code for the occurrence of the event.

- Only one status variable can be specified. If multiple STATUS subcommands are specified, COXREG uses the last specification and displays a warning.
- The keyword EVENT is optional, but the value list in parentheses must be specified.
- The value list must be enclosed in parentheses. All cases with non-negative times that do not have a code within the range specified after EVENT are classified as **censored cases**—that is, cases for which the event has not yet occurred.
- The value list can be one value, a list of values separated by blanks or commas, a range of values using the keyword THRU, or a combination.
- If missing values occur within the specified ranges, they are ignored if MISSING=EXCLUDE (the default) is specified, but they are treated as valid values for the range if MISSING=INCLUDE is specified.
- The status variable can be either numeric or string. If a string variable is specified, the EVENT values must be enclosed in apostrophes and the keyword THRU cannot be used.

Example

```
COXREG SURVIVAL WITH GROUP
  /STATUS SURVSTA (3 THRU 5, 8 THRU 10).
```

- STATUS specifies that *SURVSTA* is the status variable.
- A value between either 3 and 5, or 8 and 10, inclusive, means that the terminal event occurred.
- Values outside the specified ranges indicate censored cases.

STRATA Subcommand

STRATA identifies a stratification variable. A different baseline survival function is computed for each stratum.

- The only specification is the subcommand keyword with one, and only one, variable name.
- If you have more than one stratification variable, create a new variable that corresponds to the combination of categories of the individual variables before invoking the COXREG command.
- There is no limit to the number of levels for the strata variable.

Example

```
COXREG SURVIVAL WITH GROUP
 /STATUS SURVSTA (1)
 /STRATA=LOCATION.
```

- STRATA specifies *LOCATION* as the strata variable.
- Different baseline survival functions are computed for each value of *LOCATION*.

CATEGORICAL Subcommand

CATEGORICAL identifies covariates that are nominal or ordinal. Variables that are declared to be categorical are automatically transformed to a set of contrast variables (see "CONTRAST Subcommand," below). If a variable coded as $0 - 1$ is declared as categorical, by default, its coding scheme will be changed to deviation contrasts.

- Covariates not specified on CATEGORICAL are assumed to be at least interval, except for strings.
- Variables specified on CATEGORICAL but not on VARIABLES or any METHOD subcommand are ignored.
- Variables specified on CATEGORICAL are replaced by sets of contrast variables. If the categorical variable has n distinct values, there will be $n - 1$ contrast variables generated. The set of contrast variables associated with one categorical variable are entered or removed from the model together.
- If any one of the variables in an interaction term is specified on CATEGORICAL, the interaction term is replaced by contrast variables.
- All string variables are categorical. Only the first eight characters of each value of a string variable are used in distinguishing among values. Thus, if two values of a string variable are identical for the first eight characters, the values are treated as though they were the same.

CONTRAST Subcommand

CONTRAST specifies the type of contrast used for categorical covariates. The interpretation of the regression coefficients for categorical covariates depends on the contrasts used. The default is DEVIATION. For illustration of contrast types, see Appendix B.

- The categorical covariate is specified in parentheses following CONTRAST.
- If the categorical variable has n values, there will be $n - 1$ rows in the contrast matrix. Each contrast matrix is treated as a set of independent variables in the analysis.
- Only one variable can be specified per CONTRAST subcommand, but multiple CONTRAST subcommands can be specified.
- You can specify one of the contrast keywords in the parentheses after the variable specification to request a specific contrast type.

The following contrast types are available:

DEVIATION(refcat) *Deviations from the overall effect.* This is the default. The effect for each category of the independent variable except one is compared to the overall effect. Refcat is the category for which parameter estimates are not displayed (they must be calculated from the others). By default, refcat is the last category. To omit a category other than the last, specify the sequence number of the omitted category (which is not necessarily the same as its value) in parentheses after the keyword DEVIATION.

SIMPLE(refcat) *Each category of the independent variable except the last is compared to the last category.* To use a category other than the last as the omitted reference category, specify its sequence number (which is not necessarily the same as its value) in parentheses following the keyword SIMPLE.

DIFFERENCE *Difference or reverse Helmert contrasts.* The effects for each category of the covariate except the first are compared to the mean effect of the previous categories.

HELMERT *Helmert contrasts.* The effects for each category of the independent variable except the last are compared to the mean effects of subsequent categories.

POLYNOMIAL(metric) *Polynomial contrasts.* The first degree of freedom contains the linear effect across the categories of the independent variable, the second contains the quadratic effect, and so on. By default, the categories are assumed to be equally spaced; unequal spacing can be specified by entering a metric consisting of one integer for each category of the independent variable in parentheses after the keyword POLYNOMIAL. For example, `CONTRAST (STIMULUS) = POLYNOMIAL(1,2,4)` indicates that the three levels of *STIMULUS* are actually in the proportion 1:2:4. The default metric is always $(1,2,...,k)$, where k categories are involved. Only the relative differences between the terms of the metric matter: (1,2,4) is the same metric as (2,3,5) or (20,30,50) because, in each instance, the difference between the second and third numbers is twice the difference between the first and second.

REPEATED *Comparison of adjacent categories.* Each category of the independent variable except the first is compared to the previous category.

SPECIAL(matrix) *A user-defined contrast.* After this keyword, a matrix is entered in parentheses with $k-1$ rows and k columns, where k is the number of categories of the independent variable. The rows of the contrast matrix contain the special contrasts indicating the desired comparisons between categories. If the special contrasts are linear combinations of each other, COXREG reports the linear dependency and stops processing. If k rows are entered, the first row is discarded and only the last $k-1$ rows are used as the contrast matrix in the analysis.

INDICATOR(refcat) *Indicator variables.* Contrasts indicate the presence or absence of category membership. By default, refcat is the last category (represented in the contrast matrix as a row of zeros). To omit a category other than the last, specify the sequence number of the omitted category (which is not necessarily the same as its value) in parentheses after keyword INDICATOR.

Example

```
COXREG SURVIVAL WITH GROUP
 /STATUS SURVSTA (1)
 /STRATA=LOCATION
 /CATEGORICAL = GROUP
 /CONTRAST(GROUP)=SPECIAL(2 -1 -1
                          0  1 -1).
```

- The specification of *GROUP* on CATEGORICAL replaces the variable with a set of contrast variables.
- *GROUP* identifies whether a case is in one of the three treatment groups.
- A SPECIAL type contrast is requested. A three-column, two-row contrast matrix is entered in parentheses.

METHOD Subcommand

METHOD specifies the order of processing and the manner in which the covariates enter the model. If no METHOD subcommand is specified, the default method is ENTER.

- The subcommand keyword METHOD can be omitted.
- You can list all covariates to be used for the method on a variable list. If no variable list is specified, the default is ALL: all covariates named after WITH on the VARIABLES subcommand are used for the method.
- The keyword BY can be used between two variable names to specify an interaction term.
- Variables specified on CATEGORICAL are replaced by sets of contrast variables. The contrast variables associated with a categorical variable are entered or removed from the model together.

Three keywords are available to specify how the model is to be built:

ENTER *Forced entry.* All variables are entered in a single step. This is the default if the METHOD subcommand is omitted.

FSTEP *Forward stepwise.* The covariates specified on FSTEP are tested for entry into the model one by one based on the significance level of the score statistic. The variable with the smallest significance less than PIN is entered into the model. After each entry, variables that are already in the model are tested for possible removal based on the significance of the Wald statistic, likelihood ratio, or conditional criterion. The variable with the largest probability greater than the specified POUT value is removed and the model is reestimated. Variables in the model are then again evaluated for removal. Once no more variables satisfy the removal criteria, covariates not in the model are evaluated for entry. Model building stops when no more variables meet entry or removal criteria, or when the current model is the same as a previous one.

BSTEP *Backward stepwise.* As a first step, the covariates specified on BSTEP are entered into the model together and are tested for removal one by one. Stepwise removal and entry then follow the same process as described for FSTEP until no more variables meet entry and removal criteria, or when the current model is the same as a previous one.

- Multiple METHOD subcommands are allowed and are processed in the order in which they are specified. Each method starts with the results from the previous method. If BSTEP is used, all eligible variables are entered at the first step. All variables are then eligible for entry and removal unless they have been excluded from the METHOD variable list.

The statistic used in the test for removal can be specified by an additional keyword in parentheses following FSTEP or BSTEP. If FSTEP or BSTEP is specified by itself, the default is COND.

COND *Conditional statistic.* This is the default if FSTEP or BSTEP is specified by itself.

WALD *Wald statistic.* The removal of a covariate from the model is based on the significance of the Wald statistic.

LR *Likelihood ratio.* The removal of a covariate from the model is based on the significance of the change in the log likelihood. If LR is specified, the model must be reestimated without each of the variables in the model. This can substantially increase computational time. However, the likelihood ratio statistic is better than the Wald statistic for deciding which variables are to be removed.

Example

```
COXREG SURVIVAL WITH GROUP SMOKE DRINK
 /STATUS SURVSTA (1)
 /CATEGORICAL = GROUP SMOKE DRINK
 /METHOD ENTER GROUP
 /METHOD BSTEP (LR) SMOKE DRINK SMOKE BY DRINK.
```

- *GROUP*, *SMOKE*, and *DRINK* are specified as covariates and as categorical variables.
- The first METHOD subcommand enters GROUP into the model.
- Variables in the model at the termination of the first METHOD subcommand are included in the model at the beginning of the second METHOD subcommand.
- The second METHOD subcommand adds *SMOKE*, *DRINK*, and the interaction of *SMOKE* with *DRINK* to the previous model.
- Backward stepwise regression analysis is then done using the likelihood-ratio statistic as the removal criterion. The variable *GROUP* is not eligible for removal because it was not specified on the BSTEP subcommand.
- The procedure continues until the removal of a variable will result in a decrease in the log likelihood with a probability smaller than POUT.

MISSING Subcommand

MISSING controls missing value treatments. If MISSING is omitted, the default is EXCLUDE.

- Cases with negative values on the dependent variable are automatically treated as missing and are excluded.
- To be included in the model, a case must have nonmissing values for the dependent, status, strata, and all independent variables specified on the COXREG command.

EXCLUDE *Exclude user-missing values.* User-missing values are treated as missing. This is the default if MISSING is omitted.

INCLUDE *Include user-missing values.* User-missing values are included in the analysis.

PRINT Subcommand

By default, COXREG prints a full regression report for each step. You can use the PRINT subcommand to request specific output. If PRINT is not specified, the default is DEFAULT.

DEFAULT *Full regression output including overall model statistics and statistics for variables in the equation and variables not in the equation.* This is the default when PRINT is omitted.

SUMMARY *Summary information.* The output includes –2 log likelihood for the initial model, one line of summary for each step, and the final model printed with full detail.

CORR *Correlation/covariance matrix of parameter estimates for the variables in the model.*

BASELINE *Baseline table.* For each stratum, a table is displayed showing the baseline cumulative hazard, as well as survival, standard error, and cumulative hazard evaluated at the covariate means for each observed time point in that stratum.

CI (value) *Confidence intervals for* e^{β}. Specify the confidence level in parentheses. The requested intervals are displayed whenever a variables-in-equation table is printed. The default is 95%.

ALL *All available output.*

- Estimation histories showing the last ten iterations are printed if the solution fails to converge.

Example

```
COXREG SURVIVAL WITH GROUP
 /STATUS = SURVSTA (1)
 /STRATA = LOCATION
 /CATEGORICAL = GROUP
 /METHOD = ENTER
 /PRINT ALL.
```

- PRINT requests summary information, a correlation matrix for parameter estimates, a baseline survival table for each stratum, and confidence intervals for e^{β} with each variables-in-equation table, in addition to the default output.

CRITERIA Subcommand

CRITERIA controls the statistical criteria used in building the Cox regression models. The way in which these criteria are used depends on the method specified on the METHOD subcommand. The default criteria are noted in the description of each keyword below. Iterations will stop if any of the criteria for BCON, LCON, or ITERATE are satisfied.

BCON(value) *Change in parameter estimates for terminating iteration.* Alias PCON. Iteration terminates when the parameters change by less than the specified value. BCON defaults to 1E – 4 . To eliminate this criteria, specify a value of 0.

ITERATE(value) *Maximum number of iterations.* If a solution fails to converge after the maximum number of iterations has been reached, COXREG displays an iteration history showing the last 10 iterations and terminates the procedure. The default for ITERATE is 20.

LCON(value) *Percentage change in the log likelihood ratio for terminating iteration.* If the log likelihood decreases by less than the specified value, iteration terminates. LCON defaults to 1E – 5 . To eliminate this criterion, specify a value of 0.

PIN(value) *Probability of score statistic for variable entry.* A variable whose significance level is greater than PIN cannot enter the model. The default for PIN is 0.05.

POUT(value) *Probability of Wald, LR, or conditional LR statistic to remove a variable.* A variable whose significance is less than POUT cannot be removed. The default for POUT is 0.1.

Example

```
COXREG SURVIVAL WITH GROUP AGE BP TMRSZ
 /STATUS = SURVSTA (1)
 /STRATA = LOCATION
 /CATEGORICAL = GROUP
 /METHOD BSTEP
 /CRITERIA BCON(0) ITERATE(10) PIN(0.01) POUT(0.05).
```

- A backward stepwise Cox regression analysis is performed.
- CRITERIA alters four of the default statistical criteria that control the building of a model.
- Zero specified on BCON indicates that change in parameter estimates is not a criterion for termination. BCON can be set to 0 if only LCON and ITER are to be used.
- ITERATE specifies that the maximum number of iterations is 10. LCON is not changed and the default remains in effect. If either ITERATE or LCON is met, iterations will terminate.
- POUT requires that the probability of the statistic used to test whether a variable should remain in the model be smaller than 0.05. This is more stringent than the default value of 0.1.
- PIN requires that the probability of the score statistic used to test whether a variable should be included be smaller than 0.01. This makes it more difficult for variables to be included in the model than does the default PIN, which has a value of 0.05.

PLOT Subcommand

You can request specific plots to be produced with the PLOT subcommand. Each requested plot is produced once for each pattern specified on the PATTERN subcommand. If PLOT is not specified, the default is NONE (no plots are printed). Requested plots are displayed at the end of the final model.

- The set of plots requested is displayed for the functions at the mean of the covariates and at each combination of covariate values specified on PATTERN.
- If time-dependent covariates are included in the model, no plots are produced.
- Lines on a plot are connected as step functions.

NONE *Do not display plots.*

SURVIVAL *Plot the cumulative survival distribution.*

HAZARD *Plot the cumulative hazard function.*

LML *Plot the log-minus-log-of-survival function.*

PATTERN Subcommand

PATTERN specifies the pattern of covariate values to be used for the requested plots and coefficient tables.

- A value must be specified for each variable specified on PATTERN.
- Continuous variables that are included in the model but not named on PATTERN are evaluated at their means.
- Categorical variables that are included in the model but not named on PATTERN are evaluated at the means of the set of contrasts generated to replace them.
- You can request separate lines for each category of a variable that is in the model. Specify the name of the categorical variable after the keyword BY. The BY variable must be a categorical covariate. You cannot specify a value for the BY covariate.
- Multiple PATTERN subcommands can be specified. COXREG produces a set of requested plots for each specified pattern.
- PATTERN cannot be used when time-dependent covariates are included in the model.

OUTFILE Subcommand

OUTFILE writes an external SPSS data file. COXREG writes two types of data files. You can specify the file type to be created with one of the two keywords, followed by the file specification in parentheses.

COEFF *Write an SPSS data file containing the coefficients from the final model.*

TABLE *Write the survival table to an SPSS data file.* The file contains cumulative survival, standard error, and cumulative hazard statistics for each uncensored time within each stratum evaluated at the baseline and at the mean of the covariates. Additional covariate patterns can be requested on PATTERN.

- The specified SPSS data file must be an external file. You cannot specify an asterisk (*) to identify the working data file.
- The variables saved in the external file are listed in the output.

SAVE Subcommand

SAVE saves the temporary variables created by COXREG. The temporary variables include:

SURVIVAL *Survival function evaluated at the current case.*

SE *Standard error of the survival function.*

HAZARD *Cumulative hazard function evaluated at the current case.* Alias RESID.

LML *Log-minus-log-of-survival function.*

DFBETA *Change in the coefficient if the current case is removed.* There is one *DFBETA* for each covariate in the final model. If there are time-dependent covariates, only *DFBETA* can be requested. Requests for any other temporary variable are ignored.

PRESID *Partial residuals.* There is one residual variable for each covariate in the final model. If a covariate is not in the final model, the corresponding new variable has the system-missing value.

XBETA *Linear combination of mean corrected covariates times regression coefficients from the final model.*

- To specify variable names for the new variables, assign the new names in parentheses following each temporary variable name.
- Assigned variable names must be unique in the working data file. Scratch or system variable names cannot be used (that is, the variable names cannot begin with # or $).
- If new variable names are not specified, COXREG generates default names. The default name is composed of the first three characters of the name of the temporary variable (two for *SE*), followed by an underscore and a number to make it unique.
- A temporary variable can be saved only once on the same SAVE subcommand.

Example

```
COXREG SURVIVAL WITH GROUP
 /STATUS = SURVSTA (1)
 /STRATA = LOCATION
 /CATEGORICAL = GROUP
 /METHOD = ENTER
 /SAVE SURVIVAL HAZARD.
```

- COXREG saves cumulative survival and hazard in two new variables, *SUR_1* and *HAZ_1*, provided that neither of the two names exists in the working data file. If one does, the numeric suffixes will be incremented to make a distinction.

EXTERNAL Subcommand

EXTERNAL specifies that the data for each split-file group should be held in an external scratch file during processing. This helps conserve working space when running analyses with large data sets.

- The EXTERNAL subcommand takes no other keyword and is specified by itself.
- If time-dependent covariates exist, external data storage is unavailable, and EXTERNAL is ignored.

GENLOG

```
GENLOG varlist[BY] varlist [WITH covariate varlist]

 [/CSTRUCTURE=varname]

 [/GRESID=varlist]

 [/GLOR=varlist]

 [/MODEL={POISSON**   }]
         {MULTINOMIAL}

 [/CRITERIA=[CONVERGE({0.001**})][ITERATE({20**})][DELTA({0.5**})]
                      {n      }           {n   }         {n    }

            [CIN({95**})] [EPS({1E-8**})]
                 {n   }        {n     }

            [DEFAULT]

 [/PRINT=[FREQ**][RESID**][ADJRESID**][DEV**]
         [ZRESID][ITERATE][COV][DESIGN][ESTIM][COR]
         [ALL] [NONE]
         [DEFAULT]]

 [/PLOT={DEFAULT**                     }]
        {RESID([ADJRESID][DEV])        }
        {NORMPROB([ADJRESID][DEV])     }
        {NONE                          }

 [/SAVE=tempvar (newvar)[tempvar (newvar)...]]

 [/MISSING=[{EXCLUDE**}]]
            {INCLUDE  }

 [/DESIGN=effect[(n)] effect[(n)]... effect {BY} effect...]
                                            {* }
```

**Default if subcommand or keyword is omitted.

Overview

GENLOG is a general procedure for model fitting, hypothesis testing, and parameter estimation for any model that has categorical variables as its major components. As such, GENLOG subsumes a variety of related techniques, including general models of multiway contingency tables, logit models, logistic regression on categorical variables, and quasi-independence models.

GENLOG, following the regression approach, uses dummy coding to construct a design matrix for estimation and produces maximum-likelihood estimates of parameters by means of the Newton-Raphson algorithm. Since the regression approach uses the original parameter spaces, the parameter estimates correspond to the original levels of the categories and are therefore easier to interpret.

HILOGLINEAR, which uses an iterative proportional-fitting algorithm, is more efficient for hierarchical models and useful in model building, but it cannot produce parameter estimates for unsaturated models, does not permit specification of contrasts for parameters, and does not display a correlation matrix of the parameter estimates.

The General Loglinear Analysis and Logit Loglinear Analysis dialog boxes are both associated with the GENLOG command. In previous releases of SPSS, these dialog boxes were associated with the LOGLINEAR command. The LOGLINEAR command is now available only as a syntax command. The differences are described in the discussion of the LOGLINEAR command.

Options

Cell Weights. You can specify cell weights (such as structural zero indicators) for the model with the CSTRUCTURE subcommand.

Linear Combinations. You can compute linear combinations of observed cell frequencies, expected cell frequencies, and adjusted residuals using the GRESID subcommand.

Generalized Log-Odds Ratios. You can specify contrast variables on the GLOR subcommand and test whether the generalized log-odds ratio equals 0.

Model Assumption. You can specify POISSON or MULTINOMIAL on the MODEL subcommand to request the Poisson loglinear model or the product multinomial loglinear model.

Tuning the Algorithm. You can control the values of algorithm-tuning parameters with the CRITERIA subcommand.

Output Display. You can control the output display with the PRINT subcommand.

Optional Plots. You can request plots of adjusted or deviance residuals against observed and expected counts, or normal plots and detrended normal plots of adjusted or deviance residuals using the PLOT subcommand.

Basic Specification

The basic specification is one or more factor variables that define the tabulation. By default, GENLOG assumes a Poisson distribution and estimates the saturated model. Default output includes the factors or effects, their levels, and any labels; observed and expected frequencies and percentages for each factor and code; and residuals, adjusted residuals, and deviance residuals.

Limitations

- Maximum 10 factor variables (dependent *and* independent).
- Maximum 200 covariates.

Subcommand Order

- The variable specification must come first.
- Subcommands can be specified in any order.
- When multiple subcommands are specified, only the last specification takes effect.

Example

```
GENLOG DPREF RACE CAMP.
```

- *DPREF*, *RACE*, and *CAMP* are categorical variables.
- This is a general loglinear model because no BY keyword appears.
- The design defaults to a saturated model that includes all main effects and two-way and three-way interaction effects.

Example

```
GENLOG GSLEVEL EDUC SEX
  /DESIGN=GSLEVEL EDUC SEX.
```

- *GSLEVEL*, *EDUC*, and *SEX* are categorical variables.
- DESIGN specifies a model with main effects only.

Variable List

The variable list specifies the variables to be included in the model. GENLOG analyzes two classes of variables—categorical and continuous. Categorical variables are used to define the cells of the table. Continuous variables are used as cell covariates.

- The list of categorical variables must be specified first. Categorical variables must be numeric.
- Continuous variables can be specified only after the WITH keyword following the list of categorical variables.
- To specify a logit model, use the keyword BY (see "Logit Model" below). A variable list without the keyword BY generates a general loglinear model.
- A variable can be specified only once in the variable list—as a dependent variable immediately following GENLOG, as an independent variable following the keyword BY, or as a covariate following the keyword WITH.
- No range needs to be specified for categorical variables.

Logit Model

The logit model examines the relationships between dependent and independent factor variables.

- To separate the independent variables from the dependent variables in a logit model, use the keyword BY. The categorical variables preceding BY are the dependent variables; the categorical variables following BY are the independent variables.
- Up to 10 variables can be specified, including both dependent and independent variables.
- For the logit model, you must specify MULTINOMIAL on the MODEL subcommand.
- GENLOG displays an analysis of dispersion and two measures of association—entropy and concentration. These measures are discussed in Haberman (1982) and can be used to

quantify the magnitude of association among the variables. Both are proportional-reduction-in-error measures. The entropy statistic is analogous to Theil's entropy measure, while the concentration statistic is analogous to Goodman and Kruskal's tau-*b*. Both statistics measure the strength of association between the dependent variable and the independent variable set.

Example

```
GENLOG  GSLEVEL BY EDUC SEX
  /MODEL=MULTINOMIAL
  /DESIGN=GSLEVEL, GSLEVEL BY EDUC, GSLEVEL BY SEX.
```

- Keyword BY on the variable list specifies a logit model in which *GSLEVEL* is the dependent variable and *EDUC* and *SEX* are the independent variables.
- A logit model is multinomial.
- DESIGN specifies a model that can test for the absence of the joint effect of *SEX* and *EDUC* on *GSLEVEL*.

Cell Covariates

- Continuous variables can be used as covariates. When used, the covariates must be specified after the WITH keyword following the list of categorical variables.
- A variable cannot be named as both a categorical variable and a cell covariate.
- To enter cell covariates into a model, the covariates must be specified on the DESIGN subcommand.
- Cell covariates are not applied on a case-by-case basis. The weighted covariate mean for a cell is applied to that cell.

Example

```
GENLOG DPREF RACE CAMP WITH X
  /DESIGN=DPREF RACE CAMP X.
```

- Variable *X* is a continuous variable specified as a cell covariate. Cell covariates must be specified after the keyword WITH following the variable list. No range is defined for cell covariates.
- To include the cell covariate in the model, variable *X* is specified on DESIGN.

CSTRUCTURE Subcommand

CSTRUCTURE specifies the variable that contains values for computing cell weights, such as structural zero indicators. By default, cell weights are equal to 1.

- The specification must be a numeric variable.
- Variables specified as dependent or independent variables in the variable list cannot be specified on CSTRUCTURE.
- Cell weights are not applied on a case-by-case basis. The weighted mean for a cell is applied to that cell.

- CSTRUCTURE can be used to impose structural, or *a priori*, zeros on the model. This feature is useful in specifying a quasi-symmetry model and in excluding cells from entering into estimation.
- If multiple CSTRUCTURE subcommands are specified, the last specification takes effect.

Example

```
COMPUTE  CWT=(HUSED NE WIFED).
GENLOG HUSED WIFED WITH DISTANCE
  /CSTRUCTURE=CWT
  /DESIGN=HUSED WIFED DISTANCE.
```

- The Boolean expression assigns *CWT* the value of 1 when *HUSED* is not equal to *WIFED*, and the value of 0 otherwise.
- CSTRUCTURE imposes structural zeros on the diagonal of the symmetric crosstabulation.

GRESID Subcommand

GRESID (Generalized Residual) calculates linear combinations of observed and expected cell frequencies as well as simple, standardized, and adjusted residuals.

- The variables specified must be numeric, and they must contain coefficients of the desired linear combinations.
- Variables specified as dependent or independent variables in the variable list cannot be specified on GRESID.
- The generalized residual coefficient is not applied on a case-by-case basis. The weighted coefficient mean of the value for all cases in a cell is applied to that cell.
- Each variable specified on the GRESID subcommand contains a single linear combination.
- If multiple GRESID subcommands are specified, the last specification takes effect.

Example

```
COMPUTE GR_1=(MONTH LE 6).
COMPUTE GR_2=(MONTH GE 7).
GENLOG  MONTH WITH Z
 /GRESID=GR_1 GR_2
 /DESIGN=Z.
```

- The first variable, *GR_1*, combines the first six months into a single effect; the second variable, *GR_2*, combines the rest of the months.
- For each effect, GENLOG displays the observed and expected counts as well as the simple, standardized, and adjusted residuals.

GLOR Subcommand

GLOR (Generalized Log-Odds Ratio) specifies the population contrast variable(s). For each variable specified, GENLOG tests the null hypothesis that the generalized log-odds ratio equals 0 and displays the Wald statistic and the confidence interval. You can specify the level of the confidence interval using the CIN significance level keyword on CRITERIA. By default, the confidence level is 95%.

- The variable sum is 0 for the loglinear model and for each combined level of independent variables for the logit model.
- Variables specified as dependent or independent variables in the variable list cannot be specified on GLOR.
- The coefficient is not applied on a case-by-case basis. The weighted mean for a cell is applied to that cell.
- If multiple GLOR subcommands are specified, the last specification takes effect.

Example

```
GENLOG A B
 /GLOR=COEFF
 /DESIGN=A B.
```

- Variable *COEFF* contains the coefficients of two dichotomous factors *A* and *B*.
- If the weighted cell mean for *COEFF* is 1 when *A* equals *B* and –1 otherwise, this example tests whether the log-odds ratio equals 0, or in this case, whether variables *A* and *B* are independent.

MODEL Subcommand

MODEL specifies the assumed distribution of your data.

- You can specify only one keyword on MODEL. The default is POISSON.
- If more than one MODEL subcommand is specified, the last specification takes effect.

POISSON *The Poisson distribution.* This is the default.

MULTINOMIAL *The multinomial distribution.* For the logit model, you must specify MULTINOMIAL.

CRITERIA Subcommand

CRITERIA specifies the values used in tuning the parameters for the Newton-Raphson algorithm.

- If multiple CRITERIA subcommands are specified, the last specification takes effect.

CONVERGE(n) *Convergence criterion.* Specify a positive value for the convergence criterion. The default is 0.001.

ITERATE(n) *Maximum number of iterations.* Specify an integer. The default number is 20.

DELTA(n) *Cell delta value.* Specify a non-negative value to add to each cell frequency for the first iteration. (For the saturated model, the delta value is added for all iterations.) The default is 0.5. The delta value is used to solve mathematical problems created by 0 observations; if all of your observations are greater than 0, we recommend that you set DELTA to 0.

CIN(n) *Level of confidence interval.* Specify the percentage interval used in the test of generalized log-odds ratios and parameter estimates. The value must be between 50 and 99.99, inclusive. The default is 95.

EPS(n) *Epsilon value used for redundancy checking in design matrix.* Specify a positive value. The default is 0.00000001.

DEFAULT *Default values are used.* DEFAULT can be used to reset all criteria to default values.

Example

```
GENLOG  DPREF BY RACE ORIGIN CAMP
 /MODEL=MULTINOMIAL
 /CRITERIA=ITERATION(50) CONVERGE(.0001).
```

- ITERATION increases the maximum number of iterations to 50.
- CONVERGE lowers the convergence criterion to 0.0001.

PRINT Subcommand

PRINT controls the display of statistics.

- By default, GENLOG displays the frequency table and simple, adjusted, and deviance residuals.
- When PRINT is specified with one or more keywords, only the statistics requested by these keywords are displayed.
- When multiple PRINT subcommands are specified, the last specification takes effect.

The following keywords can be used on PRINT:

FREQ *Observed and expected cell frequencies and percentages.* This is displayed by default.

RESID *Simple residuals.* This is displayed by default.

ZRESID *Standardized residuals.*

ADJRESID *Adjusted residuals.* This is displayed by default.

DEV *Deviance residuals.* This is displayed by default.

DESIGN *The design matrix of the model.* The design matrix corresponding to the specified model is displayed.

ESTIM *The parameter estimates of the model.* The parameter estimates refer to the original categories.

COR *The correlation matrix of the parameter estimates.*

COV *The covariance matrix of the parameter estimates.*

ALL *All available output.*

DEFAULT *FREQ, RESID, ADJRESID, and DEV.* This keyword can be used to reset PRINT to its default setting.

NONE *The design and model information with goodness-of-fit statistics only.* This option overrides all other specifications on the PRINT subcommand.

Example

```
GENLOG A B
 /PRINT=ALL
 /DESIGN=A B.
```

- The DESIGN subcommand specifies a main-effects model, which tests the hypothesis of no interaction. The PRINT subcommand displays all available output for this model.

PLOT Subcommand

PLOT specifies what plots you want displayed. Plots of adjusted residuals against observed and expected counts, and normal and detrended normal plots of the adjusted residuals are displayed if PLOT is not specified or is specified without a keyword. When multiple PLOT subcommands are specified, only the last specification is executed.

DEFAULT *RESID (ADJRESID) and NORMPROB (ADJRESID).* This is the default if PLOT is not specified or is specified with no keyword.

RESID (type) *Plots of residuals against observed and expected counts.* You can specify the type of residuals to plot. ADJRESID plots adjusted residuals; DEV plots deviance residuals. ADJRESID is the default if you do not specify a type.

NORMPROB (type) *Normal and detrended normal plots of the residuals.* You can specify the type of residuals to plot. ADJRESID plots adjusted residuals; DEV plots deviance residuals. ADJRESID is the default if you do not specify a type.

NONE *No plots.*

Example

```
GENLOG  RESPONSE BY SEASON
  /MODEL=MULTINOMIAL
  /PLOT=RESID(ADJRESID,DEV)
  /DESIGN=RESPONSE SEASON(1) BY RESPONSE.
```

- This example requests plots of adjusted and deviance residuals against observed and expected counts.
- Note that if you specify `/PLOT=RESID(ADJRESID) RESID(DEV)`, only the deviance residuals are plotted. The first keyword specification, `RESID(ADJRESID)`, is ignored.

MISSING Subcommand

MISSING controls missing values. By default, GENLOG excludes all cases with system- or user-missing values for any variable. You can specify INCLUDE to include user-missing values.

EXCLUDE *Delete cases with user-missing values.* This is the default if the subcommand is omitted. You can also specify the keyword DEFAULT.

INCLUDE *Include cases with user-missing values.* Only cases with system-missing values are deleted.

Example

```
MISSING VALUES A(0).
GENLOG A B
 /MISSING=INCLUDE
 /DESIGN=B.
```

- Even though 0 was specified as missing, it is treated as a nonmissing category of *A* in this analysis.

SAVE Subcommand

SAVE saves specified temporary variables into the working data file. You can assign a new name to each temporary variable saved.

- The temporary variables you can save include *RESID* (raw residual), *ZRESID* (standardized residual), *ADJRESID* (adjusted residual), *DEV* (deviance residual), and *PRED* (predicted cell frequency). An explanatory label is assigned to each saved variable.
- A temporary variable can be saved only once on a SAVE subcommand.
- To assign a name to a saved temporary variable, specify the new name in parentheses following that temporary variable. The new name must conform to SPSS naming conventions and must be unique in the working data file. The names cannot begin with # or $.
- If you do not specify a variable name in parentheses, GENLOG assigns default names to the saved temporary variables. A default name starts with the first three characters of the name of the saved temporary variable, followed by an underscore and a unique number. For example, *RESID* will be saved as *RES_n*, where *n* is a number incremented each time a default name is assigned to a saved *RESID*.
- The saved variables are pertinent to cells in the contingency table, *not* to individual observations. In the Data Editor, all cases that define one cell receive the same value. To make sense of these values, you need to aggregate the data to obtain cell counts.

Example

```
GENLOG A B
 /SAVE PRED (PREDA_B)
 /DESIGN = A, B.
```

- SAVE saves the predicted values for two independent variables *A* and *B*.
- The saved variable is renamed *PREDA_B* and added to the working data file.

DESIGN Subcommand

DESIGN specifies the model to be fit. If DESIGN is omitted or used with no specifications, the saturated model is produced. The saturated model fits all main effects and all interaction effects.

- Only one design can be specified on the subcommand.
- To obtain main-effects models, name all of the variables listed on the variables specification.
- To obtain interactions, use the keyword BY or an asterisk (*) to specify each interaction, for example, `A BY B` or `C*D`. To obtain the single-degree-of-freedom partition of a specified factor, specify the partition in parentheses following the factor (see the example below).
- To include cell covariates in the model, first identify them on the variable list by naming them after the keyword WITH, and then specify the variable names on DESIGN.
- Effects that involve only independent variables result in redundancy. GENLOG removes these effects from the model.
- If your variable list includes a cell covariate (identified by the keyword WITH), you cannot imply the saturated model by omitting DESIGN or specifying it alone. You need to request the model explicitly by specifying all main effects and interactions on DESIGN.

Example

```
COMPUTE X=MONTH.
GENLOG MONTH WITH X
  /DESIGN X.
```

- This example tests the linear effect of the dependent variable.
- The variable specification identifies *MONTH* as a categorical variable. The keyword WITH identifies *X* as a covariate.
- DESIGN tests the linear effect of *MONTH*.

Example

```
GENLOG A B
  /DESIGN=A.

GENLOG A B
  /DESIGN=A,B.
```

- Both designs specify main-effects models.
- The first design tests the homogeneity of category probabilities for *B*; it fits the marginal frequencies on *A* but assumes that membership in any of the categories of *B* is equiprobable.
- The second design tests the independence of *A* and *B*. It fits the marginals on both *A* and *B*.

Example

```
GENLOG A  B  C
  /DESIGN=A,B,C, A BY B.
```

- This design consists of the *A* main effect, the *B* main effect, the *C* main effect, and the interaction of *A* and *B*.

Example

```
GENLOG A BY B
 /MODEL=MULTINOMIAL
 /DESIGN=A,A BY B(1).
```

- This example specifies single-degree-of-freedom partitions.
- The value 1 following *B* refers to the first category of *B*.

Example

```
GENLOG HUSED WIFED WITH DISTANCE
  /DESIGN=HUSED WIFED DISTANCE.
```

- The continuous variable *DISTANCE* is identified as a cell covariate by the keyword WITH. The cell covariate is then included in the model by naming it on DESIGN.

Example

```
COMPUTE  X=1.
GENLOG  MONTH WITH X
  /DESIGN=X.
```

- This example specifies an equiprobability model.
- The design tests whether the frequencies in the table are equal by using a constant of 1 as a cell covariate.

HILOGLINEAR

```
HILOGLINEAR {varlist} (min,max) [varlist ...]
            {ALL    }

 [/METHOD [= BACKWARD]]

 [/MAXORDER = k]

 [/CRITERIA = [CONVERGE({0.25**})] [ITERATE({20**})] [P({0.05**})]
                        {n     }            {n   }       {prob  }
              [DELTA({0.5**})] [MAXSTEPS({10**})]
                     {d    }             {n   }
              [DEFAULT] ]

 [/CWEIGHT = {varname }]
             {(matrix)}

 [/PRINT = {[FREQ**] [RESID**] [ESTIM**][ASSOCIATION**]}]
           {DEFAULT**                                  }
           {ALL                                        }
           {NONE                                       }

 [/PLOT = [{NONE**                }]
           {DEFAULT               }
           {[RESID] [NORMPROB]    }
           {ALL                   }

 [/MISSING = [{EXCLUDE**}]]
              {INCLUDE  }

 [/DESIGN = effectname effectname*effectname ...]
```

** Default if subcommand or keyword is omitted.

Example

```
HILOGLINEAR V1(1,2) V2(1,2)
  /DESIGN=V1*V2.
```

Overview

HILOGLINEAR fits hierarchical loglinear models to multidimensional contingency tables using an iterative proportional-fitting algorithm. HILOGLINEAR also estimates parameters for saturated models. These techniques are described in Everitt (1977), Bishop et al. (1975), and Goodman (1978). HILOGLINEAR is much more efficient for these models than the LOGLINEAR procedure because HILOGLINEAR uses an iterative proportional-fitting algorithm rather than the Newton-Raphson method used in LOGLINEAR.

Options

Design Specification. You can request automatic model selection using backward elimination with the METHOD subcommand. You can also specify any hierarchical design and request multiple designs using the DESIGN subcommand.

Design Control. You can control the criteria used in the iterative proportional-fitting and model-selection routines with the CRITERIA subcommand. You can also limit the order of effects in the model with the MAXORDER subcommand and specify structural zeros for cells in the tables you analyze with the CWEIGHT subcommand.

Display and Plots. You can select the display for each design with the PRINT subcommand. For saturated models, you can request tests for different orders of effects as well. With the PLOT subcommand, you can request residuals plots or normal probability plots of residuals.

Basic Specification

- The basic specification is a variable list with at least two variables followed by their minimum and maximum values.
- HILOGLINEAR estimates a saturated model for all variables in the analysis.
- By default, HILOGLINEAR displays parameter estimates, measures of partial association, goodness of fit, and frequencies for the saturated model.

Subcommand Order

- The variable list must be specified first.
- Subcommands affecting a given DESIGN must appear before the DESIGN subcommand. Otherwise, subcommands can appear in any order.
- MISSING can be placed anywhere after the variable list.

Syntax Rules

- DESIGN is optional. If DESIGN is omitted or the last specification is not a DESIGN subcommand, a default saturated model is estimated.
- You can specify multiple PRINT, PLOT, CRITERIA, MAXORDER, and CWEIGHT subcommands. The last of each type specified is in effect for subsequent designs.
- PRINT, PLOT, CRITERIA, MAXORDER, and CWEIGHT specifications remain in effect until they are overridden by new specifications on these subcommands.
- You can specify multiple METHOD subcommands, but each one affects only the next design.
- MISSING can be specified only once.

Operations

- HILOGLINEAR builds a contingency table using all variables on the variable list. The table contains a cell for each possible combination of values within the range specified for each variable.
- HILOGLINEAR assumes that there is a category for every integer value in the range of each variable. Empty categories waste space and can cause computational problems. If there

are empty categories, use the RECODE command to create consecutive integer values for categories.

- Cases with values outside the range specified for a variable are excluded.
- If the last subcommand is not a DESIGN subcommand, HILOGLINEAR displays a warning and generates the default model. This is the saturated model unless MAXORDER is specified. This model is in addition to any that are explicitly requested.
- If the model is not saturated (for example, when MAXORDER is less than the number of factors), only the goodness of fit and the observed and expected frequencies are given.
- The display uses the WIDTH subcommand defined on the SET command. If the defined width is less than 132, some portions of the display may be deleted.

Limitations

The HILOGLINEAR procedure cannot estimate all possible frequency models, and it produces limited output for unsaturated models.

- It can estimate only hierarchical loglinear models.
- It treats all table variables as nominal. (You can use LOGLINEAR to fit nonhierarchical models to tables involving variables that are ordinal.)
- It can produce parameter estimates for saturated models only (those with all possible main-effect and interaction terms).
- It can estimate partial associations for saturated models only.
- It can handle tables with no more than 10 factors.

Example

```
HILOGLINEAR V1(1,2) V2(1,2) V3(1,3) V4(1,3)
  /DESIGN=V1*V2*V3, V4.
```

- HILOGLINEAR builds a $2 \times 2 \times 3 \times 3$ contingency table for analysis.
- DESIGN specifies the generating class for a hierarchical model. This model consists of main effects for all four variables, two-way interactions among *V1*, *V2*, and *V3*, and the three-way interaction term *V1* by *V2* by *V3*.

Variable List

The required variable list specifies the variables in the analysis. The variable list must precede all other subcommands.

- Variables must be numeric and have integer values. If a variable has a fractional value, the fractional portion is truncated.
- Keyword ALL can be used to refer to all user-defined variables in the working data file.
- A range must be specified for each variable, with the minimum and maximum values separated by a comma and enclosed in parentheses.

- If the same range applies to several variables, the range can be specified once after the last variable to which it applies.
- If ALL is specified, all variables must have the same range.

METHOD Subcommand

By default, HILOGLINEAR tests the model specified on the DESIGN subcommand (or the default model) and does not perform any model selection. All variables are entered and none are removed. Use METHOD to specify automatic model selection using backward elimination for the next design specified.

- You can specify METHOD alone or with the keyword BACKWARD for an explicit specification.
- When the backward-elimination method is requested, a step-by-step output is displayed regardless of the specification on the PRINT subcommand.
- METHOD affects only the next design.

BACKWARD *Backward elimination.* Perform backward elimination of terms in the model. All terms are entered. Those that do not meet the P criterion specified on the CRITERIA subcommand (or the default P) are removed one at a time.

MAXORDER Subcommand

MAXORDER controls the maximum order of terms in the model estimated for subsequent designs. If MAXORDER is specified, HILOGLINEAR tests a model only with terms of that order or less.

- MAXORDER specifies the highest-order term that will be considered for the next design. MAXORDER can thus be used to abbreviate computations for the BACKWARD method.
- If the integer on MAXORDER is less than the number of factors, parameter estimates and measures of partial association are not available. Only the goodness of fit and the observed and expected frequencies are displayed.
- You can use MAXORDER with backward elimination to find the best model with terms of a certain order or less. This is computationally much more efficient than eliminating terms from the saturated model.

Example

```
HILOGLINEAR V1 V2 V3(1,2)
  /MAXORDER=2
  /DESIGN=V1 V2 V3
  /DESIGN=V1*V2*V3.
```

- HILOGLINEAR builds a $2 \times 2 \times 2$ contingency table for *V1*, *V2*, and *V3*.
- MAXORDER has no effect on the first DESIGN subcommand because the design requested considers only main effects.
- MAXORDER restricts the terms in the model specified on the second DESIGN subcommand to two-way interactions and main effects.

CRITERIA Subcommand

Use the CRITERIA subcommand to change the values of constants in the iterative proportional-fitting and model-selection routines for subsequent designs.

- The default criteria are in effect if the CRITERIA subcommand is omitted (see below).
- You cannot specify the CRITERIA subcommand without any keywords.
- Specify each CRITERIA keyword followed by a criterion value in parentheses. Only those criteria specifically altered are changed.
- You can specify more than one keyword on CRITERIA, and they can be in any order.

DEFAULT *Reset parameters to their default values.* If you have specified criteria other than the defaults for a design, use this keyword to restore the defaults for subsequent designs.

CONVERGE(n) *Convergence criterion.* The default is 10^{-3} times the largest cell size, or 0.25, whichever is larger.

ITERATE(n) *Maximum number of iterations.* The default is 20.

P(n) *Probability for change in chi-square if term is removed.* Specify a value between (but not including) 0 and 1 for the significance level. The default is 0.05. P is in effect only when you request BACKWARD on the METHOD subcommand.

MAXSTEPS(n) *Maximum number of steps for model selection.* Specify an integer between 1 and 99, inclusive. The default is 10.

DELTA(d) *Cell delta value.* The value of delta is added to each cell frequency for the first iteration. It is left in the cells for saturated models only. The default value is 0.5. You can specify any decimal value between 0 and 1 for *d*. HILOGLINEAR does not display parameter estimates or the covariance matrix of parameter estimates if any zero cells (either structural or sampling) exist in the expected table after delta is added.

CWEIGHT Subcommand

CWEIGHT specifies cell weights for a model. CWEIGHT is typically used to specify structural zeros in the table. You can also use CWEIGHT to adjust tables to fit new margins.

- You can specify the name of a variable whose values are cell weights, or provide a matrix of cell weights enclosed in parentheses.
- If you use a variable to specify cell weights, you are allowed only one CWEIGHT subcommand.
- If you specify a matrix, you must provide a weight for every cell in the contingency table, where the number of cells equals the product of the number of values of all variables.
- Cell weights are indexed by the values of the variables in the order in which they are specified on the variable list. The index values of the rightmost variable change the most quickly.
- You can use the notation $n*cw$ to indicate that cell weight *cw* is repeated *n* times in the matrix.

Example

```
HILOGLINEAR V1(1,2) V2(1,2) V3(1,3)
  /CWEIGHT=CELLWGT
  /DESIGN=V1*V2, V2*V3, V1*V3.
```

- This example uses the variable *CELLWGT* to assign cell weights for the table. Only one CWEIGHT subcommand is allowed.

Example

```
HILOGLINEAR V4(1,3) V5(1,3)
  /CWEIGHT=(0 1 1  1 0 1  1 1 0)
  /DESIGN=V4, V5.
```

- The HILOGLINEAR command sets the diagonal cells in the model to structural zeros. This type of model is known as a **quasi-independence model**.
- Because both *V4* and *V5* have three values, weights must be specified for nine cells.
- The first cell weight is applied to the cell in which *V4* is 1 and *V5* is 1; the second weight is applied to the cell in which *V4* is 1 and *V5* is 2; and so on.

Example

```
HILOGLINEAR V4(1,3) V5(1,3)
  /CWEIGHT=(0 3*1 0 3*1 0)
  /DESIGN=V4,V5.
```

- This example is the same as the previous example except that the *n*cw* notation is used.

Example

```
* An Incomplete Rectangular Table

DATA LIST FREE / LOCULAR RADIAL FREQ.
WEIGHT BY FREQ.
BEGIN DATA
1 1 462
1 2 130
1 3 2
1 4 1
2 1 103
2 2 35
2 3 1
2 4 0
3 5 614
3 6 138
3 7 21
3 8 14
3 9 1
4 5 443
4 6 95
4 7 22
4 8 8
4 9 5
END DATA.
HILOGLINEAR LOCULAR (1,4) RADIAL (1,9)
  /CWEIGHT=(4*1 5*0   4*1 5*0   4*0 5*1   4*0 5*1)
  /DESIGN LOCULAR RADIAL.
```

- This example uses aggregated table data as input.
- The DATA LIST command defines three variables. The values of *LOCULAR* and *RADIAL* index the levels of those variables, so that each case defines a cell in the table. The values of *FREQ* are the cell frequencies.
- The WEIGHT command weights each case by the value of the variable *FREQ*. Because each case represents a cell in this example, the WEIGHT command assigns the frequencies for each cell.
- The BEGIN DATA and END DATA commands enclose the inline data.
- The HILOGLINEAR variable list specifies two variables. *LOCULAR* has values 1, 2, 3, and 4. *RADIAL* has integer values 1 through 9.
- The CWEIGHT subcommand identifies a block rectangular pattern of cells that are logically empty. There is one weight specified for each cell of the 36-cell table.
- In this example, the matrix form needs to be used in CWEIGHT because the structural zeros do not appear in the actual data. (For example, there is no case corresponding to *LOCULAR*=1, *RADIAL*=5.)
- The DESIGN subcommand specifies main effects only for *LOCULAR* and *RADIAL*. Lack of fit for this model indicates an interaction of the two variables.
- Because there is no PRINT or PLOT subcommand, HILOGLINEAR produces the default output for an unsaturated model.

PRINT Subcommand

PRINT controls the display produced for the subsequent designs.

- If PRINT is omitted or included with no specifications, the default display is produced.
- If any keywords are specified on PRINT, only output specifically requested is displayed.
- HILOGLINEAR displays Pearson and likelihood-ratio chi-square goodness-of-fit tests for models. For saturated models, it also provides tests that the k-way effects and the k-way and higher-order effects are 0.
- Both adjusted and unadjusted degrees of freedom are displayed for tables with sampling or structural zeros. K-way and higher-order tests use the unadjusted degrees of freedom.
- The unadjusted degrees of freedom are not adjusted for zero cells, and they estimate the upper bound of the true degrees of freedom. These are the same degrees of freedom you would get if all cells were filled.
- The adjusted degrees of freedom are calculated from the number of non-zero-fitted cells minus the number of parameters that would be estimated if all cells were filled (that is, unadjusted degrees of freedom minus the number of zero-fitted cells). This estimate of degrees of freedom may be too low if some parameters do not exist because of zeros.

DEFAULT *Default displays.* This option includes FREQ and RESID output for nonsaturated models, and FREQ, RESID, ESTIM, and ASSOCIATION output for saturated models. For saturated models, the observed and expected frequencies are equal, and the residuals are zeros.

FREQ *Observed and expected cell frequencies.*

RESID *Raw and standardized residuals.*

ESTIM *Parameter estimates for a saturated model.*

ASSOCIATION *Partial associations.* You can request partial associations of effects only when you specify a saturated model. This option is computationally expensive for tables with many factors.

ALL *All available output.*

NONE *Design information and goodness-of-fit statistics only.* Use of this option overrides all other specifications on PRINT.

PLOT Subcommand

Use PLOT to request residuals plots.

- If PLOT is included without specifications, standardized residuals and normal probability plots are produced.
- No plots are displayed for saturated models.
- If PLOT is omitted, no plots are produced.

RESID *Standardized residuals by observed and expected counts.*

NORMPLOT *Normal probability plots of adjusted residuals.*

NONE *No plots.* Specify NONE to suppress plots requested on a previous PLOT subcommand. This is the default if PLOT is omitted.

DEFAULT *Default plots.* Includes RESID and NORMPLOT. This is the default when PLOT is specified without keywords.

ALL *All available plots.*

MISSING Subcommand

By default, a case with either system-missing or user-missing values for any variable named on the HILOGLINEAR variable list is omitted from the analysis. Use MISSING to change the treatment of cases with user-missing values.

- MISSING can be named only once and can be placed anywhere following the variable list.
- MISSING cannot be used without specifications.
- A case with a system-missing value for any variable named on the variable list is always excluded from the analysis.

EXCLUDE *Delete cases with missing values.* This is the default if the subcommand is omitted. You can also specify keyword DEFAULT.

INCLUDE *Include user-missing values as valid.* Only cases with system-missing values are deleted.

DESIGN Subcommand

By default, HILOGLINEAR uses a saturated model that includes all variables on the variable list. The model contains all main effects and interactions for those variables. Use DESIGN to specify a different generating class for the model.

- If DESIGN is omitted or included without specifications, the default model is estimated. When DESIGN is omitted, SPSS issues a warning message.
- To specify a design, list the highest-order terms, using variable names and asterisks (*) to indicate interaction effects.
- In a hierarchical model, higher-order interaction effects imply lower-order interaction and main effects. `V1*V2*V3` implies the three-way interaction *V1* by *V2* by *V3*, two-way interactions *V1* by *V2*, *V1* by *V3*, and *V2* by *V3*, and main effects for *V1*, *V2*, and *V3*. The highest-order effects to be estimated are the generating class.
- Any PRINT, PLOT, CRITERIA, METHOD, and MAXORDER subcommands that apply to a DESIGN subcommand must appear before it.
- All variables named on DESIGN must be named or implied on the variable list.
- You can specify more than one DESIGN subcommand. One model is estimated for each DESIGN subcommand.
- If the last subcommand on HILOGLINEAR is not DESIGN, the default model will be estimated in addition to models explicitly requested. SPSS issues a warning message for a missing DESIGN subcommand.

KM

```
KM varname [BY factor varname]

   /STATUS = varname [EVENT](vallist) [LOST(vallist)]

   [/STRATA = varname]

   [/PLOT = {NONE**                              }]
            {[SURVIVAL][LOGSURV][HAZARD]}

   [/ID  = varname]

   [/PRINT = [TABLE**][MEAN**][NONE]]

   [/PERCENTILE = [(]{25, 50, 75 }[)]]
                     {value list }

   [/TEST = [LOGRANK**][BRESLOW][TARONE]]

   [/COMPARE = [{OVERALL**}][{POOLED**}]]
                {PAIRWISE }  {STRATA  }

   [/TREND = [(METRIC)]]

   [/SAVE = tempvar[(newvar)],...]
```

**Default if subcommand or keyword is omitted.

Temporary variables created by Kaplan-Meier are:

SURVIVAL
HAZARD
SE
CUMEVENT

Example:

```
KM LENGTH BY SEXRACE
 /STATUS=EMPLOY  EVENT (1) LOST (2)
 /STRATA=LOCATION.
```

Overview

KM (alias K-M) uses the Kaplan-Meier (product-limit) technique to describe and analyze the length of time to the occurrence of an event, often known as **survival time**. KM is similar to SURVIVAL in that it produces nonparametric estimates of the survival functions. However, instead of dividing the period of time under examination into arbitrary intervals, KM evaluates the survival function at the observed event times. For analysis of survival times with covariates, including time-dependent covariates, see the COXREG command.

Options

KM Tables. You can include one factor variable on the KM command. A KM table is produced for each level of the factor variable. You can also suppress the KM tables in the output with the PRINT subcommand.

Survival Status. You can specify the code(s) indicating that an event has occurred as well as code(s) for cases lost to follow-up using the STATUS subcommand.

Plots. You can plot the survival functions on a linear or log scale or plot the hazard function for each combination of factor and stratum with the PLOT subcommand.

Test Statistics. When a factor variable is specified, you can specify one or more tests of equality of survival distributions for the different levels of the factor using the TEST subcommand. You can also specify a trend metric for the requested tests with the TREND subcommand.

Display ID and Percentiles. You can specify an ID variable on the ID subcommand to identify each case. You can also request display of percentiles in the output with the PERCENTILES subcommand.

Comparisons. When a factor variable is specified, you can use the COMPARE subcommand to compare the different levels of the factor, either pairwise or across all levels, and either pooled across all strata or within a stratum.

Add New Variables to Working Data File. You can save new variables appended to the end of the working data file with the SAVE subcommand.

Basic Specification

- The basic specification requires a survival variable and the STATUS subcommand naming a variable that indicates whether the event occurred.
- The basic specification prints one survival table followed by the mean and median survival time with standard errors and 95% confidence intervals.

Subcommand Order

- The survival variable and the factor variable (if there is one) must be specified first.
- Remaining subcommands can be specified in any order.

Syntax Rules

- Only one survival variable can be specified. To analyze multiple survival variables, use multiple KM commands.
- Only one factor variable can be specified following the BY keyword. If you have multiple factors, use the transformation language to create a single factor variable before invoking KM.
- Only one status variable can be listed on the STATUS subcommand. You must specify the value(s) indicating that the event occurred.
- Only one variable can be specified on the STRATA subcommand. If you have more than one stratum, use the transformation language to create a single variable to specify on the STRATA subcommand.

Operations

- KM deletes all cases that have negative values for the survival variable.
- KM estimates the survival function and associated statistics for each combination of factor and stratum.
- Three statistics can be computed to test the equality of survival functions across factor levels within a stratum or across all factor levels while controlling for strata. The statistics are the log rank (Mantel-Cox), the generalized Wilcoxon (Breslow), and the Tarone-Ware tests.
- When the PLOTS subcommand is specified, KM produces one plot of survival functions for each stratum, with all factor levels represented by different symbols or colors.

Limitations

- Maximum 35 factor levels (symbols) can appear in a plot.

Example

```
KM LENGTH BY SEXRACE
 /STATUS=EMPLOY EVENT (1) LOST (2)
 /STRATA=LOCATION.
```

- Survival analysis is used to examine the length of unemployment. The survival variable *LENGTH* contains the number of months a subject is unemployed. The factor variable *SEXRACE* combines sex and race factors.
- A value of 1 on the variable *EMPLOY* indicates the occurrence of the event (employment). All other observed cases are censored. A value of 2 on *EMPLOY* indicates cases lost to follow-up. Cases with other values for *EMPLOY* are known to have remained unemployed during the course of the study. KM separates the two types of censored cases in the KM table if LOST is specified.
- For each combination of *SEXRACE* and *LOCATION*, one KM table is produced, followed by the mean and median survival time with standard errors and confidence intervals.

Survival and Factor Variables

You must identify the survival and factor variables for the analysis.

- The minimum specification is one, and only one, survival variable.
- Only one factor variable can be specified using the BY keyword. If you have more than one factor, create a new variable combining all factors. There is no limit to the factor levels.

Example

```
DO IF SEX = 1.
+ COMPUTE SEXRACE = RACE.
ELSE.
+ COMPUTE SEXRACE = RACE + SEX.
END IF.
KM LENGTH BY SEXRACE
 /STATUS=EMPLOY EVENT (1) LOST (2).
```

- The two control variables, *SEX* and *RACE*, each with two values, 1 and 2, are combined into one factor variable, *SEXRACE*, with four values, 1 to 4.
- KM specifies *LENGTH* as the survival variable and *SEXRACE* as the factor variable.
- One KM table is produced for each factor level.

STATUS Subcommand

To determine whether the terminal event has occurred for a particular observation, KM checks the value of a status variable. STATUS lists the status variable and the code(s) for the occurrence of the event. The code(s) for cases lost to follow-up can also be specified.

- Only one status variable can be specified. If multiple STATUS subcommands are specified, KM uses the last specification and displays a warning.
- The keyword EVENT is optional, but the value list in parentheses must be specified. Use EVENT for clarity's sake, especially when LOST is specified.
- The value list must be enclosed in parentheses. All cases with non-negative times that do not have a code within the range specified after EVENT are classified as **censored cases**—that is, cases for which the event has not yet occurred.
- The keyword LOST and the following value list are optional. LOST cannot be omitted if the value list for lost cases is specified.
- When LOST is specified, all cases with non-negative times that have a code within the specified value range are classified as lost to follow-up. Cases lost to follow-up are treated as censored in the analysis, and the statistics do not change, but the two types of censored cases are listed separately in the KM table.
- The value lists on EVENT or LOST can be one value, a list of values separated by blanks or commas, a range of values using the keyword THRU, or a combination.
- The status variable can be either numeric or string. If a string variable is specified, the EVENT or LOST values must be enclosed in apostrophes, and the keyword THRU cannot be used.

Example

```
KM LENGTH BY SEXRACE
 /STATUS=EMPLOY  EVENT (1) LOST (3,5 THRU 8).
```

- STATUS specifies that *EMPLOY* is the status variable.
- A value of 1 for *EMPLOY* means that the event (employment) occurred for the case.
- Values of 3 and 5 through 8 for *EMPLOY* mean that contact was lost with the case. The different values code different causes for the loss of contact.
- The summary table in the output includes columns for number lost and percentage lost, as well as for number censored and percentage censored.

STRATA Subcommand

STRATA identifies a **stratification variable**—that is, a variable whose values are used to form subgroups (strata) within the categories of the factor variable. Analysis is done within each level of the strata variable for each factor level, and estimates are pooled over strata for an overall comparison of factor levels.

- The minimum specification is the subcommand keyword with one, and only one, variable name.
- If you have more than one strata variable, create a new variable to combine the levels on separate variables before invoking the KM command.
- There is no limit to the number of levels for the strata variable.

Example

```
KM LENGTH BY SEXRACE
 /STATUS=EMPLOY EVENT (1) LOST (3,5 THRU 8)
 /STRATA=LOCATION.
```

- STRATA specifies *LOCATION* as the stratification variable. Analysis of the length of unemployment is done for each location within each sex and race subgroup.

PLOT Subcommand

PLOT plots the cumulative survival distribution on a linear or logarithmic scale or plots the cumulative hazard function. A separate plot with all factor levels is produced for each stratum. Each factor level is represented by a different symbol or color. Censored cases are indicated by markers.

- When PLOT is omitted, no plots are produced. The default is NONE.
- When PLOT is specified without a keyword, the default is SURVIVAL. A plot of survival functions for each stratum is produced.
- To request specific plots, specify, following the PLOT subcommand, any combination of the keywords defined below.
- Multiple keywords can be used on the PLOT subcommand, each requesting a different plot. The effect is cumulative.

NONE *Suppress all plots.* NONE is the default if PLOT is omitted.

SURVIVAL *Plot the cumulative survival distribution on a linear scale.* SURVIVAL is the default when PLOT is specified without a keyword.

LOGSURV *Plot the cumulative survival distribution on a logarithmic scale.*

HAZARD *Plot the cumulative hazard function.*

Example

```
KM LENGTH BY SEXRACE
 /STATUS=EMPLOY EVENT (1) LOST (3,5 THRU 8)
 /STRATA=LOCATION
 /PLOT = SURVIVAL HAZARD.
```

- PLOT produces one plot of the cumulative survival distribution on a linear scale and one plot of the cumulative hazard rate for each value of *LOCATION*.

ID Subcommand

ID specifies a variable used for labeling cases. If the ID variable is a string, KM uses the string values as case identifiers in the KM table. If the ID variable is numeric, KM uses value labels or numeric values if value labels are not defined.

- ID is the first column of the KM table displayed for each combination of factor and stratum.
- If a string value or a value label exceeds 20 characters in width, KM truncates the case identifier and displays a warning.

PRINT Subcommand

By default, KM prints survival tables and the mean and median survival time with standard errors and confidence intervals if PRINT is omitted. If PRINT is specified, only the specified keyword is in effect. Use PRINT to suppress tables or the mean statistics.

TABLE *Print the KM tables.* If PRINT is not specified, TABLE, together with MEAN, is the default. Specify TABLE on PRINT to suppress the mean statistics.

MEAN *Print the mean statistics.* KM prints the mean and median survival time with standard errors and confidence intervals. If PRINT is not specified, MEAN, together with TABLE, is the default. Specify MEAN on PRINT to suppress the KM tables.

NONE *Suppress both the KM tables and the mean statistics.* Only plots and comparisons are printed.

Example

```
KM LENGTH BY SEXRACE
 /STATUS=EMPLOY EVENT (1) LOST (3,5 THRU 8)
 /STRATA=LOCATION
 /PLOT=SURVIVAL HAZARD
 /PRINT=NONE.
```

- `PRINT=NONE` suppresses both the KM tables and the mean statistics.

PERCENTILES Subcommand

PERCENTILES displays percentiles for each combination of factor and stratum. Percentiles are not displayed without the PERCENTILES subcommand. If the subcommand is specified without a value list, the default is 25, 50, and 75 for quartile display. You can specify any values between 0 and 100.

TEST Subcommand

TEST specifies the test statistic to use for testing the equality of survival distributions for the different levels of the factor.

- TEST is valid only when a factor variable is specified. If no factor variable is specified, KM issues a warning and TEST is not executed.
- If TEST is specified without a keyword, the default is LOGRANK. If a keyword is specified on TEST, only the specified test is performed.
- Each of the test statistics has a chi-square distribution with one degree of freedom.

LOGRANK *Perform the log rank (Mantel-Cox) test.*

BRESLOW *Perform the Breslow (generalized Wilcoxon) test.*

TARONE *Perform the Tarone-Ware test.*

COMPARE Subcommand

COMPARE compares the survival distributions for the different levels of the factor. Each of the keywords specifies a different method of comparison.

- COMPARE is valid only when a factor variable is specified. If no factor variable is specified, KM issues a warning and COMPARE is not executed.
- COMPARE uses whatever tests are specified on the TEST subcommand. If no TEST subcommand is specified, the log rank test is used.
- If COMPARE is not specified, the default is OVERALL and POOLED. All factor levels are compared across strata in a single test. The test statistics are displayed after the summary table at the end of output.
- Multiple COMPARE subcommands can be specified to request different comparisons.

OVERALL *Compare all factor levels in a single test.* OVERALL, together with POOLED, is the default when COMPARE is not specified.

PAIRWISE *Compare each pair of factor levels.* KM compares all distinct pairs of factor levels.

POOLED *Pool the test statistics across all strata.* The test statistics are displayed after the summary table for all strata. POOLED, together with OVERALL, is the default when COMPARE is not specified.

STRATA *Compare the factor levels for each stratum.* The test statistics are displayed for each stratum separately.

- If a factor variable has different levels across strata, you cannot request a pooled comparison. If you specify POOLED on COMPARE, KM displays a warning and ignores the request.

Example

```
KM LENGTH BY SEXRACE
 /STATUS=EMPLOY EVENT (1) LOST (3,5 THRU 8)
 /STRATA=LOCATION
 /TEST = BRESLOW
 /COMPARE = PAIRWISE.
```

- TEST specifies the Breslow test.
- COMPARE uses the Breslow test statistic to compare all distinct pairs of *SEXRACE* values and pools the test results over all strata defined by *LOCATION*.
- Test statistics are displayed at the end of output for all strata.

TREND Subcommand

TREND specifies that there is a trend across factor levels. This information is used when computing the tests for equality of survival functions specified on the TEST subcommand.

- The minimum specification is the subcommand keyword by itself. The default metric is chosen as follows:

  if g is even,

  $$(-(g-1), \ldots, -3, -1, 1, 3, \ldots, (g-1))$$

  otherwise,

  $$\left(-\frac{(g-1)}{2}, \ldots, -1, 0, 1, \ldots, \frac{(g-1)}{2}\right)$$

  where g is the number of levels for the factor variable.
- If TREND is specified but COMPARE is not, KM performs the default log rank test with the trend metric for an OVERALL POOLED comparison.
- If the metric specified on TREND is longer than required by the factor levels, KM displays a warning and ignores extra values.

Example

```
KM LENGTH BY SEXRACE
 /STATUS=EMPLOY EVENT (1) LOST (3,5 THRU 8)
 /STRATA=LOCATION
 /TREND.
```

- TREND is specified by itself. KM uses the default metric. Since *SEXRACE* has four levels, the default is (–3,–1, 1, 3).
- Even though no TEST or COMPARE subcommand is specified, KM performs the default log rank test with the trend metric and does a default OVERALL POOLED comparison.

SAVE Subcommand

SAVE saves the temporary variables created by KM. The following temporary variables can be saved:

SURVIVAL *Survival function evaluated at current case.*

SE *Standard error of the survival function.*

HAZARD *Cumulative hazard function evaluated at current case.*

CUMEVENT *Cumulative number of events.*

- To specify variable names for the new variables, assign the new names in parentheses following each temporary variable name.
- Assigned variable names must be unique in the working data file. Scratch or system variable names cannot be used (that is, variable names cannot begin with # or $).
- If new variable names are not specified, KM generates default names. The default name is composed of the first three characters of the name of the temporary variable (two for *SE*), followed by an underscore and a number to make it unique.
- A temporary variable can be saved only once on the same SAVE subcommand.

Example

```
KM LENGTH BY SEXRACE
 /STATUS=EMPLOY EVENT (1) LOST (3,5 THRU 8)
 /STRATA=LOCATION
 /SAVE SURVIVAL HAZARD.
```

- KM saves cumulative survival and cumulative hazard rates in two new variables, *SUR_1* and *HAZ_1*, provided that neither name exists in the working data file. If one does, the numeric suffixes will be incremented to make a distinction.

LOGISTIC REGRESSION

```
LOGISTIC REGRESSION [VARIABLES =] dependent var
        [WITH independent varlist [BY var [BY var] ... ]]

[/CATEGORICAL = var1, var2, ... ]

[/CONTRAST (categorical var) = [{DEVIATION [(refcat)]     }]]
                                {SIMPLE [(refcat)]        }
                                {DIFFERENCE               }
                                {HELMERT                  }
                                {REPEATED                 }
                                {POLYNOMIAL[({1,2,3...})]}
                                             {metric   }
                                {SPECIAL (matrix)         }
                                {INDICATOR [(refcat)]     }

 [/METHOD = {ENTER**             }    [{ALL     }]]
            {BSTEP [{COND}]}     {varlist}
                    {LR   }
                    {WALD}
            {FSTEP [{COND}]}
                    {LR   }
                    {WALD}

 [/SELECT = {ALL**                    }]
            {varname relation value}

 [/{NOORIGIN**}]
   {ORIGIN    }

 [/ID = [variable]]

 [/PRINT = [DEFAULT**] [SUMMARY] [CORR] [ALL] [ITER [({1})]]]
                                                     {n}

 [/CRITERIA = [BCON ({0.001**})] [ITERATE({20**})] [LCON({0.01**})]
                     {value   }           {n    }        {value }

            [PIN({0.05**})] [POUT({0.10**})] [EPS({.00000001**})]]
                 {value  }         {value  }       {value       }

 [/CLASSPLOT]

 [/MISSING = {EXCLUDE **}]
             {INCLUDE   }

 [/CASEWISE = [tempvarlist]   [OUTLIER({2**   })]]
                                       {value}

 [/SAVE = tempvar[(newname)] tempvar[(newname)]...]

 [/EXTERNAL]
```

** Default if subcommand or keyword is omitted.

Temporary variables created by LOGISTIC REGRESSION are:

PRED	*LEVER*	*COOK*
PGROUP	*LRESID*	*DFBETA*
RESID	*SRESID*	
DEV	*ZRESID*	

Example:

```
LOGISTIC REGRESSION PROMOTED WITH AGE, JOBTIME, JOBRATE.
```

Overview

LOGISTIC REGRESSION regresses a dichotomous dependent variable on a set of independent variables (Aldrich & Nelson, 1984; Fox, 1984; Hosmer & Lemeshow, 1989; McCullagh & Nelder, 1989; Agresti, 1990). Categorical independent variables are replaced by sets of contrast variables, each set entering and leaving the model in a single step.

Options

Processing of Independent Variables. You can specify which independent variables are categorical in nature on the CATEGORICAL subcommand. You can control treatment of categorical independent variables by the CONTRAST subcommand. Seven methods are available for entering independent variables into the model. You can specify any one of them on the METHOD subcommand. You can also use the keyword BY between variable names to enter interaction terms.

Selecting Cases. You can use the SELECT subcommand to define subsets of cases to be used in estimating a model.

Regression through the Origin. You can use the ORIGIN subcommand to exclude a constant term from a model.

Specifying Termination and Model-building Criteria. You can further control computations when building the model by specifying criteria on the CRITERIA subcommand.

Adding New Variables to the Working Data File. You can save the residuals, predicted values, and diagnostics generated by LOGISTIC REGRESSION in the working data file.

Output. You can use the PRINT subcommand to print optional output, use the CASEWISE subcommand to request analysis of residuals, and use the ID subcommand to specify a variable whose values or value labels identify cases in output. You can request plots of the actual and predicted values for each case with the CLASSPLOT subcommand.

Basic Specification

- The minimum specification is the VARIABLES subcommand with one dichotomous dependent variable. You must specify a list of independent variables either following the keyword WITH on the VARIABLES subcommand or on a METHOD subcommand.
- The default output includes goodness-of-fit tests for the model and a classification table for the predicted and observed group memberships. The regression coefficient, standard error of the regression coefficient, Wald statistic and its significance level, and a multiple correlation coefficient adjusted for the number of parameters (Atkinson, 1980) are displayed for each variable in the equation.

Subcommand Order

- Subcommands can be named in any order. If the VARIABLES subcommand is not specified first, a slash (/) must precede it.
- The ordering of METHOD subcommands determines the order in which models are estimated. Different sequences may result in different models.

Syntax Rules

- Only one dependent variable can be specified for each LOGISTIC REGRESSION.
- Any number of independent variables may be listed. The dependent variable may not appear on this list.
- The independent variable list is required if any of the METHOD subcommands are used without a variable list or if the METHOD subcommand is not used. The keyword TO cannot be used on any variable list.
- If you specify the keyword WITH on the VARIABLES subcommand, all independent variables must be listed.
- If the keyword WITH is used on the VARIABLES subcommand, interaction terms do not have to be specified on the variable list, but the individual variables that make up the interactions must be listed.
- Multiple METHOD subcommands are allowed.
- The minimum truncation for this command is LOGI REG.

Operations

- Independent variables specified on the CATEGORICAL subcommand are replaced by sets of contrast variables. In stepwise analyses, the set of contrast variables associated with a categorical variable is entered or removed from the model as a single step.
- Independent variables are screened to detect and eliminate redundancies.
- If the linearly dependent variable is one of a set of contrast variables, the set will be reduced by the redundant variable or variables. A warning will be issued, and the reduced set will be used.
- For the forward stepwise method, redundancy checking is done when a variable is to be entered into the model.
- When backward stepwise or direct-entry methods are requested, all variables for each METHOD subcommand are checked for redundancy before that analysis begins.

Limitations

- The dependent variable must be dichotomous for each split-file group. Specifying a dependent variable with more or less than two nonmissing values per split-file group will result in an error.

Example

```
LOGISTIC REGRESSION PASS WITH GPA, MAT, GRE.
```

- *PASS* is specified as the dependent variable.
- *GPA*, *MAT*, and *GRE* are specified as independent variables.
- LOGISTIC REGRESSION produces the default output for the logistic regression of *PASS* on *GPA*, *MAT*, and *GRE*.

VARIABLES Subcommand

VARIABLES specifies the dependent variable and, optionally, all independent variables in the model. The dependent variable appears first on the list and is separated from the independent variables by the keyword WITH.

- One VARIABLES subcommand is allowed for each Logistic Regression procedure.
- The dependent variable must be dichotomous—that is, it must have exactly two values other than system-missing and user-missing values for each split-file group.
- The dependent variable may be a string variable if its two values can be differentiated by their first eight characters.
- You can indicate an interaction term on the variable list by using the keyword BY to separate the individual variables.
- If all METHOD subcommands are accompanied by independent variable lists, the keyword WITH and the list of independent variables may be omitted.
- If the keyword WITH is used, *all* independent variables must be specified. For interaction terms, only the individual variable names that make up the interaction (for example, `X1, X2`) need to be specified. Specifying the actual interaction term (for example, `X1 BY X2`) on the VARIABLES subcommand is optional if you specify it on a METHOD subcommand.

Example

```
LOGISTIC REGRESSION PROMOTED WITH AGE,JOBTIME,JOBRATE,
    AGE BY JOBTIME.
```

- *PROMOTED* is specified as the dependent variable.
- *AGE*, *JOBTIME*, *JOBRATE*, and the interaction *AGE* by *JOBTIME* are specified as the independent variables.
- Because no METHOD is specified, all three single independent variables and the interaction term are entered into the model.
- LOGISTIC REGRESSION produces the default output.

CATEGORICAL Subcommand

CATEGORICAL identifies independent variables that are nominal or ordinal. Variables that are declared to be categorical are automatically transformed to a set of contrast variables as specified on the CONTRAST subcommand. If a variable coded as $0-1$ is declared as categorical, its coding scheme will be changed to deviation contrasts by default.

- Independent variables not specified on CATEGORICAL are assumed to be at least interval level, except for string variables.
- Any variable specified on CATEGORICAL is ignored if it does not appear either after WITH on the VARIABLES subcommand or on any METHOD subcommand.
- Variables specified on CATEGORICAL are replaced by sets of contrast variables. If the categorical variable has n distinct values, there will be $n-1$ contrast variables generated. The set of contrast variables associated with a categorical variable is entered or removed from the model as a step.
- If any one of the variables in an interaction term is specified on CATEGORICAL, the interaction term is replaced by contrast variables.
- All string variables are categorical. Only the first eight characters of each value of a string variable are used in distinguishing between values. Thus, if two values of a string variable are identical for the first eight characters, the values are treated as though they were the same.

Example

```
LOGISTIC REGRESSION PASS WITH GPA, GRE, MAT, CLASS, TEACHER
 /CATEGORICAL = CLASS,TEACHER.
```

- The dichotomous dependent variable *PASS* is regressed on the interval-level independent variables *GPA*, *GRE*, and *MAT* and the categorical variables *CLASS* and *TEACHER*.

CONTRAST Subcommand

CONTRAST specifies the type of contrast used for categorical independent variables. The interpretation of the regression coefficients for categorical variables depends on the contrasts used. The default is DEVIATION. The categorical independent variable is specified in parentheses following CONTRAST. The closing parenthesis is followed by one of the contrast-type keywords.

- If the categorical variable has n values, there will be $n-1$ rows in the contrast matrix. Each contrast matrix is treated as a set of independent variables in the analysis.
- Only one categorical independent variable can be specified per CONTRAST subcommand, but multiple CONTRAST subcommands can be specified.

The following contrast types are available. See Finn (1974) and Kirk (1982) for further information on a specific type. For illustration of contrast types, see Appendix B.

DEVIATION(refcat) *Deviations from the overall effect.* This is the default. The effect for each category of the independent variable except one is compared to the overall effect. Refcat is the category for which parameter estimates are not displayed (they must be calculated from the others). By default, refcat is the last category. To omit a category other than the last, specify the sequence number of the omitted category (which is not necessarily the same as its value) in parentheses after the keyword DEVIATION.

SIMPLE(refcat) *Each category of the independent variable except the last is compared to the last category.* To use a category other than the last as the omitted reference category, specify its sequence number (which is not necessarily the same as its value) in parentheses following the keyword SIMPLE.

DIFFERENCE *Difference or reverse Helmert contrasts.* The effects for each category of the independent variable except the first are compared to the mean effects of the previous categories.

HELMERT *Helmert contrasts.* The effects for each category of the independent variable except the last are compared to the mean effects of subsequent categories.

POLYNOMIAL(metric) *Polynomial contrasts.* The first degree of freedom contains the linear effect across the categories of the independent variable, the second contains the quadratic effect, and so on. By default, the categories are assumed to be equally spaced; unequal spacing can be specified by entering a metric consisting of one integer for each category of the independent variable in parentheses after the keyword POLYNOMIAL. For example, `CONTRAST(STIMULUS)=POLYNOMIAL(1,2,4)` indicates that the three levels of STIMULUS are actually in the proportion 1:2:4. The default metric is always $(1,2,...,k)$, where k categories are involved. Only the relative differences between the terms of the metric matter: (1,2,4) is the same metric as (2,3,5) or (20,30,50) because the difference between the second and third numbers is twice the difference between the first and second in each instance.

REPEATED *Comparison of adjacent categories.* Each category of the independent variable except the first is compared to the previous category.

SPECIAL(matrix) *A user-defined contrast.* After this keyword, a matrix is entered in parentheses with $k-1$ rows and k columns (where k is the number of categories of the independent variable). The rows of the contrast matrix contain the special contrasts indicating the desired comparisons between categories. If the special contrasts are linear combinations of each other, LOGISTIC REGRESSION reports the linear dependency and stops processing. If k rows are entered, the first row is discarded and only the last $k-1$ rows are used as the contrast matrix in the analysis.

INDICATOR(refcat) *Indicator variables.* Contrasts indicate the presence or absence of category membership. By default, refcat is the last category (represented in the contrast matrix as a row of zeros). To omit a category other than the last, specify the sequence number of the omitted category (which is not necessarily the same as its value) in parentheses after the keyword INDICATOR.

Example

```
LOGISTIC REGRESSION PASS WITH GRE, CLASS
 /CATEGORICAL = CLASS
 /CONTRAST(CLASS)=HELMERT.
```

- A logistic regression analysis of the dependent variable *PASS* is performed on the interval independent variable *GRE* and the categorical independent variable *CLASS*.
- *PASS* is a dichotomous variable representing course pass/fail status and *CLASS* identifies whether a student is in one of three classrooms. A HELMERT contrast is requested.

Example

```
LOGISTIC REGRESSION PASS WITH GRE, CLASS
 /CATEGORICAL = CLASS
 /CONTRAST(CLASS)=SPECIAL(2 -1 -1
                          0  1 -1).
```

- In this example, the contrasts are specified with the keyword SPECIAL.

METHOD Subcommand

METHOD indicates how the independent variables enter the model. The specification is the METHOD subcommand followed by a single method keyword. The keyword METHOD can be omitted. Optionally, specify the independent variables and interactions for which the method is to be used. Use the keyword BY between variable names of an interaction term.

- If no variable list is specified or if the keyword ALL is used, all the independent variables following the keyword WITH on the VARIABLES subcommand are eligible for inclusion in the model.
- If no METHOD subcommand is specified, the default method is ENTER.
- Variables specified on CATEGORICAL are replaced by sets of contrast variables. The set of contrast variables associated with a categorical variable is entered or removed from the model as a single step.
- Any number of METHOD subcommands can appear in a Logistic Regression procedure. METHOD subcommands are processed in the order in which they are specified. Each method starts with the results from the previous method. If BSTEP is used, all remaining eligible variables are entered at the first step. All variables are then eligible for entry and removal unless they have been excluded from the METHOD variable list.
- The beginning model for the first METHOD subcommand is either the constant variable (by default or if NOORIGIN is specified) or an empty model (if ORIGIN is specified).

The available METHOD keywords are:

ENTER *Forced entry.* All variables are entered in a single step. This is the default if the METHOD subcommand is omitted.

FSTEP *Forward stepwise.* The variables (or interaction terms) specified on FSTEP are tested for entry into the model one by one, based on the significance level of the score statistic. The variable with the smallest significance less than PIN is entered into the model. After each entry, variables that are already in the model are tested for possible removal, based on the significance of the conditional statistic, the Wald statistic, or the likelihood-ratio criterion. The variable with the largest probability greater than the specified POUT value is removed and the model is reestimated. Variables in the model are then evaluated again for removal. Once no more variables satisfy the removal criterion, covariates not in the model are evaluated for entry. Model building stops when no more variables meet entry or removal criteria, or when the current model is the same as a previous one.

BSTEP *Backward stepwise.* As a first step, the variables (or interaction terms) specified on BSTEP are entered into the model together and are tested for removal one by one. Stepwise removal and entry then follow the same process as described for FSTEP until no more variables meet entry or removal criteria, or when the current model is the same as a previous one.

The statistic used in the test for removal can be specified by an additional keyword in parentheses following FSTEP or BSTEP. If FSTEP or BSTEP is specified by itself, the default is COND.

COND *Conditional statistic.* This is the default if FSTEP or BSTEP is specified by itself.

WALD *Wald statistic.* The removal of a variable from the model is based on the significance of the Wald statistic.

LR *Likelihood ratio.* The removal of a variable from the model is based on the significance of the change in the log likelihood. If LR is specified, the model must be reestimated without each of the variables in the model. This can substantially increase computational time. However, the likelihood-ratio statistic is the best criterion for deciding which variables are to be removed.

Example

```
LOGISTIC REGRESSION PROMOTED WITH AGE JOBTIME JOBRATE RACE SEX AGENCY
 /CATEGORICAL RACE SEX AGENCY
 /METHOD ENTER AGE  JOBTIME
 /METHOD BSTEP (LR) RACE SEX JOBRATE AGENCY.
```

- *AGE*, *JOBTIME*, *JOBRATE*, *RACE*, *SEX*, and *AGENCY* are specified as independent variables. *RACE*, *SEX*, and *AGENCY* are specified as categorical independent variables.
- The first METHOD subcommand enters *AGE* and *JOBTIME* into the model.
- Variables in the model at the termination of the first METHOD subcommand are included in the model at the beginning of the second METHOD subcommand.
- The second METHOD subcommand adds the variables *RACE*, *SEX*, *JOBRATE*, and *AGENCY* to the previous model.
- Backward stepwise logistic regression analysis is then done with only the variables on the BSTEP variable list tested for removal using the LR statistic.
- The procedure continues until all variables from the BSTEP variable list have been removed or the removal of a variable will not result in a decrease in the log likelihood with a probability larger than POUT.

SELECT Subcommand

By default, all cases in the working data file are considered for inclusion in LOGISTIC REGRESSION. Use the optional SELECT subcommand to include a subset of cases in the analysis.

- The specification is either a logical expression or keyword ALL. ALL is the default. Variables named on VARIABLES, CATEGORICAL, or METHOD subcommands cannot appear on SELECT.
- In the logical expression on SELECT, the relation can be EQ, NE, LT, LE, GT, or GE. The variable must be numeric and the value can be any number.
- Only cases for which the logical expression on SELECT is true are included in calculations. All other cases, including those with missing values for the variable named on SELECT, are unselected.
- Diagnostic statistics and classification statistics are reported for both selected and unselected cases.
- Cases deleted from the working data file with the SELECT IF or SAMPLE command are not included among either the selected or unselected cases.

Example

```
LOGISTIC REGRESSION VARIABLES=GRADE WITH GPA,TUCE,PSI
 /SELECT SEX EQ 1 /CASEWISE=RESID.
```

- Only cases with the value 1 for *SEX* are included in the logistic regression analysis.
- Residual values generated by CASEWISE are displayed for both selected and unselected cases.

ORIGIN and NOORIGIN Subcommands

ORIGIN and NOORIGIN control whether or not the constant is included. NOORIGIN (the default) includes a constant term (intercept) in all equations. ORIGIN suppresses the constant term and requests regression through the origin. (NOCONST can be used as an alias for ORIGIN.)

- The only specification is either ORIGIN or NOORIGIN.
- ORIGIN or NOORIGIN can be specified only once per Logistic Regression procedure, and it affects all METHOD subcommands.

Example

```
LOGISTIC REGRESSION VARIABLES=PASS WITH GPA,GRE,MAT /ORIGIN.
```

- ORIGIN suppresses the automatic generation of a constant term.

ID Subcommand

ID specifies a variable whose values or value labels identify the casewise listing. By default, cases are labeled by their case number.

- The only specification is the name of a single variable that exists in the working data file.
- Only the first eight characters of the variable's value labels are used to label cases. If the variable has no value labels, the values are used.
- Only the first eight characters of a string variable are used to label cases.

PRINT Subcommand

PRINT controls the display of optional output. If PRINT is omitted, DEFAULT output (defined below) is displayed.

- The minimum specification is PRINT followed by a single keyword.
- If PRINT is used, only the requested output is displayed.

DEFAULT *Classification tables and statistics for the variables in and not in the equation at each step.* Tables and statistics are displayed for each split file and METHOD subcommand.

SUMMARY *Summary information.* Same output as DEFAULT, except that the output for each step is not displayed.

CORR *Correlation matrix of parameter estimates for the variables in the model.*

ITER(value) *Iterations at which parameter estimates are to be displayed.* The value in parentheses controls the spacing of iteration reports. If the value is *n*, the parameter estimates are displayed for every *n*th iteration starting at 0. If a value is not supplied, intermediate estimates are displayed at each iteration.

ALL *All available output.*

Example

```
LOGISTIC REGRESSION VARIABLES=PASS WITH GPA,GRE,MAT
 /METHOD FSTEP
 /PRINT CORR SUMMARY ITER(2).
```

- A forward stepwise logistic regression analysis of *PASS* on *GPA*, *GRE*, and *MAT* is specified.
- The PRINT subcommand requests the display of the correlation matrix of parameter estimates for the variables in the model (CORR), classification tables and statistics for the variables in and not in the equation for the final model (SUMMARY), and parameter estimates at every second iteration (ITER(2)).

CRITERIA Subcommand

CRITERIA controls the statistical criteria used in building the logistic regression models. The way in which these criteria are used depends on the method specified on the METHOD subcommand. The default criteria are noted in the description of each keyword below. Iterations will stop if the criterion for BCON, LCON, or ITERATE is satisfied.

BCON(value) *Change in parameter estimates to terminate iteration.* Iteration terminates when the parameters change by less than the specified value. The default is 0.001. To eliminate this criterion, specify a value of 0.

ITERATE *Maximum number of iterations.* The default is 20.

LCON(value) *Percentage change in the log likelihood ratio for termination of iterations.* If the log likelihood decreases by less than the specified value, iteration terminates. The default is 0.01. To eliminate this criterion, specify a value of 0.

PIN(value) *Probability of score statistic for variable entry.* The default is 0.05. The larger the specified probability, the easier it is for a variable to enter the model.

POUT(value) *Probability of conditional, Wald, or LR statistic to remove a variable.* The default is 0.1. The larger the specified probability, the easier it is for a variable to remain in the model.

EPS(value) *Epsilon value used for redundancy checking.* The specified value must be less than or equal to 0.05 and greater than or equal to 10^{-12}. The default is 10^{-8}. Larger values make it harder for variables to pass the redundancy check—that is, they are more likely to be removed from the analysis.

Example

```
LOGISTIC REGRESSION PROMOTED WITH AGE JOBTIME RACE
 /CATEGORICAL RACE
 /METHOD BSTEP
 /CRITERIA BCON(0.01) PIN(0.01) POUT(0.05).
```

- A backward stepwise logistic regression analysis is performed for the dependent variable *PROMOTED* and the independent variables *AGE*, *JOBTIME*, and *RACE*.
- CRITERIA alters four of the statistical criteria that control the building of a model.
- BCON specifies that if the change in the absolute value of all of the parameter estimates is less than 0.01, the iterative estimation process should stop. Larger values lower the number of iterations required. Notice that the ITER and LCON criteria remain unchanged and that if either of them is met before BCON, iterations will terminate. (LCON can be set to 0 if only BCON and ITER are to be used.)
- POUT requires that the probability of the statistic used to test whether a variable should remain in the model be smaller than 0.05. This is more stringent than the default value of 0.1.
- PIN requires that the probability of the score statistic used to test whether a variable should be included be smaller than 0.01. This makes it more difficult for variables to be included in the model than the default value of 0.05.

CLASSPLOT Subcommand

CLASSPLOT generates a classification plot of the actual and predicted values of the dichotomous dependent variable at each step.

- Keyword CLASSPLOT is the only specification.
- If CLASSPLOT is not specified, plots are not generated.

Example

```
LOGISTIC REGRESSION PROMOTED WITH JOBTIME RACE
 /CATEGORICAL RACE
 /CLASSPLOT.
```

- A logistic regression model is constructed for the dichotomous dependent variable *PROMOTED* and the independent variables *JOBTIME* and *RACE*.

- CLASSPLOT produces a classification plot for the dependent variable *PROMOTED*. The vertical axis of the plot is the frequency of the variable *PROMOTED*. The horizontal axis is the predicted probability of membership in the second of the two levels of *PROMOTED*.

CASEWISE Subcommand

CASEWISE produces a casewise listing of the values of the temporary variables created by LOGISTIC REGRESSION.

The following keywords are available for specifying temporary variables (see Fox, 1984). When CASEWISE is specified by itself, the default lists *PRED*, *PGROUP*, *RESID*, and *ZRESID*. If a list of variable names is given, only those named temporary variables are displayed.

PRED *Predicted probability.* For each case, the predicted probability of having the second of the two values of the dichotomous dependent variable.

PGROUP *Predicted group.* The group to which a case is assigned based on the predicted probability.

RESID *Difference between observed and predicted probabilities.*

DEV *Deviance values.* For each case, a log-likelihood-ratio statistic, which measures how well the model fits the case, is computed.

LRESID *Logit residual.* Residual divided by the product of *PRED* and 1–šš*PRED*.

SRESID *Studentized residual.*

ZRESID *Normalized residual.* Residual divided by the square root of the product of *PRED* and 1–šš*PRED*.

LEVER *Leverage value.* A measure of the relative influence of each observation on the model's fit.

COOK *Analog of Cook's influence statistic.*

DFBETA *Difference in beta.* The difference in the estimated coefficients for each independent variable if the case is omitted.

The following keyword is available for restricting the cases to be displayed, based on the absolute value of *SRESID*:

OUTLIER (value) *Cases with absolute values of SRESID greater than or equal to the specified value are displayed.* If OUTLIER is specified with no value, the default is 2.

Example

```
LOGISTIC REGRESSION PROMOTED WITH JOBTIME SEX RACE
 /CATEGORICAL SEX RACE
 /METHOD ENTER
 /CASEWISE SRESID LEVER DFBETA.
```

- CASEWISE produces a casewise listing of the temporary variables *SRESID*, *LEVER*, and *DFBETA*.
- There will be one *DFBETA* value for each parameter in the model. The continuous variable *JOBTIME*, the two-level categorical variable *SEX*, and the constant each require one pa-

rameter while the four-level categorical variable *RACE* requires three parameters. Thus, six values of *DFBETA* will be produced for each case.

MISSING Subcommand

LOGISTIC REGRESSION excludes all cases with missing values on any of the independent variables. For a case with a missing value on the dependent variable, predicted values are calculated if it has nonmissing values on all independent variables. The MISSING subcommand controls the processing of user-missing values. If the subcommand is not specified, the default is EXCLUDE.

EXCLUDE *Delete cases with user-missing values as well as system-missing values.* This is the default.

INCLUDE *Include user-missing values in the analysis.*

SAVE Subcommand

SAVE saves the temporary variables created by LOGISTIC REGRESSION. To specify variable names for the new variables, assign the new names in parentheses following each temporary variable name. If new variable names are not specified, LOGISTIC REGRESSION generates default names.

- Assigned variable names must be unique in the working data file. Scratch or system variable names (that is, names that begin with # or $) cannot be used.
- A temporary variable can be saved only once on the same SAVE subcommand.

Example

```
LOGISTIC REGRESSION PROMOTED WITH JOBTIME AGE
 /SAVE PRED (PREDPRO) DFBETA (DF).
```

- A logistic regression analysis of *PROMOTED* on the independent variables *JOBTIME* and *AGE* is performed.
- SAVE adds four variables to the working data file: one variable named *PREDPRO*, containing the predicted value from the specified model for each case, and three variables named *DF0*, *DF1*, and *DF2*, containing, respectively, the *DFBETA* values for each case of the constant, the independent variable *JOBTIME*, and the independent variable *AGE*.

EXTERNAL Subcommand

EXTERNAL indicates that the data for each split-file group should be held in an external scratch file during processing. This can help conserve memory resources when running complex analyses or analyses with large data sets.

- The keyword EXTERNAL is the only specification.
- Specifying EXTERNAL may result in slightly longer processing time.
- If EXTERNAL is not specified, all data are held internally and no scratch file is written.

LOGLINEAR

The syntax for LOGLINEAR is available only in a syntax window, not from the dialog box interface. See GENLOG for information on the LOGLINEAR command available from the dialog box interface.

```
LOGLINEAR varlist(min,max)...[BY] varlist(min,max)

          [WITH covariate varlist]

 [/CWEIGHT={varname }] [/CWEIGHT=(matrix)...]
           {(matrix)}

 [/GRESID={varlist }]  [/GRESID=(matrix)...]
          {(matrix)}

 [/CONTRAST (varname)={DEVIATION [(refcat)]        } [/CONTRAST...]]
                      {DIFFERENCE                   }
                      {HELMERT                      }
                      {SIMPLE [(refcat)]            }
                      {REPEATED                     }
                      {POLYNOMIAL [({1,2,3,...})]   }
                      {             {metric    }    }
                      {[BASIS] SPECIAL(matrix)      }

 [/CRITERIA=[CONVERGE({0.001**})] [ITERATE({20**})] [DELTA({0.5**})]
                      {n      }            {n   }          {n    }
            [DEFAULT]]

 [/PRINT={[FREQ**][RESID**][DESIGN][ESTIM][COR]}]
         {DEFAULT                              }
         {ALL                                  }
         {NONE                                 }

 [/PLOT={NONE**  }]
        {DEFAULT }
        {RESID   }
        {NORMPROB}

 [/MISSING=[{EXCLUDE**}]]
            {INCLUDE  }

 [/DESIGN=effect[(n)] effect[(n)]... effect BY effect...] [/DESIGN...]
```

**Default if subcommand or keyword is omitted.

Example:

```
LOGLINEAR JOBSAT (1,2) ZODIAC (1,12) /DESIGN=JOBSAT.
```

Overview

LOGLINEAR is a general procedure for model fitting, hypothesis testing, and parameter estimation for any model that has categorical variables as its major components. As such, LOGLINEAR subsumes a variety of related techniques, including general models of multiway contingency tables, logit models, logistic regression on categorical variables, and quasi-independence models.

LOGLINEAR models cell frequencies using the multinomial response model and produces maximum-likelihood estimates of parameters by means of the Newton-Raphson algorithm

(Haberman, 1978). HILOGLINEAR, which uses an iterative proportional-fitting algorithm, is more efficient for hierarchical models, but it cannot produce parameter estimates for unsaturated models, does not permit specification of contrasts for parameters, and does not display a correlation matrix of the parameter estimates.

Comparison of the GENLOG and LOGLINEAR Commands

The General Loglinear Analysis and Logit Loglinear Analysis dialog boxes are both associated with the GENLOG command. In previous releases of SPSS, these dialog boxes were associated with the LOGLINEAR command. The LOGLINEAR command is now available only as a syntax command. The differences are described below.

Distribution assumptions

- GENLOG can handle both Poisson and multinomial distribution assumptions for observed cell counts.
- LOGLINEAR assumes only multinomial distribution.

Approach

- GENLOG uses a regression approach to parameterize a categorical variable in a design matrix.
- LOGLINEAR uses contrasts to reparameterize a categorical variable. The major disadvantage of the reparameterization approach is in the interpretation of the results when there is a redundancy in the corresponding design matrix. Also, the reparameterization approach may result in incorrect degrees of freedom for an incomplete table, leading to incorrect analysis results.

Contrasts and generalized log-odds ratios (GLOR)

- GENLOG doesn't provide contrasts to reparameterize the categories of a factor. However, it offers generalized log-odds ratios (GLOR) for cell combinations. Often, comparisons among categories of factors can be derived from GLOR.
- LOGLINEAR offers contrasts to reparameterize the categories of a factor.

Deviance residual

- GENLOG calculates and displays the deviance residual and its normal probability plot, in addition to the other residuals.
- LOGLINEAR does not calculate the deviance residual.

Factor-by-covariate design

- When there is a factor-by-covariate term in the design, GENLOG generates one regression coefficient of the covariate for each combination of factor values. The estimates of these regression coefficients are calculated and displayed.
- LOGLINEAR estimates and displays the contrasts of these regression coefficients.

Partition effect

- In GENLOG, the term *partition effect* refers to the category of a factor.
- In LOGLINEAR, the term *partition effect* refers to a particular contrast.

Options

Model Specification. You can specify the model or models to be fit using the DESIGN subcommand.

Cell Weights. You can specify cell weights, such as structural zeros, for the model with the CWEIGHT subcommand.

Output Display. You can control the output display with the PRINT subcommand.

Optional Plots. You can produce plots of adjusted residuals against observed and expected counts, normal plots, and detrended normal plots with the PLOT subcommand.

Linear Combinations. You can calculate linear combinations of observed cell frequencies, expected cell frequencies, and adjusted residuals using the GRESID subcommand.

Contrasts. You can indicate the type of contrast desired for a factor using the CONTRAST subcommand.

Criteria for Algorithm. You can control the values of algorithm-tuning parameters with the CRITERIA subcommand.

Basic Specification

The basic specification is two or more variables that define the crosstabulation. The minimum and maximum values for each variable must be specified in parentheses after the variable name.

By default, LOGLINEAR estimates the saturated model for a multidimensional table. Output includes the factors or effects, their levels, and any labels; observed and expected frequencies and percentages for each factor and code; residuals, standardized residuals, and adjusted residuals; two goodness-of-fit statistics (the likelihood-ratio chi-square and Pearson's chi-square); and estimates of the parameters with accompanying *Z* values and 95% confidence intervals.

Limitations

- Maximum 10 independent (factor) variables.
- Maximum 200 covariates.

Subcommand Order

- The variables specification must come first.
- The subcommands that affect a specific model must be placed before the DESIGN subcommand specifying the model.
- All subcommands can be used more than once and, with the exception of the DESIGN subcommand, are carried from model to model unless explicitly overridden.
- If the last subcommand is not DESIGN, LOGLINEAR generates a saturated model in addition to the explicitly requested model(s).

Example

```
LOGLINEAR JOBSAT (1,2) ZODIAC (1,12) /DESIGN=JOBSAT, ZODIAC.
```

- The variable list specifies two categorical variables, *JOBSAT* and *ZODIAC*. *JOBSAT* has values 1 and 2. *ZODIAC* has values 1 through 12.
- DESIGN specifies a model with main effects only.

Example

```
LOGLINEAR DPREF (2,3) RACE CAMP (1,2).
```

- *DPREF* is a categorical variable with values 2 and 3. *RACE* and *CAMP* are categorical variables with values 1 and 2.
- This is a general loglinear model because no BY keyword appears. The design defaults to a saturated model that includes all main effects and interaction effects.

Example

```
LOGLINEAR GSLEVEL (4,8) EDUC (1,4) SEX (1,2)
  /DESIGN=GSLEVEL EDUC SEX.
```

- *GSLEVEL* is a categorical variable with values 4 through 8. *EDUC* is a categorical variable with values 1 through 4. *SEX* has values 1 and 2.
- DESIGN specifies a model with main effects only.

Example

```
LOGLINEAR GSLEVEL (4,8) BY EDUC (1,4) SEX (1,2)
  /DESIGN=GSLEVEL, GSLEVEL BY EDUC, GSLEVEL BY SEX.
```

- Keyword BY on the variable list specifies a logit model in which *GSLEVEL* is the dependent variable and *EDUC* and *SEX* are the independent variables.
- DESIGN specifies a model that can test for the absence of joint effect of *SEX* and *EDUC* on *GSLEVEL*.

Variable List

The variable list specifies the variables to be included in the model. LOGLINEAR analyzes two classes of variables: categorical and continuous. Categorical variables are used to define the cells of the table. Continuous variables are used as cell covariates. Continuous variables can be specified only after the keyword WITH following the list of categorical variables.

- The list of categorical variables must be specified first. Categorical variables must be numeric and integer.

- A range must be defined for each categorical variable by specifying, in parentheses after each variable name, the minimum and maximum values for that variable. Separate the two values with at least one space or a comma.
- To specify the same range for a list of variables, specify the list of variables followed by a single range. The range applies to all variables on the list.
- To specify a logit model, use keyword BY (see Logit Model below). A variable list without keyword BY generates a general loglinear model.
- Cases with values outside the specified range are excluded from the analysis. Non-integer values within the range are truncated for the purpose of building the table.

Logit Model

- To segregate the independent (factor) variables from the dependent variables in a logit model, use the keyword BY. The categorical variables preceding BY are the dependent variables; the categorical variables following BY are the independent variables.
- A total of 10 categorical variables can be specified. In most cases, one of them is dependent.
- A DESIGN subcommand should be used to request the desired logit model.
- LOGLINEAR displays an analysis of dispersion and two measures of association: entropy and concentration. These measures are discussed in Haberman (1982) and can be used to quantify the magnitude of association among the variables. Both are proportional reduction in error measures. The entropy statistic is analogous to Theil's entropy measure, while the concentration statistic is analogous to Goodman and Kruskal's tau-*b*. Both statistics measure the strength of association between the dependent variable and the predictor variable set.

Cell Covariates

- Continuous variables can be used as covariates. When used, the covariates must be specified after the keyword WITH following the list of categorical variables. Ranges are not specified for the continuous variables.
- A variable cannot be named as both a categorical variable and a cell covariate.
- To enter cell covariates into a model, the covariates must be specified on the DESIGN subcommand.
- Cell covariates are not applied on a case-by-case basis. The mean covariate value for a cell in the contingency table is applied to that cell.

Example

```
LOGLINEAR DPREF(2,3) RACE CAMP (1,2) WITH CONSTANT
  /DESIGN=DPREF RACE CAMP CONSTANT.
```

- Variable *CONSTANT* is a continuous variable specified as a cell covariate. Cell covariates must be specified after keyword WITH following the variable list. No range is defined for cell covariates.
- To include the cell covariate in the model, variable *CONSTANT* is specified on DESIGN.

CWEIGHT Subcommand

CWEIGHT specifies cell weights, such as structural zeros, for a model. By default, cell weights are equal to 1.

- The specification is either one numeric variable or a matrix of weights enclosed in parentheses.
- If a matrix of weights is specified, the matrix must contain the same number of elements as the product of the levels of the categorical variables. An asterisk can be used to signify repetitions of the same value.
- If weights are specified for a multiple-factor model, the index value of the rightmost factor increments the most rapidly.
- If a numeric variable is specified, only one CWEIGHT subcommand can be used on LOGLINEAR.
- To use multiple cell weights on the same LOGLINEAR command, specify all weights in matrix format. Each matrix must be specified on a separate CWEIGHT subcommand, and each CWEIGHT specification remains in effect until explicitly overridden by another CWEIGHT subcommand.
- CWEIGHT can be used to impose structural, or *a priori*, zeros on the model. This feature is useful in the analysis of symmetric tables.

Example

```
COMPUTE  CWT=1.
IF (HUSED EQ WIFED) CWT=0.
LOGLINEAR HUSED WIFED(1,4) WITH DISTANCE
  /CWEIGHT=CWT
  /DESIGN=HUSED WIFED DISTANCE.
```

- COMPUTE initially assigns *CWT* the value 1 for all cases.
- IF assigns *CWT* the value 0 when *HUSED* equals *WIFED*.
- CWEIGHT imposes structural zeros on the diagonal of the symmetric crosstabulation. Because a variable name is specified, only one CWEIGHT can be used.

Example

```
LOGLINEAR  HUSED WIFED(1,4) WITH DISTANCE
  /CWEIGHT=(0, 4*1, 0, 4*1, 0, 4*1, 0)
  /DESIGN=HUSED WIFED DISTANCE
  /CWEIGHT=(16*1)
  /DESIGN=HUSED WIFED DISTANCE.
```

- The first CWEIGHT matrix specifies the same values as variable *CWT* provided in the first example. The specified matrix is as follows:

```
0 1 1 1
1 0 1 1
1 1 0 1
1 1 1 0
```

- The same matrix can be specified in full as `(0 1 1 1 1 0 1 1 1 1 0 1 1 1 1 0)`.

- By using the matrix format on CWEIGHT rather than a variable name, a different CWEIGHT subcommand can be used for the second model.

GRESID Subcommand

GRESID (Generalized Residual) calculates linear combinations of observed cell frequencies, expected cell frequencies, and adjusted residuals.

- The specification is either a numeric variable or a matrix whose contents are coefficients of the desired linear combinations.
- If a matrix of coefficients is specified, the matrix must contain the same number of elements as the number of cells implied by the variables specification. An asterisk can be used to signify repetitions of the same value.
- Each GRESID subcommand specifies a single linear combination. Each matrix or variable must be specified on a separate GRESID subcommand. All GRESID subcommands specified are displayed for each design.

Example

```
LOGLINEAR  MONTH(1,18) WITH Z
 /GRESID=(6*1,12*0)
 /GRESID=(6*0,6*1,6*0)
 /GRESID=(12*0,6*1)
 /DESIGN=Z.
```

- The first GRESID subcommand combines the first six months into a single effect. The second GRESID subcommand combines the second six months, and the third GRESID subcommand combines the last six months.
- For each effect, LOGLINEAR displays the observed and expected counts, the residual, and the adjusted residual.

CONTRAST Subcommand

CONTRAST indicates the type of contrast desired for a factor, where a factor is any categorical dependent or independent variable. The default contrast is DEVIATION for each factor.

- The specification is CONTRAST, which is followed by a variable name in parentheses and the contrast-type keyword.
- To specify a contrast for more than one factor, use a separate CONTRAST subcommand for each specified factor. Only one contrast can be in effect for each factor on each DESIGN.
- A contrast specification remains in effect for subsequent designs until explicitly overridden by another CONTRAST subcommand.
- The design matrix used for the contrasts can be displayed by specifying keyword DESIGN on the PRINT subcommand. However, this matrix is the basis matrix that is used to determine contrasts; it is not the contrast matrix itself.
- CONTRAST can be used for a multinomial logit model, in which the dependent variable has more than two categories.

- CONTRAST can be used for fitting linear logit models. Keyword BASIS is not appropriate for such models.
- In a logit model, CONTRAST is used to transform the independent variable into a metric variable. Again, keyword BASIS is not appropriate.

The following contrast types are available. For illustration of contrast types, see Appendix B.

DEVIATION(refcat) *Deviations from the overall effect.* DEVIATION is the default contrast if the CONTRAST subcommand is not used. Refcat is the category for which parameter estimates are not displayed (they are the negative of the sum of the others). By default, refcat is the last category of the variable.

DIFFERENCE *Levels of a factor with the average effect of previous levels of a factor.* Also known as reverse Helmert contrasts.

HELMERT *Levels of a factor with the average effect of subsequent levels of a factor.*

SIMPLE(refcat) *Each level of a factor to the reference level.* By default, LOGLINEAR uses the last category of the factor variable as the reference category. Optionally, any level can be specified as the reference category enclosed in parentheses after keyword SIMPLE. The sequence of the level, not the actual value, must be specified.

REPEATED *Adjacent comparisons across levels of a factor.*

POLYNOMIAL(metric) *Orthogonal polynomial contrasts.* The default is equal spacing. Optionally, the coefficients of the linear polynomial can be specified in parentheses, indicating the spacing between levels of the treatment measured by the given factor.

[BASIS]SPECIAL(matrix) *User-defined contrast.* As many elements as the number of categories squared must be specified. If BASIS is specified before SPECIAL, a basis matrix is generated for the special contrast, which makes the coefficients of the contrast equal to the special matrix. Otherwise, the matrix specified is transposed and then used as the basis matrix to determine coefficients for the contrast matrix.

Example

```
LOGLINEAR  A(1,4) BY B(1,4)
 /CONTRAST(B)=POLYNOMIAL
 /DESIGN=A A BY B(1)
 /CONTRAST(B)=SIMPLE
 /DESIGN=A A BY B(1).
```

- The first CONTRAST subcommand requests polynomial contrasts of *B* for the first design.
- The second CONTRAST subcommand requests the simple contrast of *B*, with the last category (value 4) used as the reference category for the second DESIGN subcommand.

Example

```
* Multinomial logit model

LOGLINEAR  PREF(1,5) BY RACE ORIGIN CAMP(1,2)
  /CONTRAST(PREF)=SPECIAL(5*1, 1 1 1 1 -4, 3 -1 -1 -1 0,
       0 1 1 -2 0, 0 1 -1 0 0).
```

- LOGLINEAR builds special contrasts among the five categories of the dependent variable *PREF*, which measures preference for training camps among Army recruits. For *PREF*, 1=stay, 2=move to north, 3=move to south, 4=move to unnamed camp, and 5=undecided.
- The four contrasts are: (1) move or stay versus undecided, (2) stay versus move, (3) named camp versus unnamed, and (4) northern camp versus southern. Because these contrasts are orthogonal, SPECIAL and BASIS SPECIAL produce equivalent results.

Example

```
* Contrasts for a linear logit model

LOGLINEAR RESPONSE(1,2) BY YEAR(0,20)
 /PRINT=DEFAULT ESTIM
 /CONTRAST(YEAR)=SPECIAL(21*1, -10, -9, -8, -7, -6, -5, -4,
                          -3, -2, -1, 0, 1, 2, 3, 4, 5, 6, 7,
                           8, 9, 10, 399*1)
 /DESIGN=RESPONSE RESPONSE BY YEAR(1).
```

- *YEAR* measures years of education and ranges from 0 through 20. Therefore, allowing for the constant effect, *YEAR* has 20 estimable parameters associated with it.
- The SPECIAL contrast specifies the constant—that is, 21*1—and the linear effect of *YEAR*—that is, –10 to 10. The other 399 1's fill out the 21*21 matrix.

Example

```
* Contrasts for a logistic regression model

LOGLINEAR RESPONSE(1,2) BY TIME(1,4)
  /CONTRAST(TIME) = SPECIAL(4*1, 7 14 27 51, 8*1)
  /PRINT=ALL /PLOT=DEFAULT
  /DESIGN=RESPONSE, TIME(1) BY RESPONSE.
```

- CONTRAST is used to transform the independent variable into a metric variable.
- *TIME* represents elapsed time in days. Therefore, the weights in the contrast represent the metric of the passage of time.

CRITERIA Subcommand

CRITERIA specifies the values of some constants in the Newton-Raphson algorithm. Defaults or specifications remain in effect until overridden with another CRITERIA subcommand.

CONVERGE(n) *Convergence criterion.* Specify a value for the convergence criterion. The default is 0.001.

ITERATE(n) *Maximum number of iterations.* Specify the maximum number of iterations for the algorithm. The default number is 20.

DELTA(n) *Cell delta value.* The value of delta is added to each cell frequency for the first iteration. For saturated models, it remains in the cell. The default value is 0.5. LOGLINEAR does not display parameter estimates or correlation matrices of parameter estimates if any sampling zero cells exist in the expected table after delta is added. Parameter estimates and correlation matrices can be displayed in the presence of structural zeros.

DEFAULT *Default values are used.* DEFAULT can be used to reset the parameters to the default.

Example

```
LOGLINEAR  DPREF(2,3) BY RACE ORIGIN CAMP(1,2)
 /CRITERIA=ITERATION(50) CONVERGE(.0001).
```

- ITERATION increases the maximum number of iterations to 50.
- CONVERGE lowers the convergence criterion to 0.0001.

PRINT Subcommand

PRINT requests statistics that are not produced by default.

- By default, LOGLINEAR displays the frequency table and residuals. The parameter estimates of the model are also displayed if DESIGN is not used.
- Multiple PRINT subcommands are permitted. The specifications are cumulative.

The following keywords can be used on PRINT:

FREQ *Observed and expected cell frequencies and percentages.* This is displayed by default.

RESID *Raw, standardized, and adjusted residuals.* This is displayed by default.

DESIGN *The design matrix of the model, showing the basis matrix corresponding to the contrasts used.*

ESTIM *The parameter estimates of the model.* If you do not specify a design on the DESIGN subcommand, LOGLINEAR generates a saturated model and displays the parameter estimates for the saturated model. LOGLINEAR does not display parameter estimates or correlation matrices of parameter estimates if any sampling zero cells exist in the expected table after delta is added. Parameter estimates and a correlation matrix are displayed when structural zeros are present.

COR *The correlation matrix of the parameter estimates.* Alias COV.

ALL *All available output.*

DEFAULT *FREQ and RESID.* ESTIM is also displayed by default if the DESIGN subcommand is not used.

NONE *The design information and goodness-of-fit statistics only.* This option overrides all other specifications on the PRINT subcommand. The NONE option applies only to the PRINT subcommand.

Example

```
LOGLINEAR A(1,2) B(1,2)
 /PRINT=ESTIM
 /DESIGN=A,B,A BY B
 /PRINT=ALL
 /DESIGN=A,B.
```

- The first design is the saturated model. The parameter estimates are displayed with ESTIM specified on PRINT.
- The second design is the main-effects model, which tests the hypothesis of no interaction. The second PRINT subcommand displays all available display output for this model.

PLOT Subcommand

PLOT produces optional plots. No plots are displayed if PLOT is not specified or is specified without any keyword. Multiple PLOT subcommands can be used. The specifications are cumulative.

RESID *Plots of adjusted residuals against observed and expected counts.*

NORMPROB *Normal and detrended normal plots of the adjusted residuals.*

NONE *No plots.*

DEFAULT *RESID and NORMPROB.* Alias ALL.

Example

```
LOGLINEAR  RESPONSE(1,2) BY TIME(1,4)
  /CONTRAST(TIME)=SPECIAL(4*1, 7 14 27 51, 8*1)
  /PLOT=DEFAULT
  /DESIGN=RESPONSE TIME(1) BY RESPONSE
  /PLOT=NONE
  /DESIGN.
```

- RESID and NORMPROB plots are displayed for the first design.
- No plots are displayed for the second design.

MISSING Subcommand

MISSING controls missing values. By default, LOGLINEAR excludes all cases with system- or user-missing values on any variable. You can specify INCLUDE to include user-missing values. If INCLUDE is specified, user-missing values must also be included in the value range specification.

EXCLUDE *Delete cases with user-missing values.* This is the default if the subcommand is omitted. You can also specify keyword DEFAULT.

INCLUDE *Include user-missing values.* Only cases with system-missing values are deleted.

Example

```
MISSING VALUES A(0).
LOGLINEAR A(0,2) B(1,2) /MISSING=INCLUDE
 /DESIGN=B.
```

- Even though 0 was specified as missing, it is treated as a nonmissing category of *A* in this analysis.

DESIGN Subcommand

DESIGN specifies the model or models to be fit. If DESIGN is omitted or used with no specifications, the saturated model is produced. The saturated model fits all main effects and all interaction effects.

- To specify more than one model, use more than one DESIGN subcommand. Each DESIGN specifies one model.
- To obtain main-effects models, name all the variables listed on the variables specification.
- To obtain interactions, use keyword BY to specify each interaction, as in `A BY B` and `C BY D`. To obtain the single-degree-of-freedom partition of a specified contrast, specify the partition in parentheses following the factor (see the example below).
- To include cell covariates in the model, first identify them on the variable list by naming them after keyword WITH, and then specify the variable names on DESIGN.
- To specify an equiprobability model, name a cell covariate that is actually a constant of 1.

Example

```
* Testing the linear effect of the dependent variable

COMPUTE X=MONTH.
LOGLINEAR MONTH (1,12) WITH X
  /DESIGN X.
```

- The variable specification identifies *MONTH* as a categorical variable with values 1 through 12. Keyword WITH identifies *X* as a covariate.
- DESIGN tests the linear effect of *MONTH.*

Example

```
* Specifying main effects models

LOGLINEAR A(1,4) B(1,5)
  /DESIGN=A
  /DESIGN=A,B.
```

- The first design tests the homogeneity of category probabilities for *B*; it fits the marginal frequencies on *A*, but assumes that membership in any of the categories of *B* is equiprobable.
- The second design tests the independence of *A* and *B*. It fits the marginals on both *A* and *B*.

Example

```
* Specifying interactions

LOGLINEAR A(1,4) B(1,5) C(1,3)
  /DESIGN=A,B,C, A BY B.
```

- This design consists of the *A* main effect, the *B* main effect, the *C* main effect, and the interaction of *A* and *B*.

Example

```
* Single-degree-of-freedom partitions

LOGLINEAR A(1,4) BY B(1,5)
  /CONTRAST(B)=POLYNOMIAL
  /DESIGN=A,A BY B(1).
```

- The value 1 following *B* refers to the first partition of *B*, which is the linear effect of *B*; this follows from the contrast specified on the CONTRAST subcommand.

Example

```
* Specifying cell covariates

LOGLINEAR HUSED WIFED(1,4) WITH DISTANCE
   /DESIGN=HUSED WIFED DISTANCE.
```

- The continuous variable *DISTANCE* is identified as a cell covariate by specifying it after WITH on the variable list. The cell covariate is then included in the model by naming it on DESIGN.

Example

```
* Equiprobability model

COMPUTE  X=1.
LOGLINEAR  MONTH(1,18) WITH X
  /DESIGN=X.
```

- This model tests whether the frequencies in the 18-cell table are equal by using a cell covariate that is a constant of 1.

MANOVA: Overview

```
MANOVA dependent varlist [BY factor list (min,max)[factor list...]
                         [WITH covariate list]]

 [/WSFACTORS=varname (levels) [varname...] ]

 [/WSDESIGN]*

 [/TRANSFORM [(dependent varlist [/dependent varlist])]=
                    [ORTHONORM] [{CONTRAST}] {DEVIATION (refcat)    } ]
                                {BASIS   }   {DIFFERENCE            }
                                             {HELMERT               }
                                             {SIMPLE (refcat)       }
                                             {REPEATED              }
                                             {POLYNOMIAL [({1,2,3...})]}
                                             {             {metric  }  }
                                             {SPECIAL (matrix)      }

 [/MEASURE=newname newname...]

 [/RENAME={newname} {newname}...]
          {*      } {*      }

 [/ERROR={WITHIN           } ]
         {RESIDUAL         }
         {WITHIN + RESIDUAL}
         {n                }

 [/CONTRAST (factorname)={DEVIATION** [(refcat)]     }] †
                         {POLYNOMIAL**[({1,2,3...})] }
                         {              {metric  }   }
                         {SIMPLE [(refcat)]          }
                         {DIFFERENCE                 }
                         {HELMERT                    }
                         {REPEATED                   }
                         {SPECIAL (matrix)           }

 [/PARTITION (factorname)[=({1,1...  })]]
                            {n1,n2...}

 [/METHOD=[{UNIQUE**   }] [{CONSTANT** }] [{QR**     }]]
           {SEQUENTIAL}    {NOCONSTANT}    {CHOLESKY}

 [/{PRINT  }= [CELLINFO [([MEANS] [SSCP] [COV] [COR] [ALL])]]
   {NOPRINT}  [HOMOGENEITY [([ALL] [BARTLETT] [COCHRAN] [BOXM])]]
              [DESIGN [([OVERALL] [ONEWAY] [DECOMP] [BIAS] [SOLUTION]
                       [REDUNDANCY] [COLLINEARITY] [ALL])]]
              [PARAMETERS [([ESTIM] [ORTHO][COR][NEGSUM][EFSIZE][OPTIMAL][ALL])]]
              [SIGNIF [[(SINGLEDF)]
                       [(MULTIV**)] [(EIGEN)] [(DIMENR)]
                       [(UNIV**)] [(HYPOTH)][(STEPDOWN)] [(BRIEF)]
                       [{(AVERF**)}] [(HF)] [(GG)] [(EFSIZE)]]
                        {(AVONLY) }
             [ERROR[(STDDEV)][(COR)][(COV)][(SSCP)]]

 [/OMEANS =[VARIABLES(varlist)] [TABLES ({factor name      }] ]
                                         {factor BY factor}
                                         {CONSTANT        }

 [/PMEANS =[VARIABLES(varlist)] [TABLES ({factor name      })] [PLOT]] ]
                                         {factor BY factor}
                                         {CONSTANT        }

 [/RESIDUALS=[CASEWISE] [PLOT] ]

 [/POWER=[T({.05**})] [F({.05**})] [{APPROXIMATE}]]
            {a    }       {a    }    {EXACT      }
```

```
[/CINTERVAL=[{INDIVIDUAL}][({.95}) ]
            {JOINT     }   {a  }
            [UNIVARIATE ({SCHEFFE})]
                         {BONFER }
            [MULTIVARIATE  ({ROY       })]  ]
                            {PILLAI    }
                            {BONFER    }
                            {HOTELLING }
                            {WILKS     }
[/PCOMPS [COR] [COV] [ROTATE(rottype)]
         [NCOMP(n)] [MINEIGEN(eigencut)] [ALL] ]

[/PLOT=[BOXPLOTS] [CELLPLOTS] [NORMAL]  [ALL] ]

[/DISCRIM [RAW] [STAN] [ESTIM] [COR] [ALL]
          [ROTATE(rottype)] [ALPHA({.25**})]]
                                   {a    }

[/MISSING=[LISTWISE**] [{EXCLUDE**}] ]
                        {INCLUDE  }

[/MATRIX=[IN({file})]  [OUT({file})]]
             {[*] }         {[*] }

[/ANALYSIS [({UNCONDITIONAL**})]=[(]dependent varlist
             {CONDITIONAL    }      [WITH covariate varlist]
                                     [/dependent varlist...][)][WITH varlist] ]

[/DESIGN={factor [(n)]   }[BY factor[(n)]] [WITHIN factor[(n)]][WITHIN...]
         {[POOL(varlist)}

         [+ {factor [(n)]  }...]
            {POOL(varlist)}

         [[= n] {AGAINST} {WITHIN  }
                {VS     } {RESIDUAL}
                          {WR      }
                          {n       }

         [{factor [(n)] } ... ]
          {POOL(varlist)}

         [MWITHIN factor(n)]
         [MUPLUS]
         [CONSTANT [=n] ]
```

* WSDESIGN uses the same specification as DESIGN, with only within-subjects factors.

† DEVIATION is the default for between-subjects factors while POLYNOMIAL is the default for within-subjects factors.

** Default if subcommand or keyword is omitted.

Example 1:

```
* Analysis of Variance

MANOVA RESULT BY TREATMNT(1,4) GROUP(1,2).
```

Example 2:

```
* Analysis of Covariance

MANOVA RESULT BY TREATMNT(1,4) GROUP(1,2) WITH RAINFALL.
```

Example 3:

```
* Repeated Measures Analysis

MANOVA SCORE1 TO SCORE4 BY CLASS(1,2)
  /WSFACTORS=MONTH(4).
```

Example 4:

```
* Parallelism Test with Crossed Factors

MANOVA YIELD BY PLOT(1,4) TYPEFERT(1,3) WITH FERT
  /ANALYSIS YIELD
  /DESIGN FERT, PLOT, TYPEFERT, PLOT BY TYPEFERT,
   FERT BY PLOT + FERT BY TYPEFERT
   + FERT BY PLOT BY TYPEFERT.
```

Overview

MANOVA (multivariate analysis of variance) is a generalized procedure for analysis of variance and covariance. MANOVA is the most powerful of the analysis-of-variance procedures in SPSS and can be used for both univariate and multivariate designs. Only MANOVA allows you to perform the following tasks:

- Specify nesting of effects.
- Specify individual error terms for effects in mixed-model analyses.
- Estimate covariate-by-factor interactions to test the assumption of homogeneity of regressions.
- Obtain parameter estimates for a variety of contrast types, including irregularly spaced polynomial contrasts with multiple factors.
- Test user-specified special contrasts with multiple factors.
- Partition effects in models.
- Pool effects in models.

To simplify the presentation, reference material on MANOVA is divided into three sections: *univariate* designs with one dependent variable; *multivariate* designs with several interrelated dependent variables; and *repeated measures* designs in which the dependent variables represent the same types of measurements taken at more than one time.

If you are unfamiliar with the models, assumptions, and statistics used in MANOVA, consult Chapters 2, 3, and 4.

The full syntax diagram for MANOVA is presented here. The MANOVA sections that follow include partial syntax diagrams showing the subcommands and specifications discussed in that section. Individually, those diagrams are incomplete. Subcommands listed for univariate designs are available for any analysis, and subcommands listed for multivariate designs can be used in any multivariate analysis, including repeated measures.

MANOVA was designed and programmed by Philip Burns of Northwestern University.

MANOVA: Univariate

```
MANOVA dependent var [BY factor list (min,max)][factor list...]
                     [WITH covariate list]
 [/ERROR={WITHIN           } ]
         {RESIDUAL         }
         {WITHIN + RESIDUAL}
         {n                }
 [/CONTRAST (factor name)={DEVIATION** [(refcat)]     }]
                          {POLYNOMIAL  [({1,2,3...})] }
                          {              {metric   }  }
                          {SIMPLE [(refcat)]          }
                          {DIFFERENCE                 }
                          {HELMERT                    }
                          {REPEATED                   }
                          {SPECIAL (matrix)           }
 [/PARTITION (factor name)[=({1,1...  })]]
                             {n1,n2...}
 [/METHOD=[{UNIQUE**  }]  [{CONSTANT**}] [{QR**     }]]
           {SEQUENTIAL}    {NOCONSTANT}   {CHOLESKY}
 [/{PRINT  }= [CELLINFO [([MEANS] [SSCP] [COV] [COR] [ALL])]]
   {NOPRINT}  [HOMOGENEITY [([ALL] [BARTLETT] [COCHRAN])]]
              [DESIGN [([OVERALL] [ONEWAY] [DECOMP] [BIAS] [SOLUTION]
                       [REDUNDANCY] [COLLINEARITY])]]
              [PARAMETERS [([ESTIM][ORTHO][COR][NEGSUM][EFSIZE][OPTIMAL][ALL])]]
              [SIGNIF[(SINGLEDF)]]
              [ERROR[(STDDEV)]]                                         ]
 [/OMEANS =[VARIABLES(varlist)] [TABLES ({factor name     }] ]
                                         {factor BY factor}
                                         {CONSTANT        }
 [/PMEANS =[TABLES ({factor name     })] [PLOT]] ]
                    {factor BY factor}
                    {CONSTANT        }
 [/RESIDUALS=[CASEWISE] [PLOT] ]
 [/POWER=[T({.05**})] [F({.05**})] [{APPROXIMATE}]]
            {a     }       {a     }    {EXACT      }
 [/CINTERVAL=[{INDIVIDUAL}][({.95}) ]] [UNIVARIATE ({SCHEFFE})]
              {JOINT     }   { a}                   {BONFER }
 [/PLOT=[BOXPLOTS] [CELLPLOTS] [NORMAL]  [ALL] ]
 [/MISSING=[LISTWISE**] [{EXCLUDE**}] ]
                         {INCLUDE  }
 [/MATRIX=[IN({file})]  [OUT({file})]]
              {[*] }         {[*] }
 [/ANALYSIS=dependent var [WITH covariate list]]
 [/DESIGN={factor [(n)]    }[BY factor[(n)]] [WITHIN factor[(n)]][WITHIN...]
          {[POOL(varlist)}
          [+ {factor [(n)]  }...]
             {POOL(varlist)}
          [[= n] {AGAINST}  {WITHIN  }
                 {VS     }  {RESIDUAL}
                            {WR      }
                            {n       }
          [{factor [(n)]  } ... ]
           {POOL(varlist)}
          [MUPLUS]
          [MWITHIN factor(n)]
          [CONSTANT [=n] ]
```

** Default if subcommand or keyword is omitted.

Example:

```
MANOVA YIELD BY SEED(1,4) FERT(1,3)
  /DESIGN.
```

Overview

This section describes the use of MANOVA for univariate analyses. However, the subcommands described here can be used in any type of analysis with MANOVA. For additional subcommands used in those types of analysis, see MANOVA: Multivariate and MANOVA: Repeated Measures. For basic specification, syntax rules, and limitations of the MANOVA procedures, see MANOVA: Overview. If you are unfamiliar with the models, assumptions, and statistics used in MANOVA, consult Chapter 2.

Options

Design Specification. You can specify which terms to include in the design on the DESIGN subcommand. This allows you to estimate a model other than the default full factorial model, incorporate factor-by-covariate interactions, indicate nesting of effects, and indicate specific error terms for each effect in mixed models. You can specify a different continuous variable as a dependent variable or work with a subset of the continuous variables with the ANALYSIS subcommand.

Contrast Types. You can specify contrasts other than the default deviation contrasts on the CONTRAST subcommand. You can also subdivide the degrees of freedom associated with a factor using the PARTITION subcommand and test the significance of a specific contrast or group of contrasts.

Optional Output. You can choose from a wide variety of optional output on the PRINT subcommand or suppress output using the NOPRINT subcommand. Output appropriate to univariate designs includes cell means, design or other matrices, parameter estimates, and tests for homogeneity of variance across cells. Using the OMEANS, PMEANS, RESIDUAL, and PLOT subcommands, you can also request tables of observed and/or predicted means, casewise values and residuals for your model, and various plots useful in checking assumptions. In addition, you can request observed power values based on fixed-effect assumptions using the POWER subcommand and request simultaneous confidence intervals for each parameter estimate and regression coefficient using the CINTERVAL subcommand.

Matrix Materials. You can write matrices of intermediate results to a matrix data file, and you can read such matrices in performing further analyses using the MATRIX subcommand.

Basic Specification

- The basic specification is a variable list identifying the dependent variable, the factors (if any), and the covariates (if any).
- By default, MANOVA uses a full factorial model, which includes all main effects and all possible interactions among factors. Estimation is performed using the cell-means model and UNIQUE (regression-type) sums of squares, adjusting each effect for all other effects in the model. Parameters are estimated using DEVIATION contrasts to determine if their categories differ significantly from the mean.

Subcommand Order

- The variable list must be specified first.
- Subcommands applicable to a specific design must be specified before that DESIGN subcommand. Otherwise, subcommands can be used in any order.

Syntax Rules

- For many analyses, the MANOVA variable list and the DESIGN subcommand are the only specifications needed. If a full factorial design is desired, DESIGN can be omitted.
- All other subcommands apply only to designs that follow. If you do not enter a DESIGN subcommand or if the last subcommand is not DESIGN, MANOVA will use a full factorial model.
- Unless replaced, MANOVA subcommands other than DESIGN remain in effect for all subsequent models.
- MISSING can be specified only once.
- The following words are reserved as keywords or internal commands in the MANOVA procedure: AGAINST, CONSPLUS, CONSTANT, CONTIN, MUPLUS, MWITHIN, POOL, R, RESIDUAL, RW, VERSUS, VS, W, WITHIN, and WR. Variable names that duplicate these words should be changed before you invoke MANOVA.
- If you enter one of the multivariate specifications in a univariate analysis, MANOVA will ignore it.

Limitations

- Maximum 20 factors.
- Maximum 200 dependent variables.
- Memory requirements depend primarily on the number of cells in the design. For the default full factorial model, this equals the product of the number of levels or categories in each factor.

Example

```
MANOVA YIELD BY SEED(1,4) FERT(1,3) WITH RAINFALL
  /PRINT=CELLINFO(MEANS) PARAMETERS(ESTIM)
  /DESIGN.
```

- *YIELD* is the dependent variable; *SEED* (with values 1, 2, 3, and 4) and *FERT* (with values 1, 2, and 3) are factors; *RAINFALL* is a covariate.
- The PRINT subcommand requests the means of the dependent variable for each cell and the default deviation parameter estimates.
- The DESIGN subcommand requests the default design, a full factorial model. This subcommand could have been omitted or could have been specified in full as:

```
/DESIGN = SEED, FERT, SEED BY FERT.
```

MANOVA Variable List

The variable list specifies all variables that will be used in any subsequent analyses.

- The dependent variable must be the first specification on MANOVA.
- By default, MANOVA treats a list of dependent variables as jointly dependent, implying a multivariate design. However, you can change the role of a variable or its inclusion status in the analysis on the ANALYSIS subcommand.
- The names of the factors follow the dependent variable. Use the keyword BY to separate the factors from the dependent variable.
- Factors must have adjacent integer values, and you must supply the minimum and maximum values in parentheses after the factor name(s).
- If several factors have the same value range, you can specify a list of factors followed by a single value range in parentheses.
- Certain one-cell designs, such as univariate and multivariate regression analysis, canonical correlation, and one-sample Hotelling's T^2, do not require a factor specification. To perform these analyses, omit keyword BY and the factor list.
- Enter the covariates, if any, following the factors and their ranges. Use keyword WITH to separate covariates from factors (if any) and the dependent variable.

Example

```
MANOVA DEPENDNT BY FACTOR1 (1,3) FACTOR2, FACTOR3 (1,2).
```

- In this example, three factors are specified.
- *FACTOR1* has values 1, 2, and 3, while *FACTOR2* and *FACTOR3* have values 1 and 2.
- A default full factorial model is used for the analysis.

Example

```
MANOVA Y BY A(1,3) WITH X
  /DESIGN.
```

- In this example, the *A* effect is tested after adjusting for the effect of the covariate *X*. It is a test of equality of adjusted *A* means.
- The test of the covariate *X* is adjusted for *A*. It is a test of the pooled within-groups regression of *Y* on *X*.

ERROR Subcommand

ERROR allows you to specify or change the error term used to test all effects for which you do not explicitly specify an error term on the DESIGN subcommand. ERROR affects all terms in all subsequent designs, except terms for which you explicitly provide an error term.

WITHIN *Terms in the model are tested against the within-cell sum of squares.* This specification can be abbreviated to W. This is the default unless there is no variance within cells or unless a continuous variable is named on the DESIGN subcommand.

RESIDUAL *Terms in the model are tested against the residual sum of squares.* This specification can be abbreviated to R. This includes all terms not named on the DESIGN subcommand.

WITHIN+RESIDUAL *Terms are tested against the pooled within-cells and residual sum of squares.* This specification can be abbreviated to WR or RW. This is the default for designs in which a continuous variable appears on the DESIGN subcommand.

error number *Terms are tested against a numbered error term.* The error term must be defined on each DESIGN subcommand (for a discussion of error terms, see the DESIGN subcommand on p. 440).

- If you specify `ERROR=WITHIN+RESIDUAL` and one of the components does not exist, MANOVA uses the other component alone.
- If you specify your own error term by number and a design does not have an error term with the specified number, MANOVA does not carry out significance tests. It will, however, display hypothesis sums of squares and, if requested, parameter estimates.

Example

```
MANOVA DEP BY A(1,2) B(1,4)
  /ERROR = 1
  /DESIGN = A, B, A BY B = 1 VS WITHIN
  /DESIGN = A, B.
```

- ERROR defines error term 1 as the default error term.
- In the first design, *A* by *B* is defined as error term 1 and is therefore used to test the *A* and *B* effects. The *A* by *B* effect itself is explicitly tested against the within-cells error.
- In the second design, no term is defined as error term 1, so no significance tests are carried out. Hypothesis sums of squares are displayed for *A* and *B*.

CONTRAST Subcommand

CONTRAST specifies the type of contrast desired among the levels of a factor. For a factor with k levels or values, the contrast type determines the meaning of its $k-1$ degrees of freedom. If the subcommand is omitted or is specified with no keyword, the default is DEVIATION for between-subjects factors.

- Specify the factor name in parentheses following the subcommand CONTRAST.
- You can specify only one factor per CONTRAST subcommand, but you can enter multiple CONTRAST subcommands.
- After closing the parentheses, enter an equals sign followed by one of the contrast keywords.
- To obtain F tests for individual degrees of freedom for the specified contrast, enter the factor name followed by a number in parentheses on the DESIGN subcommand. The number refers to a partition of the factor's degrees of freedom. If you do not use the PARTITION subcommand, each degree of freedom is a distinct partition.

The following contrast types are available:

DEVIATION *Deviations from the grand mean.* This is the default for between-subjects factors. Each level of the factor except one is compared to the grand mean. One category (by default the last) must be omitted so that the effects will be independent of one another. To omit a category other than the last, specify the number of the omitted category (which is not necessarily the same as its value) in parentheses after keyword DEVIATION. For example,

```
MANOVA A BY B(2,4)
  /CONTRAST(B)=DEVIATION(1).
```

The specified contrast omits the first category, in which *B* has the value 2. Deviation contrasts are not orthogonal.

POLYNOMIAL *Polynomial contrasts.* This is the default for within-subjects factors. The first degree of freedom contains the linear effect across the levels of the factor, the second contains the quadratic effect, and so on. In a balanced design, polynomial contrasts are orthogonal. By default, the levels are assumed to be equally spaced; you can specify unequal spacing by entering a metric consisting of one integer for each level of the factor in parentheses after keyword POLYNOMIAL. For example,

```
MANOVA RESPONSE BY STIMULUS (4,6)
  /CONTRAST(STIMULUS) = POLYNOMIAL(1,2,4).
```

The specified contrast indicates that the three levels of *STIMULUS* are actually in the proportion 1:2:4. The default metric is always (1,2,...,k), where k levels are involved. Only the relative differences between the terms of the metric matter (1,2,4) is the same metric as (2,3,5) or (20,30,50) because, in each instance, the difference between the second and third numbers is twice the difference between the first and second.

DIFFERENCE *Difference or reverse Helmert contrasts.* Each level of the factor except the first is compared to the mean of the previous levels. In a balanced design, difference contrasts are orthogonal.

HELMERT *Helmert contrasts.* Each level of the factor except the last is compared to the mean of subsequent levels. In a balanced design, Helmert contrasts are orthogonal.

SIMPLE *Each level of the factor except the last is compared to the last level.* To use a category other than the last as the omitted reference category, specify its number (which is not necessarily the same as its value) in parentheses following keyword SIMPLE. For example,

```
MANOVA A BY B(2,4)
  /CONTRAST(B)=SIMPLE(1).
```

The specified contrast compares the other levels to the first level of *B*, in which *B* has the value 2. Simple contrasts are not orthogonal.

REPEATED *Comparison of adjacent levels.* Each level of the factor except the first is compared to the previous level. Repeated contrasts are not orthogonal.

SPECIAL *A user-defined contrast.* After this keyword, enter a square matrix in parentheses with as many rows and columns as there are levels in the factor. The first row represents the mean effect of the factor and is generally a vector of 1's. It represents a set of weights indicating how to collapse over the categories of this factor in estimating parameters for other factors. The other rows of the contrast matrix contain the special contrasts indicating the desired comparisons between levels of the factor. If the special contrasts are linear combinations of each other, MANOVA reports the linear dependency and stops processing.

Orthogonal contrasts are particularly useful. In a balanced design, contrasts are orthogonal if the sum of the coefficients in each contrast row is 0 and if, for any pair of contrast rows, the products of corresponding coefficients sum to 0. DIFFERENCE, HELMERT, and POLYNOMIAL contrasts always meet these criteria in balanced designs. For illustration of contrast types, see Appendix B.

Example

```
MANOVA DEP BY FAC(1,5)
  /CONTRAST(FAC)=DIFFERENCE
  /DESIGN=FAC(1) FAC(2) FAC(3) FAC(4).
```

- The factor *FAC* has five categories and therefore four degrees of freedom.
- CONTRAST requests DIFFERENCE contrasts, which compare each level (except the first) with the mean of the previous levels.
- Each of the four degrees of freedom is tested individually on the DESIGN subcommand.

PARTITION Subcommand

PARTITION subdivides the degrees of freedom associated with a factor. This permits you to test the significance of the effect of a specific contrast or group of contrasts of the factor instead of the overall effect of all contrasts of the factor. The default is a single degree of freedom for each partition.

- Specify the factor name in parentheses following the PARTITION subcommand.
- Specify an integer list in parentheses after the optional equals sign to indicate the degrees of freedom for each partition.
- Each value in the partition list must be a positive integer, and the sum of the values cannot exceed the degrees of freedom for the factor.
- The degrees of freedom available for a factor are one less than the number of levels of the factor.
- The meaning of each degree of freedom depends upon the contrast type for the factor. For example, with deviation contrasts (the default for between-subjects factors), each degree of freedom represents the deviation of the dependent variable in one level of the factor from its grand mean over all levels. With polynomial contrasts, the degrees of freedom represent the linear effect, the quadratic effect, and so on.

- If your list does not account for all the degrees of freedom, MANOVA adds one final partition containing the remaining degrees of freedom.
- You can use a repetition factor of the form *n** to specify a series of partitions with the same number of degrees of freedom.
- To specify a model that tests only the effect of a specific partition of a factor in your design, include the number of the partition in parentheses on the DESIGN subcommand (see example below).
- If you want the default single degree-of-freedom partition, you can omit the PARTITION subcommand and simply enter the appropriate term on the DESIGN subcommand.

Example

```
MANOVA OUTCOME BY TREATMNT(1,12)
  /PARTITION(TREATMNT) = (3*2,4)
  /DESIGN TREATMNT(2).
```

- The factor *TREATMNT* has 12 categories, hence 11 degrees of freedom.
- PARTITION divides the effect of *TREATMNT* into four partitions, containing respectively two, two, two, and four degrees of freedom. A fifth partition is formed to contain the remaining one degree of freedom.
- DESIGN specifies a model in which only the second partition of *TREATMNT* is tested. This partition contains the third and fourth degrees of freedom.
- Since the default contrast type for between-subjects factors is DEVIATION, this second partition represents the deviation of the third and fourth levels of *TREATMNT* from the grand mean.

METHOD Subcommand

METHOD controls the computational aspects of the MANOVA analysis. You can specify one of two different methods for partitioning the sums of squares. The default is UNIQUE.

UNIQUE *Regression approach.* Each term is corrected for every other term in the model. With this approach, sums of squares for various components of the model do not add up to the total sum of squares unless the design is balanced. This is the default if the METHOD subcommand is omitted or if neither of the two keywords is specified.

SEQUENTIAL *Hierarchical decomposition of the sums of squares.* Each term is adjusted only for the terms that precede it on the DESIGN subcommand. This is an orthogonal decomposition, and the sums of squares in the model add up to the total sum of squares.

You can control how parameters are to be estimated by specifying one of the following two keywords available on MANOVA. The default is QR.

QR *Use modified Givens rotations.* QR bypasses the normal equations and the inaccuracies that can result from creating the cross-products matrix, and it generally results in extremely accurate parameter estimates. This is the default if the METHOD subcommand is omitted or if neither of the two keywords is specified.

CHOLESKY *Use Cholesky decomposition of the cross-products matrix.* Useful for large data sets with covariates entered on the DESIGN subcommand.

You can also control whether a constant term is included in all models. Two keywords are available on METHOD. The default is CONSTANT.

CONSTANT *All models include a constant (grand mean) term, even if none is explicitly specified on the DESIGN subcommand.* This is the default if neither of the two keywords is specified.

NOCONSTANT *Exclude constant terms from models that do not include keyword CONSTANT on the DESIGN subcommand.*

Example

```
MANOVA DEP BY A B C (1,4)
  /METHOD=NOCONSTANT
  /DESIGN=A, B, C
  /METHOD=CONSTANT SEQUENTIAL
  /DESIGN.
```

- For the first design, a main effects model, the METHOD subcommand requests the model to be fitted with no constant.
- The second design requests a full factorial model to be fitted with a constant and with a sequential decomposition of sums of squares.

PRINT and NOPRINT Subcommands

PRINT and NOPRINT control the display of optional output.

- Specifications on PRINT remain in effect for all subsequent designs.
- Some PRINT output, such as CELLINFO, applies to the entire MANOVA procedure and is displayed only once.
- You can turn off optional output that you have requested on PRINT by entering a NOPRINT subcommand with the specifications originally used on the PRINT subcommand.
- Additional output can be obtained on the PCOMPS, DISCRIM, OMEANS, PMEANS, PLOT, and RESIDUALS subcommands.
- Some optional output greatly increases the processing time. Request only the output you want to see.

The following specifications are appropriate for univariate MANOVA analyses. For information on PRINT specifications appropriate for other MANOVA models, see MANOVA: Multivariate and MANOVA: Repeated Measures.

CELLINFO *Basic information about each cell in the design.*

PARAMETERS *Parameter estimates.*

HOMOGENEITY *Tests of homogeneity of variance.*

DESIGN *Design information.*

ERROR *Error standard deviations.*

CELLINFO Keyword

You can request any of the following cell information by specifying the appropriate keyword(s) in parentheses after CELLINFO. The default is MEANS.

MEANS *Cell means, standard deviations, and counts for the dependent variable and covariates.* Confidence intervals for the cell means are displayed if you have set width wide. This is the default when CELLINFO is requested with no further specification.

SSCP *Within-cell sum-of-squares and cross-products matrices for the dependent variable and covariates.*

COV *Within-cell variance-covariance matrices for the dependent variable and covariates.*

COR *Within-cell correlation matrices, with standard deviations on the diagonal, for the dependent variable and covariates.*

ALL *MEANS, SSCP, COV, COR.*

- Output from CELLINFO is displayed once before the analysis of any particular design. Specify CELLINFO only once.
- When you specify SSCP, COV, or COR, the cells are numbered for identification, beginning with cell 1.
- The levels vary most rapidly for the factor named last on the MANOVA variables specification.
- Empty cells are neither displayed nor numbered.
- A table showing the levels of each factor corresponding to each cell number is displayed at the beginning of MANOVA output.

Example

```
MANOVA DEP BY A(1,4) B(1,2) WITH COV
  /PRINT=CELLINFO(MEANS COV)
  /DESIGN.
```

- For each combination of levels of *A* and *B*, MANOVA displays separately the means and standard deviations of *DEP* and *COV*. Beginning with cell 1, it will then display the variance-covariance matrix of *DEP* and *COV* within each non-empty cell.
- A table of cell numbers will be displayed to show the factor levels corresponding to each cell.
- Keyword COV, as a parameter of CELLINFO, is not confused with variable *COV*.

PARAMETERS Keyword

Keyword PARAMETERS displays information relating to the estimated size of the effects in the model. You can specify any of the following keywords in parentheses on PARAMETERS. The default is ESTIM.

ESTIM *The estimated parameters themselves, along with their standard errors,* t *tests, and confidence intervals.* Only nonredundant parameters are displayed. This is the default if PARAMETERS is requested without further specification.

NEGSUM *The negative of the sum of parameters for each effect.* For DEVIATION main effects, this equals the parameter for the omitted (redundant) contrast. NEGSUM is displayed along with the parameter estimates.

ORTHO *The orthogonal estimates of parameters used to produce the sums of squares.*

COR *Covariance factors and correlations among the parameter estimates.*

EFSIZE *The effect size values.*

OPTIMAL *Optimal Scheffé contrast coefficients.*

ALL *ESTIM, NEGSUM, ORTHO, COR, EFSIZE, and OPTIMAL.*

SIGNIF Keyword

SIGNIF requests special significance tests, most of which apply to multivariate designs (see MANOVA: Multivariate). The following specification is useful in univariate applications of MANOVA:

SINGLEDF *Significance tests for each single degree of freedom making up each effect for analysis-of-variance tables.*

- When nonorthogonal contrasts are requested or when the design is unbalanced, the SINGLEDF effects will differ from single degree-of-freedom partitions. SINGLEDEF effects are orthogonal within an effect; single degree-of-freedom partitions are not.

Example

```
MANOVA DEP BY FAC(1,5)
  /CONTRAST(FAC)=POLY
  /PRINT=SIGNIF(SINGLEDF)
  /DESIGN.
```

- POLYNOMIAL contrasts are applied to *FAC*, testing the linear, quadratic, cubic, and quartic components of its five levels. POLYNOMIAL contrasts are orthogonal in balanced designs.
- The SINGLEDF specification on SIGNIF requests significance tests for each of these four components.

HOMOGENEITY Keyword

HOMOGENEITY requests tests for the homogeneity of variance of the dependent variable across the cells of the design. You can specify one or more of the following specifications in parentheses. If HOMOGENEITY is requested without further specification, the default is ALL.

BARTLETT *Bartlett-Box* F *test.*

COCHRAN *Cochran's* C.

ALL *Both BARTLETT and COCHRAN.* This is the default.

DESIGN Keyword

You can request the following by entering one or more of the specifications in parentheses following keyword DESIGN. If DESIGN is requested without further specification, the default is OVERALL.

The DECOMP and BIAS matrices can provide valuable information on the confounding of the effects and the estimability of the chosen contrasts. If two effects are confounded, the entry corresponding to them in the BIAS matrix will be non-zero; if they are orthogonal, the entry will be zero. This is particularly useful in designs with unpatterned empty cells. For further discussion of the matrices, see Bock (1985).

OVERALL *The overall reduced-model design matrix (not the contrast matrix).* This is the default.

ONEWAY *The one-way basis matrix (not the contrast matrix) for each factor.*

DECOMP *The upper triangular QR/CHOLESKY decomposition of the design.*

BIAS *Contamination coefficients displaying the bias present in the design.*

SOLUTION *Coefficients of the linear combinations of the cell means used in significance testing.*

REDUNDANCY *Exact linear combinations of parameters that form a redundancy.* This keyword displays a table only if QR (the default) is the estimation method.

COLLINEARITY *Collinearity diagnostics for design matrices.* These diagnostics include the singular values of the normalized design matrix (which are the same as those of the normalized decomposition matrix), condition indexes corresponding to each singular value, and the proportion of variance of the corresponding parameter accounted for by each principal component. For greatest accuracy, use the QR method of estimation whenever you request collinearity diagnostics.

ALL *All available options.*

ERROR Keyword

Generally, keyword ERROR on PRINT produces error matrices. In univariate analyses, the only valid specification for ERROR is STDDEV, which is the default if ERROR is specified by itself.

STDDEV *The error standard deviation.* Normally, this is the within-cells standard deviation of the dependent variable. If you specify multiple error terms on DESIGN, this specification will display the standard deviation for each.

OMEANS Subcommand

OMEANS (observed means) displays tables of the means of continuous variables for levels or combinations of levels of the factors.

- Use keywords VARIABLES and TABLES to indicate which observed means you want to display.
- With no specifications, the OMEANS subcommand is equivalent to requesting CELLINFO (MEANS) on PRINT.
- OMEANS displays confidence intervals for the cell means if you have set width to 132.
- Output from OMEANS is displayed once before the analysis of any particular design. This subcommand should be specified only once.

VARIABLES *Continuous variables for which you want means.* Specify the variables in parentheses after the keyword VARIABLES. You can request means for the dependent variable or any covariates. If you omit the VARIABLES keyword, observed means are displayed for the dependent variable and all covariates. If you enter the keyword VARIABLES, you must also enter the keyword TABLES, discussed below.

TABLES *Factors for which you want the observed means displayed.* List in parentheses the factors, or combinations of factors, separated with BY. Observed means are displayed for each level, or combination of levels, of the factors named (see example below). Both weighted means and unweighted means (where all cells are weighted equally, regardless of the number of cases they contain) are displayed. If you enter the keyword CONSTANT, the grand mean is displayed.

Example

```
MANOVA DEP BY A(1,3) B(1,2)
  /OMEANS=TABLES(A,B)
  /DESIGN.
```

- Because there is no VARIABLES specification on the OMEANS subcommand, observed means are displayed for all continuous variables. *DEP* is the only dependent variable here, and there are no covariates.
- The TABLES specification on the OMEANS subcommand requests tables of observed means for each of the three categories of *A* (collapsing over *B*) and for both categories of *B* (collapsing over *A*).
- MANOVA displays both weighted means, in which all cases count equally, and unweighted means, in which all cells count equally.

PMEANS Subcommand

PMEANS (predicted means) displays a table of the predicted cell means of the dependent variable, both adjusted for the effect of covariates in the cell and unadjusted for covariates. For comparison, it also displays the observed cell means.

- Output from PMEANS can be computationally expensive.
- PMEANS without any additional specifications displays a table showing for each cell the observed mean of the dependent variable, the predicted mean adjusted for the effect of covariates in that cell (ADJ. MEAN), the predicted mean unadjusted for covariates (EST. MEAN), and the raw and standardized residuals from the estimated means.

- Cells are numbered in output from PMEANS so that the levels vary most rapidly on the factor named last in the MANOVA variables specification. A table showing the levels of each factor corresponding to each cell number is displayed at the beginning of the MANOVA output.
- Predicted means are suppressed for any design in which the MUPLUS keyword appears.
- Covariates are not predicted.
- In designs with covariates and multiple error terms, use the ERROR subcommand to designate which error term's regression coefficients are to be used in calculating the standardized residuals.

For univariate analysis, the following keywords are available on the PMEANS subcommand:

TABLES *Additional tables showing adjusted predicted means for specified factors or combinations of factors.* Enter the names of factors or combinations of factors in parentheses after this keyword. For each factor or combination, MANOVA displays the predicted means (adjusted for covariates) collapsed over all other factors.

PLOT *A plot of the predicted means for each cell.*

Example

```
MANOVA DEP BY A(1,4) B(1,3)
  /PMEANS TABLES(A, B, A BY B)
  /DESIGN = A, B.
```

- PMEANS displays the default table of observed and predicted means for *DEP* and raw and standardized residuals in each of the 12 cells in the model.
- The TABLES specification on PMEANS displays tables of predicted means for *A* (collapsing over *B*), for *B* (collapsing over *A*), and all combinations of *A* and *B*.
- Because *A* and *B* are the only factors in the model, the means for *A* by *B* in the TABLES specification come from every cell in the model. They are identical to the adjusted predicted means in the default PMEANS table, which always includes all non-empty cells.
- Predicted means for *A* by *B* can be requested in the TABLES specification, even though the *A* by *B* effect is not in the design.

RESIDUALS Subcommand

Use RESIDUALS to display and plot casewise values and residuals for your models.

- Use the ERROR subcommand to specify an error term other than the default to be used to standardize the residuals.
- If a designated error term does not exist for a given design, no predicted values or residuals are calculated.
- If you specify RESIDUALS without any keyword, CASEWISE output is displayed.

The following keywords are available:

CASEWISE *A case-by-case listing of the observed, predicted, residual, and standardized residual values for each dependent variable.*

PLOT *A plot of observed values, predicted values, and case numbers versus the standardized residuals, plus normal and detrended normal probability plots for the standardized residuals (five plots in all).*

POWER Subcommand

POWER requests observed power values based on fixed-effect assumptions for all univariate and multivariate *F* tests and *t* tests. Both approximate and exact power values can be computed, although exact multivariate power is displayed only when there is one hypothesis degree of freedom. If POWER is specified by itself, with no keywords, MANOVA calculates the approximate observed power values of all *F* tests at 0.05 significance level.

The following keywords are available on the POWER subcommand:

APPROXIMATE *Approximate power values.* This is the default if POWER is specified without any keyword. Approximate power values for univariate tests are derived from an Edgeworth-type normal approximation to the noncentral beta distribution. Approximate values are normally accurate to three decimal places and are much cheaper to compute than exact values.

EXACT *Exact power values.* Exact power values for univariate tests are computed from the noncentral incomplete beta distribution.

F(a) *Alpha level at which the power is to be calculated for* F *tests.* The default is 0.05. To change the default, specify a decimal number between 0 and 1 in parentheses after *F*. The numbers 0 and 1 themselves are not allowed. *F* test at 0.05 significance level is the default when POWER is omitted or specified without any keyword.

T(a) *Alpha level at which the power is to be calculated for* t *tests.* The default is 0.05. To change the default, specify a decimal number between 0 and 1 in parentheses after *t*. The numbers 0 and 1 themselves are not allowed.

- For univariate *F* tests and *t* tests, MANOVA computes a measure of the effect size based on partial η^2:

$$\text{partial}\eta^2 = (ssh) / (ssh + sse)$$

where *ssh* is hypothesis sum of squares and *sse* is error sum of squares. The measure is an overestimate of the actual effect size. However, it is consistent and is applicable to all *F* tests and *t* tests. For a discussion of effect size measures, see Cohen (1977) or Hays (1981).

CINTERVAL Subcommand

CINTERVAL requests simultaneous confidence intervals for each parameter estimate and regression coefficient. MANOVA provides either individual or joint confidence intervals at any desired confidence level. You can compute joint confidence intervals using either Scheffé or Bonferroni intervals. Scheffé intervals are based on all possible contrasts, while Bonferroni intervals are based on the number of contrasts actually made. For a large number of contrasts,

Bonferroni intervals will be larger than Scheffé intervals. Timm (1975) provides a good discussion of which intervals are best for certain situations. Both Scheffé and Bonferroni intervals are computed separately for each term in the design. You can request only one type of confidence interval per design.

The following keywords are available on the CINTERVAL subcommand. If the subcommand is specified without any keyword, CINTERVAL automatically displays individual univariate confidence intervals at the 0.95 level.

INDIVIDUAL(a) *Individual confidence intervals.* Specify the desired confidence level in parentheses following the keyword. The desired confidence level can be any decimal number between 0 and 1. When individual intervals are requested, BONFER and SCHEFFE have no effect.

JOINT(a) *Joint confidence intervals.* Specify the desired confidence level in parentheses after the keyword. The default is 0.95. The desired confidence level can be any decimal number between 0 and 1.

UNIVARIATE(type) *Univariate confidence interval.* Specify either SCHEFFE for Scheffé intervals or BONFER for Bonferroni intervals in parentheses after the keyword. The default specification is SCHEFFE.

PLOT Subcommand

MANOVA can display a variety of plots useful in checking the assumptions needed in the analysis. Plots are produced only once in the MANOVA procedure, regardless of how many DESIGN subcommands you enter. Use the following keywords on the PLOT subcommand to request plots. If the PLOT subcommand is specified by itself, the default is BOXPLOT.

BOXPLOTS *Boxplots.* Plots are displayed for each continuous variable (dependent or covariate) named on the MANOVA variable list. Boxplots provide a simple graphical means of comparing the cells in terms of mean location and spread. The data must be stored in memory for these plots; if there is not enough memory, boxplots are not produced and a warning message is issued. This is the default if the PLOT subcommand is specified without a keyword.

CELLPLOTS *Cell statistics, including a plot of cell means versus cell variances, a plot of cell means versus cell standard deviations, and a histogram of cell means.* Plots are produced for each continuous variable (dependent or covariate) named on the MANOVA variable list. The first two plots aid in detecting heteroscedasticity (nonhomogeneous variances) and in determining an appropriate data transformation if one is needed. The third plot gives distributional information for the cell means.

NORMAL *Normal and detrended normal plots.* Plots are produced for each continuous variable (dependent or covariate) named on the MANOVA variable list. MANOVA ranks the scores and then plots the ranks against the expected normal deviate, or detrended expected normal deviate, for that rank. These plots aid in detecting non-normality and outlying observations. All data must be held in memory to compute ranks. If not enough memory is available, MANOVA displays a warning and skips the plots.

- ZCORR, an additional plot available on the PLOT subcommand, is described in MANOVA: Multivariate.
- You can request other plots on PMEANS and RESIDUALS (see respective subcommands).

MISSING Subcommand

By default, cases with missing values for any of the variables on the MANOVA variable list are excluded from the analysis. The MISSING subcommand allows you to include cases with user-missing values. If MISSING is not specified, the defaults are LISTWISE and EXCLUDE.

- The same missing-value treatment is used to process all designs in a single execution of MANOVA.
- If you enter more than one MISSING subcommand, the last one entered will be in effect for the entire procedure, including designs specified before the last MISSING subcommand.
- Pairwise deletion of missing data is not available in MANOVA.
- Keywords INCLUDE and EXCLUDE are mutually exclusive; either can be specified with LISTWISE.

LISTWISE *Cases with missing values for any variable named on the MANOVA variable list are excluded from the analysis.* This is always true in the MANOVA procedure.

EXCLUDE *Exclude both user-missing and system-missing values.* This is the default when MISSING is not specified.

INCLUDE *User-missing values are treated as valid.* For factors, you must include the missing-value codes within the range specified on the MANOVA variable list. It may be necessary to recode these values so that they will be adjacent to the other factor values. System-missing values cannot be included in the analysis.

MATRIX Subcommand

MATRIX reads and writes SPSS matrix data files. It writes correlation matrices that can be read by subsequent MANOVA procedures.

- Either IN or OUT is required to specify the matrix file in parentheses. When both IN and OUT are used on the same MANOVA procedure, they can be specified on separate MATRIX subcommands or on the same subcommand.
- The matrix materials include the N, mean, and standard deviation. Documents from the file that form the matrix are not included in the matrix data file.
- MATRIX=IN cannot be used in place of GET or DATA LIST to begin a new SPSS command file. MATRIX is a subcommand on MANOVA, and MANOVA cannot run before a working data file is defined. To begin a new command file and immediately read a matrix, first GET the matrix file, and then specify IN(*) on MATRIX.
- Records in the matrix data file read by MANOVA can be in any order, with the following exceptions: the order of split-file groups cannot be violated, and all CORR vectors must appear contiguously within each split-file group.
- When MANOVA reads matrix materials, it ignores the record containing the total number of cases. In addition, it skips unrecognized records. MANOVA does not issue a warning when it skips records.

The following two keywords are available on the MATRIX subcommand:

OUT *Write an SPSS matrix data file.* Specify either a file or an asterisk, and enclose the specification in parentheses. If you specify a file, the file is stored on disk and can be retrieved at any time. If you specify an asterisk (*) or leave the parentheses empty, the matrix file replaces the working data file but is not stored on disk unless you use SAVE or XSAVE.

IN *Read an SPSS matrix data file.* If the matrix file *is not* the current working data file, specify a file in parentheses. If the matrix file *is* the current working data file, specify an asterisk (*) or leave the parentheses empty.

Format of the SPSS Matrix Data File

The SPSS matrix data file includes two special variables created by SPSS: *ROWTYPE_* and *VARNAME_*.

- Variable *ROWTYPE_* is a short string variable having values N, MEAN, CORR (for Pearson correlation coefficients), and STDDEV.
- Variable *VARNAME_* is a short string variable whose values are the names of the variables and covariates used to form the correlation matrix. When *ROWTYPE_* is CORR, *VARNAME_* gives the variable associated with that row of the correlation matrix.
- Between *ROWTYPE_* and *VARNAME_* are the factor variables (if any) defined in the BY portion of the MANOVA variable list. (Factor variables receive the system-missing value on vectors that represent pooled values.)
- Remaining variables are the variables used to form the correlation matrix.

Split Files and Variable Order

- When split-file processing is in effect, the first variables in the matrix system file will be the split variables, followed by *ROWTYPE_*, the factor variable(s), *VARNAME_*, and then the variables used to form the correlation matrix.
- A full set of matrix materials is written for each subgroup defined by the split variable(s).
- A split variable cannot have the same variable name as any other variable written to the matrix data file.
- If a split file is in effect when a matrix is written, the same split file must be in effect when that matrix is read into another procedure.

Additional Statistics

In addition to the CORR values, MANOVA always includes the following with the matrix materials:

- The total weighted number of cases used to compute each correlation coefficient.
- A vector of N's for each cell in the data.
- A vector of MEAN's for each cell in the data.

- A vector of pooled standard deviations, STDDEV. This is the square root of the within-cells mean square error for each variable.

Example

```
GET FILE IRIS.
MANOVA SEPALLEN SEPALWID PETALLEN PETALWID BY TYPE(1,3)
  /MATRIX=OUT(MANMTX).
```

- MANOVA reads data from the SPSS data file *IRIS* and writes one set of matrix materials to the file *MANMTX*.
- The working data file is still *IRIS*. Subsequent commands are executed on the file *IRIS*.

Example

```
GET FILE IRIS.
MANOVA SEPALLEN SEPALWID PETALLEN PETALWID BY TYPE(1,3)
  /MATRIX=OUT(*).
LIST.
```

- MANOVA writes the same matrix as in the example above. However, the matrix file replaces the working data file. The LIST command is executed on the matrix file, not on the file *IRIS*.

Example

```
GET FILE=PRSNNL.
FREQUENCIES VARIABLE=AGE.

MANOVA SEPALLEN SEPALWID PETALLEN PETALWID BY TYPE(1,3)
  /MATRIX=IN(MANMTX).
```

- This example assumes that you want to perform a frequencies analysis on the file *PRSNNL* and then use MANOVA to read a different file. The file you want to read is an existing SPSS matrix data file. The external matrix file *MANMTX* is specified in parentheses after IN on the MATRIX subcommand.
- *MANMTX* does not replace *PRSNNL* as the working file.

Example

```
GET FILE=MANMTX.
MANOVA SEPALLEN SEPALWID PETALLEN PETALWID BY TYPE(1,3)
  /MATRIX=IN(*).
```

- This example assumes that you are starting a new session and want to read an existing SPSS matrix data file. GET retrieves the matrix file *MANMTX*.
- An asterisk is specified in parentheses after IN on the MATRIX subcommand to read the working data file. You can also leave the parentheses empty to indicate the default.
- If the GET command is omitted, SPSS issues an error message.
- If you specify *MANMTX* in parentheses after IN, SPSS issues an error message.

ANALYSIS Subcommand

ANALYSIS allows you to work with a subset of the continuous variables (dependent variable and covariates) you have named on the MANOVA variable list. In univariate analysis of variance, you can use ANALYSIS to allow factor-by-covariate interaction terms in your model (see the DESIGN subcommand on p. 440). You can also use it to switch the roles of the dependent variable and a covariate.

- In general, ANALYSIS gives you complete control over which continuous variables are to be dependent variables, which are to be covariates, and which are to be neither.
- ANALYSIS specifications are like the MANOVA variables specification, except that factors are not named. Enter the dependent variable and, if there are covariates, the keyword WITH and the covariates.
- Only variables listed as dependent variables or covariates on the MANOVA variable list can be entered on the ANALYSIS subcommand.
- In a univariate analysis of variance, the most important use of ANALYSIS is to omit covariates altogether from the analysis list, thereby making them available for inclusion on DESIGN (see example below and the DESIGN subcommand examples).
- For more information on ANALYSIS, refer to MANOVA: Multivariate.

Example

```
MANOVA DEP BY FACTOR(1,3) WITH COV
  /ANALYSIS DEP
  /DESIGN FACTOR, COV, FACTOR BY COV.
```

- *COV*, a continuous variable, is included on the MANOVA variable list as a covariate.
- *COV* is not mentioned on ANALYSIS, so it will not be included in the model as a dependent variable or covariate. It can, therefore, be explicitly included on the DESIGN subcommand.
- DESIGN includes the main effects of *FACTOR* and *COV* and the *FACTOR* by *COV* interaction.

DESIGN Subcommand

DESIGN specifies the effects included in a specific model. It must be the last subcommand entered for any model.

The cells in a design are defined by all of the possible combinations of levels of the factors in that design. The number of cells equals the product of the number of levels of all the factors. A design is *balanced* if each cell contains the same number of cases. MANOVA can analyze both balanced and unbalanced designs.

- Specify a list of terms to be included in the model, separated by spaces or commas.
- The default design, if the DESIGN subcommand is omitted or is specified by itself, is a full factorial model containing all main effects and all orders of factor-by-factor interaction.
- If the last subcommand specified is not DESIGN, a default full factorial design is estimated.
- To include a term for the main effect of a factor, enter the name of the factor on the DESIGN subcommand.

- To include a term for an interaction between factors, use the keyword BY to join the factors involved in the interaction.
- Terms are entered into the model in the order in which you list them on DESIGN. If you have specified SEQUENTIAL on the METHOD subcommand to partition the sums of squares in a hierarchical fashion, this order may affect the significance tests.
- You can specify other types of terms in the model, as described in the following sections.
- Multiple DESIGN subcommands are accepted. An analysis of one model is produced for each DESIGN subcommand.

Example

```
MANOVA Y BY A(1,2) B(1,2) C(1,3)
  /DESIGN
  /DESIGN A, B, C
  /DESIGN A, B, C, A BY B, A BY C.
```

- The first DESIGN produces the default full factorial design, with all main effects and interactions for factors *A*, *B*, and *C*.
- The second DESIGN produces an analysis with main effects only for *A*, *B*, and *C*.
- The third DESIGN produces an analysis with main effects and the interactions between *A* and the other two factors. The interaction between *B* and *C* is not in the design, nor is the interaction between all three factors.

Partitioned Effects: Number in Parentheses

You can specify a number in parentheses following a factor name on the DESIGN subcommand to identify individual degrees of freedom or partitions of the degrees of freedom associated with an effect.

- If you specify PARTITION, the number refers to a partition. Partitions can include more than one degree of freedom (see the PARTITION subcommand on p. 427). For example, if the first partition of *SEED* includes two degrees of freedom, the term SEED(1) on a DESIGN subcommand tests the two degrees of freedom.
- If you do not use PARTITION, the number refers to a single degree of freedom associated with the effect.
- The number refers to an individual level for a factor if that factor follows the keyword WITHIN or MWITHIN (see the sections on nested effects and pooled effects below).
- A factor has one less degree of freedom than it has levels or values.

Example

```
MANOVA YIELD BY SEED(1,4) WITH RAINFALL
  /PARTITION(SEED)=(2,1)
  /DESIGN=SEED(1) SEED(2).
```

- Factor *SEED* is subdivided into two partitions, one containing the first two degrees of freedom and the other the last degree of freedom.
- The two partitions of *SEED* are treated as independent effects.

Nested Effects: WITHIN Keyword

Use the WITHIN keyword (alias W) to nest the effects of one factor within those of another factor or an interaction term.

Example

```
MANOVA YIELD BY SEED(1,4) FERT(1,3) PLOT (1,4)
  /DESIGN = FERT WITHIN SEED BY PLOT.
```

- The three factors in this example are type of seed (*SEED*), type of fertilizer (*FERT*), and location of plots (*PLOT*).
- The DESIGN subcommand nests the effects of *FERT* within the interaction term of *SEED* by *PLOT*. The levels of *FERT* are considered distinct for each combination of levels of *SEED* and *PLOT*.

Simple Effects: WITHIN and MWITHIN Keywords

A factor can be nested within one specific level of another factor by indicating the level in parentheses. This allows you to estimate simple effects or the effect of one factor within only one level of another. Simple effects can be obtained for higher-order interactions as well.

Use WITHIN to request simple effects of between-subjects factors.

Example

```
MANOVA YIELD BY SEED(2,4) FERT(1,3) PLOT (1,4)
  /DESIGN = FERT WITHIN SEED (1).
```

- This example requests the simple effect of *FERT* within the first level of *SEED*.
- The number (*n*) specified after a WITHIN factor refers to the level of that factor. It is the ordinal position, which is not necessarily the value of that level. In this example, the first level is associated with value 2.
- The number does *not* refer to the number of partitioned effects (see "Partitioned Effects: Number in Parentheses" on p. 441).

Example

```
MANOVA YIELD BY SEED(2,4) FERT(1,3) PLOT (3,5)
 /DESIGN = FERT WITHIN PLOT(1) WITHIN SEED(2)
```

- This example requests the effect of *FERT* within the second *SEED* level of the first *PLOT* level.
- The second *SEED* level is associated with value 3 and the first *PLOT* level is associated with value 3.

Use MWITHIN to request simple effects of within-subjects factors in repeated measures analysis (see MANOVA: Repeated Measures).

Pooled Effects: Plus Sign

To pool different effects for the purpose of significance testing, join the effects with a plus sign (+). A single test is made for the combined effect of the pooled terms.

- Keyword BY is evaluated before effects are pooled together.
- Parentheses are not allowed to change the order of evaluation. For example, it is illegal to specify `(A + B) BY C`. You must specify `/DESIGN=A BY C + B BY C`.

Example

```
MANOVA Y BY A(1,3) B(1,4) WITH X
 /ANALYSIS=Y
 /DESIGN=A, B, A BY B, A BY X + B BY X + A BY B BY X.
```

- This example shows how to test homogeneity of regressions in a two-way analysis of variance.
- The + signs are used to produce a pooled test of all interactions involving the covariate *X*. If this test is significant, the assumption of homogeneity of variance is questionable.

MUPLUS Keyword

MUPLUS combines the constant term (μ) in the model with the term specified after it. The normal use of this specification is to obtain parameter estimates that represent weighted means for the levels of some factor. For example, `MUPLUS SEED` represents the constant, or overall, mean plus the effect for each level of *SEED*. The significance of such effects is usually uninteresting, but the parameter estimates represent the weighted means for each level of *SEED*, adjusted for any covariates in the model.

- MUPLUS cannot appear more than once on a given DESIGN subcommand.
- MUPLUS is the only way to get standard errors for the predicted mean for each level of the factor specified.
- Parameter estimates are not displayed by default; you must explicitly request them on the PRINT subcommand or via a CONTRAST subcommand.
- You can obtain the unweighted mean by specifying the full factorial model, excluding those terms contained by an effect, and prefixing the effect whose mean is to be found by MUPLUS.

Effects of Continuous Variables

Usually you name factors but not covariates on the DESIGN subcommand. The linear effects of covariates are removed from the dependent variable before the design is tested. However, the design can include variables measured at the interval level and originally named as covariates or as additional dependent variables.

- Continuous variables on a DESIGN subcommand must be named as dependents or covariates on the MANOVA variable list.

- Before you can name a continuous variable on a DESIGN subcommand, you must supply an ANALYSIS subcommand that does *not* name the variable. This excludes it from the analysis as a dependent variable or covariate and makes it eligible for inclusion on DESIGN.
- More than one continuous variable can be pooled into a single effect (provided that they are all excluded on an ANALYSIS subcommand) with keyword POOL(varlist). For a single continuous variable, `POOL(VAR)` is equivalent to `VAR`.
- The TO convention in the variable list for POOL refers to the order of continuous variables (dependent variables and covariates) on the original MANOVA variable list, which is not necessarily their order on the working data file. This is the only allowable use of keyword TO on a DESIGN subcommand.
- You can specify interaction terms between factors and continuous variables. If *FAC* is a factor and *COV* is a covariate that has been omitted from an ANALYSIS subcommand, `FAC BY COV` is a valid specification on a DESIGN statement.
- You cannot specify an interaction between two continuous variables. Use the COMPUTE command to create a variable representing the interaction prior to MANOVA.

Example

```
*  This example tests whether the regression of the dependent
   variable Y on the two variables X1 and X2 is the same across
   all the categories of the factors AGE and TREATMNT.

MANOVA Y BY AGE(1,5) TREATMNT(1,3) WITH X1, X2
  /ANALYSIS = Y
  /DESIGN = POOL(X1,X2),
            AGE, TREATMNT, AGE BY TREATMNT,
            POOL(X1,X2) BY AGE + POOL(X1,X2) BY TREATMNT
              + POOL(X1,X2) BY AGE BY TREATMNT.
```

- ANALYSIS excludes *X1* and *X2* from the standard treatment of covariates, so that they can be used in the design.
- DESIGN includes five terms. `POOL(X1,X2)`, the overall regression of the dependent variable on *X1* and *X2*, is entered first, followed by the two factors and their interaction.
- The last term is the test for equal regressions. It consists of three factor-by-continuous-variable interactions pooled together. `POOL(X1,X2) BY AGE` is the interaction between *AGE* and the combined effect of the continuous variables *X1* and *X2*. It is combined with similar interactions between *TREATMNT* and the continuous variables and between the *AGE* by *TREATMNT* interaction and the continuous variables.
- If the last term is not statistically significant, there is no evidence that the regression of *Y* on *X1* and *X2* is different across any combination of the categories of *AGE* and *TREATMNT.*

Error Terms for Individual Effects

The “error” sum of squares against which terms in the design are tested is specified on the ERROR subcommand. For any particular term on a DESIGN subcommand, you can specify a different error term to be used in the analysis of variance. To do so, name the term followed by keyword VS (or AGAINST) and the error term keyword.

- To test a term against only the within-cells sum of squares, specify the term followed by VS WITHIN on the DESIGN subcommand. For example, `GROUP VS WITHIN` tests the effect of the factor *GROUP* against only the within-cells sum of squares. For most analyses, this is the default error term.
- To test a term against only the residual sum of squares (the sum of squares for all terms not included in your DESIGN), specify the term followed by VS RESIDUAL.
- To test against the combined within-cells and residual sums of squares, specify the term followed by VS WITHIN+RESIDUAL.
- To test against any other sum of squares in the analysis of variance, include a term corresponding to the desired sum of squares in the design and assign it to an integer between 1 and 10. You can then test against the number of the error term. It is often convenient to test against the term before you define it. This is perfectly acceptable as long as you define the error term on the same DESIGN subcommand.

Example

```
MANOVA DEP BY A, B, C (1,3)
  /DESIGN=A VS 1,
          B WITHIN A = 1 VS 2,
          C WITHIN B WITHIN A = 2 VS WITHIN.
```

- In this example, the factors *A*, *B*, and *C* are completely nested; levels of *C* occur within levels of *B*, which occur within levels of *A*. Each factor is tested against everything within it.
- *A*, the outermost factor, is tested against the *B* within *A* sum of squares, to see if it contributes anything beyond the effects of *B* within each of its levels. The *B* within *A* sum of squares is defined as error term number 1.
- *B* nested within *A*, in turn, is tested against error term number 2, which is defined as the *C* within *B* within *A* sum of squares.
- Finally, *C* nested within *B* nested within *A* is tested against the within-cells sum of squares.

User-defined error terms are specified by simply inserting `= n` after a term, where *n* is an integer from 1 to 10. The equals sign is required. Keywords used in building a design term, such as BY or WITHIN, are evaluated first. For example, error term number 2 in the above example consists of the entire term `C WITHIN B WITHIN A`. An error-term *number,* but not an error-term *definition,* can follow keyword VS.

CONSTANT Keyword

By default, the constant (grand mean) term is included as the first term in the model.

- If you have specified NOCONSTANT on the METHOD subcommand, a constant term will not be included in any design unless you request it with the CONSTANT keyword on DESIGN.
- You can specify an error term for the constant.
- A factor named CONSTANT will not be recognized on the DESIGN subcommand.

MANOVA: Multivariate

```
MANOVA dependent varlist [BY factor list (min,max) [factor list...]]
                         [WITH covariate list]

 [/TRANSFORM [(dependent varlist [/dependent varlist])]=
                  [ORTHONORM] [{CONTRAST}] {DEVIATIONS (refcat)     }]
                              {BASIS   }   {DIFFERENCE              }
                                           {HELMERT                 }
                                           {SIMPLE (refcat)         }
                                           {REPEATED                }
                                           {POLYNOMIAL[({1,2,3...})]}
                                                        {metric   }
                                           {SPECIAL (matrix)        }

 [/RENAME={newname} {newname}...]
          {*      } {*      }

 [/{PRINT  }=[HOMOGENEITY [(BOXM)]]
   {NOPRINT} [ERROR [([COV] [COR] [SSCP] [STDDEV])]]
             [SIGNIF [([MULTIV**] [EIGEN] [DIMENR]
                      [UNIV**] [HYPOTH][STEPDOWN] [BRIEF])]]
             [TRANSFORM]                                  ]

 [/PCOMPS=[COR] [COV] [ROTATE(rottype)]
          [NCOMP(n)] [MINEIGEN(eigencut)] [ALL]]

 [/PLOT=[ZCORR]]

 [/DISCRIM [RAW] [STAN] [ESTIM] [COR] [ALL]
           [ROTATE(rottype)] [ALPHA({.25**})]]
                                    {a    }

 [/POWER=[T({.05**})] [F({.05**})] [{APPROXIMATE}]]
            {a    }      {a    }     {EXACT      }

 [/CINTERVAL=[MULTIVARIATE  ({ROY      })]]
                             {PILLAI   }
                             {BONFER   }
                             {HOTELLING}
                             {WILKS    }

 [/ANALYSIS [({UNCONDITIONAL**})]=[(]dependent varlist
              {CONDITIONAL    }       [WITH covariate varlist]
                                        [/dependent varlist...][)][WITH varlist]]

 [/DESIGN...]*
```

* The DESIGN subcommand has the same syntax as is described in MANOVA: Univariate.

**Default if subcommand or keyword is omitted.

Example:

```
MANOVA SCORE1 TO SCORE4 BY METHOD(1,3).
```

Overview

This section discusses the subcommands that are used in multivariate analysis of variance and covariance designs with several interrelated dependent variables. The discussion focuses on subcommands and keywords that do not apply, or apply in different manners, to univariate analyses. It does not contain information on all the subcommands you will need to specify the design. For subcommands not covered here, see MANOVA: Univariate.

Options

Dependent Variables and Covariates. You can specify subsets and reorder the dependent variables and covariates using the ANALYSIS subcommand. You can specify linear transformations of the dependent variables and covariates using the TRANSFORM subcommand. When transformations are performed, you can rename the variables using the RENAME subcommand and request the display of a transposed transformation matrix currently in effect using the PRINT subcommand.

Optional Output. You can request or suppress output on the PRINT and NOPRINT subcommands. Additional output appropriate to multivariate analysis includes error term matrices, Box's *M* statistic, multivariate and univariate *F* tests, and other significance analyses. You can also request predicted cell means for specific dependent variables on the PMEANS subcommand, produce a canonical discriminant analysis for each effect in your model with the DISCRIM subcommand, specify a principal components analysis of each error sum-of-squares and cross-product matrix in a multivariate analysis on the PCOMPS subcommand, display multivariate confidence intervals using the CINTERVAL subcommand, and generate a half-normal plot of the within-cells correlations among the dependent variables with the PLOT subcommand.

Basic Specification

- The basic specification is a variable list identifying the dependent variables, with the factors (if any) named after BY and the covariates (if any) named after WITH.
- By default, MANOVA produces multivariate and univariate *F* tests.

Subcommand Order

- The variable list must be specified first.
- Subcommands applicable to a specific design must be specified before that DESIGN subcommand. Otherwise, subcommands can be used in any order.

Syntax Rules

- All syntax rules applicable to univariate analysis apply to multivariate analysis. See "Syntax Rules" on p. 423 in MANOVA: Univariate.
- If you enter one of the multivariate specifications in a univariate analysis, MANOVA ignores it.

Limitations

- Maximum 20 factors.
- Memory requirements depend primarily on the number of cells in the design. For the default full factorial model, this equals the product of the number of levels or categories in each factor.

MANOVA Variable List

- Multivariate MANOVA calculates statistical tests that are valid for analyses of dependent variables that are correlated with one another. The dependent variables must be specified first.
- The factor and covariate lists follow the same rules as in univariate analyses.
- If the dependent variables are uncorrelated, the univariate significance tests have greater statistical power.

TRANSFORM Subcommand

TRANSFORM performs linear transformations of some or all of the continuous variables (dependent variables and covariates). Specifications on TRANSFORM include an optional list of variables to be transformed, optional keywords to describe how to generate a transformation matrix from the specified contrasts, and a required keyword specifying the transformation contrasts.

- Transformations apply to all subsequent designs unless replaced by another TRANSFORM subcommand.
- TRANSFORM subcommands are not cumulative. Only the transformation specified most recently is in effect at any time. You can restore the original variables in later designs by specifying SPECIAL with an identity matrix.
- You should not use TRANSFORM when you use the WSFACTORS subcommand to request repeated measures analysis; a transformation is automatically performed in repeated measures analysis (see MANOVA: Repeated Measures).
- Transformations are in effect for the duration of the MANOVA procedure only. After the procedure is complete, the original variables remain in the working data file.
- By default, the transformation matrix is not displayed. Specify the keyword TRANSFORM on the PRINT subcommand to see the matrix generated by the TRANSFORM subcommand.
- If you do not use the RENAME subcommand with TRANSFORM, the variables specified on TRANSFORM are renamed temporarily (for the duration of the procedure) as *T1*, *T2*, etc. Explicit use of RENAME is recommended.
- Subsequent references to transformed variables should use the new names. The only exception is when you supply a VARIABLES specification on the OMEANS subcommand after using TRANSFORM. In this case, specify the original names. OMEANS displays observed means of original variables (see the OMEANS subcommand on p. 432 in MANOVA: Univariate).

Variable Lists

- By default, MANOVA applies the transformation you request to all continuous variables (dependent variables and covariates).
- You can enter a variable list in parentheses following the TRANSFORM subcommand. If you do, only the listed variables are transformed.

- You can enter multiple variable lists, separated by slashes, within a single set of parentheses. Each list must have the same number of variables, and the lists must not overlap. The transformation is applied separately to the variables on each list.
- In designs with covariates, transform only the dependent variables, or, in some designs, apply the same transformation separately to the dependent variables and the covariates.

CONTRAST, BASIS, and ORTHONORM Keywords

You can control how the transformation matrix is to be generated from the specified contrasts. If none of these three keywords is specified on TRANSFORM, the default is CONTRAST.

CONTRAST *Generate the transformation matrix directly from the contrast matrix specified* (see the CONTRAST subcommand on p. 425 in MANOVA: Univariate). This is the default.

BASIS *Generate the transformation matrix from the one-way basis matrix corresponding to the specified contrast matrix.* BASIS makes a difference only if the transformation contrasts are not orthogonal.

ORTHONORM *Orthonormalize the transformation matrix by rows before use.* MANOVA eliminates redundant rows. By default, orthonormalization is not done.

- CONTRAST and BASIS are alternatives and are mutually exclusive.
- ORTHONORM is independent of the CONTRAST/BASIS choice; you can enter it before or after either of those keywords.

Transformation Methods

To specify a transformation method, use one of the following keywords available on the TRANSFORM subcommand. Note that these are identical to the keywords available for the CONTRAST subcommand (see the CONTRAST subcommand on p. 425 in MANOVA: Univariate). However, in univariate designs, they are applied to the different levels of a factor. Here they are applied to the continuous variables in the analysis. This reflects the fact that the different dependent variables in a multivariate MANOVA setup can often be thought of as corresponding to different levels of some factor.

- The transformation keyword (and its specifications, if any) must follow all other specifications on the TRANSFORM subcommand.

DEVIATION *Deviations from the mean of the variables being transformed.* The first transformed variable is the mean of all variables in the transformation. Other transformed variables represent deviations of individual variables from the mean. One of the original variables (by default the last) is omitted as redundant. To omit a variable other than the last, specify the number of the variable to be omitted in parentheses after the DEVIATION keyword. For example,

```
/TRANSFORM (A B C) = DEVIATION(1)
```

omits *A* and creates variables representing the mean, the deviation of *B* from the mean, and the deviation of *C* from the mean. A DEVIATION transformation is not orthogonal.

DIFFERENCE *Difference or reverse Helmert transformation.* The first transformed variable is the mean of the original variables. Each of the original variables except the first is then transformed by subtracting the mean of those (original) variables that precede it. A DIFFERENCE transformation is orthogonal.

HELMERT *Helmert transformation.* The first transformed variable is the mean of the original variables. Each of the original variables except the last is then transformed by subtracting the mean of those (original) variables that follow it. A HELMERT transformation is orthogonal.

SIMPLE *Each original variable, except the last, is compared to the last of the original variables.* To use a variable other than the last as the omitted reference variable, specify its number in parentheses following keyword SIMPLE. For example,

```
/TRANSFORM(A B C) = SIMPLE(2)
```

specifies the second variable, *B*, as the reference variable. The three transformed variables represent the mean of *A*, *B*, and *C*, the difference between *A* and *B*, and the difference between *C* and *B*. A SIMPLE transformation is not orthogonal.

POLYNOMIAL *Orthogonal polynomial transformation.* The first transformed variable represents the mean of the original variables. Other transformed variables represent the linear, quadratic, and higher-degree components. By default, values of the original variables are assumed to represent equally spaced points. You can specify unequal spacing by entering a metric consisting of one integer for each variable in parentheses after keyword POLYNOMIAL. For example,

```
/TRANSFORM(RESP1 RESP2 RESP3) = POLYNOMIAL(1,2,4)
```

might indicate that three response variables correspond to levels of some stimulus that are in the proportion 1:2:4. The default metric is always (1,2,...,*k*), where *k* variables are involved. Only the relative differences between the terms of the metric matter: (1,2,4) is the same metric as (2,3,5) or (20,30,50) because in each instance the difference between the second and third numbers is twice the difference between the first and second.

REPEATED *Comparison of adjacent variables.* The first transformed variable is the mean of the original variables. Each additional transformed variable is the difference between one of the original variables and the original variable that followed it. Such transformed variables are often called *difference scores.* A REPEATED transformation is not orthogonal.

SPECIAL *A user-defined transformation.* After keyword SPECIAL, enter a square matrix in parentheses with as many rows and columns as there are variables to transform. MANOVA multiplies this matrix by the vector of original variables to obtain the transformed variables (see the examples below).

Example

```
MANOVA X1 TO X3 BY A(1,4)
  /TRANSFORM(X1 X2 X3) = SPECIAL( 1  1  1,
                                  1  0 -1,
                                  2 -1 -1)
  /DESIGN.
```

- The given matrix will be post-multiplied by the three continuous variables (considered as a column vector) to yield the transformed variables. The first transformed variable will therefore equal $X1 + X2 + X3$, the second will equal $X1 - X3$, and the third will equal $2X1 - X2 - X3$.
- The variable list is optional in this example since all three interval-level variables are transformed.
- You do not need to enter the matrix one row at a time, as shown above. For example,

  ```
  /TRANSFORM = SPECIAL(1 1 1 1 0 -1 2 -1 -1)
  ```

  is equivalent to the TRANSFORM specification in the above example.
- You can specify a repetition factor followed by an asterisk to indicate multiple consecutive elements of a SPECIAL transformation matrix. For example,

  ```
  /TRANSFORM = SPECIAL (4*1 0 -1 2 2*-1)
  ```

  is again equivalent to the TRANSFORM specification above.

Example

```
MANOVA X1 TO X3, Y1 TO Y3 BY A(1,4)
 /TRANSFORM (X1 X2 X3/Y1 Y2 Y3) = SPECIAL( 1  1  1,
                                           1  0 -1,
                                           2 -1 -1)
 /DESIGN.
```

- Here the same transformation shown in the previous example is applied to *X1*, *X2*, *X3* and to *Y1*, *Y2*, *Y3*.

RENAME Subcommand

Use RENAME to assign new names to transformed variables. Renaming variables after a transformation is strongly recommended. If you transform but do not rename the variables, the names *T1*, *T2*,...,*Tn* are used as names for the transformed variables.

- Follow RENAME with a list of new variable names.
- You must enter a new name for each dependent variable and covariate on the MANOVA variable list.
- Enter the new names in the order in which the original variables appeared on the MANOVA variable list.
- To retain the original name for one or more of the interval variables, you can either enter an asterisk or reenter the old name as the new name.
- References to dependent variables and covariates on subcommands following RENAME must use the new names. The original names will not be recognized within the MANOVA

procedure. The only exception is the OMEANS subcommand, which displays observed means of the original (untransformed) variables. Use the original names on OMEANS.

- The new names exist only during the MANOVA procedure that created them. They do not remain in the working data file after the procedure is complete.

Example

```
MANOVA A, B, C, V4, V5 BY TREATMNT(1,3)
  /TRANSFORM(A, B, C) = REPEATED
  /RENAME = MEANABC, AMINUSB, BMINUSC, *, *
  /DESIGN.
```

- The REPEATED transformation produces three transformed variables, which are then assigned mnemonic names *MEANABC*, *AMINUSB*, and *BMINUSC*.
- *V4* and *V5* retain their original names.

Example

```
MANOVA WT1, WT2, WT3, WT4 BY TREATMNT(1,3) WITH COV
  /TRANSFORM (WT1 TO WT4) = POLYNOMIAL
  /RENAME = MEAN, LINEAR, QUAD, CUBIC, *
  /ANALYSIS = MEAN, LINEAR, QUAD WITH COV
  /DESIGN.
```

- After the polynomial transformation of the four *WT* variables, RENAME assigns appropriate names to the various trends.
- Even though only four variables were transformed, RENAME applies to all five continuous variables. An asterisk is required to retain the original name for *COV*.
- The ANALYSIS subcommand following RENAME refers to the interval variables by their new names.

PRINT and NOPRINT Subcommands

All of the PRINT specifications described in MANOVA: Univariate are available in multivariate analyses. The following additional output can be requested. To suppress any optional output, specify the appropriate keyword on NOPRINT.

ERROR *Error matrices.* Three types of matrices are available.

SIGNIF *Significance tests.*

TRANSFORM *Transformation matrix.* It is available if you have transformed the dependent variables with the TRANSFORM subcommand.

HOMOGENEITY *Test for homogeneity of variance.* BOXM is available for multivariate analyses.

ERROR Keyword

In multivariate analysis, error terms consist of entire matrices, not single values. You can display any of the following error matrices on a PRINT subcommand by requesting them in parentheses following the keyword ERROR. If you specify ERROR by itself, without further specifications, the default is to display COV and COR.

SSCP *Error sums-of-squares and cross-products matrix.*

COV *Error variance-covariance matrix.*

COR *Error correlation matrix with standard deviations on the diagonal.* This also displays the determinant of the matrix and Bartlett's test of sphericity, a test of whether the error correlation matrix is significantly different from an identity matrix.

SIGNIF Keyword

You can request any of the optional output listed below by entering the appropriate specification in parentheses after the keyword SIGNIF on the PRINT subcommand. Further specifications for SIGNIF are described in MANOVA: Repeated Measures.

MULTIV *Multivariate* F *tests for group differences.* MULTIV is always printed unless explicitly suppressed with the NOPRINT subcommand.

EIGEN *Eigenvalues of the* $S_h S_e^{-1}$ *matrix.* This matrix is the product of the hypothesis sums-of-squares and cross-products (SSCP) matrix and the inverse of the error SSCP matrix. To print EIGEN, request it on the PRINT subcommand.

DIMENR *A dimension-reduction analysis.* To print DIMENR, request it on the PRINT subcommand.

UNIV *Univariate* F *tests.* UNIV is always printed except in repeated measures analysis. If the dependent variables are uncorrelated, univariate tests have greater statistical power. To suppress UNIV, use the NOPRINT subcommand.

HYPOTH *The hypothesis SSCP matrix.* To print HYPOTH, request it on the PRINT subcommand.

STEPDOWN *Roy-Bargmann stepdown* F *tests.* To print STEPDOWN, request it on the PRINT subcommand.

BRIEF *Abbreviated multivariate output.* This is similar to a univariate analysis of variance table but with Wilks' multivariate *F* approximation (lambda) replacing the univariate *F*. BRIEF overrides any of the SIGNIF specifications listed above.

SINGLEDF *Significance tests for the single degree of freedom making up each effect for ANOVA tables.* Results are displayed separately corresponding to each hypothesis degree of freedom. See MANOVA: Univariate.

- If neither PRINT nor NOPRINT is specified, MANOVA displays the results corresponding to MULTIV and UNIV for a multivariate analysis not involving repeated measures.
- If you enter any specification except BRIEF or SINGLEDF for SIGNIF on the PRINT subcommand, the requested output is displayed in addition to the default.
- To suppress the default, specify the keyword(s) on the NOPRINT subcommand.

TRANSFORM Keyword

The keyword TRANSFORM specified on PRINT displays the transposed transformation matrix in use for each subsequent design. This matrix is helpful in interpreting a multivariate analysis in which the interval-level variables have been transformed with either TRANSFORM or WSFACTORS.

- The matrix displayed by this option is the transpose of the transformation matrix.
- Original variables correspond to the rows of the matrix, and transformed variables correspond to the columns.
- A **transformed variable** is a linear combination of the original variables using the coefficients displayed in the column corresponding to that transformed variable.

HOMOGENEITY Keyword

In addition to the BARTLETT and COCHRAN specifications described in MANOVA: Univariate, the following test for homogeneity is available for multivariate analyses:

BOXM *Box's* M *statistic.* BOXM requires at least two dependent variables. If there is only one dependent variable when BOXM is requested, MANOVA prints Bartlett-Box *F* test statistic and issues a note.

PLOT Subcommand

In addition to the plots described in MANOVA: Univariate, the following is available for multivariate analyses:

ZCORR *A half-normal plot of the within-cells correlations among the dependent variables.* MANOVA first transforms the correlations using Fisher's *Z* transformation. If errors for the dependent variables are uncorrelated, the plotted points should lie close to a straight line.

PCOMPS Subcommand

PCOMPS requests a principal components analysis of each error matrix in a multivariate analysis. You can display the principal components of the error correlation matrix, the error variance-covariance matrix, or both. These principal components are corrected for differences due to the factors and covariates in the MANOVA analysis. They tend to be more useful than principal components extracted from the raw correlation or covariance matrix when there are significant group differences between the levels of the factors or when a significant amount of error variance is accounted for by the covariates. You can specify any of the keywords listed below on PCOMPS.

COR *Principal components analysis of the error correlation matrix.*

COV *Principal components analysis of the error variance-covariance matrix.*

ROTATE *Rotate the principal components solution.* By default, no rotation is performed. Specify a rotation type (either VARIMAX, EQUAMAX, or QUARTIMAX) in parentheses after keyword ROTATE. To cancel a rotation specified for a previous design, enter NOROTATE in the parentheses after ROTATE.

NCOMP(n) *The number of principal components to rotate.* Specify a number in parentheses. The default is the number of dependent variables.

MINEIGEN(n) *The minimum eigenvalue for principal component extraction.* Specify a cutoff value in parentheses. Components with eigenvalues below the cutoff will not be retained in the solution. The default is 0: all components (or the number specified on NCOMP) are extracted.

ALL *COR, COV, and ROTATE.*

- You must specify either COR or COV (or both). Otherwise, MANOVA will not produce any principal components.
- Both NCOMP and MINEIGEN limit the number of components that are rotated.
- If the number specified on NCOMP is less than two, two components are rotated provided that at least two components have eigenvalues greater than any value specified on MINEIGEN.
- Principal components analysis is computationally expensive if the number of dependent variables is large.

DISCRIM Subcommand

DISCRIM produces a canonical discriminant analysis for each effect in a design. (For covariates, DISCRIM produces a canonical correlation analysis.) These analyses aid in the interpretation of multivariate effects. You can request the following statistics by entering the appropriate keywords after the subcommand DISCRIM:

RAW *Raw discriminant function coefficients.*

STAN *Standardized discriminant function coefficients.*

ESTIM *Effect estimates in discriminant function space.*

COR *Correlations between the dependent variables and the canonical variables defined by the discriminant functions.*

ROTATE *Rotation of the matrix of correlations between dependent and canonical variables.* Specify rotation type VARIMAX, EQUAMAX, or QUARTIMAX in parentheses after this keyword.

ALL *RAW, STAN, ESTIM, COR, and ROTATE.*

By default, the significance level required for the extraction of a canonical variable is 0.25. You can change this value by specifying keyword ALPHA and a value between 0 and 1 in parentheses:

ALPHA *The significance level required before a canonical variable is extracted.* The default is 0.25. To change the default, specify a decimal number between 0 and 1 in parentheses after ALPHA.

- The correlations between dependent variables and canonical functions are not rotated unless at least two functions are significant at the level defined by ALPHA.
- If you set ALPHA to 1.0, all discriminant functions are reported (and rotated, if you so request).
- If you set ALPHA to 0, no discriminant functions are reported.

POWER Subcommand

The following specifications are available for POWER in multivariate analysis. For applications of POWER in univariate analysis, see MANOVA: Univariate.

APPROXIMATE *Approximate power values.* This is the default. Approximate power values for multivariate tests are derived from procedures presented by Muller and Peterson (1984). Approximate values are normally accurate to three decimal places and are much cheaper to compute than exact values.

EXACT *Exact power values.* Exact power values for multivariate tests are computed from the non-central F distribution. Exact multivariate power values will be displayed only if there is one hypothesis degree of freedom, where all the multivariate criteria have identical power.

- For information on the multivariate generalizations of power and effect size, see Muller and Peterson (1984), Green (1978), and Huberty (1972).

CINTERVAL Subcommand

In addition to the specifications described in MANOVA: Univariate, the keyword MULTIVARIATE is available for multivariate analysis. You can specify a type in parentheses after the MULTIVARIATE keyword. The following type keywords are available on MULTIVARIATE:

ROY *Roy's largest root.* An approximation given by Pillai (1967) is used. This approximation is accurate for upper percentage points (0.95 to 1), but it is not as good for lower percentage points. Thus, for Roy intervals, the user is restricted to the range 0.95 to 1.

PILLAI *Pillai's trace.* The intervals are computed by approximating the percentage points with percentage points of the F distribution.

WILKS *Wilks' lambda.* The intervals are computed by approximating the percentage points with percentage points of the F distribution.

HOTELLING *Hotelling's trace.* The intervals are computed by approximating the percentage points with percentage points of the F distribution.

BONFER *Bonferroni intervals.* This approximation is based on Student's t distribution.

- The Wilks', Pillai's, and Hotelling's approximate confidence intervals are thought to match exact intervals across a wide range of alpha levels, especially for large sample sizes (Burns, 1984). Use of these intervals, however, has not been widely investigated.
- To obtain multivariate intervals separately for each parameter, choose individual multivariate intervals. For individual multivariate confidence intervals, the hypothesis degree

of freedom is set to 1, in which case Hotelling's, Pillai's, Wilks', and Roy's intervals will be identical and equivalent to those computed from percentage points of Hotelling's T^2 distribution. Individual Bonferroni intervals will differ and, for a small number of dependent variables, will generally be shorter.

- If you specify MULTIVARIATE on CINTERVAL, you must specify a type keyword. If you specify CINTERVAL without any keyword, the default is the same as with univariate analysis: CINTERVAL displays individual-univariate confidence intervals at the 0.95 level.

ANALYSIS Subcommand

ANALYSIS is discussed in MANOVA: Univariate as a means of obtaining factor-by-covariate interaction terms. In multivariate analyses, it is considerably more useful.

- ANALYSIS specifies a subset of the continuous variables (dependent variables and covariates) listed on the MANOVA variable list and completely redefines which variables are dependent and which are covariates.
- All variables named on an ANALYSIS subcommand must have been named on the MANOVA variable list. It does not matter whether they were named as dependent variables or as covariates.
- Factors cannot be named on an ANALYSIS subcommand.
- After keyword ANALYSIS, specify the names of one or more dependent variables and, optionally, keyword WITH followed by one or more covariates.
- An ANALYSIS specification remains in effect for all designs until you enter another ANALYSIS subcommand.
- Continuous variables named on the MANOVA variable list but omitted from the ANALYSIS subcommand currently in effect can be specified on the DESIGN subcommand. See the DESIGN subcommand on p. 440 in MANOVA: Univariate.
- You can use an ANALYSIS subcommand to request analyses of several groups of variables provided that the groups do not overlap. Separate the groups of variables with slashes and enclose the entire ANALYSIS specification in parentheses.

CONDITIONAL and UNCONDITIONAL Keywords

When several analysis groups are specified on a single ANALYSIS subcommand, you can control how each list is to be processed by specifying CONDITIONAL or UNCONDITIONAL in the parentheses immediately following the ANALYSIS subcommand. The default is UNCONDITIONAL.

UNCONDITIONAL *Process each analysis group separately, without regard to other lists.* This is the default.

CONDITIONAL *Use variables specified in one analysis group as covariates in subsequent analysis groups.*

- CONDITIONAL analysis is not carried over from one ANALYSIS subcommand to another.
- You can specify a final covariate list outside the parentheses. These covariates apply to every list within the parentheses, regardless of whether you specify CONDITIONAL or

UNCONDITIONAL. The variables on this global covariate list must not be specified in any individual lists.

Example

```
MANOVA A B C BY FAC(1,4) WITH D, E
  /ANALYSIS = (A, B / C / D WITH E)
  /DESIGN.
```

- The first analysis uses *A* and *B* as dependent variables and uses no covariates.
- The second analysis uses *C* as a dependent variable and uses no covariates.
- The third analysis uses *D* as the dependent variable and uses *E* as a covariate.

Example

```
MANOVA A, B, C, D, E BY FAC(1,4) WITH F G
  /ANALYSIS = (A, B / C / D WITH E) WITH F G
  /DESIGN.
```

- A final covariate list `WITH F G` is specified outside the parentheses. The covariates apply to every list within the parentheses.
- The first analysis uses *A* and *B*, with *F* and *G* as covariates.
- The second analysis uses *C*, with *F* and *G* as covariates.
- The third analysis uses *D*, with *E*, *F*, and *G* as covariates.
- Factoring out *F* and *G* is the only way to use them as covariates in all three analyses, since no variable can be named more than once on an ANALYSIS subcommand.

Example

```
MANOVA A B C BY FAC(1,3)
  /ANALYSIS(CONDITIONAL) = (A WITH B / C)
  /DESIGN.
```

- In the first analysis, *A* is the dependent variable, *B* is a covariate, and *C* is not used.
- In the second analysis, *C* is the dependent variable, and both *A* and *B* are covariates.

MANOVA: Repeated Measures

```
MANOVA dependent varlist [BY factor list (min,max)[factor list...]
        [WITH [varying covariate list] [(constant covariate list)]]

  /WSFACTORS = varname (levels) [varname...]

 [/WSDESIGN = [effect effect...]

 [/MEASURE = newname newname...]

 [/RENAME = newname newname...]

 [/{PRINT   }=[SIGNIF({AVERF**}) (HF) (GG) (EFSIZE)]]
   {NOPRINT}         {AVONLY }

 [/DESIGN]*
```

* The DESIGN subcommand has the same syntax as is described in MANOVA: Univariate.

** Default if subcommand or keyword is omitted.

Example:

```
MANOVA Y1 TO Y4 BY GROUP(1,2)
  /WSFACTORS=YEAR(4).
```

Overview

This section discusses the subcommands that are used in repeated measures designs, in which the dependent variables represent measurements of the same variable (or variables) at different times. This section does not contain information on all subcommands you will need to specify the design. For some subcommands or keywords not covered here, such as DESIGN, see MANOVA: Univariate. For information on optional output and the multivariate significance tests available, see MANOVA: Multivariate.

- In a simple repeated measures analysis, all dependent variables represent different measurements of the same variable for different values (or levels) of a within-subjects factor. Between-subjects factors and covariates can also be included in the model, just as in analyses not involving repeated measures.
- A **within-subjects factor** is simply a factor that distinguishes measurements made on the same subject or case, rather than distinguishing different subjects or cases.
- MANOVA permits more complex analyses, in which the dependent variables represent levels of two or more within-subjects factors.
- MANOVA also permits analyses in which the dependent variables represent measurements of several variables for the different levels of the within-subjects factors. These are known as **doubly multivariate designs**.
- A repeated measures analysis includes a within-subjects design describing the model to be tested with the within-subjects factors, as well as the usual between-subjects design describing the effects to be tested with between-subjects factors. The default for both types of design is a full factorial model.

- MANOVA always performs an orthonormal transformation of the dependent variables in a repeated measures analysis. By default, MANOVA renames them as *T1*, *T2*, and so forth.

Basic Specification

- The basic specification is a variable list followed by the WSFACTORS subcommand.
- By default, MANOVA performs special repeated measures processing. Default output includes SIGNIF(AVERF) but not SIGNIF(UNIV). In addition, for any within-subjects effect involving more than one transformed variable, the Mauchly test of sphericity is displayed to test the assumption that the covariance matrix of the transformed variables is constant on the diagonal and zero off the diagonal. The Greenhouse-Geiser epsilon and the Huynh-Feldt epsilon are also displayed for use in correcting the significance tests in the event that the assumption of sphericity is violated. These tests are discussed in the relevant chapters in this manual.

Subcommand Order

- The list of dependent variables, factors, and covariates must be first.
- WSFACTORS must be the first subcommand used after the variable list.

Syntax Rules

- The WSFACTORS (within-subjects factors), WSDESIGN (within-subjects design), and MEASURE subcommands are used only in repeated measures analysis.
- WSFACTORS is required for any repeated measures analysis.
- If WSDESIGN is not specified, a full factorial within-subjects design consisting of all main effects and interactions among within-subjects factors is used by default.
- The MEASURE subcommand is used for doubly multivariate designs, in which the dependent variables represent repeated measurements of more than one variable.
- Do not use the TRANSFORM subcommand with the WSFACTORS subcommand because WSFACTORS automatically causes an orthonormal transformation of the dependent variables.

Limitations

- Maximum 20 between-subjects factors. There is no limit on the number of measures for doubly multivariate designs.
- Memory requirements depend primarily on the number of cells in the design. For the default full factorial model, this equals the product of the number of levels or categories in each factor.

Example

```
MANOVA Y1 TO Y4 BY GROUP(1,2)
  /WSFACTORS=YEAR(4)
  /CONTRAST(YEAR)=POLYNOMIAL
  /RENAME=CONST, LINEAR, QUAD, CUBIC
  /PRINT=TRANSFORM PARAM(ESTIM)
  /WSDESIGN=YEAR
  /DESIGN=GROUP.
```

- WSFACTORS immediately follows the MANOVA variable list and specifies a repeated measures analysis in which the four dependent variables represent a single variable measured at four levels of the within-subjects factor. The within-subjects factor is called *YEAR* for the duration of the MANOVA procedure.
- CONTRAST requests polynomial contrasts for the levels of *YEAR*. Because the four variables, *Y1*, *Y2*, *Y3*, and *Y4*, in the working data file represent the four levels of *YEAR*, the effect is to perform an orthonormal polynomial transformation of these variables.
- RENAME assigns names to the dependent variables to reflect the transformation.
- PRINT requests that the transformation matrix and the parameter estimates be displayed.
- WSDESIGN specifies a within-subjects design that includes only the effect of the *YEAR* within-subjects factor. Because *YEAR* is the only within-subjects factor specified, this is the default design, and WSDESIGN could have been omitted.
- DESIGN specifies a between-subjects design that includes only the effect of the *GROUP* between-subjects factor. This subcommand could have been omitted.

MANOVA Variable List

The list of dependent variables, factors, and covariates must be specified first.

- WSFACTORS determines how the dependent variables on the MANOVA variable list will be interpreted.
- The number of dependent variables on the MANOVA variable list must be a multiple of the number of cells in the within-subjects design. If there are six cells in the within-subjects design, each group of six dependent variables represents a single within-subjects variable that has been measured in each of the six cells.
- Normally, the number of dependent variables should equal the number of cells in the within-subjects design multiplied by the number of variables named on the MEASURE subcommand (if one is used). If you have more groups of dependent variables than are accounted for by the MEASURE subcommand, MANOVA will choose variable names to label the output, which may be difficult to interpret.
- Covariates are specified after keyword WITH. You can specify either varying covariates or constant covariates, or both. **Varying covariates**, similar to dependent variables in a repeated measures analysis, represent measurements of the same variable (or variables) at different times while **constant covariates** represent variables whose values remain the same at each within-subjects measurement.
- If you use varying covariates, the number of covariates specified must be an integer multiple of the number of dependent variables.

- If you use constant covariates, you must specify them in parentheses. If you use both constant and varying covariates, constant variates must be specified after all varying covariates.

Example

```
MANOVA MATH1 TO MATH4 BY METHOD(1,2) WITH PHYS1 TO PHYS4 (SES)
  /WSFACTORS=SEMESTER(4).
```

- The four dependent variables represent a score measured four times (corresponding to the four levels of *SEMESTER*).
- The four varying covariates *PHYS1* to *PHYS4* represents four measurements of another score.
- *SES* is a constant covariate. Its value does not change over the time covered by the four levels of *SEMESTER*.
- Default contrast (POLYNOMIAL) is used.

WSFACTORS Subcommand

WSFACTORS names the within-subjects factors and specifies the number of levels for each.

- For repeated measures designs, WSFACTORS must be the first subcommand after the MANOVA variable list.
- Only one WSFACTORS subcommand is permitted per execution of MANOVA.
- Names for the within-subjects factors are specified on the WSFACTORS subcommand. Factor names must not duplicate any of the dependent variables, factors, or covariates named on the MANOVA variable list.
- If there are more than one within-subjects factors, they must be named in the order corresponding to the order of the dependent variables on the MANOVA variable list. MANOVA varies the levels of the last-named within-subjects factor most rapidly when assigning dependent variables to within-subjects cells (see example below).
- Levels of the factors must be represented in the data by the dependent variables named on the MANOVA variable list.
- Enter a number in parentheses after each factor to indicate how many levels the factor has. If two or more adjacent factors have the same number of levels, you can enter the number of levels in parentheses after all of them.
- Enter only the number of levels for within-subjects factors, not a range of values.
- The number of cells in the within-subjects design is the product of the number of levels for all within-subjects factors.

Example

```
MANOVA X1Y1 X1Y2 X2Y1 X2Y2 X3Y1 X3Y2 BY TREATMNT(1,5) GROUP(1,2)
  /WSFACTORS=X(3) Y(2)
  /DESIGN.
```

- The MANOVA variable list names six dependent variables and two between-subjects factors, *TREATMNT* and *GROUP*.

- WSFACTORS identifies two within-subjects factors whose levels distinguish the six dependent variables. *X* has three levels and *Y* has two. Thus, there are 3 × 2 = 6 cells in the within-subjects design, corresponding to the six dependent variables.
- Variable *X1Y1* corresponds to levels 1,1 of the two within-subjects factors; variable *X1Y2* corresponds to levels 1,2; *X2Y1* to levels 2,1; and so on up to *X3Y2*, which corresponds to levels 3,2. The first within-subjects factor named, *X*, varies most slowly, and the last within-subjects factor named, *Y*, varies most rapidly on the list of dependent variables.
- Because there is no WSDESIGN subcommand, the within-subjects design will include all main effects and interactions: *X*, *Y*, and *X* by *Y*.
- Likewise, the between-subjects design includes all main effects and interactions: *TREATMNT*, *GROUP*, and *TREATMNT* by *GROUP*.
- In addition, a repeated measures analysis always includes interactions between the within-subjects factors and the between-subjects factors. There are three such interactions for each of the three within-subjects effects.

CONTRAST for WSFACTORS

The levels of a within-subjects factor are represented by different dependent variables. Therefore, contrasts between levels of such a factor compare these dependent variables. Specifying the type of contrast amounts to specifying a transformation to be performed on the dependent variables.

- An orthonormal transformation is automatically performed on the dependent variables in a repeated measures analysis.
- To specify the type of orthonormal transformation, use the CONTRAST subcommand for the within-subjects factors.
- Regardless of the contrast type you specify, the transformation matrix is orthonormalized before use.
- If you do not specify a contrast type for within-subjects factors, the default contrast type is orthogonal POLYNOMIAL. Intrinsically orthogonal contrast types are recommended for within-subjects factors if you wish to examine each degree-of-freedom test. Other orthogonal contrast types are DIFFERENCE and HELMERT. MULTIV and AVERF tests are identical, no matter what contrast was specified.
- To perform non-orthogonal contrasts, you must use the TRANSFORM subcommand instead of CONTRAST. The TRANSFORM subcommand is discussed in MANOVA: Multivariate.
- When you implicitly request a transformation of the dependent variables with CONTRAST for within-subjects factors, the same transformation is applied to any covariates in the analysis. The number of covariates must be an integer multiple of the number of dependent variables.
- You can display the transpose of the transformation matrix generated by your within-subjects contrast using keyword TRANSFORM on the PRINT subcommand.

Example

```
MANOVA SCORE1 SCORE2 SCORE3 BY GROUP(1,4)
  /WSFACTORS=ROUND(3)
  /CONTRAST(ROUND)=DIFFERENCE
  /CONTRAST(GROUP)=DEVIATION
  /PRINT=TRANSFORM PARAM(ESTIM).
```

- This analysis has one between-subjects factor, *GROUP*, with levels 1, 2, 3, and 4, and one within-subjects factor, *ROUND*, with three levels that are represented by the three dependent variables.
- The first CONTRAST subcommand specifies difference contrasts for *ROUND*, the within-subjects factor.
- There is no WSDESIGN subcommand, so a default full factorial within-subjects design is assumed. This could also have been specified as `WSDESIGN=ROUND`, or simply `WSDESIGN`.
- The second CONTRAST subcommand specifies deviation contrasts for *GROUP*, the between-subjects factor. This subcommand could have been omitted because deviation contrasts are the default.
- PRINT requests the display of the transformation matrix generated by the within-subjects contrast and the parameter estimates for the model.
- There is no DESIGN subcommand, so a default full factorial between-subjects design is assumed. This could also have been specified as `DESIGN=GROUP`, or simply `DESIGN`.

PARTITION for WSFACTORS

The PARTITION subcommand also applies to factors named on WSFACTORS. (See the PARTITION subcommand on p.427 in MANOVA: Univariate.)

WSDESIGN Subcommand

WSDESIGN specifies the design for within-subjects factors. Its specifications are like those of the DESIGN subcommand, but it uses the within-subjects factors rather than the between-subjects factors.

- The default WSDESIGN is a full factorial design, which includes all main effects and all interactions for within-subjects factors. The default is in effect whenever a design is processed without a preceding WSDESIGN or when the preceding WSDESIGN subcommand has no specifications.
- A WSDESIGN specification can include main effects, factor-by-factor interactions, nested terms (term within term), terms using keyword MWITHIN, and pooled effects using the plus sign. The specification is the same as on the DESIGN subcommand but involves only within-subjects factors.
- A WSDESIGN specification cannot include between-subjects factors or terms based on them, nor does it accept interval-level variables, keywords MUPLUS or CONSTANT, or error-term definitions or references.
- The WSDESIGN specification applies to all subsequent within-subjects designs until another WSDESIGN subcommand is encountered.

Example

```
MANOVA JANLO,JANHI,FEBLO,FEBHI,MARLO,MARHI BY SEX(1,2)
  /WSFACTORS MONTH(3) STIMULUS(2)
  /WSDESIGN MONTH, STIMULUS
  /WSDESIGN
  /DESIGN SEX.
```

- There are six dependent variables, corresponding to three months and two different levels of stimulus.
- The dependent variables are named on the MANOVA variable list in such an order that the level of stimulus varies more rapidly than the month. Thus, *STIMULUS* is named last on the WSFACTORS subcommand.
- The first WSDESIGN subcommand specifies only the main effects for within-subjects factors. There is no *MONTH* by *STIMULUS* interaction term.
- The second WSDESIGN subcommand has no specifications and, therefore, invokes the default within-subjects design, which includes the main effects and their interaction.

MWITHIN Keyword for Simple Effects

You can use MWITHIN on either the WSDESIGN or the DESIGN subcommand in a model with both between- and within-subjects factors to estimate simple effects for factors nested within factors of the opposite type.

Example

```
MANOVA WEIGHT1 WEIGHT2 BY TREAT(1,2)
 /WSFACTORS=WEIGHT(2)
 /DESIGN=MWITHIN TREAT(1) MWITHIN TREAT(2)
MANOVA WEIGHT1 WEIGHT2 BY TREAT(1,2)
 /WSFACTORS=WEIGHT(2)
 /WSDESIGN=MWITHIN WEIGHT(1) MWITHIN WEIGHT(2)
 /DESIGN.
```

- The first DESIGN tests the simple effects of *WEIGHT* within each level of *TREAT*.
- The second DESIGN tests the simple effects of *TREAT* within each level of *WEIGHT*.

MEASURE Subcommand

In a doubly multivariate analysis, the dependent variables represent multiple variables measured under the different levels of the within-subjects factors. Use MEASURE to assign names to the variables that you have measured for the different levels of within-subjects factors.

- Specify a list of one or more variable names to be used in labeling the averaged results. If no within-subjects factor has more than two levels, MEASURE has no effect.

- The number of dependent variables on the DESIGN subcommand should equal the product of the number of cells in the within-subjects design and the number of names on MEASURE.
- If you do not enter a MEASURE subcommand and there are more dependent variables than cells in the within-subjects design, MANOVA assigns names (normally *MEAS.1*, *MEAS.2*, etc.) to the different measures.
- All of the dependent variables corresponding to each measure should be listed together and ordered so that the within-subjects factor named last on the WSFACTORS subcommand varies most rapidly.

Example

```
MANOVA TEMP1 TO TEMP6, WEIGHT1 TO WEIGHT6 BY GROUP(1,2)
  /WSFACTORS=DAY(3) AMPM(2)
  /MEASURE=TEMP WEIGHT
  /WSDESIGN=DAY, AMPM, DAY BY AMPM
  /PRINT=SIGNIF(HYPOTH AVERF)
  /DESIGN.
```

- There are 12 dependent variables: 6 temperatures and 6 weights, corresponding to morning and afternoon measurements on three days.
- WSFACTORS identifies the two factors (*DAY* and *AMPM*) that distinguish the temperature and weight measurements for each subject. These factors define six within-subjects cells.
- MEASURE indicates that the first group of six dependent variables correspond to *TEMP* and the second group of six dependent variables correspond to *WEIGHT*.
- These labels, *TEMP* and *WEIGHT*, are used on the output requested by PRINT.
- WSDESIGN requests a full factorial within-subjects model. Because this is the default, WSDESIGN could have been omitted.

RENAME Subcommand

Because any repeated measures analysis involves a transformation of the dependent variables, it is always a good idea to rename the dependent variables. Choose appropriate names depending on the type of contrast specified for within-subjects factors. This is easier to do if you are using one of the orthogonal contrasts. The most reliable way to assign new names is to inspect the transformation matrix.

Example

```
MANOVA LOW1 LOW2 LOW3 HI1 HI2 HI3
  /WSFACTORS=LEVEL(2) TRIAL(3)
  /CONTRAST(TRIAL)=DIFFERENCE
  /RENAME=CONST LEVELDIF TRIAL21 TRIAL312 INTER1 INTER2
  /PRINT=TRANSFORM
  /DESIGN.
```

- This analysis has two within-subjects factors and no between-subjects factors.
- Difference contrasts are requested for *TRIAL*, which has three levels.
- Because all orthonormal contrasts produce the same F test for a factor with two levels, there is no point in specifying a contrast type for *LEVEL*.

- New names are assigned to the transformed variables based on the transformation matrix. These names correspond to the meaning of the transformed variables: the mean or constant, the average difference between levels, the average effect of trial 2 compared to 1, the average effect of trial 3 compared to 1 and 2; and the two interactions between *LEVEL* and *TRIAL*.
- The transformation matrix requested by the PRINT subcommand looks like Figure 1.

Figure 1 Transformation matrix

	CONST	LEVELDIF	TRIAL1	TRIAL2	INTER1	INTER2
LOW1	0.408	0.408	-0.500	-0.289	-0.500	-0.289
LOW2	0.408	0.408	0.500	-0.289	0.500	-0.289
LOW3	0.408	0.408	0.000	0.577	0.000	0.577
HI1	0.408	-0.408	-0.500	-0.289	0.500	0.289
HI2	0.408	-0.408	0.500	-0.289	-0.500	0.289
HI3	0.408	-0.408	0.000	0.577	0.000	-0.577

PRINT Subcommand

The following additional specifications on PRINT are useful in repeated measures analysis:

SIGNIF(AVERF) *Averaged* F *tests for use with repeated measures.* This is the default display in repeated measures analysis. The averaged *F* tests in the multivariate setup for repeated measures are equivalent to the univariate (or split-plot or mixed-model) approach to repeated measures.

SIGNIF(AVONLY) *Only the averaged* F *test for repeated measures.* AVONLY produces the same output as AVERF and suppresses all other SIGNIF output.

SIGNIF(HF) *The Huynh-Feldt corrected significance values for averaged univariate* F *tests.*

SIGNIF(GG) *The Greenhouse-Geisser corrected significance values for averaged univariate* F *tests.*

SIGNIF(EFSIZE) *The effect size for the univariate* F *and* t *tests.*

- Keywords AVERF and AVONLY are mutually exclusive.
- When you request repeated measures analysis with the WSFACTORS subcommand, the default display includes SIGNIF(AVERF) but does not include the usual SIGNIF(UNIV).
- The averaged *F* tests are appropriate in repeated measures because the dependent variables that are averaged actually represent contrasts of the WSFACTOR variables. When the analysis is not doubly multivariate, as discussed above, you can specify `PRINT=SIGNIF(UNIV)` to obtain significance tests for each degree of freedom, just as in univariate MANOVA.

MATRIX—END MATRIX

This command is not available on all operating systems.

```
MATRIX

matrix statements

END MATRIX
```

The following matrix language statements can be used in a matrix program:

BREAK	DO IF	END LOOP	MSAVE	SAVE
CALL	ELSE	GET	PRINT	WRITE
COMPUTE	ELSE IF	LOOP	READ	
DISPLAY	END IF	MGET	RELEASE	

The following functions can be used in matrix language statements:

ABS	Absolute values of matrix elements
ALL	Test if all elements are positive
ANY	Test if any element is positive
ARSIN	Arc sines of matrix elements
ARTAN	Arc tangents of matrix elements
BLOCK	Create block diagonal matrix
CDFNORM	Cumulative normal distribution function
CHICDF	Cumulative chi-squared distribution function
CHOL	Cholesky decomposition
CMAX	Column maxima
CMIN	Column minima
COS	Cosines of matrix elements
CSSQ	Column sums of squares
CSUM	Column sums
DESIGN	Create design matrix
DET	Determinant
DIAG	Diagonal of matrix
EOF	Check end of file
EVAL	Eigenvalues of symmetric matrix
EXP	Exponentials of matrix elements
FCDF	Cumulative *F* distribution function
GINV	Generalized inverse
GRADE	Rank elements in matrix, using sequential integers for ties
GSCH	Gram-Schmidt orthonormal basis
IDENT	Create identity matrix
INV	Inverse

KRONECKER	Kronecker product of two matrices
LG10	Logarithms to base 10 of matrix elements
LN	Logarithms to base *e* of matrix elements
MAGIC	Create magic square
MAKE	Create a matrix with all elements equal
MDIAG	Create a matrix with the given diagonal
MMAX	Maximum element in matrix
MMIN	Minimum element in matrix
MOD	Remainders after division
MSSQ	Matrix sum of squares
MSUM	Matrix sum
NCOL	Number of columns
NROW	Number of rows
RANK	Matrix rank
RESHAPE	Change shape of matrix
RMAX	Row maxima
RMIN	Row minima
RND	Round off matrix elements to nearest integer
RNKORDER	Rank elements in matrix, averaging ties
RSSQ	Row sums of squares
RSUM	Row sums
SIN	Sines of matrix elements
SOLVE	Solve systems of linear equations
SQRT	Square roots of matrix elements
SSCP	Sums of squares and cross-products
SVAL	Singular values
SWEEP	Perform sweep transformation
T	(Synonym for TRANSPOS)
TCDF	Cumulative normal *t* distribution function
TRACE	Calculate trace (sum of diagonal elements)
TRANSPOS	Transposition of matrix
TRUNC	Truncation of matrix elements to integer
UNIFORM	Create matrix of uniform random numbers

Example:

```
MATRIX.
READ A /FILE=MATRDATA /SIZE={6,6} /FIELD=1 TO 60.
CALL EIGEN(A,EIGENVEC,EIGENVAL).
LOOP J=1 TO NROW(EIGENVAL).
+ DO IF (EIGENVAL(J) > 1.0).
+   PRINT EIGENVAL(J) / TITLE="Eigenvalue:" /SPACE=3.
+   PRINT T(EIGENVEC(:,J)) / TITLE="Eigenvector:" /SPACE=1.
+ END IF.
END LOOP.
END MATRIX.
```

Overview

The MATRIX and END MATRIX commands enclose statements that are executed by the SPSS matrix processor. Using matrix programs, you can write your own statistical routines in the compact language of matrix algebra. Matrix programs can include mathematical calculations, control structures, display of results, and reading and writing matrices as character files or SPSS data files.

As discussed below, a matrix program is for the most part independent of the rest of the SPSS session, although it can read and write SPSS data files, including the working data file.

This Syntax Reference section does not attempt to explain the rules of matrix algebra. Many textbooks, such as Hadley (1961) and O'Nan (1971), teach the application of matrix methods to statistics.

The SPSS MATRIX procedure was originally developed at the Madison Academic Computing Center, University of Wisconsin.

Terminology

A variable within a matrix program represents a **matrix**, which is simply a set of values arranged in a rectangular array of rows and columns.

- An $n \times m$ (read "n by m") matrix is one that has n rows and m columns. The integers n and m are the dimensions of the matrix. An $n \times m$ matrix contains $n \times m$ elements, or data values.
- An $n \times 1$ matrix is sometimes called a **column vector**, and a $1 \times n$ matrix is sometimes called a **row vector**. A vector is a special case of a matrix.
- A 1×1 matrix, containing a single data value, is often called a **scalar**. A scalar is also a special case of a matrix.
- An **index** to a matrix or vector is an integer that identifies a specific row or column. Indexes normally appear in printed works as subscripts, as in $'A_{31}$, but are specified in the matrix language within parentheses, as in $'A(3,1)$. The row index for a matrix precedes the column index.
- The **main diagonal** of a matrix consists of the elements whose row index equals their column index. It begins at the top left corner of the matrix; in a square matrix, it runs to the bottom right corner.
- The **transpose** of a matrix is the matrix with rows and columns interchanged. The transpose of an $n \times m$ matrix is an $m \times n$ matrix.
- A **symmetric matrix** is a square matrix that is unchanged if you flip it about the main diagonal. That is, the element in row *i*, column *j* equals the element in row *j*, column *i*. A symmetric matrix equals its transpose.
- Matrices are always rectangular, although it is possible to read or write symmetric matrices in triangular form. Vectors and scalars are considered degenerate rectangles.
- It is an error to try to create a matrix whose rows have different numbers of elements.

A matrix program does not process individual cases unless you so specify, using the control structures of the matrix language. Unlike ordinary SPSS variables, matrix variables do not have distinct values for different cases. A matrix is a single entity.

Vectors in matrix processing should not be confused with the vectors temporarily created by the VECTOR command in SPSS. The latter are shorthand for a list of SPSS variables and, like all ordinary SPSS variables, are unavailable during matrix processing.

Matrix Variables

A matrix variable is created by a matrix statement that assigns a value to a variable name.

- A matrix variable name follows the same rules as those applicable to an ordinary SPSS variable name.
- The names of matrix functions and procedures cannot be used as variable names within a matrix program. (In particular, the letter *T* cannot be used as a variable name because T is an alias for the TRANSPOS function.)
- The COMPUTE, READ, GET, MGET, and CALL statements create matrices. An index variable named on a LOOP statement creates a scalar with a value assigned to it.
- A variable name can be redefined within a matrix program without regard to the dimensions of the matrix it represents. The same name can represent scalars, vectors, and full matrices at different points in the matrix program.
- MATRIX—END MATRIX does not include any special processing for missing data. When reading a data matrix from an SPSS data file, you must therefore specify whether missing data are to be accepted as valid or excluded from the matrix.

String Variables in Matrix Programs

Matrix variables can contain short string data. Support for string variables is limited, however.

- MATRIX will attempt to carry out calculations with string variables if you so request. The results will not be meaningful.
- You must specify a format (such as A8) when you display a matrix that contains string data.

Syntax of Matrix Language

A matrix program consists of statements. Matrix statements must appear in a matrix program, between the MATRIX and END MATRIX commands. They are analogous to SPSS commands and follow the rules of the SPSS command language regarding the abbreviation of keywords; the equivalence of upper and lower case; the use of spaces, commas, and equals signs; and the splitting of statements across multiple lines. However, commas are required to separate arguments to matrix functions and procedures and to separate variable names on the RELEASE statement.

Matrix statements are composed of the following elements:

- Keywords, such as the names of matrix statements.
- Variable names.

- Explicitly written matrices, which are enclosed within braces ({ }).
- Arithmetic and logical operators.
- Matrix functions.
- The SPSS command terminator, which serves as a statement terminator within a matrix program.

Comments in Matrix Programs

Within a matrix program, you can enter comments in any of the forms recognized by SPSS: on lines beginning with the COMMENT command, on lines beginning with an asterisk, or between the characters /* and */ on a command line.

Matrix Notation in SPSS

To write a matrix explicitly:

- Enclose the matrix within braces ({ }).
- Separate the elements of each row by commas.
- Separate the rows by semicolons.
- String elements must be enclosed in either apostrophes or quotation marks, as is generally true in the SPSS command language.

Example

```
{1,2,3;4,5,6}
```

- The example represents the following matrix:

$$\begin{bmatrix} 1 & 2 & 3 \\ 4 & 5 & 6 \end{bmatrix}$$

Example

```
{1,2,3}
```

- This example represents a row vector:

$$\begin{bmatrix} 1 & 2 & 3 \end{bmatrix}$$

Example

```
{11;12;13}
```

- This example represents a column vector:

$$\begin{bmatrix} 11 \\ 12 \\ 13 \end{bmatrix}$$

Example

`{3}`

- This example represents a scalar. The braces are optional. You can specify the same scalar as `3`.

Matrix Notation Shorthand

You can simplify the construction of matrices using notation shorthand.

Consecutive Integers. Use a colon to indicate a range of consecutive integers. For example, the vector `{1,2,3,4,5,6}` can be written `{1:6}`.

Incremented Ranges of Integers. Use a second colon followed by an integer to indicate the increment. The matrix `{1,3,5,7;2,5,8,11}` can be written as `{1:7:2;2:11:3}`, where `1:7:2` indicates the integers from 1 to 7 incrementing by 2, and `2:11:3` indicates the integers from 2 to 11 incrementing by 3.

- You must use integers when specifying a range in either of these ways. Numbers with fractional parts are truncated to integers.
- If an arithmetic expression is used, it should be enclosed in parentheses.

Extraction of an Element, a Vector, or a Submatrix

You can use indexes in parentheses to extract an element from a vector or matrix, a vector from a matrix, or a submatrix from a matrix. In the following discussion, an **integer index** refers to an integer expression used as an index, which can be a scalar matrix with an integer value or an integer element extracted from a vector or matrix. Similarly, a **vector index** refers to a vector expression used as an index, which can be a vector matrix or a vector extracted from a matrix.

For example, if S is a scalar matrix, $S = \begin{bmatrix} 2 \end{bmatrix}$, R is a row vector, $R = \begin{bmatrix} 1 & 3 & 5 \end{bmatrix}$, C is a column vector, $C = \begin{bmatrix} 2 \\ 3 \\ 4 \end{bmatrix}$, and A is a 5×5 matrix, $A = \begin{bmatrix} 11 & 12 & 13 & 14 & 15 \\ 21 & 22 & 23 & 24 & 25 \\ 31 & 32 & 33 & 34 & 35 \\ 41 & 42 & 43 & 44 & 45 \\ 51 & 52 & 53 & 54 & 55 \end{bmatrix}$, then:

$R(S) = R(2) = \{3\}$

$C(S) = C(2) = \{3\}$

- An integer index extracts an element from a vector matrix.
- The distinction between a row and a column vector does not matter when an integer index is used to extract an element from it.

$A(2,3) = A(S,3) = \{23\}$

- Two integer indexes separated by a comma extract an element from a rectangular matrix.

$A(R,2)=A(1:5:2,2)=\{12; 32; 52\}$
$A(2,R)=A(2,1:5:2)=\{21, 23, 25\}$
$A(C,2)=A(2:4,2)= \{22;32;42\}$
$A(2,C)=A(2,2:4)= \{22,23,24\}$

- An integer and a vector index separated by a comma extract a vector from a matrix.
- The distinction between a row and a column vector does not matter when used as indexes in this way.

$A(2,:)=A(S,:) = \{21, 22, 23, 24, 25\}$
$A(:,2) =A(:,S)= \{12; 22; 32; 42; 52\}$

- A colon by itself used as an index extracts an entire row or column vector from a matrix.

$A(R,C)=A(R,2:4)=A(1:5:2,C)=A(1:5:2,2:4)=\{12,13,14;32,33,34;52,53,54\}$
$A(C,R)=A(C,1:5:2)=A(2:4,R)=A(2:4,1:5:2)=\{21,23,25;31,33,35;41,43,45\}$

- Two vector indexes separated by a comma extract a submatrix from a matrix.
- The distinction between a row and a column vector does not matter when used as indexes in this way.

Construction of a Matrix from Other Matrices

You can use vector or rectangular matrices to construct a new matrix, separating row expressions by semicolons and components of row expressions by commas. If a column vector V_c has n elements and matrix M has the dimensions $n \times m$, then $\{M, V_c\}$ is an $n \times (m + 1)$ matrix. Similarly, if the row vector V_r has m elements and M is the same, then $\{M;V_r\}$ is an $(n + 1) \times m$ matrix. In fact, you can paste together any number of matrices and vectors this way.

- All of the components of each column expression must have the same number of actual rows, and all of the row expressions must have the same number of actual columns.
- The distinction between row vectors and column vectors must be observed carefully when constructing matrices in this way, so that the components will fit together properly.
- Several of the matrix functions are also useful in constructing matrices; see in particular the MAKE, UNIFORM, and IDENT functions in "Matrix Functions" on p. 481.

Example

```
COMPUTE M={CORNER, COL3; ROW3}.
```

- This example constructs the matrix *M* from the matrix *CORNER*, the column vector *COL3*, and the row vector *ROW3*.
- *COL3* supplies new row components and is separated from *CORNER* by a comma.
- *ROW3* supplies column elements and is separated from previous expressions by a semicolon.
- *COL3* must have the same number of rows as *CORNER*.

- *ROW3* must have the same number of columns as the matrix resulting from the previous expressions.
- For example, if $CORNER = \begin{bmatrix} 11 & 12 \\ 21 & 22 \end{bmatrix}$, $COL3 = \begin{bmatrix} 13 \\ 23 \end{bmatrix}$, and $ROW3 = \begin{bmatrix} 31 & 32 & 33 \end{bmatrix}$,

  then: $M = \begin{bmatrix} 11 & 12 & 13 \\ 21 & 22 & 23 \\ 31 & 32 & 33 \end{bmatrix}$

Matrix Operations

You can perform matrix calculations according to the rules of matrix algebra and compare matrices using relational or logical operators.

Conformable Matrices

Many operations with matrices make sense only if the matrices involved have "suitable" dimensions. Most often, this means that they should be the same size, with the same number of rows and the same number of columns. Matrices that are the right size for an operation are said to be **conformable matrices**. If you attempt to do something in a matrix program with a matrix that is not conformable for that operation—a matrix that has the wrong dimensions—you will receive an error message, and the operation will not be performed. An important exception, where one of the matrices is a scalar, is discussed below.

Requirements for carrying out matrix operations include:

- Matrix addition and subtraction require that the two matrices be the same size.
- The relational and logical operations described below require that the two matrices be the same size.
- Matrix multiplication requires that the number of columns of the first matrix equal the number of rows of the second matrix.
- Raising a matrix to a power can be done only if the matrix is square. This includes the important operation of *inverting* a matrix, where the power is -1 .
- Conformability requirements for matrix functions are noted in "Matrix Functions" on p. 481 and in "COMPUTE Statement" on p. 480.

Scalar Expansion

When one of the matrices involved in an operation is a scalar, the scalar is treated as a matrix of the correct size in order to carry out the operation. This internal scalar expansion is performed for the following operations:

- Addition and subtraction.

- Elementwise multiplication, division, and exponentiation. Note that multiplying a matrix elementwise by an expanded scalar is equivalent to ordinary scalar multiplication: each element of the matrix is multiplied by the scalar.
- All relational and logical operators.

Arithmetic Operators

You can add, subtract, multiply, or exponentiate matrices according to the rules of matrix algebra, or you can perform elementwise arithmetic, in which you multiply, divide, or exponentiate each element of a matrix separately. The arithmetic operators are listed below.

Unary - *Sign reversal.* A minus sign placed in front of a matrix reverses the sign of each element. (The unary + is also accepted but has no effect.)

+ *Matrix addition.* Corresponding elements of the two matrices are added. The matrices must have the same dimensions, or one must be a scalar.

- *Matrix subtraction.* Corresponding elements of the two matrices are subtracted. The matrices must have the same dimensions, or one must be a scalar.

***** *Multiplication.* There are two cases. First, *scalar multiplication*: if either of the matrices is a scalar, each element of the other matrix is multiplied by that scalar. Second, *matrix multiplication*: if A is an $m \times n$ matrix and B is an $n \times p$ matrix, $A*B$ is an $m \times p$ matrix in which the element in row i, column k, is equal to $\Sigma_{j=1}^{n}$'A (i,j) × 'B (j,k) .

/ *Division.* The division operator performs elementwise division (described below). True matrix division, the inverse operation of matrix multiplication, is accomplished by taking the INV function (square matrices) or the GINV function (rectangular matrices) of the denominator and multiplying.

****** *Matrix exponentiation.* A matrix can be raised only to an integer power. The matrix, which must be square, is multiplied by itself as many times as the absolute value of the exponent. If the exponent is negative, the result is then inverted.

&* *Elementwise multiplication.* Each element of the matrix is multiplied by the corresponding element of the second matrix. The matrices must have the same dimensions, or one must be a scalar.

&/ *Elementwise division.* Each element of the matrix is divided by the corresponding element of the second matrix. The matrices must have the same dimensions, or one must be a scalar.

&** *Elementwise exponentiation.* Each element of the first matrix is raised to the power of the corresponding element of the second matrix. The matrices must have the same dimensions, or one must be a scalar.

: *Sequential integers.* This operator creates a vector of consecutive integers from the value preceding the operator to the value following it. You can specify an optional increment following a second colon. See “Matrix Notation Shorthand” on p. 473 for the principal use of this operator.

- Use these operators only with numeric matrices. The results are undefined when they are used with string matrices.

Relational Operators

The relational operators are used to compare two matrices, element by element. The result is a matrix of the same size as the (expanded) operands and containing either 1 or 0. The value of each element, 1 or 0, is determined by whether the comparison between the corresponding element of the first matrix with the corresponding element of the second matrix is true or false, 1 for true and 0 for false. The matrices being compared must be of the same dimensions unless one of them is a scalar. The relational operators are listed in Table 1.

Table 1 Relational operators in matrix programs

>	GT	Greater than
<	LT	Less than
<> or ~= (¬=)	NE	Not equal to
<=	LE	Less than or equal to
>=	GE	Greater than or equal to
=	EQ	Equal to

- The symbolic and alphabetic forms of these operators are equivalent.
- The symbols representing NE (~= or ¬=) are system dependent. In general, the tilde (~) is valid for ASCII systems while the logical-not sign (¬), or whatever symbol over the number 6 on the keyboard, is valid for IBM EBCDIC systems.
- Use these operators only with numeric matrices. The results are undefined when they are used with string matrices.

Logical Operators

Logical operators combine two matrices, normally containing values of 1 (true) or 0 (false). When used with other numerical matrices, they treat all positive values as true and all negative and 0 values as false. The logical operators are:

NOT *Reverses the truth of the matrix that follows it.* Positive elements yield 0, and negative or 0 elements yield 1.

AND *Both must be true.* The matrix *A* AND *B* is 1 where the corresponding elements of *A* and *B* are both positive, and 0 elsewhere.

OR *Either must be true.* The matrix *A* OR *B* is 1 where the corresponding element of either *A* or *B* is positive, and 0 where both elements are negative or 0.

XOR *Either must be true, but not both.* The matrix *A* XOR *B* is 1 where one, but not both, of the corresponding elements of *A* and *B* is positive, and 0 where both are positive or neither is positive.

Precedence of Operators

Parentheses can be used to control the order in which complex expressions are evaluated. When the order of evaluation is not specified by parentheses, operations are carried out in the order listed below. The operations higher on the list take precedence over the operations lower on the list.

+ - (Unary)
:
** &**
* &* &/
+ - (Addition and Subtraction)
> >= < <= <>=
NOT
AND
OR XOR

Operations of equal precedence are performed left to right of the expressions.

Examples

```
COMPUTE A = {1,2,3;4,5,6}.
COMPUTE B = A + 4.
COMPUTE C = A &** 2.
COMPUTE D = 2 &** A.
COMPUTE E = A < 5.
COMPUTE F = (C &/ 2) < B.
```

- The results of these COMPUTE statements are:

$$A = \begin{bmatrix} 1 & 2 & 3 \\ 4 & 5 & 6 \end{bmatrix} \quad B = \begin{bmatrix} 5 & 6 & 7 \\ 8 & 9 & 10 \end{bmatrix} \quad C = \begin{bmatrix} 1 & 4 & 9 \\ 16 & 25 & 36 \end{bmatrix}$$

$$D = \begin{bmatrix} 2 & 4 & 8 \\ 16 & 32 & 64 \end{bmatrix} \quad E = \begin{bmatrix} 1 & 1 & 1 \\ 1 & 0 & 0 \end{bmatrix} \quad F = \begin{bmatrix} 1 & 1 & 1 \\ 0 & 0 & 0 \end{bmatrix}$$

MATRIX and Other SPSS Commands

A matrix program is a single procedure within an SPSS session.

- No working data file is needed to run a matrix program. If one exists, it is ignored during matrix processing unless you specifically reference it (with an asterisk) on the GET, SAVE, MGET, or MSAVE statements.
- Variables defined in the SPSS working data file are unavailable during matrix processing, except with the GET or MGET statements.
- Matrix variables are unavailable after the END MATRIX command, unless you use SAVE or MSAVE to write them to the working data file.

- You cannot run a matrix program from a syntax window if split-file processing is in effect. If you save the matrix program into a syntax file, however, you can use the INCLUDE command to run the program even if split-file processing is in effect.

Matrix Statements

Table 2 lists all the statements that are accepted within a matrix program. Most of them have the same name as an analogous SPSS command and perform an exactly analogous function. Use only these statements between the MATRIX and END MATRIX commands. Any command not recognized as a valid matrix statement will be rejected by the matrix processor.

Table 2 Valid matrix statements

BREAK	ELSE IF	MSAVE
CALL	END IF	PRINT
COMPUTE	END LOOP	READ
DISPLAY	GET	RELEASE
DO IF	LOOP	SAVE
ELSE	MGET	WRITE

Exchanging Data with SPSS Data Files

Matrix programs can read and write SPSS data files.

- The GET and SAVE statements read and write ordinary (case-oriented) SPSS data files, treating each case as a row of a matrix and each ordinary variable as a column.
- The MGET and MSAVE statements read and write matrix-format SPSS data files, respecting the structure defined by SPSS when it creates the file. These statements are discussed below.
- Case weighting in an SPSS data file is ignored when the file is read into a matrix program.

Using a Working Data File

You can use the GET statement to read a case-oriented working data file into a matrix variable. The result is a rectangular data matrix in which cases have become rows and variables have become columns. Special circumstances can affect the processing of this data matrix.

Split-file Processing. After a SPLIT FILE command in SPSS, a matrix program executed with the INCLUDE command will read one split-file group with each execution of a GET statement. This enables you to process the subgroups separately within the matrix program.

Case Selection. When a subset of cases is selected for processing, as the result of a SELECT IF, SAMPLE, or N OF CASES command, only the selected cases will be read by the GET statement in a matrix program.

Temporary Transformations. The entire matrix program is treated as a single procedure by the SPSS system. Temporary transformations—those preceded by the TEMPORARY command—entered immediately before a matrix program are in effect throughout that program (even if you GET the working data file repeatedly) and are no longer in effect at the end of the matrix program.

Case Weighting. Case weighting in a working data file is ignored when the file is read into a matrix program.

MATRIX and END MATRIX Commands

The MATRIX command, when encountered in an SPSS session, invokes the matrix processor, which reads matrix statements until the END MATRIX or FINISH command is encountered.

- MATRIX is a procedure and cannot be entered inside a transformation structure such as DO IF or LOOP.
- The MATRIX procedure does not require a working data file.
- Comments are removed before subsequent lines are passed to the matrix processor.
- Macros are expanded before subsequent lines are passed to the matrix processor.

The END MATRIX command terminates matrix processing and returns control to the SPSS command processor.

- The contents of matrix variables are lost after an END MATRIX command.
- The working data file, if present, becomes available again after an END MATRIX command.

COMPUTE Statement

The COMPUTE statement carries out most of the calculations in the matrix program. It closely resembles the COMPUTE command in the SPSS transformation language.

- The basic specification is the target variable, an equals sign, and the assignment expression. Values of the target variable are calculated according to the specification on the assignment expression.
- The target variable must be named first, and the equals sign is required. Only one target variable is allowed per COMPUTE statement.
- Expressions that extract portions of a matrix, such as *M*(1,:) or *M*(1:3,4), are allowed to assign values. (See "Matrix Notation Shorthand" on p. 473.) The target variable must be specified as a variable.
- Matrix functions must specify at least one argument enclosed in parentheses. If an expression has two or more arguments, each argument must be separated by a comma. For a complete discussion of the functions and their arguments, see "Matrix Functions" on p. 481.

String Values on COMPUTE Statements

Matrix variables, unlike those in the SPSS transformation language, are not checked for data type (numeric or string) when you use them in a COMPUTE statement.

- Numerical calculations with matrices containing string values will produce meaningless results.

- One or more elements of a matrix can be set equal to string constants by enclosing the string constants in apostrophes or quotation marks on a COMPUTE statement.
- String values can be copied from one matrix to another with the COMPUTE statement.
- There is no way to display a matrix that contains both numeric and string values, if you compute one for some reason.

Example

```
COMPUTE LABELS={"Observe", "Predict", "Error"}.
PRINT LABELS /FORMAT=A7.
```

- *LABELS* is a row vector containing three string values.

Arithmetic Operations and Comparisons

The expression on a COMPUTE statement can be formed from matrix constants and variables, combined with the arithmetic, relational, and logical operators discussed above. Matrix constructions and matrix functions are also allowed.

Examples

```
COMPUTE PI = 3.14159265.
COMPUTE RSQ = R * R.
COMPUTE FLAGS = EIGENVAL >= 1.
COMPUTE ESTIM = {OBS, PRED, ERR}.
```

- The first statement computes a scalar. Note that the braces are optional on a scalar constant.
- The second statement computes the square of the matrix *R*. *R* can be any square matrix, including a scalar.
- The third statement computes a vector named *FLAGS*, which has the same dimension as the existing vector *EIGENVAL*. Each element of *FLAGS* equals 1 if the corresponding element of *EIGENVAL* is greater than or equal to 1, and 0 if the corresponding element is less than 1.
- The fourth statement constructs a matrix *ESTIM* by concatenating the three vectors or matrices *OBS*, *PRED*, and *ERR*. The component matrices must have the same number of rows.

Matrix Functions

The following functions are available in the matrix program. Except where noted, each takes one or more numeric matrices as arguments and returns a matrix value as its result. The arguments must be enclosed in parentheses, and multiple arguments must be separated by commas.

On the following list, matrix arguments are represented by names beginning with *M*. Unless otherwise noted, these arguments can be vectors or scalars. Arguments that must be vectors are represented by names beginning with *V*, and arguments that must be scalars are represented by names beginning with *S*.

ABS(M) *Absolute value.* Takes a single argument. Returns a matrix having the same dimensions as the argument, containing the absolute values of its elements.

ALL(M)

Test for all elements non-zero. Takes a single argument. Returns a scalar: 1 if all elements of the argument are non-zero and 0 if any element is zero.

ANY(M)

Test for any element non-zero. Takes a single argument. Returns a scalar: 1 if any element of the argument is non-zero and 0 if all elements are zero.

ARSIN(M)

Inverse sine. Takes a single argument, whose elements must be between -1 and 1. Returns a matrix having the same dimensions as the argument, containing the inverse sines (arc sines) of its elements. The results are in radians and are in the range from $-\pi/2$ to $\pi/2$.

ARTAN(M)

Inverse tangent. Takes a single argument. Returns a matrix having the same dimensions as the argument, containing the inverse tangents (arc tangents) of its elements, in radians. To convert radians to degrees, multiply by $180/\pi$, which you can compute as 45/ARTAN(1). For example, the statement `COMPUTE DEGREES=ARTAN(M)*45/ARTAN(1)` returns a matrix containing inverse tangents in degrees.

BLOCK(M1,M2,...)

Create a block diagonal matrix. Takes any number of arguments. Returns a matrix with as many rows as the sum of the rows in all the arguments, and as many columns as the sum of the columns in all the arguments, with the argument matrices down the diagonal and zeros elsewhere. For example, if:

$$A = \begin{bmatrix} 1 & 1 & 1 \\ 1 & 1 & 1 \end{bmatrix}, \quad B = \begin{bmatrix} 2 & 2 \\ 2 & 2 \end{bmatrix}, \quad C = \begin{bmatrix} 3 & 3 & 3 \\ 3 & 3 & 3 \\ 3 & 3 & 3 \\ 3 & 3 & 3 \end{bmatrix}, \text{ and } D = \begin{bmatrix} 4 & 4 & 4 \end{bmatrix},$$

then:

$$\text{BLOCK}(A, B, C, D) = \begin{bmatrix} 1 & 1 & 1 & 0 & 0 & 0 & 0 & 0 & 0 & 0 & 0 \\ 1 & 1 & 1 & 0 & 0 & 0 & 0 & 0 & 0 & 0 & 0 \\ 0 & 0 & 0 & 2 & 2 & 0 & 0 & 0 & 0 & 0 & 0 \\ 0 & 0 & 0 & 2 & 2 & 0 & 0 & 0 & 0 & 0 & 0 \\ 0 & 0 & 0 & 0 & 0 & 3 & 3 & 3 & 0 & 0 & 0 \\ 0 & 0 & 0 & 0 & 0 & 3 & 3 & 3 & 0 & 0 & 0 \\ 0 & 0 & 0 & 0 & 0 & 3 & 3 & 3 & 0 & 0 & 0 \\ 0 & 0 & 0 & 0 & 0 & 3 & 3 & 3 & 0 & 0 & 0 \\ 0 & 0 & 0 & 0 & 0 & 0 & 0 & 0 & 4 & 4 & 4 \end{bmatrix}$$

CDFNORM(M)

Standard normal cumulative distribution function of elements. Takes a single argument. Returns a matrix having the same dimensions as the argument, containing the values of the cumulative normal distribution function for each of its elements. If an element of the argument is x, the corresponding element of the result is a number between 0 and 1, giving the proportion of a normal distribution that is less than x. For example, `CDFNORM({-1.96,0,1.96})` results in, approximately, {.025,.5,.975}.

CHICDF(M,S)

Chi-square cumulative distribution function of elements. Takes two arguments, a matrix of chi-square values and a scalar giving the degrees of freedom (which must be positive). Returns a matrix having the same dimensions as the first argument, containing the values of the cumulative chi-square distribution function for each of its elements. If an element of the first argument is x and the second argument is S, the corresponding element of the result is a number between 0 and 1, giving the proportion of a chi-square distribution with S degrees of freedom that is less than x. If x is not positive, the result is 0.

CHOL(M)

Cholesky decomposition. Takes a single argument, which must be a symmetric positive-definite matrix (a square matrix, symmetric about the main diagonal, with positive eigenvalues). Returns a matrix having the same dimensions as the argument. If M is a symmetric positive-definite matrix and B=CHOL(M), then T(B)*B=M, where T is the transpose function defined below.

CMAX(M)

Column maxima. Takes a single argument. Returns a row vector with the same number of columns as the argument. Each column of the result contains the maximum value of the corresponding column of the argument.

CMIN(M)

Column minima. Takes a single argument. Returns a row vector with the same number of columns as the argument. Each column of the result contains the minimum value of the corresponding column of the argument.

COS(M)

Cosines. Takes a single argument. Returns a matrix having the same dimensions as the argument, containing the cosines of the elements of the argument. Elements of the argument matrix are assumed to be measured in radians. To convert degrees to radians, multiply by $\pi/180$, which you can compute as ARTAN(1)/45. For example, the statement `COMPUTE COSINES=COS(DEGREES*ARTAN(1)/45)` returns cosines from a matrix containing elements measured in degrees.

CSSQ(M)

Column sums of squares. Takes a single argument. Returns a row vector with the same number of columns as the argument. Each column of the result contains the sum of the squared values of the elements in the corresponding column of the argument.

CSUM(M)

Column sums. Takes a single argument. Returns a row vector with the same number of columns as the argument. Each column of the result contains the sum of the elements in the corresponding column of the argument.

DESIGN(M)

Main-effects design matrix from the columns of a matrix. Takes a single argument. Returns a matrix having the same number of rows as the argument, and as many columns as the sum of the numbers of unique values in each column of the argument. Constant columns in the argument are skipped with a warning message. The result contains 1 in the row(s) where the value in question occurs in the argument and 0 otherwise. For example, if:

$$A = \begin{bmatrix} 1 & 2 & 8 \\ 1 & 3 & 8 \\ 2 & 6 & 5 \\ 3 & 3 & 8 \\ 3 & 6 & 5 \end{bmatrix}, \text{ then: DESIGN}(A) = \begin{bmatrix} 1 & 0 & 0 & 1 & 0 & 0 & 1 & 0 \\ 1 & 0 & 0 & 0 & 1 & 0 & 1 & 0 \\ 0 & 1 & 0 & 0 & 0 & 1 & 0 & 1 \\ 0 & 0 & 1 & 0 & 1 & 0 & 1 & 0 \\ 0 & 0 & 1 & 0 & 0 & 1 & 0 & 1 \end{bmatrix}$$

The first three columns of the result correspond to the three distinct values 1, 2, and 3 in the first column of A; the fourth through sixth columns of the result correspond to the three distinct values 2, 3, and 6 in the second column of A; and the last two columns of the result correspond to the two distinct values 8 and 5 in the third column of A.

DET(M)

Determinant. Takes a single argument, which must be a square matrix. Returns a scalar, which is the determinant of the argument.

DIAG(M)

Diagonal of a matrix. Takes a single argument. Returns a column vector with as many rows as the minimum of the number of rows and the number of columns in the argument. The ith element of the result is the value in row i, column i, of the argument.

EOF(file)

End of file indicator, normally used after a READ statement. Takes a single argument, which must be either a filename in apostrophes or quotation marks, or a file handle defined on a FILE HANDLE command that precedes the matrix program. Returns a scalar equal to 1 if the last attempt to read that file encountered the last record in the file, and equal to 0 if the last attempt did not encounter the last record in the file. Calling the EOF function causes a REREAD specification on the READ statement to be ignored on the next attempt to read the file.

EVAL(M)

Eigenvalues of a symmetric matrix. Takes a single argument, which must be a symmetric matrix. Returns a column vector with the same number of rows as the argument, containing the eigenvalues of the argument in decreasing numerical order.

EXP(M)

Exponentials of matrix elements. Takes a single argument. Returns a matrix having the same dimensions as the argument, in which each element equals e raised to the power of the corresponding element in the argument matrix.

FCDF(M,S1,S2) *Cumulative* F *distribution function of elements.* Takes three arguments, a matrix of F values and two scalars giving the degrees of freedom (which must be positive). Returns a matrix having the same dimensions as the first argument M, containing the values of the cumulative F distribution function for each of its elements. If an element of the first argument is x and the second and third arguments are $S1$ and $S2$, the corresponding element of the result is a number between 0 and 1, giving the proportion of an F distribution with $S1$ and $S2$ degrees of freedom that is less than x. If x is not positive, the result is 0.

GINV(M) *Moore-Penrose generalized inverse of a matrix.* Takes a single argument. Returns a matrix with the same dimensions as the transpose of the argument. If A is the generalized inverse of a matrix M, then $M*A*M=M$ and $A*M*A=A$. Both $A*M$ and $M*A$ are symmetric.

GRADE(M) *Ranks elements in a matrix.* Takes a single argument. Uses sequential integers for ties.

GSCH(M) *Gram-Schmidt orthonormal basis for the space spanned by the column vectors of a matrix.* Takes a single argument, in which there must be as many linearly independent columns as there are rows. (That is, the rank of the argument must equal the number of rows.) Returns a square matrix with as many rows as the argument. The columns of the result form a basis for the space spanned by the columns of the argument.

IDENT(S1 [,S2]) *Create an identity matrix.* Takes either one or two arguments, which must be scalars. Returns a matrix with as many rows as the first argument and as many columns as the second argument, if any. If the second argument is omitted, the result is a square matrix. Elements on the main diagonal of the result equal 1, and all other elements equal 0.

INV(M) *Inverse of a matrix.* Takes a single argument, which must be square and nonsingular (that is, its determinant must not be 0). Returns a square matrix having the same dimensions as the argument. If A is the inverse of M, then $M*A=A*M=I$, where I is the identity matrix.

KRONECKER(M1,M2) *Kronecker product of two matrices.* Takes two arguments. Returns a matrix whose row dimension is the product of the row dimensions of the arguments and whose column dimension is the product of the column dimensions of the arguments. The Kronecker product of two matrices A and B takes the form of an array of scalar products:

$$\begin{array}{lll} A(1,1)*B & A(1,2)*B & ...A(1,N)*B \\ A(2,1)*B & A(2,2)*B & ...A(2,N)*B \\ ... & & \\ A(M,1)*B & A(M,2)*B & ...A(M,N)*B \end{array}$$

LG10(M) *Base 10 logarithms of the elements.* Takes a single argument, all of whose elements must be positive. Returns a matrix having the same dimensions as the argument, in which each element is the logarithm to base 10 of the corresponding element of the argument.

LN(M) *Natural logarithms of the elements.* Takes a single argument, all of whose elements must be positive. Returns a matrix having the same dimensions as the argument, in which each element is the logarithm to base *e* of the corresponding element of the argument.

MAGIC(S) *Magic square.* Takes a single scalar, which must be 3 or larger, as an argument. Returns a square matrix with *S* rows and *S* columns containing the integers from 1 through S^2. All the row sums and all the column sums are equal in the result matrix. (The result matrix is only one of several possible magic squares.)

MAKE(S1,S2,S3) *Create a matrix, all of whose elements equal a specified value.* Takes three scalars as arguments. Returns an $S1 \times S2$ matrix, all of whose elements equal *S3*.

MDIAG(V) *Create a square matrix with a specified main diagonal.* Takes a single vector as an argument. Returns a square matrix with as many rows and columns as the dimension of the vector. The elements of the vector appear on the main diagonal of the matrix, and the other matrix elements are all 0.

MMAX(M) *Maximum element in a matrix.* Takes a single argument. Returns a scalar equal to the numerically largest element in the argument *M*.

MMIN(M) *Minimum element in a matrix.* Takes a single argument. Returns a scalar equal to the numerically smallest element in the argument *M*.

MOD(M,S) *Remainders after division by a scalar.* Takes two arguments, a matrix and a scalar (which must not be 0). Returns a matrix having the same dimensions as *M*, each of whose elements is the remainder after the corresponding element of *M* is divided by *S*. The sign of each element of the result is the same as the sign of the corresponding element of the matrix argument *M*.

MSSQ(M) *Matrix sum of squares.* Takes a single argument. Returns a scalar that equals the sum of the squared values of all the elements in the argument.

MSUM(M) *Matrix sum.* Takes a single argument. Returns a scalar that equals the sum of all the elements in the argument.

NCOL(M) *Number of columns in a matrix.* Takes a single argument. Returns a scalar that equals the number of columns in the argument.

NROW(M) *Number of rows in a matrix.* Takes a single argument. Returns a scalar that equals the number of rows in the argument.

RANK(M) *Rank of a matrix.* Takes a single argument. Returns a scalar that equals the number of linearly independent rows or columns in the argument.

RESHAPE(M,S1,S2) *Matrix of different dimensions.* Takes three arguments, a matrix and two scalars, whose product must equal the number of elements in the matrix. Returns a matrix whose dimensions are given by the scalar arguments. For example, if *M* is any matrix with exactly 50 elements, then `RESHAPE(M, 5, 10)` is a matrix with 5 rows and 10 columns. Elements are assigned to the reshaped matrix in order by row.

RMAX(M) *Row maxima.* Takes a single argument. Returns a column vector with the same number of rows as the argument. Each row of the result contains the maximum value of the corresponding row of the argument.

RMIN(M) *Row minima.* Takes a single argument. Returns a column vector with the same number of rows as the argument. Each row of the result contains the minimum value of the corresponding row of the argument.

RND(M) *Elements rounded to the nearest integers.* Takes a single argument. Returns a matrix having the same dimensions as the argument. Each element of the result equals the corresponding element of the argument rounded to an integer.

RNKORDER(M) *Ranking of matrix elements in ascending order.* Takes a single argument. Returns a matrix having the same dimensions as the argument M. The smallest element of the argument corresponds to a result element of 1, and the largest element of the argument to a result element equal to the number of elements, except that ties (equal elements in M) are resolved by assigning a rank equal to the arithmetic mean of the applicable ranks. For example, if:

$$M = \begin{bmatrix} -1 & -21.7 & 8 \\ 0 & 3.91 & -21.7 \\ 8 & 9 & 10 \end{bmatrix}, \text{ then: RNKORDER}(M) = \begin{bmatrix} 3 & 1.5 & 6.5 \\ 4 & 5 & 1.5 \\ 6.5 & 8 & 9 \end{bmatrix}$$

RSSQ(M) *Row sums of squares.* Takes a single argument. Returns a column vector having the same number of rows as the argument. Each row of the result contains the sum of the squared values of the elements in the corresponding row of the argument.

RSUM(M) *Row sums.* Takes a single argument. Returns a column vector having the same number of rows as the argument. Each row of the result contains the sum of the elements in the corresponding row of the argument.

SIN(M) *Sines.* Takes a single argument. Returns a matrix having the same dimensions as the argument, containing the sines of the elements of the argument. Elements of the argument matrix are assumed to be measured in radians. To convert degrees to radians, multiply by $\pi/180$, which you can compute as ARTAN(1)/45. For example, the statement `COMPUTE SINES=SIN(DEGREES*ARTAN(1)/45)` computes sines from a matrix containing elements measured in degrees.

SOLVE(M1,M2) *Solution of systems of linear equations.* Takes two arguments, the first of which must be square and nonsingular (its determinant must be nonzero), and the second of which must have the same number of rows as the first. Returns a matrix with the same dimensions as the second argument. If $M1*X=M2$, then X=SOLVE($M1$, $M2$). In effect, this function sets its result X equal to INV($M1$)*$M2$.

SQRT(M)

Square roots of elements. Takes a single argument, whose elements must not be negative. Returns a matrix having the same dimensions as the arguments, whose elements are the positive square roots of the corresponding elements of the argument.

SSCP(M)

Sums of squares and cross-products. Takes a single argument. Returns a square matrix having as many rows (and columns) as the argument has columns. SSCP(M) equals T(M)*M, where T is the transpose function defined below.

SVAL(M)

Singular values of a matrix. Takes a single argument. Returns a column vector containing as many rows as the minimum of the numbers of rows and columns in the argument, containing the singular values of the argument in decreasing numerical order. The singular values of a matrix M are the square roots of the eigenvalues of T(M)*M, where T is the transpose function discussed below.

SWEEP(M,S)

Sweep transformation of a matrix. Takes two arguments, a matrix and a scalar, which must be less than or equal to both the number of rows and the number of columns of the matrix. In other words, the pivot element of the matrix, which is $M(S,S)$, must exist. Returns a matrix of the same dimensions as M. Suppose that $S=\{k\}$ and A=SWEEP(M,S). If $M(k,k)$ is not 0, then

$$A(k,k) = 1/M(k,k)$$
$$A(i,k) = -M(i,k)/M(k,k) \quad \text{for } i \text{ not equal to } k$$
$$A(k,j) = M(k,j)/(M(k,k)) \quad \text{for } j \text{ not equal to } k$$
$$A(i,j) = (M(k,k)*M(i,j) - M(i,k)*M(k,j))/M(k,k) \quad \text{for } i,j \text{ not equal to } k$$

and if $M(k,k)$ equals 0, then

$$A(i,k) = A(k,i) = 0 \quad \text{for all } i$$
$$A(i,j) = M(i,j) \quad \text{for } i,j \text{ not equal to } k$$

TCDF(M,S)

Cumulative t *distribution function of elements.* Takes two arguments, a matrix of t values and a scalar giving the degrees of freedom (which must be positive). Returns a matrix having the same dimensions as M, containing the values of the cumulative t distribution function for each of its elements. If an element of the first argument is x and the second argument is S, then the corresponding element of the result is a number between 0 and 1, giving the proportion of a t distribution with S degrees of freedom that is less than x.

TRACE(M)

Sum of the main diagonal elements. Takes a single argument. Returns a scalar, which equals the sum of the elements on the main diagonal of the argument.

TRANSPOS(M)

Transpose of the matrix. Takes a single argument. Returns the transpose of the argument. TRANSPOS can be shortened to T.

TRUNC(M) *Truncation of elements to integers.* Takes a single argument. Returns a matrix having the same dimensions as the argument, whose elements equal the corresponding elements of the argument truncated to integers.

UNIFORM(S1,S2) *Uniformly distributed pseudo-random numbers between 0 and 1.* Takes two scalars as arguments. Returns a matrix with the number of rows specified by the first argument and the number of columns specified by the second argument, containing pseudo-random numbers uniformly distributed between 0 and 1.

CALL Statement

Closely related to the matrix functions are the matrix procedures, which are invoked with the CALL statement. Procedures, similarly to functions, accept arguments enclosed in parentheses and separated by commas. They return their result in one or more of the arguments as noted in the individual descriptions below. They are implemented as procedures rather than as functions so that they can return more than one value or (in the case of SETDIAG) modify a matrix without making a copy of it.

EIGEN(M,var1,var2) *Eigenvectors and eigenvalues of a symmetric matrix.* Takes three arguments: a symmetric matrix and two valid variable names to which the results are assigned. If *M* is a symmetric matrix, the statement `CALL EIGEN(M, A, B)` will assign to *A* a matrix having the same dimensions as *M*, containing the eigenvectors of *M* as its columns, and will assign to *B* a column vector having as many rows as *M*, containing the eigenvalues of *M* in descending numerical order. The eigenvectors in *A* are ordered to correspond with the eigenvalues in *B*; thus, the first column corresponds to the largest eigenvalue, the second to the second largest, and so on.

SETDIAG(M,V) *Set the main diagonal of a matrix.* Takes two arguments, a matrix and a vector. Elements on the main diagonal of *M* are set equal to the corresponding elements of *V*. If *V* is a scalar, all the diagonal elements are set equal to that scalar. Otherwise, if *V* has fewer elements than the main diagonal of *M*, remaining elements on the main diagonal are unchanged. If *V* has more elements than are needed, the extra elements are not used. See also MDIAG on p. 486.

SVD(M,var1,var2,var3) *Singular value decomposition of a matrix.* Takes four arguments: a matrix and three valid variable names to which the results are assigned. If *M* is a matrix, the statement `CALL SVD(M,U,Q,V)` will assign to *Q* a diagonal matrix of the same dimensions as *M*, and to *U* and *V* unitary matrices (matrices whose inverses equal their transposes) of appropriate dimensions, such that $M=U*Q*T(V)$, where T is the transpose function defined above. The singular values of *M* are in the main diagonal of *Q*.

PRINT Statement

The PRINT statement displays matrices or matrix expressions. Its syntax is as follows:

```
PRINT [matrix expression]
      [/FORMAT="format descriptor"]
      [/TITLE="title"]
      [/SPACE={NEWPAGE}]
              {n      }
      [{/RLABELS=list of quoted names}]
       {/RNAMES=vector of names      }

      [{/CLABELS=list of quoted names}]
       {/CNAMES=vector of names      }
```

Matrix Expression

Matrix expression is a single matrix variable name or an expression that evaluates to a matrix. PRINT displays the specified matrix.

- The matrix specification must precede any other specifications on the PRINT statement. If no matrix is specified, no data will be displayed, but the TITLE and SPACE specifications will be honored.
- You can specify a matrix name, a matrix raised to a power, or a matrix function (with its arguments in parentheses) by itself, but you must enclose other matrix expressions in parentheses. For example, `PRINT A`, `PRINT INV(A)`, and `PRINT B**DET(T(C)*D)` are all legal, but `PRINT A+B` is not. You must specify `PRINT (A+B)`.
- Constant expressions are allowed.
- A matrix program can consist entirely of PRINT statements, without defining any matrix variables.

FORMAT Keyword

FORMAT specifies a single format descriptor for display of the matrix data.

- All matrix elements are displayed with the same format.
- You can use any printable numeric format (for numeric matrices) or string format (for string matrices) as defined in FORMATS in the *SPSS Base System Syntax Reference Guide*.
- The matrix processor will choose a suitable numeric format if you omit the FORMAT specification, but a string format such as A8 is essential when displaying a matrix containing string data.
- String values exceeding the width of a string format are truncated.
- See "Scaling Factor in Displays" on p. 492 for default formatting of matrices containing large or small values.

TITLE Keyword

TITLE specifies a title for the matrix displayed. The title must be enclosed in quotation marks or apostrophes. If it exceeds the maximum display width, it is truncated. The slash preceding

TITLE is required, even if it is the only specification on the PRINT statement. If you omit the TITLE specification, the matrix name or expression from the PRINT statement is used as a default title.

SPACE Keyword

SPACE controls output spacing before printing the title and the matrix. You can specify either a positive number or the keyword NEWPAGE. The slash preceding SPACE is required, even if it is the only specification on the PRINT statement.

NEWPAGE *Start a new page before printing the title.*

n *Skip* n *lines before displaying the title.*

RLABELS Keyword

RLABELS allows you to supply row labels for the matrix.

- The labels must be separated by commas.
- Enclose individual labels in quotation marks or apostrophes if they contain imbedded commas or if you wish to preserve lowercase letters. Otherwise, quotation marks or apostrophes are optional.
- If too many names are supplied, the extras are ignored. If not enough names are supplied, the last rows remain unlabeled.

RNAMES Keyword

RNAMES allows you to supply the name of a vector or a vector expression containing row labels for the matrix.

- Either a row vector or a column vector can be used, but the vector must contain string data.
- If too many names are supplied, the extras are ignored. If not enough names are supplied, the last rows remain unlabeled.

CLABELS Keyword

CLABELS allows you to supply column labels for the matrix.

- The labels must be separated by commas.
- Enclose individual labels in quotation marks or apostrophes if they contain imbedded commas or if you wish to preserve lowercase letters. Otherwise, quotation marks or apostrophes are optional.
- If too many names are supplied, the extras are ignored. If not enough names are supplied, the last columns remain unlabeled.

CNAMES Keyword

CNAMES allows you to supply the name of a vector or a vector expression containing column labels for the matrix.

- Either a row vector or a column vector can be used, but the vector must contain string data.
- If too many names are supplied, the extras are ignored. If not enough names are supplied, the last columns remain unlabeled.

Scaling Factor in Displays

When a matrix contains very large or very small numbers, it may be necessary to use scientific notation to display the data. If you do not specify a display format, the matrix processor chooses a power-of-ten multiplier that will allow the largest value to be displayed, and it displays this multiplier on a heading line before the data. The multiplier is not displayed for each element in the matrix. The displayed values, multiplied by the power of ten that is indicated in the heading, equal the actual values (possibly rounded).

- Values that are very small, relative to the multiplier, are displayed as 0.
- If you explicitly specify a scientific-notation format (Ew.d), each matrix element is displayed using that format. This permits you to display very large and very small numbers in the same matrix without losing precision.

Example

```
COMPUTE M = {.0000000001357, 2.468, 3690000000}.
PRINT M /TITLE "Default format".
PRINT M /FORMAT "E13" /TITLE "Explicit exponential format".
```

- The first PRINT subcommand uses the default format with 10^9 as the multiplier for each element of the matrix. This results in the following output:

```
Default format
  10 ** 9  X
    .000000000    .000000002   3.690000000
```

  Note that the first element is displayed as 0 and the second is rounded to one significant digit.

- An explicitly specified exponential format on the second PRINT subcommand allows each element to be displayed with full precision, as the following output shows:

```
Explicit exponential format
 1.3570000E-10 2.4680000E+00 3.6900000E+09
```

Matrix Control Structures

The matrix language includes two structures that allow you to alter the flow of control within a matrix program.

- The DO IF statement tests a logical expression to determine whether one or more subsequent matrix statements should be executed.

- The LOOP statement defines the beginning of a block of matrix statements that should be executed repeatedly until a termination criterion is satisfied or a BREAK statement is executed.

These statements closely resemble the DO IF and LOOP commands in the SPSS transformation language. In particular, these structures can be nested within one another as deeply as the available memory allows.

DO IF Structures

A DO IF structure in a matrix program affects the flow of control exactly as the analogous commands affect an SPSS transformation program, except that missing-value considerations do not arise in a matrix program. The syntax of the DO IF structure is as follows:

```
DO IF [(]logical expression[)]
  matrix statements
[ELSE IF [(]logical expression[)]]
  matrix statements
[ELSE IF...]
  .
  .
  .
[ELSE]
 matrix statements
END IF.
```

- The DO IF statement marks the beginning of the structure, and the END IF statement marks its end.
- The ELSE IF statement is optional and can be repeated as many times as desired within the structure.
- The ELSE statement is optional. It can be used only once and must follow any ELSE IF statements.
- The END IF statement must follow any ELSE IF and ELSE statements.
- The DO IF and ELSE IF statements must contain a logical expression, normally one involving the relational operators EQ, GT, and so on. However, the matrix language allows any expression that evaluates to a scalar to be used as the logical expression. Scalars greater than 0 are considered true, and scalars less than or equal to 0 are considered false.

A DO IF structure affects the flow of control within a matrix program as follows:

- If the logical expression on the DO IF statement is true, the statements immediately following the DO IF are executed up to the next ELSE IF or ELSE in the structure. Control then passes to the first statement following the END IF for that structure.
- If the expression on the DO IF statement is false, control passes to the first ELSE IF, where the logical expression is evaluated. If this expression is true, statements following the ELSE IF are executed up to the next ELSE IF or ELSE statement, and control passes to the first statement following the END IF for that structure.

- If the expressions on the DO IF and the first ELSE IF statements are both false, control passes to the next ELSE IF, where that logical expression is evaluated. If none of the expressions is true on any of the ELSE IF statements, statements following the ELSE statement are executed up to the END IF statement, and control falls out of the structure.
- If none of the expressions on the DO IF statement or the ELSE IF statements is true and there is no ELSE statement, control passes to the first statement following the END IF for that structure.

LOOP Structures

A LOOP structure in a matrix program affects the flow of control exactly as the analogous commands affect an SPSS transformation program, except that missing-value considerations do not arise in a matrix program. Its syntax is as follows:

```
LOOP [varname=n TO m [BY k]] [IF [(]logical expression[)]]

matrix statements

[BREAK]

matrix statements

END LOOP [IF [(]logical expression[)]]
```

The matrix statements specified between LOOP and END LOOP are executed repeatedly until one of the following four conditions is met:

- A logical expression on the IF clause of the LOOP statement is evaluated as false.
- An index variable used on the LOOP statement passes beyond its terminal value.
- A logical expression on the IF clause of the END LOOP statement is evaluated as true.
- A BREAK statement is executed within the loop structure (but outside of any nested loop structures).

Index Clause on the LOOP Statement

An index clause on a LOOP statement creates an index variable whose name is specified immediately after the keyword LOOP. The variable is assigned an initial value of *n*. Each time through the loop the variable is tested against the terminal value *m* and incremented by the increment value *k* if *k* is specified, or by 1 if *k* is not specified. When the index variable is greater than *m* for positive increments or less than *m* for negative increments, control passes to the statement after the END LOOP statement.

- Both the index clause and the IF clause are optional. If both are present, the index clause must appear first.
- The index variable must be a scalar with a valid matrix variable name.
- The initial value, *n*, the terminal value, *m*, and the increment, *k* (if present), must be scalars or matrix expressions evaluating to scalars. Non-integer values are truncated to integers before use.
- If the keyword BY and the increment *k* are absent, an increment of 1 is used.

IF Clause on the LOOP Statement

The logical expression is evaluated before each iteration of the loop structure. If it is false, the loop terminates and control passes to the statement after END LOOP.

- The IF clause is optional. If both the index clause and the IF clause are present, the index clause must appear first.
- As in the DO IF structure, the logical expression of the IF clause is evaluated as a scalar, with positive values being treated as true and zero or negative values as false.

IF Clause on the END LOOP Statement

When an IF clause is present on an END LOOP statement, the logical expression is evaluated after each iteration of the loop structure. If it is true, the loop terminates and control passes to the statement following the END LOOP statement.

- The IF clause is optional.
- As on the LOOP statement, the logical expression of the IF clause is evaluated as a scalar, with positive values being treated as true and zero or negative values as false.

BREAK Statement

The BREAK statement within a loop structure transfers control immediately to the statement following the (next) END LOOP statement. It is normally placed within a DO IF structure inside the LOOP structure to exit the loop when specified conditions are met.

Example

```
LOOP LOCATION = 1, NROW(VEC).
+  DO IF (VEC(LOCATION) = TARGET).
+     BREAK.

+  END IF.
END LOOP.
```

- This loop searches for the (first) location of a specific value, TARGET, in a vector, *VEC.*
- The DO IF statement checks whether the vector element indexed by *LOCATION* equals the target.
- If so, the BREAK statement transfers control out of the loop, leaving *LOCATION* as the index of TARGET in *VEC.*

READ Statement: Reading Character Data

The READ statement reads data into a matrix or submatrix from a character-format file—that is, a file containing ordinary numbers or words in readable form. The syntax for the READ statement is:

```
READ  variable reference
  [/FILE = file reference]
   /FIELD = c1 TO c2 [BY w]
  [/SIZE = size expression]
  [/MODE = {RECTANGULAR}]
           {SYMMETRIC  }
  [/REREAD]
  [/FORMAT = format descriptor]
```

- The file can contain values in freefield or fixed-column format. The data can appear in any of the field formats supported by DATA LIST.
- More than one matrix can be read from a single input record by rereading the record.
- If the end of the file is encountered during a READ operation (that is, fewer values are available than the number of elements required by the specified matrix size), a warning message is displayed and the contents of the unread elements of the matrix are unpredictable.

Variable Specification

The variable reference on the READ statement is a matrix variable name, with or without indexes.

For a name without indexes:

- READ creates the specified matrix variable.
- The matrix need not exist when READ is executed.
- If the matrix already exists, it is replaced by the matrix read from the file.
- You must specify the size of the matrix using the SIZE specification.

For an indexed name:

- READ creates a submatrix from an existing matrix.
- The matrix variable named must already exist.
- You can define any submatrix with indexes; for example, `M(:,I)`. To define an entire existing matrix, specify `M(:,:)`.
- The SIZE specification can be omitted. If specified, its value must match the size of the specified submatrix.

FILE Specification

FILE designates the character file containing the data. It can be an actual filename in apostrophes or quotation marks, or a file handle defined on a FILE HANDLE command that precedes the matrix program.

- The filename or handle must specify an existing file containing character data, not an SPSS data file or a specially formatted file of another kind, such as a spreadsheet file.

- The FILE specification is required on the first READ statement in a matrix program (first in order of appearance, not necessarily in order of execution). If you omit the FILE specification from a later READ statement, the statement uses the most recently named file (in order of appearance) on a READ statement in the same matrix program.

FIELD Specification

FIELD specifies the column positions of a fixed-format record where the data for matrix elements are located.

- The FIELD specification is required.
- Startcol is the number of the leftmost column of the input area.
- Endcol is the number of the rightmost column of the input area.
- Both startcol and endcol are required, and both must be constants. For example, `FIELD = 9 TO 72` specifies that values to be read appear between columns 9 and 72 (inclusive) of each input record.
- The BY clause, if present, indicates that each value appears within a fixed set of columns on the input record; that is, one value is separated from the next by its column position rather than by a space or comma. Width is the width of the area designated for each value. For example, `FIELD = 1 TO 80 BY 10` indicates that there are eight possible values per record and that one will appear between columns 1 and 10 (inclusive), another between columns 11 and 20, and so on up to columns 71 and 80. The BY value must evenly divide the length of the field. That is, endcol – startcol + 1 must be a multiple of width.
- You can use the FORMAT specification (see p. 498) to supply the same information as the BY clause of the FIELD specification. If you omit the BY clause and do not specify a format on the FORMAT specification, READ assumes that values are separated by blanks or commas within the designated field.

SIZE Specification

The SIZE specification is a matrix expression that, when evaluated, specifies the size of the matrix to be read.

- The expression should evaluate to a two-element row or column vector. The first element designates the number of rows in the matrix to be read; the second gives the number of columns.
- Values of the SIZE specification are truncated to integers if necessary.
- The size expression may be a constant, such as `{5;5}`, or a matrix variable name, such as `MSIZE`, or any valid expression, such as `INFO(1,:)`.
- If you use a scalar as the size expression, a column vector containing that number of rows is read. Thus, `SIZE=1` reads a scalar, and `SIZE=3` reads a 3×1 column vector.

You must include a SIZE specification whenever you name an entire matrix (rather than a submatrix) in the READ statement. If you specify a submatrix, the SIZE specification is optional but, if included, must agree with the size of the specified submatrix.

MODE Specification

MODE specifies the format of the matrix to be read in. It can be either rectangular or symmetric. If the MODE specification is omitted, the default is RECTANGULAR.

RECTANGULAR *Matrix is completely represented in file.* Each row begins on a new record, and all entries in that row are present on that and (possibly) succeeding records. This is the default if the MODE specification is omitted.

SYMMETRIC *Elements of the matrix below the main diagonal are the same as those above it.* Only matrix elements on and below the main diagonal are read; elements above the diagonal are set equal to the corresponding symmetric elements below the diagonal. Each row is read beginning on a new record, although it may span more than one record. Only a single value is read from the first record, two values are read from the second, and so on.

- If SYMMETRIC is specified, the matrix processor first checks that the number of rows and the number of columns are the same. If the numbers, specified either on SIZE or on the variable reference, are not the same, an error message is displayed and the command is not executed.

REREAD Specification

The REREAD specification indicates that the current READ statement should begin with the last record read by a previous READ statement.

- REREAD has no further specifications.
- REREAD cannot be used on the first READ statement to read from a file.
- If you omit REREAD, the READ statement begins with the first record following the last one read by the previous READ statement.
- The REREAD specification is ignored on the first READ statement following a call to the EOF function for the same file.

FORMAT Specification

FORMAT specifies how the matrix processor should interpret the input data. Format descriptor can be any valid SPSS data format, such as F6, E12.2, or A6, or it can be a type code; for example, F, E, or A. (See FORMATS in the *SPSS Base System Syntax Reference Guide*.)

- If you omit the FORMAT specification, the default is F.
- You can specify the width of fixed-size data fields with either a FORMAT specification or a BY clause on a FIELD specification. You can include it in both places only if you specify the same value.
- If you do not include either a FORMAT or a BY clause on FIELD, READ expects values separated by blanks or commas.
- An additional way of specifying the width is to supply a repetition factor without a width (for example, 10F, 5COMMA, or 3E). The field width is then calculated by dividing the

width of the whole input area on the FIELD specification by the repetition factor. A format with a digit for the repetition factor must be enclosed in quotes.

- Only one format can be specified. A specification such as `FORMAT='5F2.0 3F3.0 F2.0'` is invalid.

WRITE Statement: Writing Character Data

WRITE writes the value of a matrix expression to an external file. The syntax of the WRITE statement is:

```
WRITE  matrix expression
   [/OUTFILE = file reference]
    /FIELD = startcol TO endcol [BY width]
   [/MODE = {RECTANGULAR}]
            {TRIANGULAR }
   [/HOLD]
   [/FORMAT = format descriptor]
```

Matrix Expression Specification

Specify any matrix expression that evaluates to the value(s) to be written.

- The matrix specification must precede any other specifications on the WRITE statement.
- You can specify a matrix name, a matrix raised to a power, or a matrix function (with its arguments in parentheses) by itself, but you must enclose other matrix expressions in parentheses. For example, `WRITE A`, `WRITE INV(A)`, or `WRITE B**DET(T(C)*D)` is legal, but `WRITE A+B` is not. You must specify `WRITE (A+B)`.
- Constant expressions are allowed.

OUTFILE Specification

OUTFILE designates the character file to which the matrix expression is to be written. The file reference can be an actual filename in apostrophes or quotation marks, or a file handle defined on a FILE HANDLE command that precedes the matrix program. The filename or file handle must be a valid file specification.

- The OUTFILE specification is required on the first WRITE statement in a matrix program (first in order of appearance, not necessarily in order of execution).
- If you omit the OUTFILE specification from a later WRITE statement, the statement uses the most recently named file (in order of appearance) on a WRITE statement in the same matrix program.

FIELD Specification

FIELD specifies the column positions of a fixed-format record to which the data should be written.

- The FIELD specification is required.
- The start column, *c1*, is the number of the leftmost column of the output area.

- The end column, *c2*, is the number of the rightmost column of the output area.
- Both *c1* and *c2* are required, and both must be constants. For example, `FIELD = 9 TO 72` specifies that values should be written between columns 9 and 72 (inclusive) of each output record.
- The BY clause, if present, indicates how many characters should be allocated to the output value of a single matrix element. The value *w* is the width of the area designated for each value. For example, `FIELD = 1 TO 80 BY 10` indicates that up to eight values should be written per record, and that one should go between columns 1 and 10 (inclusive), another between columns 11 and 20, and so on up to columns 71 and 80. The value on the BY clause must evenly divide the length of the field. That is, $c2 - c1 + 1$ must be a multiple of *w*.
- You can use the FORMAT specification (see below) to supply the same information as the BY clause. If you omit the BY clause from the FIELD specification and do not specify a format on the FORMAT specification, WRITE uses freefield format, separating matrix elements by single blank spaces.

MODE Specification

MODE specifies the format of the matrix to be written. If MODE is not specified, the default is RECTANGULAR.

RECTANGULAR *Write the entire matrix.* Each row starts a new record, and all the values in that row are present in that and (possibly) subsequent records. This is the default if the MODE specification is omitted.

TRIANGULAR *Write only the lower triangular entries and the main diagonal.* Each row begins a new record and may span more than one record. This mode may save file space.

- A matrix written with `MODE = TRIANGULAR` must be square, but it need not be symmetric. If it is not, values in the upper triangle are not written.
- A matrix written with `MODE = TRIANGULAR` may be read with `MODE = SYMMETRIC`.

HOLD Specification

HOLD causes the last line written by the current WRITE statement to be held so that the next WRITE to that file will write on the same line. Use HOLD to write more than one matrix on a line.

FORMAT Specification

FORMAT indicates how the internal (binary) values of matrix elements should be converted to character format for output.

- The format descriptor is any valid SPSS data format, such as F6, E12.2, or A6, or it can be a format type code, such as F, E, or A. It specifies how the written data are encoded and, if a width is specified, how wide the fields containing the data are. (See FORMATS in the *SPSS Base System Syntax Reference Guide* for valid formats.)

- If you omit the FORMAT specification, the default is F.
- The data field widths may be specified either here or after BY on the FIELD specification. You may specify the width in both places only if you give the same value.
- An additional way of specifying the width is to supply a repetition factor without a width (for example, 10F or 5COMMA). The field width is then calculated by dividing the width of the whole output area on the FIELD specification by the repetition factor. A format with a digit for the repetition factor must be enclosed in quotes.
- If the field width is not specified in any of these ways, then the freefield format is used: matrix values are written separated by one blank, and each value occupies as many positions as necessary to avoid the loss of precision. Each row of the matrix is written starting with a new output record.
- Only one format descriptor can be specified. Do *not* try to specify more than one format; for example, `'5F2.0 3F3.0 F2.0'` is invalid as a FORMAT specification on WRITE.

GET Statement: Reading SPSS Data Files

GET reads matrices from an external SPSS data file or from the working data file. The syntax of GET is as follows:

```
GET variable reference
  [/FILE={file reference}]
         {*             }
  [/VARIABLES = variable list]
  [/NAMES = names vector]
  [/MISSING = {ACCEPT}]
              {OMIT  }
              {value }
  [/SYSMIS = {OMIT }]
             {value}
```

Variable Specification

The variable reference on the GET statement is a matrix variable name with or without indexes.

For a name without indexes:

- GET creates the specified matrix variable.
- The size of the matrix is determined by the amount of data read from the SPSS data file or the working file.
- If the matrix already exists, it is replaced by the matrix read from the file.

For an indexed name:

- GET creates a submatrix from an existing matrix.
- The matrix variable named must already exist.
- You can define any submatrix with indexes; for example, `M(:,I)`. To define an entire existing matrix, specify `M(:,:)`.
- The indexes, along with the size of the existing matrix, specify completely the size of the submatrix, which must agree with the dimensions of the data read from the SPSS data file.
- The specified submatrix is replaced by the matrix elements read from the SPSS data file.

FILE Specification

FILE designates the SPSS data file to be read. Use an asterisk, or simply omit the FILE specification, to designate the current working data file.

- The file reference can be either a filename enclosed in apostrophes or quotation marks, or a file handle defined on a FILE HANDLE command that precedes the matrix program.
- If you omit the FILE specification, the working data file is used.
- In a matrix program executed with the INCLUDE command, if a SPLIT FILE command is in effect, a GET statement that references the working data file will read a single split-file group of cases. (A matrix program cannot be executed from a syntax window if a SPLIT FILE command is in effect.)

VARIABLES Specification

VARIABLES specifies a list of variables to be read from the SPSS data file.

- The variable list is entered much the same as in other SPSS procedures except that the variable names *must* be separated by commas.
- The keyword TO can be used to reference consecutive variables on the SPSS data file.
- The variable list can consist of the keyword ALL to get all the variables in the SPSS data file. ALL is the default if the VARIABLES specification is omitted.
- All variables read from the SPSS data file should be numeric. If a string variable is specified, a warning message is issued and the string variable is skipped.

Example

```
GET M /VARIABLES = AGE, RESIDE, INCOME TO HEALTH.
```

- The variables *AGE*, *RESIDE*, and *INCOME* to *HEALTH* from the working data file will form the columns of the matrix *M*.

NAMES Specification

NAMES specifies a vector to store the variable names from the SPSS data file.

- If you omit the NAMES specification, the variable names are not available to the MATRIX procedure.
- In place of a vector name, you can use a matrix expression that evaluates to a vector, such as `A(N+1,:)`.

MISSING Specification

MISSING specifies how missing values declared for the SPSS data file should be handled.

- The MISSING specification is required if the SPSS data file contains missing values for any variable being read.

- If you omit the MISSING specification and a missing value is encountered for a variable being read, an error message is displayed and the GET statement is not executed.

The following keywords are available on the MISSING specification. There is no default.

ACCEPT *Accept user-missing values for entry.* If the system-missing value exists for a variable to be read, you must specify SYSMIS to indicate how the system-missing value should be handled (see "SYSMIS Specification," below).

OMIT *Skip an entire observation when a variable with a missing value is encountered.*

value *Recode all missing values encountered (including the system-missing value) to the specified value for entry.* The replacement value can be any numeric constant.

SYSMIS Specification

SYSMIS specifies how system-missing values should be handled when you have specified ACCEPT on MISSING.

- The SYSMIS specification is ignored unless ACCEPT is specified on MISSING.
- If you specify ACCEPT on MISSING but omit the SYSMIS specification, and a system-missing value is encountered for a variable being read, an error message is displayed and the GET statement is not executed.

The following keywords are available on the SYSMIS specification. There is no default.

OMIT *Skip an entire observation when a variable with a system-missing value is encountered.*

value *Recode all system-missing values encountered to the specified value for entry.* The replacement value can be any numeric constant.

Example

```
GET SCORES
  /VARIABLES = TEST1,TEST2,TEST3
  /NAMES = VARNAMES
  /MISSING = ACCEPT
  /SYSMIS = -1.0.
```

- A matrix named *SCORES* is read from the working data file.
- The variables *TEST1*, *TEST2*, and *TEST3* form the columns of the matrix, while the cases in the working file form the rows.
- A vector named *VARNAMES*, whose three elements contain the variable names *TEST1*, *TEST2*, and *TEST3*, is created.
- User-missing values defined in the working data file are accepted into the matrix *SCORES*.
- System-missing values in the working data file are converted to the value −1 in the matrix *SCORES*.

SAVE Statement: Writing SPSS Data Files

SAVE writes matrices to an SPSS data file or to the current working data file. The rows of the matrix expression become cases, and the columns become variables. The syntax of the SAVE statement is as follows:

```
SAVE matrix expression
  [/OUTFILE = {file reference}]
              {*             }
  [/VARIABLES = variable list]
  [/NAMES = names vector]
  [/STRINGS = variable list]
```

Matrix Expression Specification

The matrix expression following the keyword SAVE is any matrix language expression that evaluates to the value(s) to be written to an SPSS data file.

- The matrix specification must precede any other specifications on the SAVE statement.
- You can specify a matrix name, a matrix raised to a power, or a matrix function (with its arguments in parentheses) by itself, but you must enclose other matrix expressions in parentheses. For example, `SAVE A`, `SAVE INV(A)`, or `SAVE B**DET(T(C)*D)` is legal, but `SAVE A+B` is not. You must specify `SAVE (A+B)`.
- Constant expressions are allowed.

OUTFILE Specification

OUTFILE designates the file to which the matrix expression is to be written. It can be an actual filename in apostrophes or quotation marks, or a file handle defined on a FILE HANDLE command that precedes the matrix program. The filename or handle must be a valid file specification.

- To save a matrix expression as the working data file, specify an asterisk (*). If there is no working data file, one will be created; if there is one, it is replaced by the saved matrices.
- The OUTFILE specification is required on the first SAVE statement in a matrix program (first in order of appearance, not necessarily in order of execution). If you omit the OUTFILE specification from a later SAVE statement, the statement uses the most recently named file (in order of appearance) on a SAVE statement in the same matrix program.
- If more than one SAVE statement writes to the working data file in a single matrix program, the dictionary of the new working data file is written on the basis of the information given by the first such SAVE. All the subsequently saved matrices are appended to the new working data file as additional cases. If the number of columns differs, an error occurs.
- When you execute a matrix program with the INCLUDE command, the SAVE statement creates a new SPSS data file at the end of the matrix program's execution, so any attempt to GET the data file obtains the original data file, if any.
- When you execute a matrix program from a syntax window, SAVE creates a new SPSS data file immediately, but the file remains open, so you cannot GET it until after the END MATRIX statement.

VARIABLES Specification

You can provide variable names for the SPSS data file with the VARIABLES specification. The variable list is a list of valid SPSS variable names separated by commas.

- You can use the TO convention, as shown in the example below.
- You can also use the NAMES specification, discussed below, to provide variable names.

Example

```
SAVE {A,B,X,Y} /OUTFILE=*
  /VARIABLES = A,B,X1 TO X50,Y1,Y2.
```

- The matrix expression on the SAVE statement constructs a matrix from two column vectors *A* and *B* and two matrices *X* and *Y*. All four matrix variables must have the same number of rows so that this matrix construction will be valid.
- The VARIABLES specification provides descriptive names so that the SPSS variable names in the new working data file will resemble the names used in the matrix program.

NAMES Specification

As an alternative to the explicit list on the VARIABLES specification, you can specify a name list with a matrix expression that evaluates to a vector containing string values. The elements of this vector are used as names for the variables.

- The NAMES specification on SAVE is designed to complement the NAMES specification on the GET statement. Names extracted from an SPSS data file can be used in a new data file by specifying the same vector name on both NAMES specifications.
- If you specify both VARIABLES and NAMES, a warning message is displayed and the VARIABLES specification is used.
- If you omit both the VARIABLES and NAMES specifications, or if you do not specify names for all columns of the matrix, the MATRIX procedure creates default names. The names have the form *COLn*, where *n* is the column number.

STRINGS Specification

The STRINGS specification provides the names of variables that contain short string data rather than numeric data.

- By default, all variables are assumed to be numeric.
- The variable list specification following STRINGS consists of a list of SPSS variable names separated by commas. The names must be among those used by SAVE.

MGET Statement: Reading SPSS Matrix Data Files

MGET reads an SPSS matrix-format data file. MGET puts the data it reads into separate matrix variables. It also names these new variables automatically. The syntax of MGET is as follows:

```
MGET [ [/] FILE = file reference]
  [/TYPE = {COV    }]
           {CORR   }
           {MEAN   }
           {STDDEV }
           {N      }
           {COUNT  }
```

- Since MGET assigns names to the matrices it reads, do not specify matrix names on the MGET statement.

FILE Specification

FILE designates an SPSS matrix-format data file. (See MATRIX DATA in the *SPSS Base System Syntax Reference Guide* for a discussion of matrix-format data files.) To designate the working data file (if it is a matrix-format data file), use an asterisk, or simply omit the FILE specification.

- The file reference can be either a filename enclosed in apostrophes or quotation marks, or a file handle defined on a FILE HANDLE command that precedes the matrix program.
- The same matrix-format SPSS data file can be read more than once.
- If you omit the FILE specification, the current working data file is used.
- MGET ignores the SPLIT FILE command in SPSS when reading the working data file. It does honor the split-file groups that were in effect when the matrix-format data file was created.
- The maximum number of split-file groups that can be read is 99.
- The maximum number of cells that can be read is 99.

TYPE Specification

TYPE specifies the rowtype(s) to read from the matrix-format data file.

- By default, records of all rowtypes are read.
- If the matrix-format data file does not contain rows of the requested type, an error occurs.

Valid keywords on the TYPE specification are:

COV *A matrix of covariances.*

CORR *A matrix of correlation coefficients.*

MEAN *A vector of means.*

STDDEV *A vector of standard deviations.*

N *A vector of numbers of cases.*

COUNT *A vector of counts.*

Names of Matrix Variables from MGET

- The MGET statement automatically creates matrix variable names for the matrices it reads.
- All new variables created by MGET are reported to the user.
- If a matrix variable already exists with the same name that MGET chose for a new variable, the new variable is not created and a warning is issued. The RELEASE statement can be used to get rid of a variable. A COMPUTE statement followed by RELEASE can be used to change the name of an existing matrix variable.

MGET constructs variable names in the following manner:

- The first two characters of the name identify the row type. If there are no cells and no split file groups, these two characters constitute the name:

CV	A covariance matrix (rowtype COV)
CR	A correlation matrix (rowtype CORR)
MN	A vector of means (rowtype MEAN)
SD	A vector of standard deviations (rowtype STDDEV)
NC	A vector of numbers of cases (rowtype N)
CN	A vector of counts (rowtype COUNT)

- Characters 3–5 of the variable name identify the cell number or the split-group number. Cell identifiers consist of the letter *F* and a two-digit cell number. Split-group identifiers consist of the letter *S* and a two-digit split-group number; for example, *MNF12* or *SDS22*.
- If there are both cells and split groups, characters 3–5 identify the cell, and characters 6–8 identify the split group. The same convention for cell or split-file numbers is used; for example, *CRF12S21*.
- After the name is constructed as described above, any leading zeros are removed from the cell number and the split-group number; for example, *CNF2S99* or *CVF2S1*.

MSAVE Statement: Writing SPSS Matrix Data Files

The MSAVE statement writes matrix expressions to an SPSS matrix-format data file that can be used as matrix input to other SPSS procedures. (See MATRIX DATA in the *SPSS Base System Syntax Reference Guide* for discussion of matrix-format data files.) The syntax of MSAVE is as follows:

```
MSAVE matrix expression
   /TYPE = {COV    }
           {CORR   }
           {MEAN   }
           {STDDEV }
           {N      }
           {COUNT  }
  [/OUTFILE = {file reference}]
              {*             }
  [/VARIABLES = variable list]
  [/SNAMES = variable list]
  [/SPLIT = split vector]
  [/FNAMES = variable list]
  [/FACTOR = factor vector]
```

- Only one matrix-format data file can be saved in a single matrix program.
- Each MSAVE statement writes records of a single rowtype. Therefore, several MSAVE statements will normally be required to write a complete matrix-format data file.
- Most specifications are retained from one MSAVE statement to the next, so that it is not necessary to repeat the same specifications on a series of MSAVE statements. The exception is the FACTOR specification, as noted below.

Example

```
MSAVE M /TYPE=MEAN /OUTFILE=CORRMAT /VARIABLES=V1 TO V8.
MSAVE S /TYPE STDDEV.
MSAVE MAKE(1,8,24) /TYPE N.
MSAVE C /TYPE CORR.
```

- The series of MSAVE statements save the matrix variables *M*, *S*, and *C*, which contain, respectively, vectors of means and standard deviations and a matrix of correlation coefficients. The SPSS matrix-format data file thus created is suitable for use in a procedure such as FACTOR.
- The first MSAVE statement saves *M* as a vector of means. This statement specifies OUTFILE, a previously defined file handle, and VARIABLES, a list of variable names to be used in the SPSS data file.
- The second MSAVE statement saves *S* as a vector of standard deviations. Note that the OUTFILE and VARIABLES specifications do not have to be repeated.
- The third MSAVE statement saves a vector of case counts. The matrix function MAKE constructs an eight-element vector with values equal to the case count (24 in this example).
- The last MSAVE statement saves *C*, an 8×8 matrix, as the correlation matrix.

Matrix Expression Specification

- The matrix expression must be specified first on the MSAVE statement.
- The matrix expression specification can be any matrix language expression that evaluates to the value(s) to be written to the matrix-format file.
- You can specify a matrix name, a matrix raised to a power, or a matrix function (with its arguments in parentheses) by itself, but you must enclose other matrix expressions in parentheses. For example, `MSAVE A`, `SAVE INV(A)`, or `MSAVE B**DET(T(C)*D)` is legal, but `MSAVE N * WT` is not. You must specify `MSAVE (N * WT)`.
- Constant expressions are allowed.

TYPE Specification

TYPE specifies the rowtype to write to the matrix-format data file. Only a single rowtype can be written by any one MSAVE statement.Valid keywords on the TYPE specification are:

COV *A matrix of covariances.*

CORR *A matrix of correlation coefficients.*

MEAN *A vector of means.*

STDDEV *A vector of standard deviations.*

N *A vector of numbers of cases.*

COUNT *A vector of counts.*

OUTFILE Specification

OUTFILE designates the SPSS matrix-format data file to which the matrices are to be written. It can be an asterisk, an actual filename in apostrophes or quotation marks, or a file handle defined on a FILE HANDLE command that precedes the matrix program. The filename or handle must be a valid file specification.

- The OUTFILE specification is required on the first MSAVE statement in a matrix program.
- To save a matrix expression as the working data file (replacing any working data file created before the matrix program), specify an asterisk (*).
- Since only one matrix-format data file can be written in a single matrix program, any OUTFILE specification on the second and later MSAVE statements in one matrix program must be the same as that on the first MSAVE statement.

VARIABLES Specification

You can provide variable names for the matrix-format data file with the VARIABLES specification. The variable list is a list of valid SPSS variable names separated by commas. You can use the TO convention.

- The VARIABLES specification names only the data variables in the matrix. Split-file variables and grouping or factor variables are named on the SNAMES and FNAMES specifications.
- The names in the VARIABLES specification become the values of the special variable *VARNAME_* in the matrix-format data file, for rowtypes of CORR and COV.
- You cannot specify the reserved names *ROWTYPE_* and *VARNAME_* on the VARIABLES specification.
- If you omit the VARIABLES specification, the default names *COL1*, *COL2*,..., are used.

FACTOR Specification

To write an SPSS matrix-format data file with factor or group codes, you must use the FACTOR specification to provide a row matrix containing the values of each of the factors or group variables for the matrix expression being written by the current MSAVE statement.

- The factor vector must have the same number of columns as there are factors in the matrix data file being written. You can use a scalar when the groups are defined by a single variable. For example, `FACTOR=1` indicates that the matrix data being written are for the value 1 of the factor variable.
- The values of the factor vector are written to the matrix-format data file as values of the factors in the file.

- To create a complete matrix-format data file with factors, you must execute an MSAVE statement for every combination of values of the factors or grouping variables (in other words, for every group). If split-file variables are also present, you must execute an MSAVE statement for every combination of factor codes within every combination of values of the split-file variables.

Example

```
MSAVE M11 /TYPE=MEAN /OUTFILE=CORRMAT /VARIABLES=V1 TO V8
    /FNAMES=SEX, GROUP /FACTOR={1,1}.
MSAVE S11 /TYPE STDDEV.
MSAVE MAKE(1,8,N(1,1)) /TYPE N.
MSAVE C11 /TYPE CORR.

MSAVE M12 /TYPE=MEAN /FACTOR={1,2}.
MSAVE S12 /TYPE STDDEV.
MSAVE MAKE(1,8,N(1,2)) /TYPE N.
MSAVE C12 /TYPE CORR.

MSAVE M21 /TYPE=MEAN /FACTOR={2,1}.
MSAVE S21 /TYPE STDDEV.
MSAVE MAKE(1,8,N(2,1)) /TYPE N.
MSAVE C21 /TYPE CORR.

MSAVE M22 /TYPE=MEAN /FACTOR={2,2}.
MSAVE S22 /TYPE STDDEV.
MSAVE MAKE(1,8,N(2,2)) /TYPE N.
MSAVE C22 /TYPE CORR.
```

- The first four MSAVE statements provide data for a group defined by the variables *SEX* and *GROUP*, with both factors having the value 1.
- The second, third, and fourth groups of four MSAVE statements provide the corresponding data for the other groups, in which *SEX* and *GROUP*, respectively, equal 1 and 2, 2 and 1, and 2 and 2.
- Within each group of MSAVE statements, a suitable number-of-cases vector is created with the matrix function MAKE.

FNAMES Specification

To write an SPSS matrix-format data file with factor or group codes, you can use the FNAMES specification to provide variable names for the grouping or factor variables.

- The variable list following the keyword FNAMES is a list of valid SPSS variable names, separated by commas.
- If you omit the FNAMES specification, the default names *FAC1*, *FAC2*,..., are used.

SPLIT Specification

To write an SPSS matrix-format data file with split-file groups, you must use the SPLIT specification to provide a row matrix containing the values of each of the split-file variables for the matrix expression being written by the current MSAVE statement.

- The split vector must have the same number of columns as there are split-file variables in the matrix data file being written. You can use a scalar when there is only one split-file variable. For example, `SPLIT=3` indicates that the matrix data being written are for the value 3 of the split-file variable.
- The values of the split vector are written to the matrix-format data file as values of the split-file variable(s).
- To create a complete matrix-format data file with split-file variables, you must execute MSAVE statements for every combination of values of the split-file variables. (If factor variables are present, you must execute MSAVE statements for every combination of factor codes within every combination of values of the split-file variables.)

SNAMES Specification

To write an SPSS matrix-format data file with split-file groups, you can use the SNAMES specification to provide variable names for the split-file variables.

- The variable list following the keyword SNAMES is a list of valid SPSS variable names, separated by commas.
- If you omit the SNAMES specification, the default names *SPL1*, *SPL2*,..., are used.

DISPLAY Statement

DISPLAY provides information on the matrix variables currently defined in a matrix program, and on usage of internal memory by the matrix processor. Two keywords are available on DISPLAY:

DICTIONARY *Display variable name and row and column dimensions for each matrix variable currently defined.*

STATUS *Display the status and size of internal tables.* This display is intended as a debugging aid when writing large matrix programs that approach the memory limitations of your system.

If you enter the DISPLAY statement with no specifications, both DICTIONARY and STATUS information is displayed.

RELEASE Statement

Use the RELEASE statement to release the work areas in memory assigned to matrix variables that are no longer needed.

- Specify a list of currently defined matrix variables. Variable names on the list must be separated by commas.
- RELEASE discards the contents of the named matrix variables. Releasing a large matrix when it is no longer needed makes memory available for additional matrix variables.
- All matrix variables are released when the END MATRIX statement is encountered.

Macros Using the Matrix Language

Macro expansion (see DEFINE—END DEFINE in the *SPSS Base System Syntax Reference Guide*) occurs before command lines are passed to the matrix processor. Therefore, previously defined macro names can be used within a matrix program. If the macro name expands to one or more valid matrix statements, the matrix processor will execute those statements. Similarly, you can define an entire matrix program, including the MATRIX and END MATRIX commands, as a macro, but you cannot define a macro within a matrix program, since DEFINE and END DEFINE are not valid matrix statements.

NLR

```
MODEL PROGRAM parameter=value [parameter=value ...]
transformation commands

[DERIVATIVES
transformation commands]

[CLEAR MODEL PROGRAMS]
```

Procedure CNLR (Constrained Nonlinear Regression):

```
[CONSTRAINED FUNCTIONS
transformation commands]

CNLR dependent var

 [/FILE=file]    [/OUTFILE=file]

 [/PRED=varname]

 [/SAVE [PRED] [RESID[(varname)]] [DERIVATIVES] [LOSS]]

 [/CRITERIA=[ITER n] [MITER n] [CKDER {0.5**}]
                                      {n    }
            [ISTEP {1E+20**}] [FPR n] [LFTOL n]
                   {n      }
            [LSTOL n] [STEPLIMIT {2**}] [NFTOL n]
                                 {n  }
            [FTOL n] [OPTOL n] [CRSHTOL {.01**}]]
                                        {n    }

 [/BOUNDS=expression, expression, ...]

 [/LOSS=varname]

 [/BOOTSTRAP [=n]]
```

Procedure NLR (Nonlinear Regression):

```
NLR dependent var

 [/FILE=file]    [/OUTFILE=file]

 [/PRED=varname]

 [/SAVE [PRED] [RESID [(varname)] [DERIVATIVES]]

 [/CRITERIA=[ITER {100**}] [CKDER {0.5**}]
                  {n    }         {n    }
            [SSCON {1E-8**}]  [PCON {1E-8**}]  [RCON {1E-8**}]]
                   {n     }         {n     }         {n     }
```

**Default if subcommand or keyword is omitted.

Example:

```
MODEL PROGRAM A=.6.
COMPUTE PRED=EXP(A*X).

NLR Y.
```

Overview

Nonlinear regression is used to estimate parameter values and regression statistics for models that are not linear in their parameters. SPSS has two procedures for estimating nonlinear equations. CNLR (constrained nonlinear regression), which uses a sequential quadratic programming algorithm, is applicable for both constrained and unconstrained problems. NLR (nonlinear regression), which uses a Levenberg-Marquardt algorithm, is applicable only for unconstrained problems.

CNLR is more general. It allows linear and nonlinear constraints on any combination of parameters. It will estimate parameters by minimizing any smooth loss function (objective function), and can optionally compute bootstrap estimates of parameter standard errors and correlations. The individual bootstrap parameter estimates can optionally be saved in a separate SPSS data file.

Both programs estimate the values of the parameters for the model and, optionally, compute and save predicted values, residuals, and derivatives. Final parameter estimates can be saved in an SPSS data file and used in subsequent analyses.

CNLR and NLR use much of the same syntax. Some of the following sections discuss features common to both procedures. In these sections, the notation [C]NLR means that either the CNLR or NLR procedure can be specified. Sections that apply only to CNLR or only to NLR are clearly identified.

Options

The Model. You can use any number of transformation commands under MODEL PROGRAM to define complex models.

Derivatives. You can use any number of transformation commands under DERIVATIVES to supply derivatives.

Adding Variables to Working Data File. You can add predicted values, residuals, and derivatives to the working data file with the SAVE subcommand.

Writing Parameter Estimates to a New Data File. You can save final parameter estimates as an external SPSS data file using the OUTFILE subcommand and retrieve them in subsequent analyses using the FILE subcommand.

Controlling Model-building Criteria. You can control the iteration process used in the regression with the CRITERIA subcommand.

Additional CNLR Controls. For CNLR, you can impose linear and nonlinear constraints on the parameters with the BOUNDS subcommand. Using the LOSS subcommand, you can specify a loss function for CNLR to minimize and, using the BOOTSTRAP subcommand, provide bootstrap estimates of the parameter standard errors, confidence intervals, and correlations.

Basic Specification

The basic specification requires three commands: MODEL PROGRAM, COMPUTE (or any other computational transformation command), and [C]NLR.

- The MODEL PROGRAM command assigns initial values to the parameters and signifies the beginning of the model program.
- The computational transformation command generates a new variable to define the model. The variable can take any legitimate name, but if the name is not *PRED*, the PRED subcommand will be required.
- The [C]NLR command provides the regression specifications. The minimum specification is the dependent variable.
- By default, the residual sum of squares and estimated values of the model parameters are displayed for each iteration. Statistics generated include regression and residual sums of squares and mean squares, corrected and uncorrected total sums of squares, R^2, parameter estimates with their asymptotic standard errors and 95% confidence intervals, and an asymptotic correlation matrix of the parameter estimates.

Command Order

- The model program, beginning with the MODEL PROGRAM command, must precede the [C]NLR command.
- The derivatives program (when used), beginning with the DERIVATIVES command, must follow the model program but precede the [C]NLR command.
- The constrained functions program (when used), beginning with the CONSTRAINED FUNCTIONS command, must immediately precede the CNLR command. The constrained functions program cannot be used with the NLR command.
- The CNLR command must follow the block of transformations for the model program and the derivatives program when specified; the CNLR command must also follow the constrained functions program when specified.
- Subcommands on [C]NLR can be named in any order.

Syntax Rules

- The FILE, OUTFILE, PRED, and SAVE subcommands work the same way for both CNLR and NLR.
- The CRITERIA subcommand is used by both CNLR and NLR, but iteration criteria are different. Therefore, the CRITERIA subcommand is documented separately for CNLR and NLR.
- The BOUNDS, LOSS, and BOOTSTRAP subcommands can be used only with CNLR. They cannot be used with NLR.

Operations

- By default, the predicted values, residuals, and derivatives are created as temporary variables. To save these variables, use the SAVE subcommand.

Weighting Cases

- If case weighting is in effect, [C]NLR uses case weights when calculating the residual sum of squares and derivatives. However, the degrees of freedom in the ANOVA table are always based on unweighted cases.
- When the model program is first invoked for each case, the weight variable's value is set equal to its value in the working data file. The model program may recalculate that value. For example, to effect a robust estimation, the model program may recalculate the weight variable's value as an inverse function of the residual magnitude. [C]NLR uses the weight variable's value after the model program is executed.

Missing Values

Cases with missing values for any of the dependent or independent variables named on the [C]NLR command are excluded.

- Predicted values, but not residuals, can be calculated for cases with missing values on the dependent variable.
- [C]NLR ignores cases that have missing, negative, or zero weights. The procedure displays a warning message if it encounters any negative or zero weights at any time during its execution.
- If a variable used in the model program or the derivatives program is omitted from the independent variable list on the [C]NLR command, the predicted value and some or all of the derivatives may be missing for every case. If this happens, SPSS generates an error message.

Example

```
MODEL PROGRAM A=.5 B=1.6.
COMPUTE PRED=A*SPEED**B.

DERIVATIVES.
COMPUTE D.A=SPEED**B.
COMPUTE D.B=A*LN(SPEED)*SPEED**B.

NLR STOP.
```

- MODEL PROGRAM assigns values to the model parameters *A* and *B*.
- COMPUTE generates the variable *PRED* to define the nonlinear model using parameters *A* and *B* and the variable *SPEED* from the working data file. Because this variable is named *PRED*, the PRED subcommand is not required on NLR.
- DERIVATIVES indicates that calculations for derivatives are being supplied.
- The two COMPUTE statements on the DERIVATIVES transformations list calculate the derivatives for the parameters *A* and *B*. If either one had been omitted, NLR would have calculated it numerically.
- NLR specifies *STOP* as the dependent variable. It is not necessary to specify *SPEED* as the independent variable since it has been used in the model and derivatives programs.

MODEL PROGRAM Command

The MODEL PROGRAM command assigns initial values to the parameters and signifies the beginning of the model program. The model program specifies the nonlinear equation chosen to model the data. There is no default model.

- The model program is required and must precede the [C]NLR command.
- The MODEL PROGRAM command must specify all parameters in the model program. Each parameter must be individually named. Keyword TO is not allowed.
- Parameters can be assigned any acceptable SPSS variable name. However, if you intend to write the final parameter estimates to a file with the OUTFILE subcommand, do not use the name *SSE* or *NCASES* (see the OUTFILE subcommand on p. 520).
- Each parameter in the model program must have an assigned value. The value can be specified on MODEL PROGRAM or read from an existing parameter data file named on the FILE subcommand.
- Zero should be avoided as an initial value because it provides no information on the scale of the parameters. This is especially true for CNLR.
- The model program must include at least one command that uses the parameters and the independent variables (or preceding transformations of these) to calculate the predicted value of the dependent variable. This predicted value defines the nonlinear model. There is no default model.
- By default, the program assumes that *PRED* is the name assigned to the variable for the predicted values. If you use a different variable name in the model program, you must supply the name on the PRED subcommand (see the PRED subcommand on p. 521).
- In the model program, you can assign a label to the variable holding predicted values and also change its print and write formats, but you should not specify missing values for this variable.
- You can use any computational commands (such as COMPUTE, IF, DO IF, LOOP, END LOOP, END IF, RECODE, or COUNT) or output commands (WRITE, PRINT, or XSAVE) in the model program, but you cannot use input commands (such as DATA LIST, GET, MATCH FILES, or ADD FILES).
- Transformations in the model program are used only by [C]NLR, and they do not affect the working data file. The parameters created by the model program do not become a part of the working data file. Permanent transformations should be specified before the model program.

Caution

The selection of good initial values for the parameters in the model program is very important to the operation of [C]NLR. The selection of poor initial values can result in no solution, a local rather than a general solution, or a physically impossible solution.

Example

```
MODEL PROGRAM A=10 B=1 C=5 D=1.
COMPUTE PRED= A*exp(B*X) + C*exp(D*X).
```

- The MODEL PROGRAM command assigns starting values to the four parameters *A, B, C,* and *D.*
- COMPUTE defines the model to be fit as the sum of two exponentials.

DERIVATIVES Command

The optional DERIVATIVES command signifies the beginning of the derivatives program. The derivatives program contains transformation statements for computing some or all of the derivatives of the model. The derivatives program must follow the model program but precede the [C]NLR command.

If the derivatives program is not used, [C]NLR numerically estimates derivatives for all the parameters. Providing derivatives reduces computation time and, in some situations, may result in a better solution.

- The DERIVATIVES command has no further specifications but must be followed by the set of transformation statements that calculate the derivatives.
- You can use any computational commands (such as COMPUTE, IF, DO IF, LOOP, END LOOP, END IF, RECODE, or COUNT) or output commands (WRITE, PRINT, or XSAVE) in the derivatives program, but you cannot use input commands (such as DATA LIST, GET, MATCH FILES, or ADD FILES).
- To name the derivatives, specify the prefix *D.* before each parameter name. For example, the derivative name for the parameter *PARM1* must be *D.PARM1.*
- Once a derivative has been calculated by a transformation, the variable for that derivative can be used in subsequent transformations.
- You do not need to supply all of the derivatives. Those that are not supplied will be estimated by the program. During the first iteration of the nonlinear estimation procedure, derivatives calculated in the derivatives program are compared with numerically calculated derivatives. This serves as a check on the supplied values (see the CRITERIA subcommand on p. 523).
- Transformations in the derivatives program are used by [C]NLR only and do not affect the working data file.
- For NLR, the derivative of each parameter must be computed with respect to the predicted function. (For computation of derivatives in CNLR, see the LOSS subcommand on p. 527.)

Example

```
MODEL PROGRAM A=1, B=0, C=1, D=0
COMPUTE PRED = Ae^Bx + Ce^Dx
DERIVATIVES.
COMPUTE D.A = exp (B * X).
COMPUTE D.B = A * exp (B * X) * X.
COMPUTE D.C = exp (D * X).
COMPUTE D.D = C * exp (D * X) * X.
```

- The derivatives program specifies derivatives of the PRED function for the sum of the two exponentials in the model described by the following equation:

$$Y = Ae^{Bx} + Ce^{Dx}$$

Example

```
DERIVATIVES.
COMPUTE D.A = exp (B * X).
COMPUTE D.B = A * X * D.A.
COMPUTE D.C = exp (D * X).
COMPUTE D.D = C * X * D.C.
```

- This is an alternative way to express the same derivatives program specified in the previous example.

CONSTRAINED FUNCTIONS Command

The optional CONSTRAINED FUNCTIONS command signifies the beginning of the constrained functions program, which specifies nonlinear constraints. The constrained functions program is specified after the model program and the derivatives program (when used). It can only be used with, and must precede, the CNLR command. For more information, see the BOUNDS subcommand on p. 526.

Example

```
MODEL PROGRAM A=.5 B=1.6.
COMPUTE PRED=A*SPEED**B.

CONSTRAINED FUNCTIONS.
COMPUTE CF=A-EXP(B).

CNLR STOP
  /BOUNDS CF LE 0.
```

CLEAR MODEL PROGRAMS Command

CLEAR MODEL PROGRAMS deletes all transformations associated with the model program, the derivative program, and/or the constrained functions program previously submitted. It is primarily used in interactive mode to remove temporary variables created by these programs without affecting the working data file or variables created by other transformation programs or temporary programs. It allows you to specify new models, derivatives, or constrained functions without having to run [C]NLR.

It is not necessary to use this command if you have already executed the [C]NLR procedure. Temporary variables associated with the procedure are automatically deleted.

CNLR/NLR Command

Either the CNLR or the NLR command is required to specify the dependent and independent variables for the nonlinear regression.

- For either CNLR or NLR, the minimum specification is a dependent variable.
- Only one dependent variable can be specified. It must be a numeric variable in the working data file and cannot be a variable generated by the model or the derivatives program.

OUTFILE Subcommand

OUTFILE stores final parameter estimates for use on a subsequent [C]NLR command. The only specification on OUTFILE is the target file. Some or all of the values from this file can be read into a subsequent [C]NLR procedure with the FILE subcommand. The parameter data file created by OUTFILE stores the following variables:

- All the split-file variables. OUTFILE writes one case of values for each split-file group in the working data file.
- All the parameters named on the MODEL PROGRAM command.
- The labels, formats, and missing values of the split-file variables and parameters defined for them previous to their use in the [C]NLR procedure.
- The sum of squared residuals (named *SSE*). *SSE* has no labels or missing values. The print and write format for *SSE* is F10.8.
- The number of cases on which the analysis was based (named *NCASES*). *NCASES* has no labels or missing values. The print and write format for *NCASES* is F8.0.

When OUTFILE is used, the model program cannot create variables named *SSE* or *NCASES*.

Example

```
MODEL PROGRAM A=.5 B=1.6.
COMPUTE PRED=A*SPEED**B.
NLR STOP /OUTFILE=PARAM.
```

- OUTFILE generates a parameter data file containing one case for four variables: *A*, *B*, *SSE*, and *NCASES*.

FILE Subcommand

FILE reads starting values for the parameters from a parameter data file created by an OUTFILE subcommand from a previous [C]NLR procedure. When starting values are read from a file, they do not have to be specified on the MODEL PROGRAM command. Rather, the MODEL PROGRAM command simply names the parameters that correspond to the parameters in the data file.

- The only specification on FILE is the file that contains the starting values.
- Some new parameters may be specified for the model on the MODEL PROGRAM command while others are read from the file specified on the FILE subcommand.
- You do not have to name the parameters on MODEL PROGRAM in the order in which they occur in the parameter data file. In addition, you can name a partial list of the variables contained in the file.
- If the starting value for a parameter is specified on MODEL PROGRAM, the specification overrides the value read from the parameter data file.
- If split-file processing is in effect, the starting values for the first subfile are taken from the first case of the parameter data file. Subfiles are matched with cases in order until the starting value file runs out of cases. All subsequent subfiles use the starting values for the last case.

- To read starting values from a parameter data file and then replace those values with the final results from [C]NLR, specify the same file on the FILE and OUTFILE subcommands. The input file is read completely before anything is written in the output file.

Example

```
MODEL PROGRAM A B C=1 D=3.
COMPUTE PRED=A*SPEED**B + C*SPEED**D.
NLR STOP /FILE=PARAM /OUTFILE=PARAM.
```

- MODEL PROGRAM names four of the parameters used to calculate *PRED*, but assigns values to only *C* and *D*. The values of *A* and *B* are read from the existing data file *PARAM*.
- After NLR computes the final estimates of the four parameters, OUTFILE writes over the old input file. If, in addition to these new final estimates, the former starting values of *A* and *B* are still desired, specify a different file on the OUTFILE subcommand.

PRED Subcommand

PRED identifies the variable holding the predicted values.

- The only specification is a variable name, which must be identical to the variable name used to calculate predicted values in the model program.
- If the model program names the variable *PRED*, the PRED subcommand can be omitted. Otherwise, the PRED subcommand is required.
- The variable for predicted values is not saved in the working data file unless the SAVE subcommand is used.

Example

```
MODEL PROGRAM A=.5 B=1.6.
COMPUTE PSTOP=A*SPEED**B.
NLR STOP /PRED=PSTOP.
```

- COMPUTE in the model program creates a variable named *PSTOP* to temporarily store the predicted values for the dependent variable *STOP*.
- PRED identifies *PSTOP* as the variable used to define the model for the NLR procedure.

SAVE Subcommand

SAVE is used to save the temporary variables for the predicted values, residuals, and derivatives created by the model and the derivatives programs.

- The minimum specification is a single keyword.
- The variables to be saved must have unique names on the working data file. If a naming conflict exists, the variables are not saved.
- Temporary variables, for example, variables created after a TEMPORARY command and parameters specified by the model program, are not saved in the working data file. They will not cause naming conflicts.

The following keywords are available and can be used in any combination and in any order. The new variables are always appended to the working data file in the order in which these keywords are presented here:

PRED *Save the predicted values.* The variable's name, label, and formats are those specified for it (or assigned by default) in the model program.

RESID [(varname)] *Save the residuals variable.* You can specify a variable name in parentheses following the keyword. If no variable name is specified, the name of this variable is the same as the specification you use for this keyword. For example, if you use the three-character abbreviation RES, the default variable name will be *RES*. The variable has the same print and write format as the predicted values variable created by the model program. It has no variable label and no user-defined missing values. It is system-missing for any case in which either the dependent variable is missing or the predicted value cannot be computed.

DERIVATIVES *Save the derivative variables.* The derivative variables are named with the prefix *D.* to the first six characters of the parameter names. Derivative variables use the print and write formats of the predicted values variable and have no value labels or user-missing values. Derivative variables are saved in the same order as the parameters named on MODEL PROGRAM. Derivatives are saved for all parameters, whether or not the derivative was supplied in the derivatives program.

LOSS *Save the user-specified loss function variable.* This specification is available only with CNLR and only if the LOSS subcommand has been specified.

Asymptotic standard errors of predicted values and residuals, and special residuals used for outlier detection and influential case analysis are not provided by the [C]NLR procedure. However, for a squared loss function, the asymptotically correct values for all these statistics can be calculated using the SAVE subcommand with [C]NLR and then using the REGRESSION procedure. In REGRESSION, the dependent variable is still the same, and derivatives of the model parameters are used as independent variables. Casewise plots, standard errors of prediction, partial regression plots, and other diagnostics of the regression are valid for the nonlinear model.

Example

```
MODEL PROGRAM A=.5 B=1.6.
COMPUTE PSTOP=A*SPEED**B.
NLR STOP /PRED=PSTOP
  /SAVE=RESID(RSTOP) DERIVATIVES PRED.
REGRESSION VARIABLES=STOP D.A D.B /ORIGIN
  /DEPENDENT=STOP /ENTER D.A D.B /RESIDUALS.
```

- The SAVE subcommand creates the residuals variable *RSTOP* and the derivative variables *D.A* and *D.B*.
- Because the PRED subcommand identifies *PSTOP* as the variable for predicted values in the nonlinear model, keyword PRED on SAVE adds the variable *PSTOP* to the working data file.

- The new variables are added to the working data file in the following order: *PSTOP, RSTOP, D.A,* and *D.B.*
- The subcommand RESIDUALS for REGRESSION produces the default analysis of residuals.

CRITERIA Subcommand

CRITERIA controls the values of the cutoff points used to stop the iterative calculations in [C]NLR.

- The minimum specification is any of the criteria keywords and an appropriate value. The value can be specified in parentheses after an equals sign, a space, or a comma. Multiple keywords can be specified in any order. Defaults are in effect for keywords not specified.
- Keywords available for CRITERIA differ between CNLR and NLR and are discussed separately. However, with both CNLR and NLR, you can specify the critical value for derivative checking.

Checking Derivatives for CNLR and NLR

Upon entering the first iteration, [C]NLR always checks any derivatives calculated on the derivatives program by comparing them with numerically calculated derivatives. For each comparison, it computes an agreement score. A score of 1 indicates agreement to machine precision; a score of 0 indicates definite disagreement. If a score is less than 1, either an incorrect derivative was supplied or there were numerical problems in estimating the derivative. The lower the score, the more likely it is that the supplied derivatives are incorrect. Highly correlated parameters may cause disagreement even when a correct derivative is supplied. Be sure to check the derivatives if the agreement score is not 1.

During the first iteration, [C]NLR checks each derivative score. If any score is below 1, it begins displaying a table to show the worst (lowest) score for each derivative. If any score is below the critical value, the program stops.

To specify the critical value, use the following keyword on CRITERIA:

CKDER n *Critical value for derivative checking.* Specify a number between 0 and 1 for *n*. The default is 0.5. Specify 0 to disable this criterion.

Iteration Criteria for CNLR

The CNLR procedure uses NPSOL (Version 4.0) Fortran Package for Nonlinear Programming (Gill et al., 1986). The CRITERIA subcommand of CNLR gives the control features of NPSOL. The following section summarizes the NPSOL documentation.

CNLR uses a sequential quadratic programming algorithm, with a quadratic programming subproblem to determine the search direction. If constraints or bounds are specified, the first step is to find a point that is feasible with respect to those constraints. Each major iteration sets up a quadratic program to find the search direction, p. Minor iterations are used to solve this subproblem. Then, the major iteration determines a steplength α by a line search, and the function is evaluated at the new point. An optimal solution is found when the optimality tolerance criterion is met.

The CRITERIA subcommand has the following keywords when used with CNLR:

ITER n *Maximum number of major iterations.* Specify any positive integer for n. The default is $\max(50, 3(p + m_L) + 10m_N)$, where p is the number of parameters, m_L is the number of linear constraints, and m_N is the number of nonlinear constraints. If the search for a solution stops because this limit is exceeded, CNLR issues a warning message.

MINORITERATION n *Maximum number of minor iterations.* Specify any positive integer. This is the number of minor iterations allowed within each major iteration. The default is $\max(50, 3(n + m_L + m_N))$.

CRSHTOL n *Crash tolerance.* CRSHTOL is used to determine if initial values are within their specified bounds. Specify any value between 0 and 1. The default value is 0.01. A constraint of the form $a'X \geq l$ is considered a valid part of the working set if $|a'X - l| < \text{CRSHTOL}(1 + |l|)$.

STEPLIMIT n *Step limit.* The CNLR algorithm does not allow changes in the length of the parameter vector to exceed a factor of n. The limit prevents very early steps from going too far from good initial estimates. Specify any positive value. The default value is 2.

FTOLERANCE n *Feasibility tolerance.* This is the maximum absolute difference allowed for both linear and nonlinear constraints for a solution to be considered feasible. Specify any value greater than 0. The default value is the square root of your machine's epsilon.

LFTOLERANCE n *Linear feasibility tolerance.* If specified, this overrides FTOLERANCE for linear constraints and bounds. Specify any value greater than 0. The default value is the square root of your machine's epsilon.

NFTOLERANCE n *Nonlinear feasibility tolerance.* If specified, this overrides FTOLERANCE for nonlinear constraints. Specify any value greater than 0. The default value is the square root of your machine's epsilon.

LSTOLERANCE n *Line search tolerance.* This value must be between 0 and 1 (but not including 1). It controls the accuracy required of the line search that forms the innermost search loop. The default value, 0.9, specifies an inaccurate search. This is appropriate for many problems, particularly if nonlinear constraints are involved. A smaller positive value, corresponding to a more accurate line search, may give better performance if there are no nonlinear constraints, all (or most) derivatives are supplied in the derivatives program, and the data fit in memory.

OPTOLERANCE n *Optimality tolerance.* If an iteration point is a feasible point and the next step will not produce a relative change in either the parameter vector or the objective function of more than the square root of OPTOLERANCE, an optimal solution has been found. OPTOLERANCE can also be thought of as the number of significant digits in the objective function at the solution. For example, if OPTOLERANCE=10^{-6}, the objective function should have approximately six significant digits of accuracy. Specify any number between the FPRECISION value and 1. The default value for OPTOLERANCE is epsilon**0.8.

FPRECISION n *Function precision.* This is a measure of the accuracy with which the objective function can be checked. It acts as a relative precision when the function is large, and an absolute precision when the function is small. For example, if the objective function is larger than 1, and six significant digits are desired, FPRECISION should be 1'E – 6 . If, however, the objective function is of the order 0.001, FPRECISION should be 1'E – 9 to get six digits of accuracy. Specify any number between 0 and 1. The choice of FPRECISION can be very complicated for a badly scaled problem. Chapter 8 of Gill et al. (1981) gives some scaling suggestions. The default value is epsilon**0.9.

ISTEP n *Infinite step size.* This value is the magnitude of the change in parameters that is defined as infinite. That is, if the change in the parameters at a step is greater than ISTEP, the problem is considered unbounded, and estimation stops. Specify any positive number. The default value is 1'E + 20 .

Iteration Criteria for NLR

The NLR procedure uses an adaptation of subroutine LMSTR from the MINPACK package by Garbow et al. Because the NLR algorithm differs substantially from CNLR, the CRITERIA subcommand for NLR has a different set of keywords.

NLR computes parameter estimates using the Levenberg-Marquardt method. At each iteration, NLR evaluates the estimates against a set of control criteria. The iterative calculations continue until one of five cutoff points is met, at which point the iterations stop and the reason for stopping is displayed.

The CRITERIA subcommand has the following keywords when used with NLR:

ITER n *Maximum number of major and minor iterations allowed.* Specify any positive integer for *n*. The default is 100 iterations per parameter. If the search for a solution stops because this limit is exceeded, NLR issues a warning message.

SSCON n *Convergence criterion for the sum of squares.* Specify any non-negative number for *n*. The default is 1'E – 8 . If successive iterations fail to reduce the sum of squares by this proportion, the procedure stops. Specify 0 to disable this criterion.

PCON n *Convergence criterion for the parameter values.* Specify any non-negative number for *n*. The default is 1'E – 8 . If successive iterations fail to change any of the parameter values by this proportion, the procedure stops. Specify 0 to disable this criterion.

RCON n *Convergence criterion for the correlation between the residuals and the derivatives.* Specify any non-negative number for *n*. The default is 1'E – 8 . If the largest value for the correlation between the residuals and the derivatives equals this value, the procedure stops because it lacks the information it needs to estimate a direction for its next move. This criterion is often referred to as a gradient convergence criterion. Specify 0 to disable this criterion.

Example

```
MODEL PROGRAM A=.5 B=1.6.
COMPUTE PRED=A*SPEED**B.
NLR STOP /CRITERIA=ITER(80) SSCON=.000001.
```

- CRITERIA changes two of the five cutoff values affecting iteration, ITER and SSCON, and leaves the remaining three, PCON, RCON, and CKDER, at their default values.

BOUNDS Subcommand

The BOUNDS subcommand can be used to specify both linear and nonlinear constraints. It can be used only with CNLR; it cannot be used with NLR.

Simple Bounds and Linear Constraints

BOUNDS can be used to impose bounds on parameter values. These bounds can involve either single parameters or a linear combination of parameters and can be either equalities or inequalities.

- All bounds are specified on the same BOUNDS subcommand and separated by semicolons.
- The only variables allowed on BOUNDS are parameter variables (those named on MODEL PROGRAM).
- Only * (multiplication), + (addition), – (subtraction), = or EQ, >= or GE, and <= or LE can be used. When two relational operators are used (as in the third bound in the example below), they must both be in the same direction.

Example

```
/BOUNDS 5 >= A;
        B >= 9;
        .01 <= 2*A + C <= 1;
        D + 2*E = 10
```

- BOUNDS imposes bounds on the parameters *A, B, C,* and *D.* Specifications for each parameter are separated by a semicolon.

Nonlinear Constraints

Nonlinear constraints on the parameters can also be specified with the BOUNDS subcommand. The constrained function must be calculated and stored in a variable by a constrained functions program directly preceding the CNLR command. The constraint is then specified on the BOUNDS subcommand.

In general, nonlinear bounds will not be obeyed until an optimal solution has been found. This is different from simple and linear bounds, which are satisfied at each iteration. The constrained functions must be smooth near the solution.

Example

```
MODEL PROGRAM A=.5 B=1.6.
COMPUTE PRED=A*SPEED**B.

CONSTRAINED FUNCTIONS.
COMPUTE DIFF=A-10**B.

CNLR STOP /BOUNDS DIFF LE 0.
```

- The constrained function is calculated by a constrained functions program and stored in variable *DIFF*. The constrained functions program immediately precedes CNLR.
- BOUNDS imposes bounds on the function (less than or equal to 0).
- CONSTRAINED FUNCTIONS variables and parameters named on MODEL PROGRAM cannot be combined in the same BOUNDS expression. For example, you *cannot* specify $(DIFF + A) >= 0$ on the BOUNDS subcommand.

LOSS Subcommand

LOSS specifies a loss function for CNLR to minimize. By default, CNLR minimizes the sum of squared residuals. LOSS can be used only with CNLR; it cannot be used with NLR.

- The loss function must first be computed in the model program. LOSS is then used to specify the name of the computed variable.
- The minimizing algorithm may fail if it is given a loss function that is not smooth, such as the absolute value of residuals.
- If derivatives are supplied, the derivative of each parameter must be computed with respect to the loss function, rather than the predicted value. The easiest way to do this is in two steps: first compute derivatives of the model, and then compute derivatives of the loss function with respect to the model and multiply by the model derivatives.
- When LOSS is used, the usual summary statistics are not computed. Standard errors, confidence intervals, and correlations of the parameters are available only if the BOOTSTRAP subcommand is specified.

Example

```
MODEL PROGRAM  A=1 B=1.
COMPUTE PRED=EXP(A+B*T)/(1+EXP(A+B*T)).
COMPUTE LOSS=-W*(Y*LN(PRED)+(1-Y)*LN(1-PRED)).

DERIVATIVES.
COMPUTE D.A=PRED/(1+EXP(A+B*T)).
COMPUTE D.B=T*PRED/(1+EXP(A+B*T)).
COMPUTE D.A=(-W*(Y/PRED - (1-Y)/(1-PRED)) * D.A).
COMPUTE D.B=(-W*(Y/PRED - (1-Y)/(1-PRED)) * D.B).

CNLR Y /LOSS=LOSS.
```

- The second COMPUTE command in the model program computes the loss functions and stores its values in the variable *LOSS*, which is then specified on the LOSS subcommand.
- Because derivatives are supplied in the derivatives program, the derivatives of all parameters are computed with respect to the loss function, rather than the predicted value.

BOOTSTRAP Subcommand

BOOTSTRAP provides bootstrap estimates of the parameter standard errors, confidence intervals, and correlations. BOOTSTRAP can be used only with CNLR; it cannot be used with NLR.

Bootstrapping is a way of estimating the standard error of a statistic, using repeated samples from the original data set. This is done by sampling with replacement to get samples of the same size as the original data set.

- The minimum specification is the subcommand keyword. Optionally, specify the number of samples to use for generating bootstrap results.
- By default, BOOTSTRAP generates bootstrap results based on $10*p*(p+1)/2$ samples, where p is the number of parameters. That is, 10 samples are drawn for each statistic (standard error or correlation) to be calculated.
- When BOOTSTRAP is used, the nonlinear equation is estimated for each sample. The standard error of each parameter estimate is then calculated as the standard deviation of the bootstrapped estimates. Parameter values from the original data are used as starting values for each bootstrap sample. Even so, bootstrapping is computationally expensive.
- If the OUTFILE subcommand is specified, a case is written to the output file for each bootstrap sample. The first case in the file will be the actual parameter estimates, followed by the bootstrap samples. After the first case is eliminated (using SELECT IF), other SPSS procedures (such as FREQUENCIES) can be used to examine the bootstrap distribution.

Example

```
MODEL PROGRAM A=.5 B=1.6.
COMPUTE PSTOP=A*SPEED**B.
CNLR STOP /BOOTSTRAP /OUTFILE=PARAM.
GET FILE=PARAM.
LIST.
COMPUTE ID=$CASENUM.
SELECT IF (ID > 1).
FREQUENCIES A B /FORMAT=NOTABLE /HISTOGRAM.
```

- CNLR generates the bootstrap standard errors, confidence intervals, and parameter correlation matrix. OUTFILE saves the bootstrap estimates in the file *PARAM.*
- GET retrieves the system file *PARAM.*
- LIST lists the different sample estimates along with the original estimate. *NCASES* in the listing (see the OUTFILE subcommand on p. 520) refers to the number of distinct cases in the sample because cases are duplicated in each bootstrap sample.
- FREQUENCIES generates histograms of the bootstrapped parameter estimates.

PROBIT

```
PROBIT response-count varname OF observation-count varname
      WITH varlist [BY varname(min,max)]

 [/MODEL={PROBIT**}]
         {LOGIT   }
         {BOTH    }

 [/LOG=[{10**  }]
        {2.718 }
        {value }
        {NONE  }

 [/CRITERIA=[{OPTOL   }({epsilon**0.8})][P({0.15**})][STEPLIMIT({0.1**})]
             {CONVERGE} {n           }     {p     }            {n    }

            [ITERATE({max(50,3(p+1)**})]]
                     {n               }

 [/NATRES[=value]]

 [/PRINT={[CI**] [FREQ**] [RMP**]} [PARALL] [NONE] [ALL]]
         {DEFAULT**              }

 [/MISSING=[{EXCLUDE**}]  ]
            {INCLUDE  }
```

**Default if subcommand or keyword is omitted.

Example:

```
PROBIT  R OF N BY ROOT(1,2) WITH X
  /MODEL = BOTH.
```

Overview

PROBIT can be used to estimate the effects of one or more independent variables on a dichotomous dependent variable (such as dead or alive, employed or unemployed, product purchased or not). The program is designed for dose-response analyses and related models, but PROBIT can also estimate logistic regression models.

Options

The Model. You can request a probit or logit response model, or both, for the observed response proportions with the MODEL subcommand.

Transform Predictors. You can control the base of the log transformation applied to the predictors or request no log transformation with the LOG subcommand.

Natural Response Rates. You can instruct PROBIT to estimate the natural response rate (threshold) of the model or supply a known natural response rate to be used in the solution with the NATRES subcommand.

Algorithm Control Parameters. You can specify values of algorithm control parameters, such as the limit on iterations, using the CRITERIA subcommand.

Statistics. By default, PROBIT calculates frequencies, fiducial confidence intervals, and the relative median potency. It also produces a plot of the observed probits or logits against the values of a single independent variable. Optionally, you can use the PRINT subcommand to request a test of the parallelism of regression lines for different levels of the grouping variable or to suppress any or all of these statistics.

Basic Specification

- The basic specification is the response-count variable, keyword OF, the observation-count variable, keyword WITH, and at least one independent variable.
- PROBIT calculates maximum-likelihood estimates for the parameters of the default probit response model and automatically displays estimates of the regression coefficient and intercept terms, their standard errors, a covariance matrix of parameter estimates, and a Pearson chi-square goodness-of-fit test of the model.

Subcommand Order

- The variable specification must be first.
- Subcommands can be named in any order.

Syntax Rules

- The variables must include a response count, an observation count, and at least one predictor. A categorical grouping variable is optional.
- All subcommands are optional and each can appear only once.
- Generally, data should not be entered for individual observations. PROBIT expects predictor values, response counts, and the total number of observations as the input case.
- If the data are available only in a case-by-case form, use AGGREGATE first to compute the required response and observation counts.

Operations

- The transformed response variable is predicted as a linear function of other variables using the nonlinear-optimization method. Note that the previous releases used the iteratively weighted least-squares method, which has a different way of transforming the response variables. See "MODEL Subcommand" on p. 533.
- If individual cases are entered in the data, PROBIT skips the plot of transformed response proportions and predictor values.
- If individual cases are entered, the degrees of freedom for the chi-square goodness-of-fit statistic are based on the individual cases.

Limitations

- Only one prediction model can be tested on a single PROBIT command, although both probit and logit response models can be requested for that prediction.
- Confidence limits, the plot of transformed response proportions and predictor values, and computation of relative median potency are necessarily limited to single-predictor models.

Example

```
PROBIT  R OF N BY ROOT(1,2) WITH X
  /MODEL = BOTH.
```

- This example specifies that both the probit and logit response models be applied to the response frequency *R*, given *N* total observations and the predictor *X*.
- By default, the predictor is log transformed.

Example

```
* Using data in a case-by-case form

DATA LIST FREE / PREPARTN DOSE RESPONSE.
BEGIN DATA
1 1.5 0
 ...
4 20.0 1
END DATA.
COMPUTE SUBJECT = 1.
PROBIT RESPONSE OF SUBJECT BY PREPARTN(1,4) WITH DOSE.
```

- This dose-response model (Finney, 1971) illustrates a case-by-case analysis. A researcher tests four different preparations at varying doses and observes whether each subject responds. The data are individually recorded for each subject, with 1 indicating a response and 0 indicating no response. The number of observations is always 1 and is stored in variable *SUBJECT*.
- PROBIT warns that the data are in a case-by-case form and that the plot is therefore skipped.
- Degrees of freedom for the goodness-of-fit test are based on individual cases, not dosage groups.
- PROBIT displays predicted and observed frequencies for all individual input cases.

Example

```
* Aggregating case-by-case data

DATA LIST FREE/PREPARTN DOSE RESPONSE.
BEGIN DATA
     1.00      1.50      .00
     ...
     4.00     20.00     1.00
END DATA.
AGGREGATE OUTFILE=*
  /BREAK=PREPARTN DOSE
  /SUBJECTS=N(RESPONSE)
  /NRESP=SUM(RESPONSE).
PROBIT NRESP OF SUBJECTS BY PREPARTN(1,4) WITH DOSE.
```

- This example analyzes the same dose-response model as the previous example, but the data are first aggregated.
- AGGREGATE summarizes the data by cases representing all subjects who received the same preparation (*PREPARTN*) at the same dose (*DOSE*).
- The number of cases having a nonmissing response is recorded in the aggregated variable *SUBJECTS*.
- Because *RESPONSE* is coded 0 for no response and 1 for a response, the sum of the values gives the number of observations with a response.
- PROBIT requests a default analysis.
- The parameter estimates for this analysis are the same as those calculated for individual cases in the example above. The chi-square test, however, is based on the number of dosages.

Variable Specification

The variable specification on PROBIT identifies the variables for response count, observation count, groups, and predictors. The variable specification is required.

- The variables must be specified first. The specification must include the response-count variable, followed by the keyword OF and then the observation-count variable.
- If the value of the response-count variable exceeds that of the observation-count variable, a procedure error occurs and PROBIT is not executed.
- At least one predictor (covariate) must be specified following the keyword WITH. The number of predictors is limited only by available workspace. All predictors must be continuous variables.
- You can specify a grouping variable (factor) after the keyword BY. Only one variable can be specified. It must be numeric and can contain only integer values. You must specify, in parentheses, a range indicating the minimum and maximum values for the grouping variable. Each integer value in the specified range defines a group.
- Cases with values for the grouping variable that are outside the specified range are excluded from the analysis.
- Keywords BY and WITH can appear in either order. However, both must follow the response- and observation-count variables.

Example

```
PROBIT R OF N WITH X.
```

- The number of observations having the measured response appears in variable *R*, and the total number of observations is in *N*. The predictor is *X*.

Example

```
PROBIT  R OF N BY ROOT(1,2) WITH X.

PROBIT  R OF N WITH X BY ROOT(1,2).
```

- Because keywords BY and WITH can be used in either order, these two commands are equivalent. Each command specifies *X* as a continuous predictor and *ROOT* as a categorical grouping variable.
- Groups are identified by the levels of variable *ROOT*, which may be 1 or 2.
- For each combination of predictor and grouping variables, the variable *R* contains the number of observations with the response of interest, and *N* contains the total number of observations.

MODEL Subcommand

MODEL specifies the form of the dichotomous-response model. Response models can be thought of as transformations (*T*) of response rates, which are proportions or probabilities (*p*). Note the difference in the transformations between the current version and the previous versions.

- A **probit** is the inverse of the cumulative standard normal distribution function. Thus, for any proportion, the probit transformation returns the value below which that proportion of standard normal deviates is found. For the probit response model, the program uses $T(p) = \text{PROBIT}(p)$. Hence:

$$T(0.025) = \text{PROBIT}(0.025) = -1.96$$
$$T(0.400) = \text{PROBIT}(0.400) = -0.25$$
$$T(0.500) = \text{PROBIT}(0.500) = 0.00$$
$$T(0.950) = \text{PROBIT}(0.950) = 1.64$$

- A **logit** is simply the natural log of the odds ratio, $p/(1-p)$. In the Probit procedure, the response function is given as $T(p) = \log_e(p/(1-p))$. Hence:

$$T(0.025) = \text{LOGIT}(0.025) = -3.66$$
$$T(0.400) = \text{LOGIT}(0.400) = -0.40$$
$$T(0.500) = \text{LOGIT}(0.500) = 0.00$$
$$T(0.950) = \text{LOGIT}(0.950) = 2.94$$

You can request one or both of the models on the MODEL subcommand. The default is PROBIT if the subcommand is not specified or is specified with no keyword.

PROBIT *Probit response model.* This is the default.

LOGIT *Logit response model.*

BOTH *Both probit and logit response models.* PROBIT displays all the output for the logit model followed by the output for the probit model.

- If subgroups and multiple-predictor variables are defined, PROBIT estimates a separate intercept, a_j, for each subgroup and a regression coefficient, b_i, for each predictor.

LOG Subcommand

LOG specifies the base of the logarithmic transformation of the predictor variables or suppresses the default log transformation.

- LOG applies to all predictors.
- To transform only selected predictors, use COMPUTE commands before the Probit procedure. Then specify NONE on the LOG subcommand.
- If LOG is omitted, a logarithm base of 10 is used.
- If LOG is used without a specification, the natural logarithm base e (2.718) is used.
- If you have a control group in your data and specify NONE on the LOG subcommand, the control group is included in the analysis. See "NATRES Subcommand" on p. 535.

You can specify one of the following on LOG:

value *Logarithm base to be applied to all predictors.*

NONE *No transformation of the predictors.*

Example

```
PROBIT R OF N BY ROOT (1,2) WITH X
  /LOG = 2.
```

- LOG specifies a base-2 logarithmic transformation.

CRITERIA Subcommand

Use CRITERIA to specify the values of control parameters for the PROBIT algorithm. You can specify any or all of the keywords below. Defaults remain in effect for parameters that are not changed.

OPTOL(n) *Optimality tolerance.* Alias CONVERGE. If an iteration point is a feasible point and the next step will not produce a relative change in either the parameter vector or the log-likelihood function of more than the square root of n, an optimal solution has been found. OPTOL can also be thought of as the number of significant digits in the log-likelihood function at the solution. For example, if OPTOL=10^{-6}, the log-likelihood function should have approximately six significant digits of accuracy. The default value is machine epsilon**0.8.

ITERATE(n) *Iteration limit.* Specify the maximum number of iterations. The default is max $(50, 3(p+1))$, where p is the number of parameters in the model.

P(p) *Heterogeneity criterion probability.* Specify a cutoff value between 0 and 1 for the significance of the goodness-of-fit test. The cutoff value determines whether a heterogeneity factor is included in calculations of confidence levels for effective levels of a predictor. If the significance of chi-square is greater than the cutoff, the heterogeneity factor is not included. If you specify 0, this criterion is disabled; if you specify 1, a heterogeneity factor is automatically included. The default is 0.15.

STEPLIMIT(n) *Step limit.* The PROBIT algorithm does not allow changes in the length of the parameter vector to exceed a factor of *n*. This limit prevents very early steps from going too far from good initial estimates. Specify any positive value. The default value is 0.1.

CONVERGE(n) *Alias of OPTOL.*

NATRES Subcommand

You can use NATRES either to supply a known natural response rate to be used in the solution or to instruct PROBIT to estimate the natural (or threshold) response rate of the model.

- To supply a known natural response rate as a constraint on the model solution, specify a value less than 1 on NATRES.
- To instruct PROBIT to estimate the natural response rate of the model, you can indicate a control group by giving a 0 value to any of the predictor variables. PROBIT displays the estimate of the natural response rate and the standard error and includes the estimate in the covariance/correlation matrix as NAT RESP.
- If no control group is indicated and NATRES is specified without a given value, PROBIT estimates the natural response rate from the entire data and informs you that no control group has been provided. The estimate of the natural response rate and the standard error are displayed and NAT RESP is included in the covariance/correlation matrix.
- If you have a control group in your data and specify NONE on the LOG subcommand, the control group is included in the analysis.

Example

```
DATA LIST FREE / SOLUTION DOSE NOBSN NRESP.
BEGIN DATA
1  5 100 20
1 10  80 30
1  0 100 10
...
END DATA.

PROBIT NRESP OF NOBSN BY SOLUTION(1,4) WITH DOSE
  /NATRES.
```

- This example reads four variables and requests a default analysis with an estimate of the natural response rate.
- The predictor variable, *DOSE*, has a value of 0 for the third case.
- The response count (10) and the observation count (100) for this case establish the initial estimate of the natural response rate.

- Because the default log transformation is performed, the control group is not included in the analysis.

Example

```
DATA LIST FREE / SOLUTION DOSE NOBSN NRESP.
BEGIN DATA
1  5 100 20
1 10  80 30
1  0 100 10
  ...
END DATA.

PROBIT NRESP OF NOBSN BY SOLUTION(1,4) WITH DOSE
  /NATRES = 0.10.
```

- This example reads four variables and requests an analysis in which the natural response rate is set to 0.10. The values of the control group are ignored.
- The control group is excluded from the analysis because the default log transformation is performed.

PRINT Subcommand

Use PRINT to control the statistics calculated by PROBIT.

- PROBIT always displays the plot (for a single-predictor model) and the parameter estimates and covariances for the probit model.
- If PRINT is used, the requested statistics are calculated and displayed in addition to the parameter estimates and plot.
- If PRINT is not specified or is specified without any keyword, FREQ, CI, and RMP are calculated and displayed in addition to the parameter estimates and plot.

DEFAULT *FREQ, CI, and RMP.* This is the default if PRINT is not specified or is specified by itself.

FREQ *Frequencies.* Display a table of observed and predicted frequencies with their residual values. If observations are entered on a case-by-case basis, this listing can be quite lengthy.

CI *Fiducial confidence intervals.* Print Finney's (1971) fiducial confidence intervals for the levels of the predictor needed to produce each proportion of responses. PROBIT displays this default output for single-predictor models only. If a categorical grouping variable is specified, PROBIT produces a table of confidence intervals for each group. If the Pearson chi-square goodness-of-fit test is significant ($p < 0.15$ by default), PROBIT uses a heterogeneity factor to calculate the limits.

RMP *Relative median potency.* Display the relative median potency (RMP) of each pair of groups defined by the grouping variable. PROBIT displays this default output for single-predictor models only. For any pair of groups, the RMP is the ratio of the stimulus tolerances in those groups. **Stimulus tolerance** is the value of the predictor necessary to produce a 50% response rate. If the derived model for one predictor and two groups estimates that a predictor value of 21 produces a 50% response rate in the first group, and that a predictor value of 15 produces a 50% response rate in the second group, the relative median potency would be 21/15 = 1.40. In biological assay analyses, RMP measures the comparative strength of preparations.

PARALL *Parallelism test.* Produce a test of the parallelism of regression lines for different levels of the grouping variable. This test displays a chi-square value and its associated probability. It requires an additional pass through the data and, thus, additional processing time.

NONE *Display only the unconditional output.* This option can be used to override any other specification on the PRINT subcommand for PROBIT.

ALL *All available output.* This is the same as requesting FREQ, CI, RMP, and PARALL.

MISSING Subcommand

PROBIT always deletes cases having a missing value for any variable. In the output, PROBIT indicates how many cases it rejected because of missing data. This information is displayed with the DATA Information that prints at the beginning of the output. You can use the MISSING subcommand to control the treatment of user-missing values.

EXCLUDE *Delete cases with user-missing values.* This is the default. You can also make it explicit by using the keyword DEFAULT.

INCLUDE *Include user-missing values.* PROBIT treats user-missing values as valid. Only cases with system-missing values are rejected.

SURVIVAL

```
SURVIVAL TABLES=survival varlist
                [BY varlist (min, max)...][BY varlist (min, max)...]

  /INTERVALS=THRU n BY a [THRU m BY b ...]

  /STATUS=status variable({min, max}) FOR {ALL             }
                         {value    }        {survival varlist}
 [/STATUS=...]

 [/PLOT  ({ALL     })={ALL              } BY {ALL    }      BY {ALL    }]
          {LOGSURV }   {survival varlis}     {varlist}         {varlist}
          {SURVIVAL}
          {HAZARD  }
          {DENSITY }

 [/PRINT={TABLE**}]
         {NOTABLE}

 [/COMPARE={ALL**           } BY {ALL**  }    BY {ALL**  }]
           {survival varlist}    {varlist}       {varlist}

 [/CALCULATE=[{EXACT**    }] [PAIRWISE] [COMPARE] ]
              {CONDITIONAL}
              {APPROXIMATE}

 [/MISSING={GROUPWISE**}  [INCLUDE] ]
           {LISTWISE   }

 [/WRITE=[{NONE**}] ]
          {TABLES}
          {BOTH  }
```

**Default if subcommand or keyword is omitted.

Example:

```
SURVIVAL TABLES=MOSFREE BY TREATMNT(1,3)
  /STATUS = PRISON (1) FOR MOSFREE
  /INTERVAL=THRU 24 BY 3.
```

Overview

SURVIVAL produces actuarial life tables, plots, and related statistics for examining the length of time to the occurrence of an event, often known as **survival time**. Cases can be classified into groups for separate analyses and comparisons. Time intervals can be calculated with the SPSS date- and time-conversion functions, for example, CTIME.DAYS or YRMODA (see the *SPSS Base System Syntax Reference Guide*). For a closely related alternative nonparametric analysis of survival times using the product-limit Kaplan-Meier estimator, see the KM command. For an analysis of survival times with covariates, including time-dependent covariates, see the COXREG command.

Options

Life Tables. You can list the variables to be used in the analysis, including any control variables on the TABLES subcommand. You can also suppress the life tables in the output with the PRINT subcommand.

Intervals. SURVIVAL reports the percentage alive at various times after the initial event. You can select the time points for reporting with the INTERVALS subcommand.

Plots. You can plot the survival functions for all cases or separately for various subgroups with the PLOT subcommand.

Comparisons. When control variables are listed on the TABLES subcommand, you can compare groups based on the Wilcoxon (Gehan) statistic using the COMPARE subcommand. You can request pairwise or approximate comparisons with the CALCULATE subcommand.

Writing a File. You can write the life tables, including the labeling information, to a file with the WRITE subcommand.

Basic Specification

- The basic specification requires three subcommands: TABLES, INTERVALS, and STATUS. TABLES identifies at least one survival variable from the working data file, INTERVALS divides the time period into intervals, and STATUS names a variable that indicates whether the event occurred.
- The basic specification prints one or more life tables, depending on the number of survival and control variables specified.

Subcommand Order

- TABLES must be first.
- Remaining subcommands can be named in any order.

Syntax Rules

- Only one TABLES subcommand can be specified, but multiple survival variables can be named. A survival variable cannot be specified as a control variable on any subcommands.
- Only one INTERVALS subcommand can be in effect on a SURVIVAL command. The interval specifications apply to all the survival variables listed on TABLES. If multiple INTERVALS subcommands are used, the last specification supersedes all previous ones.
- Only one status variable can be listed on each STATUS subcommand. To specify multiple status variables, use multiple STATUS subcommands.
- You can specify multiple control variables on one BY keyword. Use a second BY keyword to specify second-order control variables to interact with the first-order control variables.
- All variables, including survival variables, control variables, and status variables, must be numeric. SURVIVAL does not process string variables.

Operations

- SURVIVAL computes time intervals according to specified interval widths, calculates the survival functions for each interval, and builds one life table for each group of survival variables. The life table is displayed unless explicitly suppressed.
- When the PLOT subcommand is specified, SURVIVAL plots the survival functions for all cases or separately for various groups.
- When the COMPARE subcommand is specified, SURVIVAL compares survival-time distributions of different groups based on the Wilcoxon (Gehan) statistic.

Limitations

- Maximum 20 survival variables.
- Maximum 100 control variables total on the first- and second-order control-variable lists combined.
- Maximum 20 THRU and BY specifications on INTERVALS.
- Maximum 35 values can appear on a plot.

Example

```
SURVIVAL TABLES=MOSFREE BY TREATMNT(1,3)
  /STATUS = PRISON (1) FOR MOSFREE
  /INTERVALS = THRU 24 BY 3.
```

- The survival analysis is used to examine the length of time between release from prison and return to prison for prisoners in three treatment programs. The variable *MOSFREE* is the length of time in months a prisoner stayed out of prison. The variable *TREATMNT* indicates the treatment group for each case.
- A value of 1 on the variable *PRISON* indicates a terminal outcome—that is, cases coded as 1 have returned to prison. Cases with other non-negative values for *PRISON* have not returned. Because we don't know their final outcome, such cases are called censored.
- Life tables are produced for each of the three subgroups. INTERVALS specifies that the survival experience be described every three months for the first two years.

TABLES Subcommand

TABLES identifies the survival and control variables to be included in the analysis.

- The minimum specification is one or more survival variables.
- To specify one or more first-order control (or factor) variables, use keyword BY followed by the control variable(s). First-order control variables are processed in sequence. For example, `BY A(1,3) B(1,2)` results in five groups (*A* = 1, *A* = 2, *A* = 3, *B* = 1, and *B* = 2).
- You can specify one or more second-order control variables following a second BY keyword. Separate life tables are generated for each combination of values of the first-order

and second-order controls. For example, `BY A(1,3) BY B(1,2)` results in six groups (*A* = 1 *B* = 1, *A* = 1 *B* = 2, *A* = 2 *B* = 1, *A* = 2 *B* = 2, *A* = 3 *B* = 1, and *A* = 3 *B* = 2).

- Each control variable must be followed by a value range in parentheses. These values must be integers separated by a comma or a blank. Non-integer values in the data are truncated, and the case is assigned to a subgroup based on the integer portion of its value on the variable. To specify only one value for a control variable, use the same value for the minimum and maximum.
- To generate life tables for all cases combined, as well as for control variables, use COMPUTE to create a variable that has the same value for all cases. With this variable as a control, tables for the entire set of cases, as well as for the control variables, will be produced.

Example

```
SURVIVAL TABLES = MOSFREE BY TREATMNT(1,3) BY RACE(1,2)
  /STATUS = PRISON(1)
  /INTERVAL = THRU 24 BY 3.
```

- *MOSFREE* is the survival variable, and *TREATMNT* is the first-order control variable. The second BY defines *RACE* as a second-order control group having a value of 1 or 2.
- Six life tables with the median survival time are produced, one for each pair of values for the two control variables.

INTERVALS Subcommand

INTERVALS determines the period of time to be examined and how the time will be grouped for the analysis. The interval specifications apply to all the survival variables listed on TABLES.

- SURVIVAL always uses 0 as the starting point for the first interval. Do not specify the 0. The INTERVALS specification *must* begin with keyword THRU.
- Specify the terminal value of the time period after the keyword THRU. The final interval includes any observations that exceed the specified terminal value.
- The grouping increment, which follows keyword BY, must be in the same units as the survival variable.
- The period to be examined can be divided into intervals of varying lengths by repeating the THRU and BY keywords. The period must be divided in ascending order. If the time period is not a multiple of the increment, the endpoint of the period is adjusted upward to the next even multiple of the grouping increment.
- When the period is divided into intervals of varying lengths by repeating the THRU and BY specifications, the adjustment of one period to produce even intervals changes the starting point of subsequent periods. If the upward adjustment of one period completely overlaps the next period, no adjustment is made and the procedure terminates with an error.

Example

```
SURVIVAL TABLES = MOSFREE BY TREATMNT(1,3)
  /STATUS = PRISON(1) FOR MOSFREE
  /INTERVALS = THRU 12 BY 1 THRU 24 BY 3.
```

- INTERVALS produces life tables computed from 0 to 12 months at one-month intervals and from 13 to 24 months at three-month intervals.

Example

```
SURVIVAL  ONSSURV BY TREATMNT (1,3)
  /STATUS = OUTCOME (3,4) FOR ONSSURV
  /INTERVALS = THRU 50 BY 6.
```

- On the INTERVALS subcommand, the value following BY (6) does not divide evenly into the period to which it applies (50). Thus, the endpoint of the period is adjusted upward to the next even multiple of the BY value, resulting in a period of 54 with 9 intervals of 6 units each.

Example

```
SURVIVAL  ONSSURV BY TREATMNT (1,3)
  /STATUS = OUTCOME (3,4) FOR ONSSURV
  /INTERVALS = THRU 50 BY 6 THRU 100 BY 10 THRU 200 BY 20.
```

- Multiple THRU and BY specifications are used on the INTERVAL subcommand to divide the period of time under examination into intervals of different lengths.
- The first THRU and BY specifications are adjusted to produce even intervals as in the previous example. As a result, the following THRU and BY specifications are automatically readjusted to generate 5 intervals of 10 units (through 104), followed by 5 intervals of 20 units (through 204).

STATUS Subcommand

To determine whether the terminal event has occurred for a particular observation, SURVIVAL checks the value of a status variable. STATUS lists the status variable associated with each survival variable and the codes which indicate that a terminal event occurred.

- Specify a status variable followed by a value range enclosed in parentheses. The value range identifies the codes that indicate that the terminal event has taken place. All cases with non-negative times that do not have a code in the value range are classified as **censored cases**, which are cases for which the terminal event has not yet occurred.
- If the status variable does not apply to all the survival variables, specify FOR and the name of the survival variable(s) to which the status variable applies.
- Each survival variable on TABLES must have an associated status variable identified by a STATUS subcommand.
- Only one status variable can be listed on each STATUS subcommand. To specify multiple status variables, use multiple STATUS subcommands.
- If FOR is omitted on the STATUS specification, the status-variable specification applies to all of the survival variables not named on another STATUS subcommand.
- If more than one STATUS subcommand omits keyword FOR, the final STATUS subcommand without FOR applies to all survival variables not specified by FOR on other STATUS subcommands. No warning is printed.

Example

```
SURVIVAL  ONSSURV BY TREATMNT (1,3)
  /INTERVALS = THRU 50 BY 5, THRU 100 BY 10
  /STATUS = OUTCOME (3,4) FOR ONSSURV.
```

- STATUS specifies that a code of 3 or 4 on *OUTCOME* means that the terminal event for the survival variable *ONSSURV* occurred.

Example

```
SURVIVAL TABLES = NOARREST MOSFREE BY TREATMNT(1,3)
  /STATUS = ARREST (1) FOR NOARREST
  /STATUS = PRISON (1)
  /INTERVAL=THRU 24 BY 3.
```

- STATUS defines the terminal event for *NOARREST* as a value of 1 for *ARREST*. Any other value for *ARREST* is considered censored.
- The second STATUS subcommand defines the value of 1 for *PRISON* as the terminal event. Keyword FOR is omitted. Thus, the status-variable specification applies to *MOSFREE*, which is the only survival variable not named on another STATUS subcommand.

PLOT Subcommand

PLOT produces plots of the cumulative survival distribution, the hazard function, and the probability density function. The PLOT subcommand can plot only the survival functions generated by the TABLES subcommand; PLOT cannot eliminate control variables.

- When specified by itself, the PLOT subcommand produces all available plots for each survival variable. Points on each plot are identified by values of the first-order control variables. If second-order controls are used, a separate plot is generated for every value of the second-order control variables.
- To request specific plots, specify, in parentheses following PLOT, any combination of the keywords defined below.
- Optionally, generate plots for only a subset of the requested life tables. Use the same syntax as used on the TABLES subcommand for specifying survival and control variables, omitting the value ranges. Each survival variable named on PLOT must have as many control levels as were specified for that variable on TABLES. However, only one control variable needs to be present for each level. If a required control level is missing on the PLOT specification, the default BY ALL is used for that level. Keyword ALL can be used to refer to an entire set of survival or control variables.
- To determine the number of plots that will be produced, multiply the number of functions plotted by the number of survival variables times the number of first-order controls times the number of distinct values represented in all of the second-order controls.

ALL *Plot all available functions.* ALL is the default if PLOT is used without specifications.

LOGSURV *Plot the cumulative survival distribution on a logarithmic scale.*

SURVIVAL *Plot the cumulative survival distribution on a linear scale.*

HAZARD *Plot the hazard function.*

DENSITY *Plot the density function.*

Example

```
SURVIVAL TABLES = NOARREST MOSFREE BY TREATMNT(1,3)
  /STATUS = ARREST (1) FOR NOARREST
  /STATUS = PRISON (1) FOR MOSFREE
  /INTERVALS = THRU 24 BY 3
  /PLOT (SURVIVAL,HAZARD) = MOSFREE.
```

- Separate life tables are produced for each of the survival variables (*NOARREST* and *MOSFREE*) for each of the three values of the control variable *TREATMNT*.
- PLOT produces plots of the cumulative survival distribution and the hazard rate for *MOSFREE* for the three values of *TREATMNT* (even though *TREATMNT* is not included on the PLOT specification).
- Because plots are requested only for the survival variable *MOSFREE*, no plots are generated for variable *NOARREST*.

PRINT Subcommand

By default, SURVIVAL prints life tables. PRINT can be used to suppress the life tables.

TABLE *Print the life tables.* This is the default.

NOTABLE *Suppress the life tables.* Only plots and comparisons are printed. The WRITE subcommand, which is used to write the life tables to a file, can be used when NOTABLE is in effect.

Example

```
SURVIVAL TABLES = MOSFREE BY TREATMNT(1,3)
  /STATUS = PRISON (1) FOR MOSFREE
  /INTERVALS = THRU 24 BY 3
  /PLOT (ALL)
  /PRINT = NOTABLE.
```

- PRINT NOTABLE suppresses the printing of life tables.

COMPARE Subcommand

COMPARE compares the survival experience of subgroups defined by the control variables. At least one first-order control variable is required for calculating comparisons.

- When specified by itself, the COMPARE subcommand produces comparisons using the TABLES variable list.
- Alternatively, specify the survival and control variables for the comparisons. Use the same syntax as used on the TABLES subcommand for specifying survival and control variables, omitting the value ranges. Only variables that appear on the TABLES subcommand can be listed on COMPARE, and their role as survival, first-order, and second-order control

variables cannot be altered. Keyword TO can be used to refer to a group of variables, and keyword ALL can be used to refer to an entire set of survival or control variables.

- By default, COMPARE calculates exact comparisons between subgroups. Use the CALCULATE subcommand to obtain pairwise comparisons or approximate comparisons.

Example

```
SURVIVAL TABLES = MOSFREE BY TREATMNT(1,3)
  /STATUS = PRISON (1) FOR MOSFREE
  /INTERVAL = THRU 24 BY 3
  /COMPARE.
```

- COMPARE computes the Wilcoxon (Gehan) statistic, degrees of freedom, and observed significance level for the hypothesis that the three survival curves based on the values of *TREATMNT* are identical.

Example

```
SURVIVAL TABLES=ONSSURV,RECSURV BY TREATMNT(1,3)
  /STATUS = RECURSIT(1,9) FOR RECSURV
  /STATUS = STATUS(3,4) FOR ONSSURV
  /INTERVAL = THRU 50 BY 5 THRU 100 BY 10
  /COMPARE = ONSSURV BY TREATMNT.
```

- COMPARE requests a comparison of *ONSSURV* by *TREATMNT*. No comparison is made of *RECSURV* by *TREATMNT*.

CALCULATE Subcommand

CALCULATE controls the comparisons of survival for subgroups specified on the COMPARE subcommand.

- The minimum specification is the subcommand keyword by itself. EXACT is the default.
- Only one of the keywords EXACT, APPROXIMATE, and CONDITIONAL can be specified. If more than one keyword is used, only one is in effect. The order of precedence is APPROXIMATE, CONDITIONAL, and EXACT.
- The keywords PAIRWISE and COMPARE can be used with any of the EXACT, APPROXIMATE, or CONDITIONAL keywords.
- If CALCULATE is used without the COMPARE subcommand, CALCULATE is ignored. However, if the keyword COMPARE is specified on CALCULATE and the COMPARE subcommand is omitted, SPSS generates an error message.
- Data can be entered into SURVIVAL for each individual case or aggregated for all cases in an interval. The way in which data are entered determines whether an exact or an approximate comparison is most appropriate. See "Using Aggregated Data" on p. 546.

EXACT *Calculate exact comparisons.* This is the default. You can obtain exact comparisons based on the survival experience of each observation with individual data. While this method is the most accurate, it requires that all of the data be in memory simultaneously. Thus, exact comparisons may be impractical for large samples. It is also inappropriate when individual data are not available and data aggregated by interval must be used.

APPROXIMATE *Calculate approximate comparisons only.* Approximate comparisons are appropriate for aggregated data. The approximate-comparison approach assumes that all events occur at the midpoint of the interval. With exact comparisons, some of these midpoint ties can be resolved. However, if interval widths are not too great, the difference between exact and approximate comparisons should be small.

CONDITIONAL *Calculate approximate comparisons if memory is insufficient.* Approximate comparisons are produced only if there is insufficient memory available for exact comparisons.

PAIRWISE *Perform pairwise comparisons.* Comparisons of all pairs of values of the first-order control variable are produced along with the overall comparison.

COMPARE *Produce comparisons only.* Survival tables specified on the TABLES subcommand are not computed, and requests for plots are ignored. This allows all available workspace to be used for comparisons. The WRITE subcommand cannot be used when this specification is in effect.

Example

```
SURVIVAL TABLES = MOSFREE BY TREATMNT(1,3)
  /STATUS = PRISON (1) FOR MOSFREE
  /INTERVAL = THRU 24 BY 3
  /COMPARE /CALCULATE = PAIRWISE.
```

- PAIRWISE on CALCULATE computes the Wilcoxon (Gehan) statistic, degrees of freedom, and observed significance levels for each pair of values of *TREATMNT*, as well as for an overall comparison of survival across all three *TREATMNT* subgroups: group 1 with group 2, group 1 with group 3, and group 2 with group 3.
- All comparisons are exact comparisons.

Example

```
SURVIVAL TABLES = MOSFREE BY TREATMNT(1,3)
  /STATUS = PRISON (1) FOR MOSFREE
  /INTERVAL = THRU 24 BY 3
  /COMPARE /CALCULATE = APPROXIMATE COMPARE.
```

- APPROXIMATE on CALCULATE computes the Wilcoxon (Gehan) statistic, degrees of freedom, and probability for the overall comparison of survival across all three *TREATMNT* subgroups using the approximate method.
- Because keyword COMPARE is specified on CALCULATE, survival tables are not computed.

Using Aggregated Data

When aggregated survival information is available, the number of censored and uncensored cases at each time point must be entered. Up to two records can be entered for each interval, one for censored cases and one for uncensored cases. The number of cases included on each record is used as the weight factor. If control variables are used, there will be up to two records (one for censored and one for uncensored cases) for each value of the control variable in each interval. These records must contain the value of the control variable and the number of cases that belong in the particular category as well as values for survival time and status.

Example

```
DATA LIST   / SURVEVAR 1-2 STATVAR 4 SEX 6 COUNT 8.
VALUE LABELS  STATVAR 1 'DECEASED' 2 'ALIVE'
              /SEX 1 'FEMALE' 2 'MALE'.
BEGIN DATA
 1 1 1 6
 1 1 1 1
 1 2 2 2
 1 1 2 1
 2 2 1 1
 2 1 1 2
 2 2 2 1
 2 1 2 3
   ...
END DATA.
WEIGHT COUNT.
SURVIVAL TABLES = SURVEVAR BY SEX (1,2)
  /INTERVALS = THRU 10 BY 1
  /STATUS = STATVAR (1) FOR SURVEVAR.
```

- This example reads aggregated data and performs a SURVIVAL analysis when a control variable with two values is used.
- The first data record has a code of 1 on the status variable *STATVAR*, indicating it is an uncensored case, and a code of 1 on *SEX*, the control variable. The number of cases for this interval is 6, the value of the variable *COUNT*. Intervals with weights of 0 do not have to be included.
- *COUNT* is not used in SURVIVAL but is the weight variable. In this example, each interval requires four records to provide all the data for each *SURVEVAR* interval.

MISSING Subcommand

MISSING controls missing-value treatments. The default is GROUPWISE.

- Negative values on the survival variables are automatically treated as missing data. In addition, cases outside the value range on a control variable are excluded.
- GROUPWISE and LISTWISE are mutually exclusive. However, each can be used with INCLUDE.

GROUPWISE *Exclude missing values groupwise.* Cases with missing values on a variable are excluded from any calculation involving that variable. This is the default.

LISTWISE *Exclude missing values listwise.* Cases missing on any variables named on TABLES are excluded from the analysis.

INCLUDE *Include user-missing values.* User-missing values are included in the analysis.

WRITE Subcommand

WRITE writes data in the survival tables to a file. This file can be used for further analyses or to produce graphics displays.

- When WRITE is omitted, the default is NONE. No output file is created.
- When WRITE is used, a PROCEDURE OUTPUT command must precede the SURVIVAL command. The OUTFILE subcommand on PROCEDURE OUTPUT specifies the output file.
- When WRITE is specified without a keyword, the default is TABLES.

NONE *Do not write procedure output to a file.* This is the default when WRITE is omitted.

TABLES *Write survival-table data records.* All survival-table statistics are written to a file.

BOTH *Write out survival-table data and label records.* Variable names, variable labels, and value labels are written out along with the survival table statistics.

Format

WRITE writes five types of records. Keyword TABLES writes record types 30, 31, and 40. Keyword BOTH writes record types 10, 20, 30, 31, and 40. The format of each record type is described in Table 1 through Table 5.

Table 1 Record type 10, produced only by keyword BOTH

Columns	Content	Format
1–2	Record type (10)	F2.0
3–7	Table number	F5.0
8–15	Name of survival variable	A8
16–55	Variable label of survival variable	A40
56	Number of BY's (0, 1, or 2)	F1.0
57–60	Number of rows in current survival table	F4.0

- One type-10 record is produced for each life table.
- Column 56 specifies the number of orders of control variables (0, 1, or 2) that have been applied to the life table.
- Columns 57–60 specify the number of rows in the life table. This number is the number of intervals in the analysis that show subjects entering; intervals in which no subjects enter are not noted in the life tables.

Table 2 Record type 20, produced by keyword BOTH

Columns	Content	Format
1–2	Record type (20)	F2.0
3–7	Table number	F5.0
8–15	Name of control variable	A8
16–55	Variable label of control variable	A40
56–60	Value of control variable	F5.0
61–80	Value label for this value	A20

- One type-20 record is produced for each control variable in each life table.
- If only first-order controls have been placed in the survival analysis, one type-20 record will be produced for each table. If second-order controls have also been applied, two type-20 records will be produced per table.

Table 3 Record type 30, produced by both keywords TABLES and BOTH

Columns	Content	Format
1–2	Record type (30)	F2.0
3–7	Table number	F5.0
8–13	Beginning of interval	F6.2
14–21	Number entering interval	F8.2
22–29	Number withdrawn in interval	F8.2
30–37	Number exposed to risk	F8.2
38–45	Number of terminal events	F8.2

- Information on record type 30 continues on record type 31. Each pair of type-30 and type-31 records contains the information from one line of the life table.

Table 4 Record type 31, continuation of record type 30

Columns	Content	Format
1–2	Record type (31)	F2.0
3–7	Table number	F5.0
8–15	Proportion terminating	F8.6
16–23	Proportion surviving	F8.6
24–31	Cumulative proportion surviving	F8.6
32–39	Probability density	F8.6
40–47	Hazard rate	F8.6
48–54	S.E. of cumulative proportion surviving	F7.4
55–61	S.E. of probability density	F7.4
62–68	S.E. of hazard rate	F7.4

- Record type 31 is a continuation of record type 30.
- As many type-30 and type-31 record pairs are output for a table as it has lines (this number is noted in columns 57-60 of the type-10 record for the table).

Table 5 Record type 40, produced by both keywords TABLES and BOTH

Columns	Content	Format
1–2	Record type (40)	F2.0

- Type-40 records indicate the completion of the series of records for one life table.

Record Order

The SURVIVAL output file contains records for each of the life tables specified on the TABLES subcommand. All records for a given table are produced together in sequence. The records for the life tables are produced in the same order as the tables themselves. All life tables for the first survival variable are written first. The values of the first- and second-order control variables rotate, with the values of the first-order controls changing more rapidly.

Example

```
PROCEDURE OUTPUT OUTFILE = SURVTBL.
SURVIVAL TABLES = MOSFREE BY TREATMNT(1,3)
  /STATUS = PRISON (1) FOR MOSFREE
  /INTERVAL = THRU 24 BY 3
  /WRITE = BOTH.
```

- WRITE generates a procedure output file called *SURVTBL*, containing life tables, variable names and labels, and value labels stored as record types 10, 20, 30, 31, and 40.

Appendix A
Matrix Macros

You can use the matrix language (see Chapter 13) with the macro facility (see the *SPSS Base System Syntax Reference Guide*) to create fairly sophisticated statistical programs. Three SPSS command syntax files are included with Advanced Statistics as examples of the power and flexibility of the matrix language combined with the macro facility. Using these command syntax files, you can perform:

- Canonical correlation and redundancy analysis, with the CANCORR macro.
- Discriminant holdout classification (sometimes referred to as "jackknifing"), with the DISCLASS macro.
- Ridge regression and ridge trace plots, with the RIDGEREG macro.

With the CANCORR macro, you can estimate coefficients that maximize the correlation between two sets of variables. You can also calculate canonical scores for each of the cases. Using the DISCLASS macro, you can obtain more realistic estimates of the true classification error rate in discriminant analysis. Each case is classified using a discriminant function from whose calculation the case has been omitted. The RIDGEREG macro is useful when independent variables are nearly collinear and the usual least-squares estimators result in unstable coefficients with large variances.

Limitations of Macros

The three macros described in this appendix define command syntax that you can enter and run in a syntax window, just like regular SPSS commands. However, macros in general are not as robust or stable as fully supported SPSS statistical procedures, and these macros have not been as extensively tested as regular SPSS commands. If you use these macros, keep the following in mind:

- The syntax must be entered exactly as documented. Enter subcommands in the order shown. Do not include a slash before the first subcommand, do not abbreviate keywords, and do not omit equals signs or add them if they are not displayed.
- The command names are, in fact, macro calls. Once the macro is defined in a session, SPSS attempts to call the macro whenever it encounters the name, even if it is not at the beginning of a command. For example, if you have run the CANCORR macro and you also have a variable named *cancorr*, any command specifications (or

dialog box choices) that refer to the variable *cancorr* will cause SPSS to run the macro, resulting in numerous error messages. This remains in effect for the duration of the SPSS session.

- Because these macros utilize MATRIX routines that can require a good deal of memory, you may not be able to run the macros with a large number of variables.
- Error checking is not comprehensive, and a large number of uninformative error messages may be generated in some instances.
- The macros create new files in the current directory. If you do not have write permission to the current directory (this may happen if you are running SPSS on a network), you must change the current directory before you can run the macros. To change the current directory, open or save a data file in a directory for which you have write permission (for example, a directory on your own hard disk).

Using the Macros

All three macros should be located in the same directory as the SPSS executable program or with sample data. To use the macros:

1. From the SPSS menus choose:

   File
   New ▸
   SPSS Syntax

   This opens a new syntax window that you can use to enter command syntax.

2. Use the INCLUDE command to access the macro. Specify the directory location and the filename of the macro in apostrophes or quotes, as in:

```
include file 'c:\spsswin\cancorr.sps'.
```

3. After the INCLUDE command, enter the syntax for the command, as in:

```
include file 'c:\spsswin\cancorr.sps'.
cancorr set1=var1 var2 var3
 /set2=var4 var5 var6.
```

4. Highlight the commands, as shown in Figure A.1, and click on the Run Syntax tool.

Figure A.1 Using a macro in a syntax window

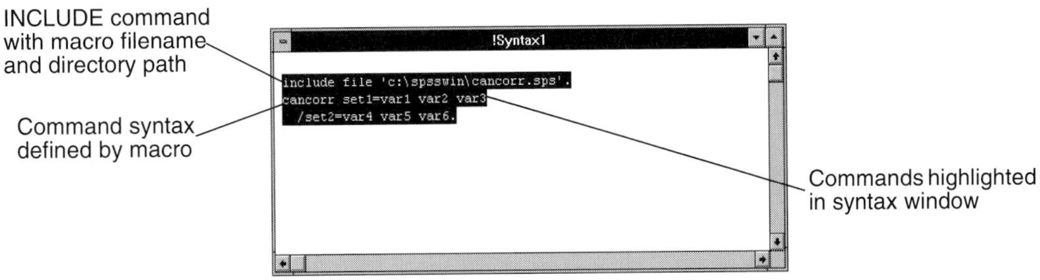

INCLUDE command with macro filename and directory path

Command syntax defined by macro

!Syntax1

```
include file 'c:\spsswin\cancorr.sps'.
cancorr set1=var1 var2 var3
  /set2=var4 var5 var6.
```

Commands highlighted in syntax window

Canonical Correlation

Canonical correlation analysis is a statistical technique for studying the association between two sets of variables, such as scores on four different reading tests and scores on three different aptitude tests. The goal is to form linear combinations of the variables that have maximum correlation. You want to estimate coefficients in such a way that the linear combination of reading scores correlates as highly as possible with the linear combination of aptitude scores. The correlation between the two linear combinations is known as the **canonical correlation**, and the coefficients used to form the linear combinations are known as the **canonical coefficients**. For each case, the values of the linear combinations are known as the **canonical scores**.

Canonical correlation analysis is not restricted to finding a single set of coefficients that results in maximum correlation. Instead, after you find the set of coefficients that results in maximum correlation, you find the set of coefficients that is orthogonal to the first and has the next largest correlation, and so on. The number of pairs of linear combinations you can form is determined by the smaller number of variables in the two sets.

As an example, consider data on CEO characteristics and company performance obtained from *Forbes* (5/28/90). (See the SPSS Base system documentation.) You want to examine the relationship between the two sets of variables. The first set is CEO age (*age*), years with the firm (*firmyrs*), years as the CEO (*ceoyrs*), and total salary (*total*). The second set is profits (*profits*), sales (*sales*), and Forbes' rank of growth (*growthrk*).

Figure A.2 contains the correlations among the individual variables in the first set, the correlations among the variables in the second set, and the correlations between the variables in the two sets.

Figure A.2 Correlation matrices from CANCORR macro

Correlations for Set-1

	AGE	FIRMYRS	CEOYRS	TOTAL
AGE	1.0000	.3562	.4270	-.0114
FIRMYRS	.3562	1.0000	.2542	-.1274
CEOYRS	.4270	.2542	1.0000	.0226
TOTAL	-.0114	-.1274	.0226	1.0000

Correlations for Set-2

	PROFITS	SALES	GROWTHRK
PROFITS	1.0000	.8473	.0700
SALES	.8473	1.0000	.1893
GROWTHRK	.0700	.1893	1.0000

Correlations Between Set-1 and Set-2

	PROFITS	SALES	GROWTHRK
AGE	.0282	.1099	-.1206
FIRMYRS	.0608	.1084	.2090
CEOYRS	-.1630	-.1413	-.2092
TOTAL	.1372	.0670	-.2050

Because the smaller number of variables in the set is three, you can form three pairs of linear combinations of the variables. Figure A.3 contains the standardized and raw coefficients that are used to form the linear combinations. If you take the first set of coefficients for each set and form the appropriate linear combination, you will obtain a correlation coefficient of 0.401. This value is shown in Figure A.4, where it is labeled the first canonical correlation. Figure A.4 also shows the canonical correlation for each pair of linear combinations.

Figure A.3 Standardized and raw canonical coefficients

Standardized Canonical Coefficients for Set-1

	1	2	3
AGE	.326	.745	.770
FIRMYRS	-.753	.151	-.014
CEOYRS	.628	-.733	.022
TOTAL	.323	.618	-.626

Raw Canonical Coefficients for Set-1

	1	2	3
AGE	.057	.131	.135
FIRMYRS	-.069	.014	-.001
CEOYRS	.089	-.103	.003
TOTAL	.000	.000	.000

Standardized Canonical Coefficients for Set-2

	1	2	3
PROFITS	-.309	.296	-1.863
SALES	.169	.742	1.787
GROWTHRK	-.992	-.283	-.064

Raw Canonical Coefficients for Set-2

	1	2	3
PROFITS	-.001	.001	-.005
SALES	.000	.000	.000
GROWTHRK	-.005	-.001	.000

Figure A.4 Canonical correlations

```
Canonical Correlations
1       .401
2       .259
3       .194
```

To examine the relationship between the canonical scores and the individual variables, you can look at the canonical loadings and cross-loadings shown in Figure A.5. The output labeled *Canonical Loadings for Set-1* contains the correlation coefficients between the variables in set 1 and the canonical scores for set 1. (Remember, canonical scores are the linear combinations of values of the set-1 variables.) The output labeled *Cross Loadings for Set-1* contains the correlations between the individual variables in set 1 and the canonical scores for set 2.

Figure A.5 Canonical loadings and cross-loadings

```
Canonical Loadings for Set-1
              1         2         3
     AGE    .323      .479      .782
 FIRMYRS   -.518      .151      .346
  CEOYRS    .583     -.362      .333
   TOTAL    .429      .573     -.632

Cross Loadings for Set-1
              1         2         3
     AGE    .130      .124      .152
 FIRMYRS   -.208      .039      .067
  CEOYRS    .234     -.094      .065
   TOTAL    .172      .149     -.123

Canonical Loadings for Set-2
                1         2         3
  PROFITS    -.236      .905     -.354
    SALES    -.281      .939      .196
GROWTHRK     -.982     -.122      .144

Cross Loadings for Set-2
                1         2         3
  PROFITS    -.095      .234     -.069
    SALES    -.113      .243      .038
GROWTHRK     -.394     -.032      .028
```

Using the results shown in Figure A.6, you can test the null hypothesis that successive groups of canonical correlations are 0. The first line corresponds to the test that all canonical correlations are 0. The next line is for the test that the second and third canonical correlations are 0. The last line is for the test that the third canonical correlation is 0.

Figure A.6 Significance tests for canonical correlation

```
Test that remaining correlations are zero:
     Wilks'      Chi-SQ         DF        Sig.
1      .753     21.830     12.000       .039
2      .898      8.299      6.000       .217
3      .962      2.954      2.000       .228
```

Figure A.7 describes the proportion of variance explained by each of the canonical variables. You see that there does not seem to be much of a relationship between the two sets of variables we were considering.

Figure A.7 Proportion of variance explained by canonical variables

```
            Redundancy Analysis:


Proportion of Variance of Set-1 Explained by Its Own Can. Var.
              Prop Var
CV1-1             .224
CV1-2             .178
CV1-3             .310


Proportion of Variance of Set-1 Explained by Opposite Can. Var.
              Prop Var
CV2-1             .036
CV2-2             .012
CV2-3             .012


Proportion of Variance of Set-2 Explained by Its Own Can. Var.
              Prop Var
CV2-1             .366
CV2-2             .572
CV2-3             .061


Proportion of Variance of Set-2 Explained by Opposite Can. Var.
              Prop Var
CV1-1             .059
CV1-2             .038
CV1-3             .002
```

Using the CANCORR Macro

The macro for canonical correlation is contained in the file *cancorr.sps*. This command syntax file should be located in the same directory as the SPSS executable program. The syntax defined by the macro is:

```
CANCORR SET1=variable list
 /SET2=variable list.
```

The variable lists specified for SET1 and SET2 must be numeric variables.

The canonical correlation macro creates two new data files in the current directory:

- The working data file is replaced with *cc__tmp1.sav*. This file contains the contents of the working data file immediately prior to running the macro.
- Canonical score variables, along with the contents of the original data file, are saved in a file named *cc__tmp2.sav*. The names of the canonical score variables are *s1__cv1*, *s2__cv1*, *s1__cv2*, *s2__cv2*, *s1__cv3*, *s2__cv3*, etc. The number of canonical score variables created is twice the smaller number of variables specified on SET1 or SET2. For example, if SET1 specifies four variables and SET2 specifies three variables, six canonical score variables will be created.

The filenames created by the macro use double underscores to minimize the risk of overwriting existing files. Each time you run the macro, the contents of these two data files are overwritten.

Discriminant Holdout Classification

In discriminant analysis, you can obtain estimates of the true misclassification rate by classifying each of the cases using a discriminant function from whose calculation the case has been excluded. Consider, for example, data on survival predictors for newborn infants (Van Vliet & Gupta, 1973).

Figure A.8 contains the group classification table obtained from the Discriminant Analysis procedure. Compare this to the classification table obtained from the DISCLASS macro shown in Figure A.9. You see that the estimated overall misclassification rate obtained from the DISCLASS macro is higher. Using the holdout technique, 66.7% of the cases are correctly classified, while 83.3% of the cases are correctly classified using the usual Discriminant Analysis procedure.

Figure A.8 Classification table from Discriminant Analysis procedure

```
CLASSIFICATION RESULTS -

                          NO. OF    PREDICTED GROUP MEMBERSHIP
   ACTUAL GROUP            CASES          1           2
 --------------------     ------    --------    --------

GROUP         1             26          22           4
 DIE                                  84.6%       15.4%

GROUP         2             22           4          18
 SURVIVE                              18.2%       81.8%

PERCENT OF "GROUPED" CASES CORRECTLY CLASSIFIED:  83.33%
```

Figure A.9 Classification table from DISCLASS macro

	Count / Col Pct	PRED 1.00	2.00	Row Total
SURVIVAL	1.00	17 70.8	9 37.5	26 54.2
	2.00	7 29.2	15 62.5	22 45.8
	Column Total	24 50.0	24 50.0	48 100.0

Page 1 of 1

With the DISCLASS macro, you can also obtain a casewise listing of observed and predicted group memberships as well as the estimated probabilities of group membership. This listing is shown in Figure A.10. *D1* and *D2* are the squared Mahalanobis distances to the centroids of the two groups. The posterior probabilities (*P1* and *P2*) are based on the distances.

Figure A.10 Casewise output from DISCLASS macro

Squared Distances (D) and Probabilities (P) of Group Membership

Actual	Predict	Error	D1	D2	P1	P2
1	2	***	10.8	10.3	.439	.561
1	1		11.4	15.1	.860	.140
1	1		4.5	7.9	.846	.154
1	1		5.2	8.1	.806	.194
1	1		17.2	17.8	.573	.427
1	2	***	9.6	8.3	.341	.659
1	1		4.2	5.6	.668	.332
1	1		7.0	10.4	.845	.155
1	2	***	4.8	4.4	.443	.557
1	2	***	8.6	8.5	.486	.514
1	1		5.0	11.6	.966	.034
1	1		9.3	16.9	.978	.022
1	1		4.3	9.6	.932	.068
1	1		10.3	11.0	.579	.421
1	2	***	13.6	13.4	.480	.520
1	1		6.3	12.6	.959	.041
1	1		5.7	13.2	.977	.023
1	1		2.1	6.8	.913	.087
1	2	***	13.7	9.4	.102	.898
1	1		5.5	6.3	.601	.399
1	2	***	6.0	4.8	.361	.639
1	1		19.7	23.2	.856	.144
1	2	***	10.5	9.3	.347	.653
1	2	***	7.7	3.4	.101	.899
1	1		4.6	5.3	.595	.405
1	1		6.0	6.0	.504	.496
2	1	***	8.6	10.1	.679	.321
2	1	***	63.9	69.0	.929	.071
2	2		5.2	4.9	.467	.533
2	1	***	10.3	10.3	.501	.499
2	2		8.3	5.9	.239	.761
2	2		10.4	6.9	.147	.853
2	2		8.1	4.7	.160	.840
2	2		6.1	5.2	.384	.616
2	2		30.4	28.2	.252	.748
2	2		94.3	84.5	.007	.993
2	2		12.9	5.7	.026	.974
2	2		20.8	10.2	.005	.995
2	2		6.8	5.3	.323	.677
2	1	***	10.4	18.3	.982	.018
2	1	***	7.5	9.2	.706	.294
2	1	***	5.9	13.9	.982	.018
2	2		4.3	4.3	.492	.508
2	1	***	77.8	80.1	.765	.235
2	2		5.5	5.0	.433	.567
2	2		5.6	3.4	.249	.751
2	2		7.6	3.7	.125	.875
2	2		9.3	4.8	.096	.904

Using the DISCLASS Macro

The macro for discriminant holdout classification is contained in the file *disclass.sps*. This command syntax file should be located in the same directory as the SPSS executable program. The syntax defined by the macro is:

```
DISCLASS VARS=variable list
 /GROUPS=grouping variable
 /HOLDOUT=NO
 /CASE=n
 /PRIORS (value list).
```

The discriminant holdout classification macro creates two new data files in the current directory:

- The working data file is replaced with *dc__tmp1.sav*. This file contains the contents of the working data file immediately prior to running the macro.
- The grouping variable and predicted group membership (variable *pred*) are saved in a file named *dc__tmp2.sav*. If you request casewise output with the CASE specification, group membership probabilities (variables *p1*, *p2*, *p3*, etc.) and squared Mahalanobis distances (variables *d1*, *d2*, *d3*, etc.) are also saved.

The filenames created by the macro use double underscores to minimize the risk of overwriting existing files. Each time you run the macro, the contents of these two data files are overwritten.

Required Specifications

VARS and GROUPS specifications are required; all other specifications are optional. The variables specified for VARS must be numeric. The grouping variable specified with GROUPS must contain consecutive integer values. (You can use the AUTORECODE command to recode string or numeric variables into consecutive integers.)

Optional Specifications

The following additional specifications are optional:

- `HOLDOUT=NO` specifies that holdout probabilities for classification should not be computed. By default, holdout probabilities are computed `(HOLDOUT=YES)`.
- `CASE=n` specifies a number of groups for which you want casewise listings of actual and predicted groups, squared Mahalanobis distance, and posterior probabilities of group classification. By default, casewise output is not displayed.
- `PRIORS(value list)` specifies prior probabilities for each group. The value list must be enclosed in parentheses, and values must be separated by commas. The values must sum to 1, and the number of values must be equal to the number of categories of the grouping variable. You can specify decimals or fractions. By default, all group probabilities are equal.

Ridge Regression

Whenever you have independent variables that are strongly interrelated and you use the usual method of least squares to estimate a regression model, you can obtain estimates that are unstable and have large variances. One possible consequence of this is regression coefficients that have the wrong sign.

The least-squares estimators are unbiased and they have the smallest variances of any unbiased estimators. (An **unbiased estimator** is one whose expected value is the population parameter. That is, if you take repeated samples from the same population and calculate the estimates, the average of the sample estimates will approach the population value.) Because the variance of the unbiased estimator can be quite large, one strategy for decreasing the variance is to allow the estimators to be biased. This means that, for example, if the true population value of a parameter is 10, we will allow the biased estimator to have an expected value close to 10, but not exactly 10. We expect to gain from this a smaller variance and a more stable estimate.

Ridge regression (Hoerl & Kennard, 1970) is one of the methods proposed for obtaining biased estimators of regression coefficients. Instead of solving the usual equation

$$\beta = (X'X)^{-1}X'Y$$

Equation A.1

we solve the equation

$$\beta(ridge) = (X'X + kI)^{-1}X'Y$$

Equation A.2

The constant k, which can range between 0 and 1, is known as the **bias parameter**. As k increases, the bias in the parameter estimates increases and the variance decreases. The residual sum of squares for the regression also increases with increasing k, so R^2 decreases. You want to choose a value of k so that the decrease in variance is greater than the increase in the bias. See Montgomery and Peck (1982) for further discussion.

One way to select a value of k, suggested by Hoerl and Kennard (1970), is to examine what is known as a **ridge trace**. In a ridge trace, you plot the values of the standardized coefficients for different values of the bias parameter k. If there is multicollinearity, the ridge estimates will initially fluctuate and then stabilize. Figure A.11 contains the ridge trace plot produced by the RIDGEREG macro for the acetylene data in Montgomery and Peck (1982). The bias parameter varies from 0.000 to 0.4 in increments of 0.005. You see that the coefficients stabilize at a k value of approximately 0.05. A plot of R^2 against the values of k is shown in Figure A.12. You see that the reduction in R^2 is not very large in the range we have considered.

Figure A.11 Ridge trace plot

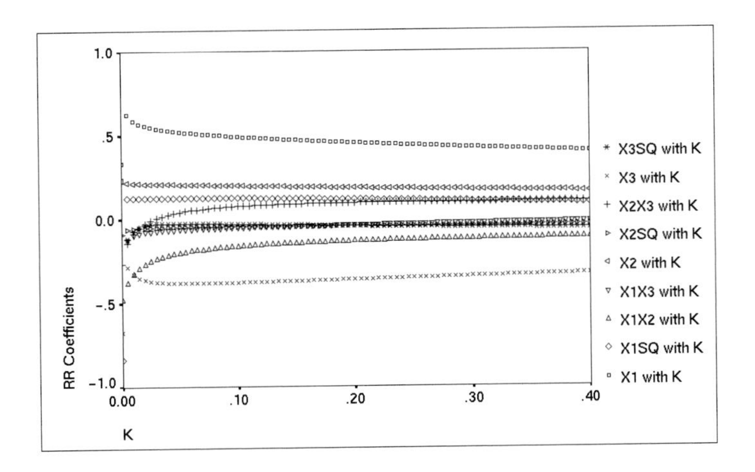

Figure A.12 Plot of R^2 versus K

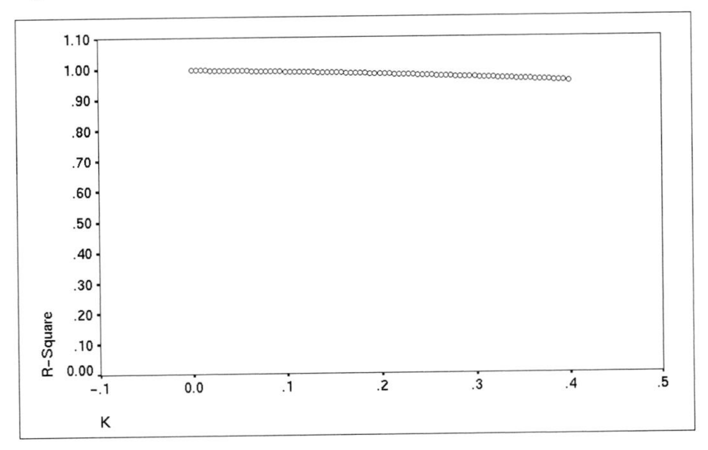

The regression estimates and an analysis-of-variance table when a bias parameter of 0.032 is used are shown in Figure A.13.

Figure A.13 Regression estimates and ANOVA table from RIDGEREG macro

```
****** Ridge Regression with k = .032 ******

Mult R       .997188219
RSquare      .994384344
Adj RSqu     .985960859
SE          1.409847641

         ANOVA table
                df          SS          MS
Regress      9.000    2111.783     234.643
Residual     6.000      11.926       1.988


      F value          Sig F
   118.0490479       .0000047

-------------Variables in the Equation----------------
                     B          SEB          Beta            T           Sig T
X1          6.41538363     .55157077     .53916097   11.63111597     .00002433
X2          2.51888121     .41405871     .21169391    6.08339143     .00089704
X3         -4.44496177     .55416490    -.37354745   -8.02100923     .00020052
X1X2       -3.13292744     .78638216    -.23285964   -3.98397575     .00725026
X1X3        -.89532968     .23892754    -.06750540   -3.74728535     .00954022
X2X3         .17443263     .83199867     .01229720     .20965493     .84087455
X1SQ        1.87399467     .81618959     .12487716    2.29602864     .06143404
X2SQ        -.51711533     .40684163    -.04806298   -1.27104820     .25076357
X3SQ        -.30873645     .72389616    -.02666682    -.42649273     .68462664
Constant   35.01555919     .73806473     .00000000   47.44239579     .00000001
```

Using the RIDGEREG Macro

The macro for ridge regression is contained in the file *ridgereg.sps*. This command syntax file should be located in the same directory as the SPSS executable program. The syntax defined by the macro is:

```
RIDGEREG DEP=dependent variable
 /ENTER variable list
 /START=value
 /STOP=value
 /INC=value
 /K=value.
```

The ridge regression macro creates two new data files in the current directory:

- The working data file is replaced with *rr__tmp1.sav*. This file contains the contents of the working data file immediately prior to running the macro.
- The values of R^2 (variable *rsq*) and *Beta* values for each independent variable (using the original independent variable names) for each value of the bias parameter (variable *k*) are saved in a file named *rr__tmp2.sav*. The number of cases is determined by the number of bias parameter values defined by START, STOP, and INC. The default values produce a file with 21 cases. If you use *k* to specify a bias parameter value, the file contains only one case.

The filenames created by the macro use double underscores to minimize the risk of overwriting existing files. Each time you run the macro, the contents of these two data files are overwritten.

Required Specifications

DEP and ENTER specifications are required. All variables specified must be numeric variables. If you do not use any of the optional specifications, the default START, STOP, and INC values are used to generate a ridge trace plot of standardized regression coefficients versus values of the bias parameter k, a plot of the model R^2 versus k, and a table listing the R^2 and *Beta* values for each k value.

Optional Specifications

The following optional specifications are also available:

- START specifies a starting bias parameter value for ridge trace mode. The value must be between 0 and 1. The default value is 0.
- STOP specifies an ending bias parameter value for ridge trace mode. The value must be greater than the START value but cannot exceed 1. The default value is 1.
- INC specifies the increment for the bias parameter value in ridge trace mode. The value must be less than the difference between the START and STOP values. The default is 0.05.
- K specifies a single value for the bias parameter and overrides any specifications on START, STOP, and INC. The value must be between 0 and 1. If you specify K, a full set of standard regression output is produced for the specified k value.

Appendix B
Categorical Variable Coding Schemes

In many SPSS procedures, you can request automatic replacement of a categorical independent variable with a set of contrast variables, which will then be entered or removed from an equation as a block. You can specify how the set of contrast variables is to be coded, usually on the CONTRAST subcommand. This appendix explains and illustrates how different contrast types requested on CONTRAST actually work.

Deviation

Deviation from the grand mean. In matrix terms, these contrasts have the form:

```
mean     (  1/k      1/k     ...    1/k     1/k )
df(1)    ( 1-1/k    -1/k     ...   -1/k    -1/k )
df(2)    (  -1/k    1-1/k    ...   -1/k    -1/k )
.                   .
.                   .
df(k-1)  (  -1/k    -1/k     ...   1-1/k   -1/k )
```

where k is the number of categories for the independent variable and the last category is omitted by default. For example, the deviation contrasts for an independent variable with three categories are as follows:

```
(  1/3    1/3    1/3 )
(  2/3   -1/3   -1/3 )
( -1/3    2/3   -1/3 )
```

To omit a category other than the last, specify the number of the omitted category in parentheses after the DEVIATION keyword. For example, the following subcommand obtains the deviations for the first and third categories and omits the second:

```
/CONTRAST(FACTOR)=DEVIATION(2)
```

Suppose that *factor* has three categories. The resulting contrast matrix will be

```
( 1/3    1/3    1/3 )
( 2/3   -1/3   -1/3 )
( -1/3  -1/3    2/3 )
```

Simple

Simple contrasts. Compares each level of a factor to the last. The general matrix form is

```
mean    ( 1/k   1/k   ...   1/k   1/k )
df(1)   (  1     0    ...    0    -1 )
df(2)   (  0     1    ...    0    -1 )
.              .
.              .
df(k-1) (  0     0    ...    1    -1 )
```

where k is the number of categories for the independent variable. For example, the simple contrasts for an independent variable with four categories are as follows:

```
( 1/4   1/4   1/4   1/4 )
(  1     0     0    -1 )
(  0     1     0    -1 )
(  0     0     1    -1 )
```

To use another category instead of the last as a reference category, specify in parentheses after the SIMPLE keyword the sequence number of the reference category, which is not necessarily the value associated with that category. For example, the following CONTRAST subcommand obtains a contrast matrix that omits the second category:

```
/CONTRAST(FACTOR) = SIMPLE(2)
```

Suppose that *factor* has four categories. The resulting contrast matrix will be

```
( 1/4   1/4   1/4   1/4 )
(  1    -1     0     0 )
(  0    -1     1     0 )
(  0    -1     0     1 )
```

Helmert

Helmert contrasts. Compares categories of an independent variable with the mean of the subsequent categories. The general matrix form is

mean	(1/*k*	1/*k*	...	1/*k*	1/*k*)
df(1)	(1	-1/(*k*-1)	...	-1/(*k*-1)	-1/(*k*-1))
df(2)	(0	1	...	-1/(*k*-2)	-1/(*k*-2))
.		.			
.		.			
df(*k*-2)	(0	0	1	-1/2	-1/2)
df(k-1)	(0	0	...	1	-1)

where *k* is the number of categories of the independent variable. For example, an independent variable with four categories has a Helmert contrast matrix of the following form:

(1/4	1/4	1/4	1/4)
(1	-1/3	-1/3	-1/3)
(0	1	-1/2	-1/2)
(0	0	1	-1)

Difference

Difference or reverse Helmert contrasts. Compares categories of an independent variable with the mean of the previous categories of the variable. The general matrix form is

mean	(	1/*k*	1/*k*	1/*k*	...	1/*k*)
df(1)	(	-1	1	0	...	0)
df(2)	(	-1/2	-1/2	1	...	0)
.			.			
.			.			
df(k-1)	(	-1/(*k*-1)	-1/(*k*-1)	-1/(*k*-1)	...	1)

where *k* is the number of categories for the independent variable. For example, the difference contrasts for an independent variable with four categories are as follows:

(1/4	1/4	1/4	1/4)
(-1	1	0	0)
(-1/2	-1/2	1	0)
(-1/3	-1/3	-1/3	1)

Polynomial

Orthogonal polynomial contrasts. The first degree of freedom contains the linear effect across all categories; the second degree of freedom, the quadratic effect; the third degree of freedom, the cubic; and so on for the higher-order effects.

You can specify the spacing between levels of the treatment measured by the given categorical variable. Equal spacing, which is the default if you omit the metric, can be specified as consecutive integers from 1 to *k*, where *k* is the number of categories. If the variable *drug* has three categories, the subcommand

```
/CONTRAST(DRUG)=POLYNOMIAL
```

is the same as

```
/CONTRAST(DRUG)=POLYNOMIAL(1,2,3)
```

Equal spacing is not always necessary, however. For example, suppose that *drug* represents different dosages of a drug given to three groups. If the dosage administered to the second group is twice that to the first group and the dosage administered to the third group is three times that to the first group, the treatment categories are equally spaced, and an appropriate metric for this situation consists of consecutive integers:

```
/CONTRAST(DRUG)=POLYNOMIAL(1,2,3)
```

If, however, the dosage administered to the second group is four times that given the first group, and the dosage given the third group is seven times that to the first, an appropriate metric is:

```
/CONTRAST(DRUG)=POLYNOMIAL(1,4,7)
```

In either case, the result of the contrast specification is that the first degree of freedom for *drug* contains the linear effect of the dosage levels and the second degree of freedom contains the quadratic effect.

Polynomial contrasts are especially useful in tests of trends and for investigating the nature of response surfaces. You can also use polynomial contrasts to perform nonlinear curve-fitting, such as curvilinear regression.

Repeated

Compares adjacent levels of an independent variable. The general matrix form is

mean	($1/k$	$1/k$	$1/k$	...	$1/k$	$1/k$)
df(1)	(1	-1	0	...	0	0)
df(2)	(0	1	-1	...	0	0)
.		.				
.		.				
df(k-1)	(0	0	0	...	1	-1)

where k is the number of categories for the independent variable. For example, the repeated contrasts for an independent variable with four categories are as follows:

(1/4	1/4	1/4	1/4)
(1	-1	0	0)
(0	1	-1	0)
(0	0	1	-1)

These contrasts are useful in profile analysis and wherever difference scores are needed.

Special

A user-defined contrast. Allows entry of special contrasts in the form of square matrices with as many rows and columns as there are categories of the given independent variable. For MANOVA and LOGLINEAR, the first row entered is always the mean, or constant, effect and represents the set of weights indicating how to average other independent variables, if any, over the given variable. Generally, this contrast is a vector of ones.

The remaining rows of the matrix contain the special contrasts indicating the desired comparisons between categories of the variable. Usually, orthogonal contrasts are the most useful. Orthogonal contrasts are statistically independent and are nonredundant. Contrasts are orthogonal if:

- For each row, contrast coefficients sum to zero.
- The products of corresponding coefficients for all pairs of disjoint rows also sum to zero.

For example, suppose that *treatment* has four levels and that you want to compare the various levels of treatment with each other. An appropriate special contrast is

(1	1	1	1)	weights for mean calculation
(3	-1	-1	-1)	compare 1st with 2nd through 4th
(0	2	-1	-1)	compare 2nd with 3rd and 4th
(0	0	1	-1)	compare 3rd with 4th

which you specify by means of the following CONTRAST subcommand for MANOVA, LOGISTIC REGRESSION, and COXREG:

```
/CONTRAST(TREATMNT)=SPECIAL( 1  1  1  1
                             3 -1 -1 -1
                             0  2 -1 -1
                             0  0  1 -1 )
```

For LOGLINEAR, you need to specify:

```
/CONTRAST(TREATMNT)=BASIS SPECIAL( 1  1  1  1
                                   3 -1 -1 -1
                                   0  2 -1 -1
                                   0  0  1 -1 )
```

Each row except the means row sums to zero. Products of each pair of disjoint rows sum to zero as well:

Rows 2 and 3: $(3)(0) + (-1)(2) + (-1)(-1) + (-1)(-1) = 0$

Rows 2 and 4: $(3)(0) + (-1)(0) + (-1)(1) + (-1)(-1) = 0$

Rows 3 and 4: $(0)(0) + (2)(0) + (-1)(1) + (-1)(-1) = 0$

The special contrasts need not be orthogonal. However, they must not be linear combinations of each other. If they are, the procedure reports the linear dependency and ceases processing. Helmert, difference, and polynomial contrasts are all orthogonal contrasts.

Indicator

Indicator variable coding. Also known as dummy coding, this is not available in LOGLINEAR or MANOVA. The number of new variables coded is $k-1$. Cases in the reference category are coded 0 for all $k-1$ variables. A case in the ith category is coded 0 for all indicator variables except the ith, which is coded 1.

Bibliography

Agresti, A. 1984. *Analysis of ordinal categorical data*. New York: John Wiley and Sons.

_____. 1990. *Categorical data analysis*. New York: John Wiley and Sons.

Aldrich, J. H., and F. D. Nelson. 1984. *Linear probability, logit, and probit models*. Beverly Hills, Calif.: Sage Publications.

Andrews, D. F., R. Gnanadesikan, and J. L. Warner. 1973. Methods for assessing multivariate normality. In: *Multivariate Analysis III*, P. R. Krishnaiah, ed. New York: Academic Press.

Ashford, J. R., and R. D. Sowden. 1970. Multivariate probit analysis. *Biometrics*, 26: 535–546.

Atkinson, A. C. 1980. A note on the generalized information criterion for choice of a model. *Biometrika*, 67: 413–418.

Bacon, L. 1980. Unpublished data.

Bancroft, T. A. 1968. *Topics in intermediate statistical methods*. Ames: Iowa State University Press.

Barnard, R. M. 1973. Field-dependent independence and selected motor abilities. Ph.D. diss., School of Education, New York University.

Bartolucci, A. A., and M. D. Fraser. 1977. Comparative step-up and composite tests for selecting prognostic indicators associated with survival. *Biometrical Journal*, 19: 437–448.

Belsey, D. A., E. Kuh, and R. E. Welsch. 1980. *Regression diagnostics: Identifying influential data and sources of collinearity*. New York: John Wiley and Sons.

Benedetti, J. K., and M. B. Brown. 1978. Strategies for the selection of log-linear models. *Biometrics*, 34: 680–686.

Bishop, Y. M. M., and S. E. Fienberg. 1969. Incomplete two-dimensional contingency tables. *Biometrics*, 25: 119–128.

Bishop, Y. M. M., S. E. Fienberg, and P. W. Holland. 1975. *Discrete multivariate analysis: Theory and practice*. Cambridge, Mass.: MIT Press.

Blossfeld, H. P., A. Hamerle, and K. U. Mayer. 1989. *Event history analysis*. Hillsdale, N.J.: Lawrence Erlbaum Associates.

Bock, R. D. 1985. *Multivariate statistical methods in behavioral research*. Mooresville, Ind.: Scientific Software, Inc.

Brown, B. W., Jr. 1980. Prediction analyses for binary data. In: *Biostatistics Casebook*, R. G. Miller, B. Efron, B. W. Brown, and L. E. Moses, eds. New York: John Wiley and Sons.

Brown, M. B., and J. K. Benedetti. 1977. Sampling behavior of tests for correlation in two-way contingency tables. *Journal of the American Statistical Association*, 72: 309–315.

Burns, P. R. 1984. Multiple comparison methods in MANOVA. In: *Proceedings of the 7th SPSS Users and Coordinators Conference*. Chicago: ISSUE, Inc., 33–66.

_____. 1984. *SPSS-6000 MANOVA update manual*. Chicago: Vogelback Computing Center.

Churchill, G. A., Jr. 1979. *Marketing research: Methodological foundations*. Hinsdale, Ill.: Dryden Press.

Cochran, W. G., and G. M. Cox. 1957. *Experimental designs*. 2nd ed. New York: John Wiley and Sons.

Cohen, J. 1960. A coefficient of agreement for nominal scales. *Educational and Psychological Measurement*, 20: 37–46.

_____. 1977. *Statistical power analysis for the behavioral sciences*. New York: Academic Press.

Conover, W. J. 1980. *Practical nonparametric statistics*. 2nd ed. New York: John Wiley and Sons.

Cooley, W. W., and P. R. Lohnes. 1971. *Multivariate data analysis*. New York: John Wiley and Sons.

Cox, D. R., and E. J. Snell. 1968. A general definition of residuals. *Journal of the Royal Statistical Society Series B*, 30: 248–275.

Crowley, J., and B. E. Storer. 1983. Comment. *Journal of the American Statistical Association*, 78: 277–281.

Daniel, C., and F. Wood. 1980. *Fitting equations to data*. Rev. ed. New York: John Wiley and Sons.

Davies, O. L. 1954. *Design and analysis of industrial experiments*. New York: Hafner Press.

Delany, M. F., and C. T. Moore. 1987. American alligator food habits in Florida. Unpublished manuscript.

Draper, N. R., and H. Smith. 1981. *Applied regression analysis*. New York: John Wiley and Sons.

Duncan, O. D. 1966. Path analysis: Sociological examples. *American Journal of Sociology*, 72: 1–16.

Elashoff, J. 1981. Data for the panel session in software for repeated measures analysis of variance. *Proceedings of the Statistical Computing Section*, American Statistical Association.

Everitt, B. S. 1977. *The analysis of contingency tables*. New York: Halsted Press.

_____. 1978. *Graphical techniques for multivariate data*. New York: North-Holland.

Eysenck, M. W. 1977. *Human memory: Theory, research and individual differences*. New York: Pergamon Press.

Finn, J. D. 1974. *A general model for multivariate analysis*. New York: Holt, Rinehart and Winston.

Finney, D. J. 1971. *Probit analysis*. Cambridge: Cambridge University Press.

Fisher, R. A. 1936. The use of multiple measurements in taxonomic problems. *Annals of Eugenics*, 7: 179–188.

Fox, J. 1984. *Linear statistical models and related methods: With applications to social research*. New York: John Wiley and Sons.

Freund, R. J. 1980. The case of the missing cell. *The American Statistician*, 34: 94–98.

Friereich, E. J., et al. 1963. The effect of 6-mercaptopurine on the duration of steroid-induced remission in acute leukemia. *Blood*, 21: 699–716.

Frome, E. L. 1983. The analysis of rates using Poisson regression models. *Biometrics*, 39: 665–674.

Gentleman, R., and J. Crowley. 1991. Graphical methods for censored data. *Journal of the American Statistical Association*, 86: 678–683.

Gilbert, E. S. 1968. On discrimination using qualitative variables. *Journal of the American Statistical Association*, 63: 1399–1412.

Gill, P. E., W. M. Murray, and M. H. Wright. 1981. *Practical optimization*. London: Academic Press.

Gill, P. E., W. M. Murray, M. A. Saunders, and M. H. Wright. 1984. Procedures for optimization problems with a mixture of bounds and general linear constraints. *ACM Transactions on Mathematical Software*, 10:3, 282–296.

_____. 1986. User's guide for NPSOL (version 4.0): A FORTRAN package for nonlinear programming. *Technical Report SOL 86-2*. Department of Operations Research, Stanford University.

Goldstein, M., and W. R. Dillon. 1978. *Discrete discriminant analysis*. New York: John Wiley and Sons.

Goodman, L. A. 1964. Simple methods of analyzing three-factor interaction in contingency tables. *Journal of the American Statistical Association*, 59: 319–352.

_____. 1978. *Analyzing qualitative/categorical data*. Cambridge, Mass.: Abt Books.

_____. 1984. *The analysis of cross-classified data having ordered categories*. Cambridge, Mass.: Harvard University Press.

Goodnight, J. H. 1979. A tutorial on the SWEEP operator. *The American Statistician*, 33: 149–158.

Green, P. E. 1978. *Analyzing multivariate data*. Hinsdale, Ill.: Dryden Press.

Greenhouse, S. W., and S. Geisser. 1959. On methods in analysis of profile data. *Psychometrika*, 24: 95–112.

Haberman, S. J. 1973. The analysis of residuals in cross-classified tables. *Biometrics*, 29: 205–220.

_____. 1978. *Analysis of qualitative data*. Vol. 1. New York: Academic Press.

_____. 1979. *Analysis of qualitative data*. Vol. 2. New York: Academic Press.

_____. 1982. Analysis of dispersion of multinomial responses. *Journal of the American Statistical Association*, 77: 568–580.

Hadley, G. 1961. *Linear algebra*. Reading, Mass.: Addison-Wesley.

Hand, D. J. 1981. *Discrimination and classification*. New York: John Wiley and Sons.

Hauck, W. W., and A. Donner. 1977. Wald's test as applied to hypotheses in logit analysis. *Journal of the American Statistical Association*, 72: 851–853.

Hays, W. L. 1981. *Statistics for the social sciences*. 3rd ed. New York: Holt, Rinehart and Winston.

Heck, D. L. 1960. Charts of some upper percentage points of the distribution of the largest characteristic root. *Annals of Mathematical Statistics*, 31: 625–642.

Hicks, C. R. 1973. *Fundamental concepts in the design of experiments*. 2nd ed. New York: Holt, Rinehart and Winston.

Hinds, M. A., and G. A. Milliken. 1982. Statistical methods to use nonlinear models to compare silage treatments. Unpublished paper.

Hoaglin, D. C., F. Mosteller, and J. W. Tukey. 1983. *Understanding robust and exploratory data analysis*. New York: John Wiley and Sons.

Hoerl, A. E., and R. W. Kennard. 1970. Ridge regression: Applications to nonorthogonal problems. *Technometrics*, 12: 69–82.

_____. 1970. Ridge regression: Biased estimation of nonorthogonal problems. *Technometrics*, 12: 55–67.

Hosmer, D. W., and S. Lemeshow. 1989. *Applied logistic regression*. New York: John Wiley and Sons.

Huberty, C. J. 1972. Multivariate indices of strength of association. *Multivariate Behavioral Research*, 7: 523–526.

Huynh, H., and L. S. Feldt. 1976. Estimation of the Box correction for degrees of freedom from sample data in randomized block and split-plot designs. *Journal of Educational Statistics*, 1: 69–82.

Huynh, H., and G. K. Mandevill. 1979. Validity conditions in repeated measures designs. *Psychological Bulletin*, 86: 964–973.

Judge, G. G., W. E. Griffiths, R. C. Hill, H. Lutkepohl, and T. C. Lee. 1985. *The theory and practice of econometrics*. 2nd ed. New York: John Wiley and Sons.

Kalbfleisch, J. D., and R. L. Prentice. 1980. *The statistical analysis of failure time data*. New York: John Wiley and Sons.

Kaplan, E. L., and P. Meier. 1958. Nonparametric estimation from incomplete observations. *Journal of the American Statistical Association*, 53: 457–481.

Kendall, M. G., and A. Stuart. 1973. *The advanced theory of statistics*. Vol. 2. New York: Hafner Press.

Kennedy, J. J. 1970. The eta coefficient in complex anova designs. *Educational and Psychological Measurement*, 30: 885–889.

Kirk, R. E. 1982. *Experimental design*. 2nd ed. Monterey, Calif.: Brooks/Cole.

Kleinbaum, D. G., L. L. Kupper, and H. Morgenstern. 1982. *Epidemiological research: Principles and quantitative methods*. Belmont, Calif.: Wadsworth, Inc.

Kleinbaum, D. G., L. L. Kupper, and K. E. Muller. 1988. *Applied regression analysis and other multivariable methods*. Boston: PWS-KENT Publishing Company.

Koch, G., S. Atkinson, and M. Stokes. 1986. Poisson regression. In: *Encyclopedia of Statistical Sciences*, Vol. 7, S. Kotz and N. Johnson, eds. New York: John Wiley and Sons.

Koch, G. G., J. R. Landis, J. L. Freeman, D. H. Freeman, and R. G. Lehnen. 1977. A general methodology for the analysis of experiments with repeated measurement of categorical data. *Biometrics*, 33: 133–158.

Kvalseth, T. O. 1985. Cautionary note about R squared. *The American Statistician*, 39:4, 279–285.

Lachenbruch, P. A. 1975. *Discriminant analysis*. New York: Hafner Press.

Lawless, J. F., and K. Singhal. 1978. Efficient screening of nonnormal regression models. *Biometrics*, 34: 318–327.

Lee, E. T. 1992. *Statistical methods for survival data analysis*. New York: John Wiley and Sons.

Lee, E., and M. Desu. 1972. A computer program for comparing k samples with right-censored data. *Computer Programs in Biomedicine*, 2: 315–321.

Magidson, J. 1981. Qualitative variance, entropy, and correlation ratios for nominal dependent variables. *Social Science Research*, 10: 177–194.

Mantel, N., and W. Haenszel. 1959. Statistical aspects of the analysis of data from retrospective studies of disease. *Journal of the National Cancer Institute*, 22: 719–748.

McCullagh, P., and J. A. Nelder. 1989. *Generalized linear models*. 2nd ed. London: Chapman and Hall.

Miller, R. G. 1981. *Simultaneous statistical inference*. 2nd ed. New York: Springer-Verlag.

Milliken, G. A. 1987. A tutorial on nonlinear modeling with an application from pharmacokinetics. Unpublished manuscript.

Milliken, G. W., and D. E. Johnson. 1984. *Analysis of messy data*. Belmont, Calif.: Lifetime Learning Publications.

Montgomery, D. C., and E. A. Peck. 1982. *Introduction to linear regression analysis*. New York: John Wiley and Sons.

Morrison, D. F. 1976. *Multivariate statistical methods*. 2nd ed. New York: McGraw-Hill.
Mudholkar, G. S., Y. P. Chaubey, and C. C. Lin. 1976. Some approximations for the noncentral-F distribution. *Technometrics*, 18: 351–358.
Muller, K. E., and B. L. Peterson. 1984. Practical methods for computing power in testing the multivariate general linear hypothesis. *Computational Statistics and Data Analysis*, 2: 143–158.
Neter, J., W. Wasserman, and R. Kutner. 1985. *Applied linear statistical models*. 2nd ed. Homewood, Ill.: Richard D. Irwin, Inc.
Norusis, M. J., and SPSS Inc. 1989. *SPSS advanced statistics user's guide*. Chicago: SPSS Inc.
O'Brien, R. G. 1983. General Scheffé tests and optimum subeffects for linear models. Presented at the annual meeting of the American Statistical Association, August, 1983.
Olsen, C. L. 1976. On choosing a test statistic in multivariate analysis of variance. *Psychological Bulletin*, 83: 579–586.
O'Nan, Michael. 1971. *Linear algebra*. New York: Harcourt Brace Jovanovich.
Pillai, K. C. S. 1967. Upper percentage points of the largest root of a matrix in multivariate analysis. *Biometrika*, 54: 189–193.
Prentice, R. L., and P. Marek. 1979. A quantitative discrepancy between censored data rank tests. *Biometrics*, 35: 861–867.
Rao, C. R. 1973. *Linear statistical inference and its applications*. 2nd ed. New York: John Wiley and Sons.
Roy, J., and R. E. Bargmann. 1958. Tests of multiple independence and the associated confidence bounds. *Annals of Mathematical Statistics*, 29: 491–503.
Schoenfeld, D. 1982. Partial residuals for the proportional hazards regression model. *Biometrika*, 69: 239–241.
Searle, S. R. 1971. *Linear models*. New York: John Wiley and Sons.
Snedecor, G. W., and W. G. Cochran. 1967. *Statistical methods*. 6th ed. Ames: Iowa State University Press.
Stablein, D. M., W. H. Carter, Jr., and J. W. Novak. 1981. Analysis of survival data with nonproportional hazard functions. *Controlled Clinical Trials*, 2: 149–159.
Tatsuoka, M. M. 1971. *Multivariate analysis*. New York: John Wiley and Sons.
Therneau, T., Grambsch, P., and Fleming, T. R. 1990. Martingale-based residuals for survival models. *Biometrika*, 77: 147–160.
Timm, N. H. 1975. *Multivariate analysis with applications in education and psychology*. Monterey, Calif.: Brooks/Cole.
Tukey, J. W. 1977. *Exploratory data analysis*. Reading, Mass.: Addison-Wesley.
Van Vliet, P. K. J., and J. M. Gupta. 1973. THAM v. sodium bicarbonate in idiopathic respiratory distress syndrome. *Archives of Disease in Childhood*, 48: 249–255.
Wald, A. 1943. Tests of statistical hypotheses concerning several parameters with applications to problems of estimation. *Transcripts of American Mathematical Society*, 54: 426–482.
Winer, B. J., D. R. Brown, and K. M. Michels. 1991. *Statistical principles in experimental design*. 3rd ed. New York: McGraw-Hill.
Witkin, H. A., and others. 1954. *Personality through perception*. New York: Harper and Brothers.
Wright, S. 1960. Path coefficients and path regressions: Alternative or complementary concepts? *Biometrics*, 16: 189–202.

Subject Index

Syntax Index